The Human Connectome

Kostas N. Fountoulakis

The Human Connectome

How the Brain Works

Springer

Kostas N. Fountoulakis
School of Medicine
Aristotle University of Thessalonik
Thessaloniki, Greece

ISBN 978-3-031-10350-6 ISBN 978-3-031-10351-3 (eBook)
https://doi.org/10.1007/978-3-031-10351-3

This Springer imprint is published by the registered company Springer Nature Switzerland AG
The registered company address is: Gewerbestrasse 11, 6330 Cham, Switzerland

To my parents, Nikos and Olga
To the parents that adopted me, Giannis and Zoumboulia
To Katerina
To Nikolas and Ioannis

Foreword

It is with great personal pleasure that I write the foreword to the book written by the prominent neuroscientist and psychiatrist Dr. Kostas Fountoulakis, Professor, Faculty of Medicine, Aristotle University of Thessaloniki, entitled "The Human Connectome: How the Brain Works" The word "human" is of major importance, granted that Man is of unique complexity, both in qualitative and quantitative terms, and that her/his uniqueness pertains mainly to her/his brain. Socrates, despite the minimal technology of his era, understood this and used to say: *«ουδενί άλλω πείθομαι ή τω λόγω»*, which means: "I am not convinced by anything other than Reason." Socrates was referring to the use of human logic in interpreting the phenomena. Indeed, humans, their society, culture, and civilization, all together, are the most complex system known in our universe. An indicative example of this astonishing complexity is that the number of synapses in a single human brain is about three times greater than the number of the stars in our galaxy. Therefore, understanding the structure and function of the human brain is neither simple nor easy, and composing a comprehensible book on the subject "human brain" is even more difficult. Below are my collective thoughts on the human brain as a short introduction to this wonderful book.

Aristotle rightly argued that humans consist of three components: *Physis*, *Ethos*, and *Logos*. It is exactly what we nowadays respectively call Genetics, Environment (Epigenetics), and Logic. Mankind has indeed stood out from all the other beings living on our planet, and this is because of the emergence of Reason through the enormous and rapid evolution of the human brain, and through the addition of the intellectual functions that have distinguished *Homo sapiens*. It is, among others, logical thinking, the will to act accordingly, the imaginary travel through space and time, and the control of feelings that liberated our species from the tyranny of the "here and now," as well as the ability to anticipate and wait for reward, that have given humans their uniqueness. However, there is another function of the human brain that, although not unique to our species, is exceptionally well developed. This is empathy, a spontaneous, mutual ability of a person to participate both cognitively and emotionally, in the thoughts and feelings of another individual(s).

Empathy is the basis of social bonding, sociability, and social life. It starts with the mutual feeling of pleasure in the communication between a mother and her child. Although the immense importance of empathy in the survival and progress of our species has been known since antiquity, its biological basis has been clarified, to some extent, only recently. The famous Hierocles' concentric circles, from the Hellenistic period, gradually increase in diameter, starting with the mutual love between the child and the mother, followed by the expanding love for the father, the siblings and the rest of the family, the members of the tribe, the national compatriots, and, finally, for humanity as a whole. The development of empathy has helped us to survive and evolve as a species, as well as to grow intellectually as individuals. Pro-sociality, morality, and an inner sense of justice are all products of empathy and interlinked social life, which finally led to the development of conventional justice.

Science, Technology, and Politics are interrelated and constitute products of human reason, that is, the brain. The first seeks the truth, while the second and the third use this truth for the benefit of the human, both as an individual and as a social whole. All three disciplines are based on real data, on Aristotle's logic of induction and deduction, and on the observation of natural and/or experimental phenomena through the senses. Scientific knowledge has increased exponentially after the Second World War, beyond the multidimensional technological advances, and offers us a better understanding of our self-awareness. This occurs through a new enlightenment that recognizes the biological basis of our life and behavior and acknowledges that rationalism plays a major role in our thought processes and decision making. This new enlightenment also realizes the radical importance of upbringing and nurturing, which is, in other words, paideia and education, and provides an appropriate way and guidance to the highly desirable states of happiness and eudaimonia. In my opinion, the book "The Human Neural Network: The Way the Human Brain Works" constitutes a necessary textbook for someone who would like to understand the human species, as well as her/himself.

As for the brain's overall neural network, I would refer to the pioneer of Neurology and Psychiatry, F.J. Gall, who was the first to develop the theory of "organology," which suggests that the human mind is composed of independent "entities" located in the brain. Today, we know that the brain neural network consists of multiple neural subnetworks that serve both separate and overlapping functions. The number of neural circuits in these subnetworks is reflected in the number of synapses in the corresponding region of the brain. Interestingly, the neural circuits in the brain regions that subserve vision and hearing, speech, and high mental and executive functions develop quite early in life, peaking in number, respectively, at 4, 9, and 36 months of age. Thus, important brain areas and functions, such as logic, empathy, love, social consciousness, moral intelligence, and sense of justice, develop in early childhood. Progressively, there is an increase in brain white matter, which represents the "wiring" of the brain that interconnects the brain subnetworks, a process that is completed at 25–27 years of life. Therefore, this is when the whole, mature, fully integrated neural network of the human brain is completed, and all brain subnetworks are being interconnected and integrated. This means that the development and integration of a fully grown adult individual takes about a third of her/his life. This is something that we need to take into consideration when we read this book.

Before composing this Foreword, I read Professor Fountoulakis' book. I started with the useful Introduction and the rich historical background of our major intellectual steps in the understanding of the brain. I continued with its macroscopic functional anatomy and the description of its microstructure. I read about the mechanisms of communication between brain cells, about the brain's anatomical and functional organization, and the functions of arousal and sleep, as well as about the higher mental functions that differentiate our species from all others as human. Finally, I read about the processes of thinking and self-consciousness, the formation of personality and its components character and temperament, the exercise of free will, and the so important human sociobiology. The writing style of the book is fluent and highly understandable; the presentation of the brain is systematic, without being tedious for the reader, while its modular structure is flowing and didactic. This book is a treasure for the educated reader who is interested in learning more about the human brain.

University Research Institute of Maternal and Child Health & Precision Medicine
National and Kapodistrian University of Athens
Athens, Greece

Georgios P. Chrousos

Introduction

It is a great challenge for anybody to try to describe the way the human brain works. Even worse, trying to explain how the human mind works is even harder and the heterogeneity of the audience makes the attempt even more difficult. However, what is important to have in mind since the beginning is that there is a conceptual difference between the words "brain" and "mind." Whether this corresponds or not to the scientific evidence so far is obviously a matter of debate. What is certain is that the brain exists either alive or dead while the mind does not exist in a dead brain. This conceptual difference exists in the vocabulary, as well as in the ideas and the approach the overwhelming majority of mental health professionals have, and the current book has mainly them as the target audience.

The word "brain" denotes what is inside the skull, as the Greek equivalent "encephalos" also means. The word "mind" comes from the Latin word "mens" which is related to the Greek words "μένος" and "μήνις" which both mean "anger." It is also related to the Greek verb "μένω" which means "I stay" and in this way it reflects states during which the mind is trapped because of the dominance of anger and intense feelings. It should also be mentioned that in ancient Greek the word "μήνις" means also the "crescent" and thus, there is a relationship between the states of the mind and the phases of the moon. The word for month is "μήνας" in Greek and "mensis" in Latin. This group of words includes also "menses" and "menstrual" which attribute a temporal/cyclic as well as affective component. The word "μαινάς" was used for women who participated in orgies during ceremonies for the God Dionysus. The word "menescus" has a similar origin (Fig. 1). The Greek word for mind is "νους" and comes from the ancient word "νόος" which is of unknown origin; however it is related to the Latin "nos" which means "we" while other versions of the word denote property ("mine"). The words "noesis," "nootropics," and others come from it.

Essentially, the brain is the anatomic structure, the organ within which the processes that give birth to the mind take place. For centuries, the human body and especially the brain were "taboos," areas forbidden for scientific research. This is true to some extent even today concerning the study of behavior, of course not in terms of real restrictions and prohibitions, but because of the pressure from ideological approaches and attempts to understand and interpret human behavior in a voluntaristic and anti-scientific frame.

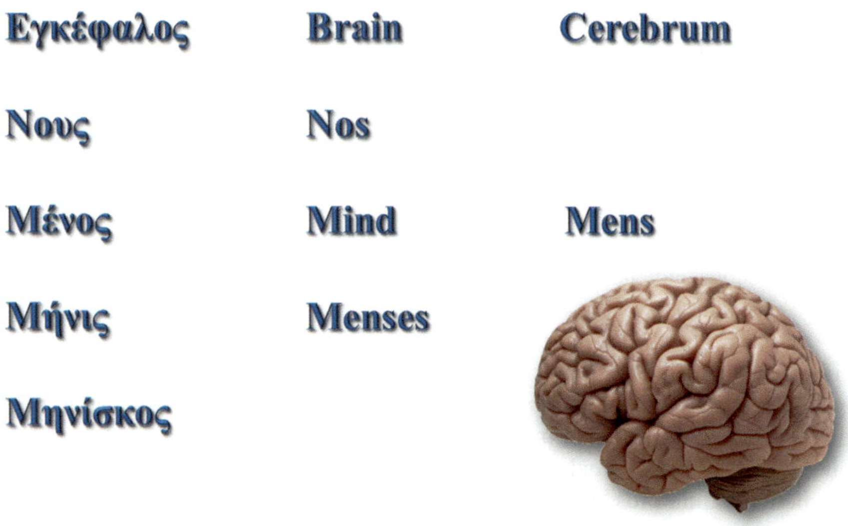

Fig. 1

The present book cannot avoid its involvement in debates of this type completely, since as the description of brain function will proceed, inevitably we will enter into reals as personality and free will. My intention is to preserve the strict scientific approach concerning all the topics this book will discuss. Equal distances are not in my intention and I never try to keep them. There are cases where I clearly adopt a certain position and cases where I reject all of them or others where I conclude that no suggestion has sufficient scientific documentation.

It is clear that the current book targets mainly the mental health care professionals as audience, and to a lesser extent the other health professionals. It is written in my view concerning the training and educational needs of Psychiatrists and Psychologists and to a lesser degree of Neurologists and Neuroscientists in general, although these disciplines will also find in its contents large pieces missing from their own training.

On the other hand, this book constitutes the end result of many years-long effort that goes beyond a Psychiatrist's standard clinical training and education, even that of a Psychiatrist that follows an academic career. As I am writing this introductory text, which I should note, I am writing after having already completed the entire book, I realized that this effort started almost 18 years ago, when as a young lecturer I was asked to teach the topic "Systems of neuronic structures" to medical students and in the frame of the lesson "general Psychiatry."

This topic was based in part on my doctorate dissertation which was on biological and psychophysiological indices in depression. But the surprise that gave a huge boost to things happened when the late professor Juan Lopez-Ibor, former President of the World Psychiatric Association, asked me to give a speech during his celebration ceremony of Emeritus at the University of Madrid in November 2012. Until that time I had several speeches and papers on biological psychiatry and psychopharmacology, but Lopez-Ibor asked me something quite different: to talk on "the human connectome," a term I was hearing for the first time and almost drove me into panic. So, the "human neural network" was the topic, something that forced me for the first time to build a presentation much more comprehensive and deeper than usual.

In the following years I continued to add pieces on the structure. But the most important was that I had to teach undergraduate medical students or to give lectures to heterogenous audiences composed of various different mental health professionals and even harder was that I was obliged to discuss on clinical matters with colleagues from other specialties of medicine, which had a completely different and manichaeistic-dualistic model in mind concerning human behavior and mental health. This interaction was continuously bringing fundamental questions to the surface, usually by students. We usually bypass this kind of questions by recruiting banal statements of the "biopsychosocial type." But these questions are so simple and so fundamental that they tend to be radical. Additionally, it was all the more obvious that many things we were discussing or teaching were not exactly that way, while some difficult questions had been "swept under the carpet."

Thus, only when I thought that I could give an answer to difficult questions, or when I exhausted (completely in my opinion) the search for answers, and I was in the position to describe the landscape of our knowledge and our understanding, even in those cases for which no real answers are available, only then I decided to start writing a book, and I started with the writing of a relevant chapter for my edited book *Psychobiology of Behavior* as a test; then I moved beyond.

So, the current book does not avoid any question; however it cannot answer all of them. It does reply to every question on the basis of where our level of knowledge and understanding is, without fear or passion and more importantly, without any kind of ideological or similar biases and on the basis of scientific evidence alone. I want also to make clear that science does not simply ask for the elimination of the supernatural from its dialectics. It additionally demands the adoption of a strict logical reasoning and development of arguments, in accord with the saying by Plato "Μηδείς αγεωμέτρητος εισίτω μοι τη θύρα" (meaning that "nobody who does not know geometry should enter"), but also in accord with the principles of analytic philosophy of Bertrand Russell and Ludwig Wittgenstein as well as with the principles of science as these were formulated primarily by Carl Popper and secondarily by Imre Lakatos.

However, it is certain that this book demands basic knowledge, and by stepping on it, goes a step further. At least basic knowledge of biology and human anatomy is needed, and secondary basic knowledge concerning the human behavior. Reading will likely be laborious at places, and probably the reader will need to go back and forth among chapters. Although every effort has been taken so that this need is minimized, it was impossible to completely extinguish it, and this is because the various structures, functions but also our knowledge are intertwined in many dimensions and demand the existence of elements from other chapters at almost every level of every topic discussion.

There was also a huge effort, and eventually rather disproportionately big after taking into consideration the volume of the text, so that the book is enriched with images and pictures of high scientific value but also high aesthetics, to the extent this was possible. Here, a huge problem both in terms of cost and in terms of clarifications was copyright. The fact that many images circulate for years widely on the internet does not mean they are not copyright protected, and their reproduction in a book would be problematic. All the images included in the current book are either public domain or permission was officially obtained (and the cost was significant) or were created by me from scratch.

Finally, I would like to thank all my associates during the last 20 years, my undergraduate and postgraduate students, my mentors, and especially Hagop Akiskal (whose loss this year was a very sad event), Hans-Jurgen Moeller, Siegfried Kasper, and Monte Buchsbaum who taught me how real teaching is.

Thessaloniki, March 2021

Kostas N. Fountoulakis

Contents

About the Author

Kostas N. Fountoulakis, MD is Professor of Psychiatry at Aristotle University of Thessaloniki, and director of the 3rd Department of Psychiatry at AHEPA University Hospital, in Thessaloniki, Greece.

He is currently zonal 8 representative (South Europe) in the Board of the World Psychiatric Association, National Representative for Mental Health in the World Health Organization, and Vice-President of the Hellenic Psychiatric Association. He chairs the Mental Health Section of the Institute of Scientific Research of the Greek Medical Association. For the years 2020–2021 he served as director of Cochrane Greece.

Dr. Fountoulakis' areas of clinical and research interest are reflected in the topics that he teaches: general psychiatry, biological psychiatry, psychopharmacology, mood disorders, schizophrenia, and personality disorders. He has coauthored more than 400 papers and more than 300 of them are published in international journals such as the *LANCET, BMJ, Am J Psychiatry, British Journal of Psychiatry, Biological Psychiatry, International Journal of Neuropsychopharmacology, Journal of Affective Disorders, Schizophrenia Research, Psychiatry Research, Bipolar Disorders,* and the *Annals of General Psychiatry* among others, approximately 15,000 citations and $h = 62$ (Publish or Perish).

During the years 2016–2018 he was ranked by expertscape.com within the top 25 world experts concerning bipolar disorder (0.068%) and among the top 50 concerning suicidality (0.097%). In 2021 he was placed among the 0.37% for year and 0.57% for the whole career of all scientists of all disciplines by the Stanford ranking.

He has authored or coauthored a number of chapters in books including the Mood disorders chapter for the Wiki project of the World Psychiatric Association (WPA). He has authored the book *Bipolar Disorders: An Evidence-Based Guide to Manic Depression* (Springer 2015), co-edited the WPA book *Advances in Psychiatry vol 3* (Springer 2018) and the book *Psychobiology of Behavior* (Springer 2019), and wrote the *Philosophical and historical roots and the current face of Psychiatry* which is in print (Springer 2022).

Dr. Fountoulakis served for 20 years as Editor of *Annals of General Psychiatry (IF = 2.5)*. Currently he serves as field editor of CNS Spectrums (IF = 3.9). In the past he also served as Section Editor of *Current Opinion in Psychiatry* as well as guest editor in other international journals.

He has received a number of national and international research awards, including the 2012 Kraepelin-Alzheimer medal of the University of Munich and the 2015 Excellence in Education Award of the WFSBP. Since 2014 he is honorary member of the WPA.

History

1

Contents

1.1 Antiquity and the Prescientific Era

In *Ancient Egypt*, history starts with Imhotep (Fig. 1.1) who lived during the Third Dynasty of the Old Kingdom under King Djoser (Zozer) (3000 BCE) and was an archpriest himself. His name means "the one who brings tranquility." The oldest surviving medical document is *The Edwin Smith Surgical Papyrus*, which was bought in the Egyptian city of Luxor in 1862 by the archeologist Edwin Smith (1822–1906). It was written approximately a thousand years after the death of Imhotep, but many historians believe that it contains much of his teachings (Breasted 1930). The papyrus (Fig. 1.2) includes information pertaining to brain injury and its treatment (Okasha and Okasha 2016).

Ancient Egypt
- Teachings of Imhotep
- The heart is the seat of mental life
- Analogy to the Nile Valley
- Dryness floods > equilibrium
- Early ideas pertaining to morality, sin, and guilt

K. N. Fountoulakis, *The Human Connectome*,
https://doi.org/10.1007/978-3-031-10351-3_1

Fig. 1.1 Imhotep, status possibly from the Hellenistic period in Egypt (between 664 and 30 BCE, Metropolitan Museum of Art)

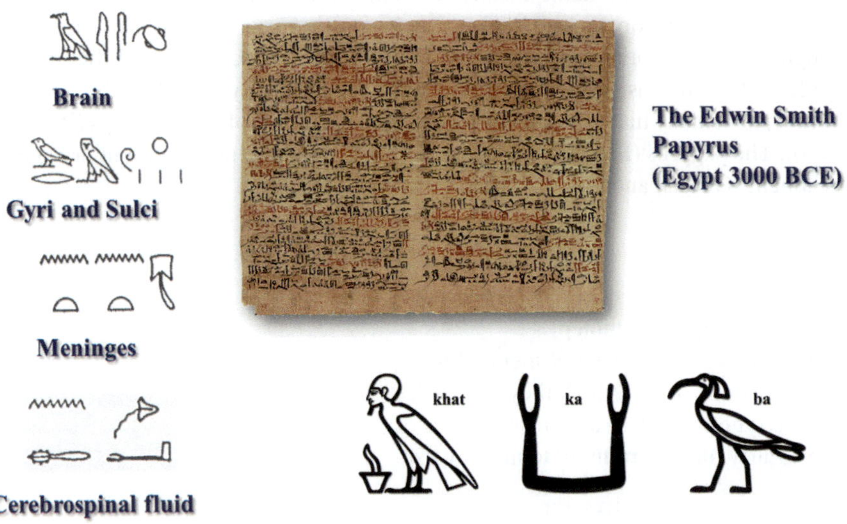

Fig. 1.2 The *Edwin Smith Papyrus* (3000 BCE) and the hieroglyphic writing of the parts of the brain and the soul according to the ancient Egyptian mythology

The Ancient Egyptians *considered the heart, not the brain*, to be the seat of mental life, and they conceptualized the body and the mind as an *analogue of the river Nile Valley*. According to their beliefs, humans were comprised by the "khat" (body), the "ka" (which protects the body), and the "ba" (spirit and destiny). They believed that a system of channels ("metu") was connecting the heart with the rest of the body by carrying air, liquids, and solid matter. The conceptualization as an analogue to the Nile environment pointed to the importance of moisture but also to the balance of elements and mechanisms, leading to the first idea of the "equilibrium." While the Nile Valley was suffering from too little (dryness) or too much water (floods), as well as from extremes of temperature, the human body and mind were suffering in a similar way. Reasoning by analogy, as the channels of the river should be kept open, but not too open, so should the "metu." In this frame, *bloodletting* was practiced both with the use of leeches and with *venesection*, while *cathartics* were used for the bowels.

Egyptian mythology suggested that after death, the heart which during life registers all good and bad deeds will be weighed against a feather to assess *sin and guilt* (Fig. 1.3). As a result, during mummification, the heart was treated with respect and was never removed from the body, in contrast to the brain which was extracted with the use of a hook through the nostrils and then discarded (Woollam 1958), while other internal organs were kept in sacred jars.

Fig. 1.3 At left, Ani (the deceased) and his wife Tutu enter the assemblage of gods. At the center, the jackal-headed god of the dead, Anubis, weights the heart of the deceased in the Duat using the feather of Maat as the measure in balance, observed by the goddesses Renenutet and Meskhenet, the god Shay, and Ani's own ba. At right, the monster Ammut, who will devour Ani's soul if he is unworthy, awaits the verdict, while the god Thoth prepares to record it. At the top are the gods acting as judges: Hu and Sia, Hathor, Horus, Isis and Nephthys, Nut, Geb, Tefnut, Shu, Atum, and Ra-Horakhty. The baboon sitting atop the scales represents Thoth, the god of writing, and is overseeing the procedures. The winged figure to the left of the scales is Ani's ba or physical soul

At the dawn of humanity, it was widely believed that *supernatural phenomena* were the cause of mental illness. These included *spiritual or demonic possession*, sorcery, the evil eye, or an angry deity (Roccatagliata 1973). Treatments were mystical but also aggressive and brutal.

Ancient Prescientific Beliefs

- Supernational conceptualization of mental life and health
- Evil spirits
- Skull trephination
- Religious and magical rituals

The first documented attempts to treat mental illness date back as early as 5000 BC. This is what the discovery of trephined skulls in regions that were home to ancient world cultures suggest (Fig. 1.4). The first report on *trephination* was made in 1863–1865 by the American anthropologist Ephraim George Squier (1821–1888) who observed a rectangular hole in a prehistoric skull during his visit to the Cuzco district of Peru (Fig. 1.5). In 1867, Paul Broca (1824–1880) confirmed that this hole was deliberately made.

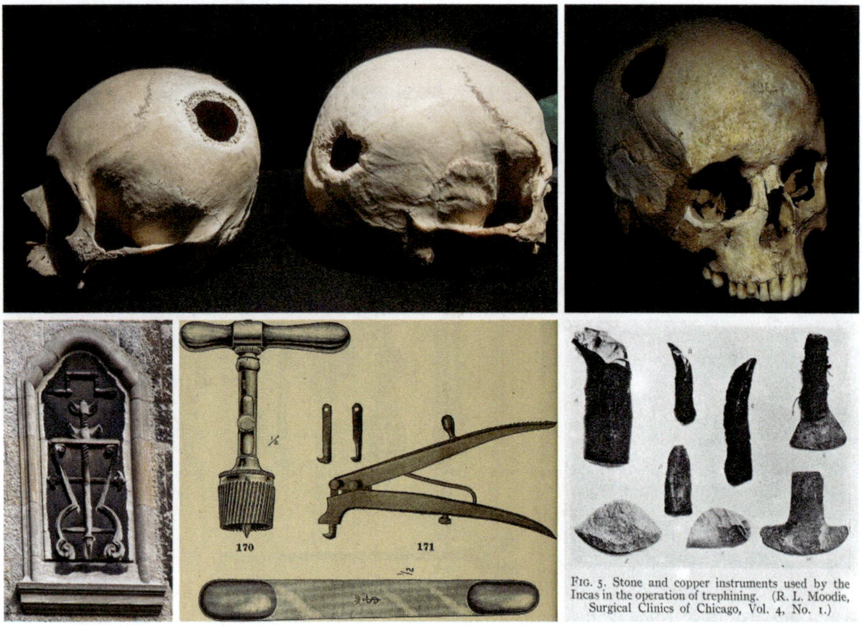

Fig. 1.4 Up: images of trephined skulls, which were found during excavations in various areas that hosted ancient civilizations. Down left: decoration of the external wall of the Montreal Neurological Institute, with depictions of trephination tools. Center: set of skull trephination tools from the book *List of staple medical and surgical supplies selected to meet war conditions* (1917). Right: ancient tools of the Incas (The Wellcome Collection)

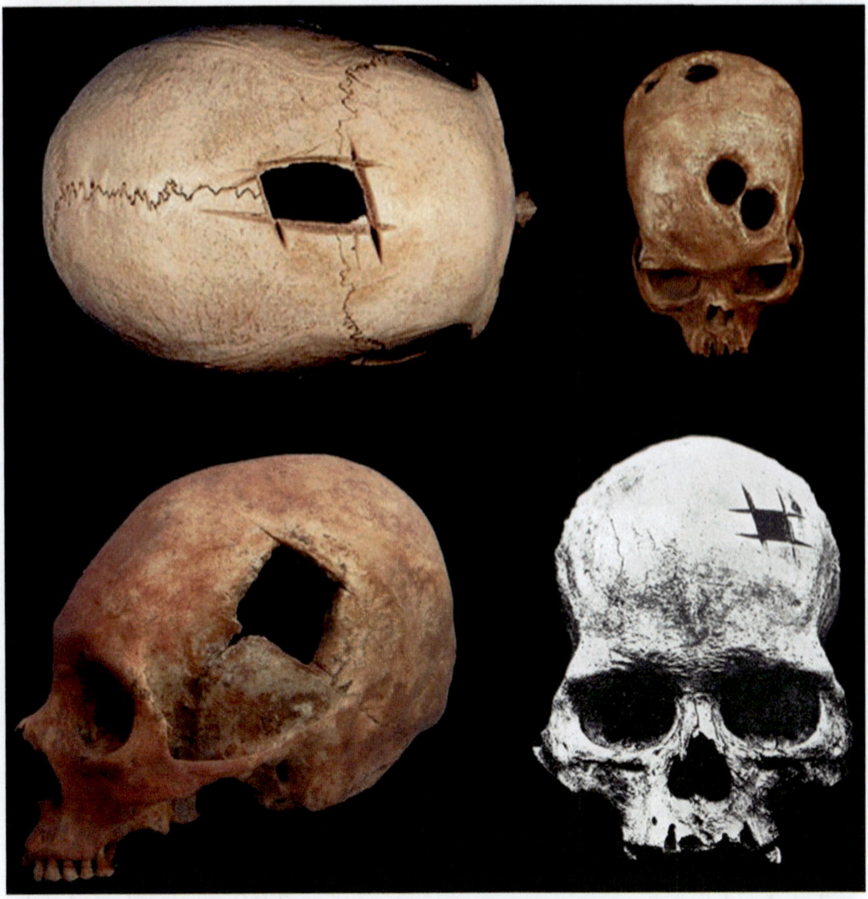

Fig. 1.5 Prehistoric skulls with marks of trephination, including the skull found by E.G Squier, at Cuzco Peru (bottom row, right)

Of prime importance was the observation concerning the presence of signs of infection in the porous structure of the bone surrounding the opening. This means that the patient had survived the operation, at least for a short period of time (O'Callaghan and Carroll 1982). The method of trephining includes the chipping of a hole, or trephine, into the skull using crude stone instruments but later more sophisticated tools (Fig. 1.6). The idea was that evil spirits would leave the head of the patient through this hole. The method survived until the early twentieth century in some cultures, as a cure for a number of health problems including skull fractures, migraines, and mental illness (Fig. 1.7).

In most ancient societies, magical powers, evil deities, or demons were considered to be the cause of mental illness, and the "treatment" concerned religious or magical rituals performed mainly by priests. Christianity named these deities "demons" (from the Greek word *δαίμων*) which means a deity of lower order and

Fig. 1.6 The bronze-cutting instrument, known as Antze's tumi, which depicts the operation of trephining (from Asenjo 1963). (Courtesy of Charles C. Thomas, Publisher, Springfield, Illinois)

attributed a definitely negative connotation to the word. However, even after the spreading of Christianity, these beliefs were dominant in many societies, even among members of the high socioeconomic classes of Europe. The history of civilization denotes that the belief in demons was changing in complex ways.

Fig. 1.7 The Cure of Folly, by Hieronymus Bosch (1450–1516), depicts a trephination operation. Painting displayed in the Museo del Prado in Madrid, completed around 1494 or later

1.2 The Dawn of the Scientific Era

The dawn of the scientific era started in *Ancient Greece*. Before that, there is a wealth of myths and stories concerning mental illness and its treatment. Melampus (name meaning "black feet") was a mythical figure who supposedly had cured the mania of Ifinoe and Ifianassa, the daughters of King Proetos of Argos. He also cured Ificlos, son of King Fylacos of Fylacae from his psychogenic impotency. In Iliad, Homer mentions the manic episode of Ajax, followed by deep depression and shame which results in the suicide of the hero. Interestingly, Ulysses had faked a mental disorder in order to avoid going to Troy. The hidden meaning of the myth of Jason and Medea and the Argonauts has been the focus of work of many scholars (Fig. 1.8). It is interesting that their names suggest a relationship with medicine and treatment. Jason is the English version of the name *Ιάσων* (*iason* meaning the healer) while Medea corresponds to *Μήδεια* (pronounced Midia, which comes from the Greek verb *μέδομαι* meaning take care, study, improvise, and predict). Medea relates also to the word Medusa (the gorgon) and to medicine (through Latin). He was of barbaric origin and considered a witch priestess. One interesting aspect of the myth is that Jason left her to marry *Γλαύκα* (Glauca, meaning owl, the bird of wisdom). After this Medea went to Thebes where she healed Hercules from the curse of Hera (bursts of anger similar to acute manic episodes during which he killed Iphitus, Hercules' best friend).

Prescientific Era in Greece
- Melampus
- Jason and Medea
- The god Apollo
- Centaur Chiron
- Asclepius

The myth of Jason and Medea and the Argonauts received a number of interpretations with the most dominant being that of the exploratory travelling of Ancient Greeks in the black sea and the Caucasus area, in search for gold. However, one cannot leave unnoticed the connotation of names: The "healer" goes east to bring eastern metaphysical tradition of how to take care and govern, only to soon abandon the newly acquired habits for the sake of knowledge. The early depictions of Jason, Media, and the dragon-guard of the golden fleece seem too similar to images of Adam, Eve, and the serpent (Fig. 1.8).

In another line of myths, Apollo, son of Zeus, was the God of medicine. He was the teacher of centaur Chiron (half human, half horse) who further enriched the healing practice with the discovery of medicinal herbs and hymns. According to Hesiod, in turn, Chiron taught Asclepius (Greek, *Ασκληπειός*; Latin, Aesculapius) who was the son of Apollo and the nymph Coronis (Edelstein et al. 1998). The name "coronis" (in Greek *κορωνίς*) is similar to Latin and French "coronis" and belongs

Fig. 1.8 Jason and Medea images on a Greek amphora from the National Archaeological Museum of Naples, c. fourth-century BC. The depiction of the serpent is quite similar to the modern symbol of medicine and the scene resembles the Old Testament scene of Adam, Eve, and the snake. The two stories share many common archetypal elements

to the same group of words with "crown" and "horn," meaning "something on the top of." This was an allegory, meaning that Asclepius was the ultimate of medical charisma. There seems to be a real person who was highly skilled in medicine, carried the name of Asclepius, and lived in Thessaly (near Pelion, the mountain of the centaurs) at the era of the Trojan War. According to myths, his wife was *Ηπιόνη* (Epione meaning the one who calms the pain). He had five daughters: *Υγεία* (Hygeia that is health) the goddess of health, cleanliness, and sanitation; *Ιασώ* or *Ιησώ* (Iaso that is Healer, conveniently similar to *Ιησούς*-Jesus which however is of a Hebrew origin) the goddess of recovery from illness; *Ακεσώ* (Aceso that is the therapist) the goddess of the healing process; *Αίγλη* (Aegli that is Prestige) the goddess of beauty, splendor, glory, magnificence, and adornment; and *Πανάκεια* (Panacea) the goddess of universal remedy. Later and after the third century BC, *Τελέσφορος* (Telesphorus meaning the one who brings a successful result) was added as a son, probably with the adoption of a Celtic god. Possibly but not proven beyond doubt, Romans added a sixth daughter called Meditrina (possibly meaning "healer") which represented the magic spells and herbs. Asclepius was associated with the Roman and Etruscan deity Vediovis (probably meaning the God on Earth). According to the myths, Asclepius managed to bring a man back from the dead; this provoked the anger of Zeus who killed him with a bolt of lightning. Like Imhotep, also Asclepius rose to become the demigod at the center of a healing cult in Greece. A number of temples, the *asclepieia* were dedicated to his worship and served as early hospitals (Fig. 1.9) (Majno 1975).

In one of these temples in the Aegean island of Kos, it is believed that *Hippocrates* may had begun his career. In these temples, treatment included a ritual purification followed by offerings or sacrifices to the God. Afterward, the patients would spend the night in the *άβατον* (avaton) or *άδυτον* (adyton); both names mean the place where nobody can enter. It constituted the most sacred place of the temple, and there, the God would contact the patients through their dreams. The priests would interpret these dreams or visions and would accordingly prescribe the appropriate therapy. An interesting detail concerns the Aesculapian snakes, a particular type of nonvenomous snake which were allowed to slither around freely on the floor in the rooms where the patients slept. In general, snakes were considered to be sacred because they emerge from the ground, and thus they carry the healing powers of Mother Earth (Gaea mother, the prototype representation of the female deity in Greek mythology). An additional reason snakes were considered sacred is that they shed their skin which is then rejuvenated; thus they are a symbol of death and resurrection and eternal life (Wickkiser 2008). Snakes were an important element of ancient Athenian worship; they represented their bonds with the land (crawl onto the ground), and they were the symbol of Athena, goddess of wisdom; a sacred snake was kept in the Parthenon before the Persian wars. It is interesting that a similar role (thought with a negative connotation) is reserved for snakes in the Genesis

Fig. 1.9 Design of the Asklepieion of Athens based on ancient ruins, from Caton, R: The temples and ritual of Asklepios at Epidaurus and Athens. Two lectures delivered at the Royal Institution of Great Britain. London, C. J. Clay, 1900

in the Old Testament. This is the origin of *Asclepius' rod* which was a snake-entwined staff, remaining until today as the symbol of medicine, denoting the therapeutic power the physician takes from Mother Earth. One similar symbol is the Caduceus (Κηρύκειον), the rod of the God Hermes, which is depicted with two snakes in a very interesting image resembling the DNA double helix and which is often used mistakenly as the symbol of medicine. However, the symbolism here is quite different, since the image comes from a myth which says that the God Hermes separated with his rod two snakes fighting to each other, thus establishing himself as the sacred messenger of peace.

The Dawn of the Scientific Era
- Thales of Miletus
- Hippocrates
- Polybus
- Corpus Hippocraticum
- The brain is the seat of mental processes
- Theory of the four humors

However, as said before, *the dawn of the scientific era started in Ancient Greece*. Maybe the first human ever to specifically reject mysticism was *Thales of Miletus* (652–588 BC; Fig. 1.10) in Ionia and then Greek east coast of Asia Minor, today west Turkey, who argued that gods are not responsible for the weather or any events in nature. He could be considered to be the first scientist on the face of the Earth. In this frame, he suggested that mental disorders are natural occurring events and should be studied and treated according to the scientific way.

The *Corpus Hippocraticum* (in Greek *Ιπποκρατική Συλλογή*) was created by the scholars of the library of Alexandria around 300 BCE, after the then ruler of Egypt Ptolemy Soter (367–283 BCE) ordered them to collect books containing all of human knowledge, by compiling all classical medical documents they could find (Canfora 1990; Sigerist 1961). These texts are probably the work of unknown authors, not of Hippocrates himself (Fig. 1.11).

In them, the *concept of the four essential body humors of the human body* is included. These humors include the following:

- Black bile (from the liver)
- Yellow bile (from the spleen)
- Phlegm (from the brain)
- Blood (from the heart)

There was the belief that these humors were secreted by different organs and were possessing different qualities and characteristics. The most important is that it was believed that the properties of the humors varied with the seasons. The early versions of this theory might had appeared in Egypt or Mesopotamia, but the essence of this approach is ascribed to the *school of Kos* and to Polybus which was a pupil

Fig. 1.10 Thales of Miletus Illustration from "Illustrerad verldshistoria utgifven av E. Wallis. volume I": Thales

and son-in-law to *Hippocrates* (fourth century BCE; Fig. 1.12) and author of the book *Περί φύσεως ανθρώπου* (*Peri physeos anthropou* meaning *On the Nature of Man*). This theory is in fact a microcosmic form of the macrocosmic theory of the four elements (earth, water, air, fire) and the four qualities (dry, wet, cold, hot) of the universe as first proposed by Empedocles (fifth century BC). Essentially, it was a model with two factors, humidity and temperature, as was the model of the universe (Fountoulakis 2015b) with wider metaphysical and philosophical consequences, as the deity for fire was Zeus and of the air was Hera, while the deity for the Earth was Ades (*Αιδωνεύς*) and of water (and sperm) was Nestis (*Νήστις*) (death vs. life).

It is important to note that *the brain was considered to be the seat of life*. In these "Hippocratic texts," some of which are only remotely related to Hippocrates himself or his school, it is clearly hypothesized that behind all diseases, there exists a specific *disequilibrium among the body humors*. In terms of Pythagorean philosophy, and in accord with previous Egyptian ideas, health was identified with *eucrasia* (in Greek *ευκρασία*, good admixture and balance), while in contrast, illness was

Fig. 1.11 Hippocrates (460–370 BCE), by unknown artist from Young Persons' Cyclopedia of Persons and Places, 1881

Fig. 1.12 The mind/body according to Pythagoras (tetractys)

identified with *dyscrasia* (in Greek *δυσκρασία*, bad admixture or imbalance). Different patterns of body fluids were believed to underlie different "idiosyncrasies" (in Greek *ιδιοσυγκρασίες* meaning unique or characteristic admixtures of elements and temperaments). It is interesting to mention that it was thought that yellow bile was responsible for the choleric temperament (irritable, hostile, and given to rage) while phlegm for the phlegmatic (indolent, irresolute, and timid) (Fountoulakis 2015a).

In general, during the times of Ancient Greece and the Roman Empire, religion, science, and philosophy were interplexed in the way mental illness was conceived. The psychological vs. somatic approach has its roots in that time period and essentially it derives from the differences in the philosophical views of *Plato* vs. those of *Aristotle's* (384–322 BCE; Fig. 1.13). Plato (424–348 BCE, Fig. 1.14) was probably the first who spoke on subconscious conflicts and mental biases. Both him and Socrates (470–399 BCE) were believing that mental symptoms (corresponding to axis I in previous DSM editions) are not related to personality (axis II), but they also proposed that one kind of mental illness is biological while another one is closely related to morality. These ideas are clearly presented in "Phaedrus" and "Timaeus" by Plato which are part of the "Socratic dialogues" (Seeskin 2008; Ahonen 2019; Ahonen 2014).

Fig. 1.13 Aristotle (384–322 BCE). Statue in the campus of the Aristotle University of Thessaloniki Greece

Fig. 1.14 "The School of Athens" by Raphael (Raffaello Sanzio da Urbino; 1483–1520) c. 1511. At the center of the picture, Plato points toward heaven and Aristotle to the Earth in a symbolic scene of the conflicting philosophical approaches. Around them gather other philosophers including the Greek Socrates, Zeno of Citium, Epicurus, Pythagoras, Antisthenes, Anaximander, Heraclitus, and Diogenes of Sinope but also Averroes, Zoroaster, and possibly other more contemporary figures

Plato and Aristotle

- Reason
- Morality values
- Psychological vs. biological approach
- Subconscious conflicts
- Mental biases
- The brain vs. the heart as the seat of mental life
- Experience vs. inherent characteristics
- Herophilus of Alexandria
- Galen

Plato was the first who considered logic and strict arguments as an absolute prerequisite for the philosophical thinking. In this way, he was the first who conceived what in the early twentieth century was named "analytical philosophy" and was introduced by Bertrand Russell (1872–1970). More specifically, Plato considered mathematics and geometry as a prerequisite and cornerstone for philosophy. He was so convinced that mathematics were absolutely necessary in order someone to become a philosopher and that he hung the inscription *Μηδείς αγεωμέτρητος εισίτω μου την στέγην*, which means that nobody who is not familiar with mathematics should enter into the academy. While Plato was more a philosopher and argued on the supremacy of the brain, Aristotle was more an early scientist rather than a philosopher and stressed the need for experimental verification and the existence of observable empirical data in order to avoid subjective misinterpretations. Aristotle was also an accomplished animal dissector, and it is reported that he had an experience of dissection with 49 animal species. This experience led to the observation that the heart was full of activity as well as blood while the brain was still and bloodless (Clarke 1963). According to Aristotle, the role of the brain was minor and supportive to that of the heart, mainly by assisting in the preservation of the homeostasis around the heart. In his book *Περί Ψυχής* (De Anima; On the Soul), he argues that experience rather than inherent ideas and traits shapes the human mind.

A different story was that of *Herophilus* (335–280 BCE), a Greek physician who lived in Alexandria of Egypt during the Hellenistic period. He was the first anatomist and he is said to had had performed more than 600 anatomic dissections in alive prisoners. He was much interested in the brain and considered it to be the site of intellect. He believed that neural transmissions occurred by means of pneuma (in Greek *πνεύμα* meaning the spirit) and that humoral disequilibrium prevented pneuma from reaching the brain and thus resulting in mental disorders (Debru and Von Staden 1991; Dobson 1925).

1.3 From the Middle Ages to the Renaissance and the Early Modern Era

In 1543, *Andreas Vesalius* (1514–1564; Fig. 1.15) published *De humani corporis fabrica* (*On the Workings of the Human Body*; Fig. 1.16) with Book VII dealing exclusively with the brain (O'Malley 1964). Vesalius realized that a problem with Galen's anatomy was that it was based entirely on animal dissection, and thus his assumptions were often not in accord with the actual human anatomy. One case of such a false belief was the existence of the "rete mirabile" in humans. This structure was supposed to mediate the communication between the soul and the brain, and it exists in some animals as a network of veins around the pituitary, but it does not exist in humans. On the other hand, Vesalius pointed out that human ventricles are almost identical in shape to those of other primates, although there is an immense difference in cognitive abilities. Vesalius was a dedicated Christian and died after a pilgrimage in Jerusalem on the Greek island of Zante and buried in the cemetery of the Church of Santa Maria delle Grazie, which was completely destroyed by the earthquake of 1953 (Fig. 1.17).

Fig. 1.15 Portrait of Vesalius created by to Jan van Calcar from page xii of *De humani corporis fabrica* (1534 edition)

Renaissance and the Early Modern Age

- Andreas Vesalius
- Rene Descartes
- Thomas Willis
- Precise anatomical brain studies
- Cartesian dualism of mind (soul) vs. brain (body)
- "Automata"
- "Soul of the beasts" or "corporeal soul"
- Hierarchical organization of the brain
- Localization of brain function

While there is not a specific year which distinguishes between the renaissance and the early modern period, it is generally considered that *René Descartes* (1596–1650; Fig. 1.18) was born during the last decades of the renaissance and died during the first years of the early modern era. In a pivotal moment in the history of modern psychiatry and the neurosciences, in his books, *The Description of the*

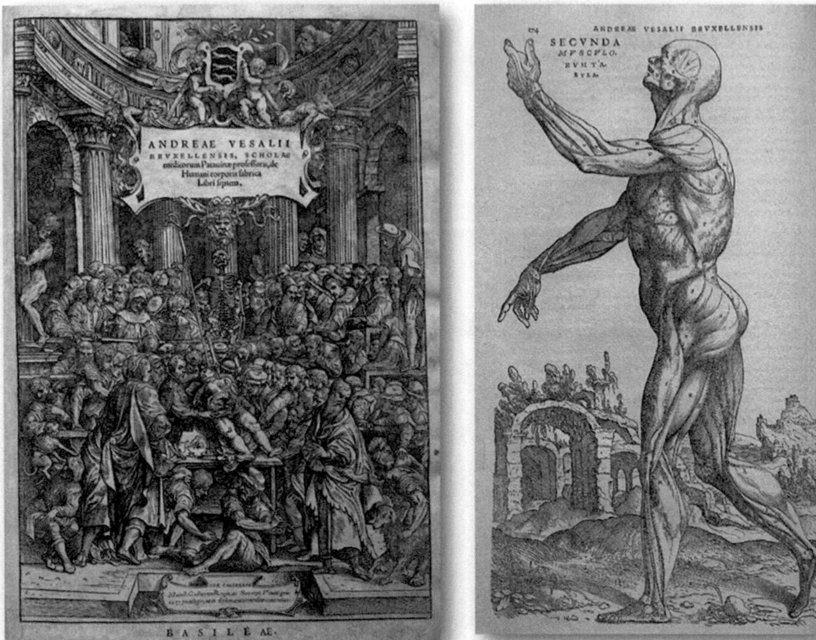

Fig. 1.16 Images from the book *De humani corporis fabrica* (*On the functioning of the human body*), by Andreas Vesalius, 1534

Human Body (1647) and *Passions of the Soul* (1649), he proposed the mind-body dichotomy (Cartesian dualism) (Descartes et al. 1984).

This was essentially a rejection of the Aristotelian approach which was dominant at that time, especially in religious circles, and in accord with the ideas of Plato (Mendler 2014). His assumption suggested that the mind was distinct from matter but could influence matter (through an unknown way). He also thought that "animal spirits" travel through the nerves. Throughout his life, Descartes experienced a threatening atmosphere because of his radical ideas, and this was the reason he chose to delay the publication of his ideas, with his cardinal work being published only months before his death. Because of these fears, he changed 24 different addresses in 13 different towns, and in general, he kept his address secret.

It is believed that Descartes was impressed and inspired by the spectacular animated fountains in the Royal Gardens of Paris, with the moving figures of Neptune, Diana, and other mythological entities. These mechanical wonders in a way inspired him on how to consider the human body. Overall Descartes was greatly impressed by the "automatons" of the time, which were manufactured mainly as toys, but also efforts were made to use them for practical as well as military purposes. Such devices were on display throughout Paris, and it was a fashion that lasted for centuries (Fig. 1.19). He modeled the human body after these devices and suggested that the brain/body and the mind are completely separate (Cottingham 1978). At that

Fig. 1.17 The monastery of Santa Maria delle Grazie, on the Greek island of Zakynthos, c. 1950, before the devastating earthquake of 1953

time, there was no knowledge of how the senses (vision, hearing, smell, etc.) function, and thus he was mainly influenced by theology and physics (Watson 1982; Cunning 2014a).

According to his theory, the two substances, body and mind, are essentially distinct and each of them could exist apart from the other. However, he gave priority to the mind by suggesting that it can exist without the body but the body cannot exist without the mind (Dicker 2013). This was essentially a compromise between the Catholic church and the scientific community, allowing medicine to progress in issues pertaining to the physical body but at the same time keeping the realm of human behavior (and mental health) still in the "jurisdiction" of religion. This is known as *Cartesian dualism* (*or mind-body dualism*) and exerted a profound influence on Western philosophy and culture (Urban 2018; Van Gijn 2005; Haas 1992; Shampo and Kyle 1975).

However, Descartes could not answer the riddle of the mind-body interaction which was evident through basic functions (e.g., hunger, sexual desire, etc.) as well as the effect of substances on behavior (e.g., alcohol) (Cunning 2014b). Thus, he proposed that the pineal gland was serving as connector between them (Berhouma 2013; Lokhorst and Kaitaro 2001; Smith 1998; Shiraev 2010) and that essentially the pineal gland is the *siege de l'ame* (seat of the soul) for a number of reasons including the "rete mirabile" and the fact that the pineal gland is a singular structure (Finger 1995). The proposal that the pineal gland was under the control of God was in accord with the widely spread belief that God was everywhere and was filling the

Fig. 1.18 Portrait of Rene Descartes (1596–1650) painted by Frans Hals the Elder

space among people and therefore every connection and communication among people was mediated by God.

Descartes and his contemporaries Johannes Kepler (1571–1630) and Galileo Galilei (1564–1642) *expelled the religious "final cause" from the physical universe* (res extensa), but as a compromise, Descartes kept the mind (or res cogitans) under the jurisdiction of religion and the Church and supported the beliefs concerning the immortality of the soul, as well as the validity of a divine determinism concerning behavior and mental life (Janiak 2015; Duncan 2000; Shapiro 1999; Dicker 2013).

Descartes remained a faithful Catholic through his entire life, and this had a profound influence on his ideas and work (Rodis-Lewis 1999). In spite of all his efforts, nothing could stop the Catholic church from prohibiting his books in 1663 (Mendler 2014). It should be noted, however, that Cartesian "dualism" is probably not in accord with the Christian teaching and dogma and reflects more oriental religions which consider the body to constitute the shell or even the prison of the soul.

It is interesting what within this largely theological frame, Descartes believed that mathematics and mechanics could explain the function of the mind (Kogge and Richter 2013). In accord with his dualistic approach to the body/mind problem, Descartes denied that animals had "soul," "reason," and "intelligence" and considered them to be similar to the automatic machines of the time ("automata") (Kogge and Richter 2013). This was widely accepted and one of the consequences it was

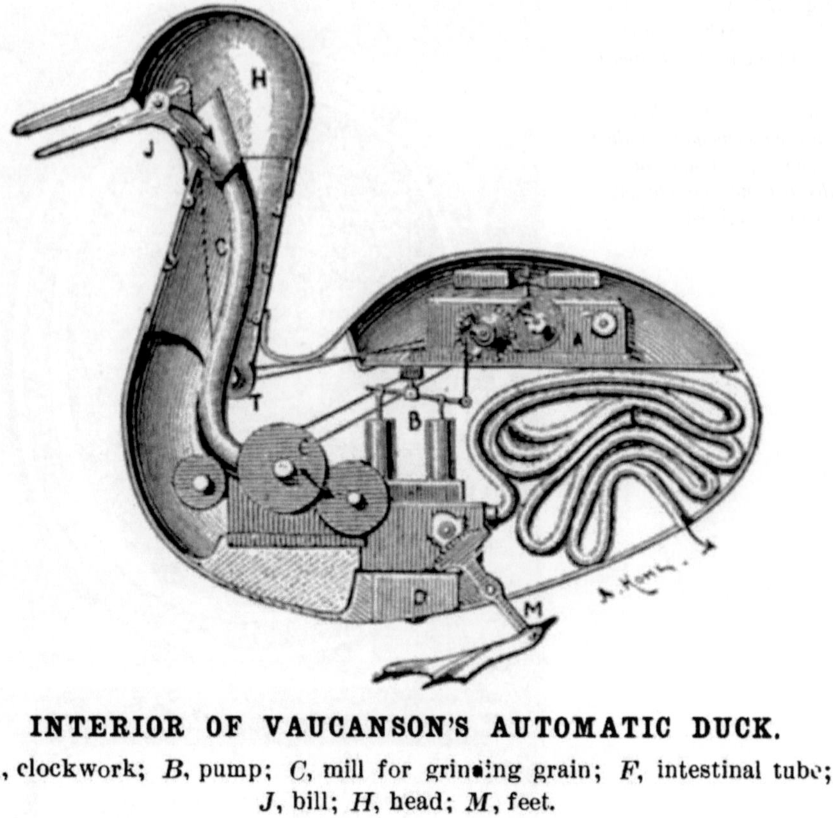

Fig. 1.19 An example of an "automaton" was the Canard Digérateur, or Digesting Duck was an automaton in the form of a duck, built by Jacques de Vaucanson in 1739. The picture is an imagery design of how this duck would look like, from the January 1899 edition of *Scientific American*

that it justified the cruel treatment of animals until the emergence of the theories of Charles Darwin in the mid-nineteenth century.

In the long term, Dualism was problematic, caused, and still causes damage, merely because it was so successful and was eventually adopted by the church and the lay culture. It is still a dominant approach for many philosophers and mental health professionals even in a vague and unsystematic way (Ventriglio and Bhugra 2015; Sessa 2009; Nierenberg 2009; George et al. 2000).

Thomas Willis (1621–1675; Fig. 1.20) is considered by many to be the father of neurology as he was the one to have coined the term *neurology* itself and gave to the term *psychology* its current meaning. In contrast to Descartes, he adopted a *monistic approach* to the mind and brain and was the first to propose a hierarchical organization of the brain into levels of function, from back and low to frontal and higher. He published the book *De anima brutorum* with his ideas concerning the mind-body problem in 1672 (Finger 2001).

Fig. 1.20 Thomas Willis (1621–1675) by unknown artist from his book *Opera omnia, nitidius quàm vnquam hactenus edita, plurimùm emendata, indice rerum copiosissimo, ac distinctione characterum exornata*, c.1694

Willis was a faithful Christian too and friend to the Archbishop of Canterbury with whom he frequently discussed his ideas and theories. Subsequently, he was very careful with issues pertaining to the mind and soul. Interestingly, in his work, *De anima brutorum*, he distinguished the immortal human soul from the "soul of brutes" (animals) which he also calls "corporeal," "body," or "material" soul. He suggested that humans possess both kinds of souls but animals only the inferior and mortal one.

The important implication of Willis' approach was that his ideas permitted animals to think and act somewhat rationally; thus, they were not simple automata without a soul. By this, Willis succeeded in a remarkable achievement, that is, to secure several issues simultaneously; he was free to research the brain and its functions and free to study aspects of human behavior, animals were not mere soulless things (automata), and at the same time, the Christian Church was fully satisfied with all these.

Starting in the late eighteenth century, Franz Josef Gall (1758–1828; Fig. 1.21) accepted the theory of Johann Lavater (1741–1801) that the physical shape of the body and face could directly reflect specific psychological characteristics and

Fig. 1.21 Franz Josef Gall (1758–1828)

eventually founded "phrenology." Phrenology was controversial at that time and today is considered to be an absolute "pseudoscience." However, it should be noted that Gall's work overall promoted psychology as a separate discipline, and it would be unfair to remember him exclusively because of phrenology.

Initially, Gall developed the theory of "organology" (mind is composed of independent entities hosted by the brain) and the method of "Cranioscopy" (assessing personality and psychological characteristics by studying the external shape of the skull; Fig. 1.22). Therefore, he is considered the first to theorize the localization of brain function (although in reality, it was Emanuel Swedenborg). Because of the opposition by the Catholic church, he was prohibited to give lectures in his home country Austria, so he was obliged to move to Paris; however, also there, he was treated as a threat to the French culture (Finger 2001).

The term "phrenology" was given to Gall's theories by his research partner Johann Gaspar Spurzheim (1776–1832) (Finger and Eling 2019). By the mid-nineteenth century, phrenology was no longer an appealing theory, and its supporters were waning dramatically. The problem with Gall's ideas was that they were based on the study of the skull and never of the brain itself. He often neglected opposing data and was cherry-picking concerning observations. However, through his work, he significantly contributed to the methods of anatomical study of the brain which previously involved unsystematic, crude, and random slicing of the tissue (Livianos-Aldana et al. 2007; Temkin 1947; Van Wyhe 2002).

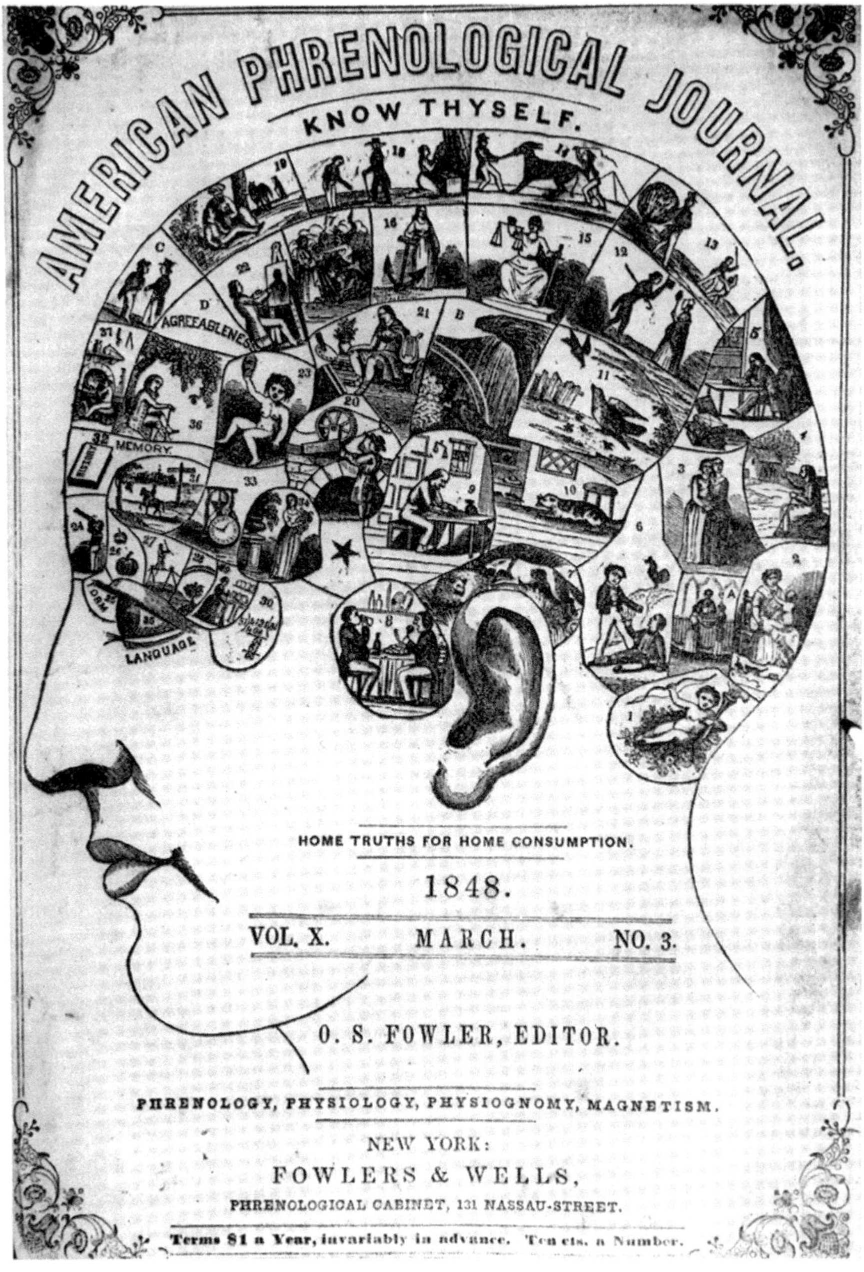

Fig. 1.22 Picture of the 1848 edition of *American Phrenological Journal* published by Fowlers and Wells, New York City. It shows the localization of the various psychological domains on the skull

1.4 The Great Revolutions and the Extinction of the Supernatural

And then the era of the great revolutions came. The first was launched by *Charles Robert Darwin* (1809–1882; Fig. 1.23) who published the theory of evolution in his book *On the Origin of Species* in 1860 (Darwin 1860), and he continued working during the 1860s and 1870s.

The Great Revolutions and the Extinction of the Supernatural

- Charles Darwin
- Sigmund Freud
- Korbinian Brodmann
- Theory of evolution
- Evolutionary approach to mental function
- Modeling of mental life
- Charting of the brain

He proposed a common ancestry for all species of life, and while at that time this is an exceptionally radical idea, today is considered to be a foundational concept

Fig. 1.23 Charles Robert Darwin (1809–1882)

and a triumph of science, a unifying theory of the life sciences (Coyne 2009). Central to his theory was the concept of *natural selection*, a process through which evolution leads to more fit species. Overall his work made the study of animals a model for the study of elements of human behavior (Kandel 2000; Fountoulakis 2019). His book *The Expression of the Emotions in Man and Animals* (Darwin 1872) triggered an interest in emotions from an evolutionary perspective (Fig. 1.24). In that book, Darwin stressed the *universal nature of emotions* and the connection of mental states to the neurological organization of movement. Central to his understanding was a shared human and animal ancestry, in sharp contrast to the contemporary claims that there were divinely created human muscles to express uniquely human feelings. Darwin's original suggestion was that emotions evolved via natural selection and therefore have cross-culturally universal counterparts, a proposal confirmed almost a century later. Furthermore, *animals undergo emotions comparable to those of humans*.

A few decades later, a young Viennese physician named *Sigmund Freud* (1856–1939; Fig. 1.25) begun treating neurotic patients with hypnosis. Later, he developed the "free thought association" and the "interpretation of dreams." In 1896, he coined the term *psychoanalysis* to define this new method of treatment. The term comes from the Greek word *ψυχή* (*psichi*, meaning soul) and *ανα* (*ana*, meaning again) and the verb *λύω* (*lio*, meaning open, free, solve, and disassemble). Gradually, he started developing a theory on the structure and functioning of human mental life with special emphasis on the "unconscious" which was a concept known to philosophers since the seventeenth century, but it was introduced to psychiatry for the first time by Freud (Brown 1999). The twentieth century begun with the publication of *The Interpretation of Dreams* by Sigmund Freud, in 1900. It triggered

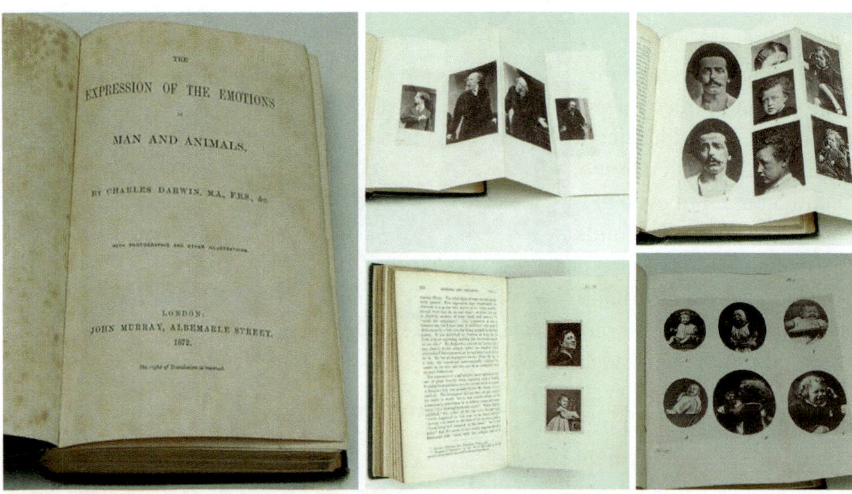

Fig. 1.24 The book *The Expression of the Emotions in Man and Animals* by Charles Darwin (1872). (After permission by Botterweg Auctions Amsterdam)

Fig. 1.25 Sigmund Freud (1856–1939)

a revolution which spread beyond psychiatry to philosophy, sociology, and lay culture, often with malignant effects (Lieberman 2016; Torrey 1992).

It should be recognized however that the major contribution of psychoanalysis was that it spread a rejection of the supernatural in the wider human culture concerning the interpretation of behavior. However, on the other hand, it was not scientific in nature, not even in terms of the early nineteenth century.

This constituted a fundamental problem, as it quickly has shown with the work of Carl Gustav Jung (1875–1961) which included strong metaphysical and supernatural elements. Jung influenced a wide range of disciplines from psychiatry to philosophy and anthropology and to religious studies. In 1910, he was the first head of Freud's newly founded International Psychoanalytic Association (Fig. 1.26). However, soon Jung developed his own and very distinct theory, that of analytical psychology, which focused on the development of self as the main task of human psychological development (Jung 1916).

His theory included a number of novel concepts, including persona, the archetypal phenomena, the collective unconscious, the psychological complex, and extraversion and introversion. The concept of "collective unconscious" probably came

Fig. 1.26 Third International Psychoanalytical Congress, 1911, in Weimar Germany. Sigmund Freud, Carl Jung, Karl Abraham, Otto Rank, and Sándor Ferenczi in the middle. What is interesting in this photo is the pyramidal shape of the gathering with Freud in the middle, whose figure emerges as taller and dominant in comparison to others. Body heights do not match, pointing to the careful setting of the photo

directly from Hegel's notion of Volksgeist and was of prime importance since it was considered to include memories, ideas, and other psychological components, all inherited. By focusing on the collective unconscious, Jung de-emphasized the importance of the libido, and this caused a bitter conflict with Freud in 1912 when his book on the psychology of the unconscious was published. The next year it seems that he experienced some kind of a psychotic episode with visual and auditory hallucinations, and from then on, his thoughts included a more spiritual essence with a mystical and occult perspective and with an interest in paranormal phenomena and alchemy (Jung and Pauli 1955; Jung 1960). What Jung seemed to have eventually embraced, a form of refined metaphysical approach, was exactly what Freud was trying to eliminate from psychiatry and psychology.

The theories by Darwin and Freud had as a common characteristic and as major contribution to human thinking the elimination of the metaphysical and the supernatural from the life sciences and mental life. After them, the total spectrum of life and human experience could be discussed in terms very close to those of the natural sciences. Their wide popularity led to deep social transformations, often in contrast to the principles these theories had, and it is ironic that they obtained an ideological quality and were determined in terms of a social movement ("Darwinism," "Freudianism") with their founders obtaining a status of a "quasi-prophet." This fact proves of course that emotional antirationalism is what characterizes the humankind much more than reason; in the overwhelming majority of cases, reason is exhausted in the solving of practical everyday problems.

In 1909, *Korbinian Brodmann* (1868–1918; Fig. 1.27) published his work on cortical cytoarchitecture, and later he described what is today known as Broadmann's brain areas. In his studies, he described the structure and cellular composition of 52 specifically defined areas of the cortex, which are included into 11 histological groups.

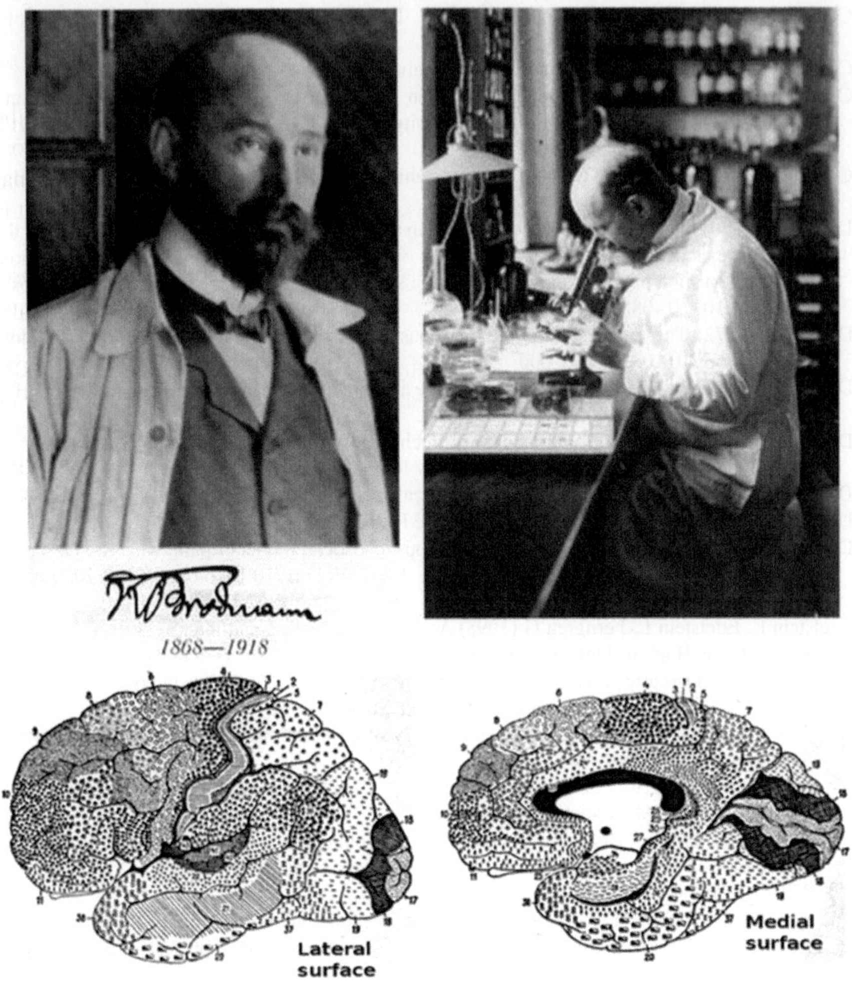

Fig. 1.27 Up: Korbinian Brodmann (1868–1918). Down: Broadmann's areas in the brain

References

Ahonen M (2014) Plato on madness and mental disorders. In: Mental disorders in ancient philosophy. Springer International Publishing, Cham, pp 35–67. https://doi.org/10.1007/978-3-319-03431-7_3

Ahonen M (2019) Ancient philosophers on mental illness. Hist Psychiatry 30(1):3–18. https://doi.org/10.1177/0957154X18803508

Berhouma M (2013) Beyond the pineal gland assumption: a neuroanatomical appraisal of dualism in Descartes' philosophy. Clin Neurol Neurosurg 115(9):1661–1670. https://doi.org/10.1016/j.clineuro.2013.02.023

Breasted J (1930) The Edwin Smith surgical papyrus. University of Chicago Press, Chicago

Brown P (1999) Pierre Janet (1959–1947). In: Freeman H (ed) A century of psychiatry. Mosby, London

Canfora L (1990) The vanished library. University of California Press, Berkeley

Clarke E (1963) Aristotelian concepts of the form and function of the brain. Bull Hist Med 37:1–14

Cottingham J (1978) 'A brute to the brutes?': Descartes' treatment of animals. Philosophy 53(206):551–559

Coyne J (2009) Why evolution is true. Oxford University Press, Oxford

Cunning D (2014a) The Cambridge companion to Descartes' *meditations*. Cambridge companions to philosophy. Cambridge University Press, Cambridge. https://doi.org/10.1017/cco9781139088220

Cunning D (2014b) The Cambridge companion to Descartes' meditations. Cambridge University Press

Darwin C (1860) On the origin of species by means of natural selection: or the preservation of the favoured races in the struggle for life. On the origin of species by means of natural selection: or the preservation of the favoured races in the struggle for life. John Murray, London. https://doi.org/10.1037/14088-000

Darwin C (1872) The expression of the emotions in man and animals. John Murray, London. https://doi.org/10.1037/10001-000

Debru A, Von Staden H (1991) [Herophilus, or the art of medicine in ancient Alexandria]. Rev Hist Sci Paris 44(3–4):435–445

Descartes R, Cottingham J, Stoothoff R, Murdochtrans D (1984) The philosophical writings of René Descartes (1641). Cambridge University Press, Cambridge

Dicker G (2013) Descartes: an analytic and historical introduction. Oxford University Press, Oxford

Dobson JF (1925) Herophilus of Alexandria. Proc R Soc Med 18:19–32

Duncan G (2000) Mind-body dualism and the biopsychosocial model of pain: what did Descartes really say? J Med Philos 25(4):485–513. https://doi.org/10.1076/0360-5310(200008)25:4;1-A;FT485

Edelstein E, Edelstein L, Ferngren G (1998) Asclepius: collection and interpretation of the testimonies. Johns Hopkins University Press, Baltimore

Finger S (1995) Descartes and the pineal gland in animals: a frequent misinterpretation. J Hist Neurosci 4(3–4):166–182. https://doi.org/10.1080/09647049509525637

Finger S (2001) Origins of neuroscience: a history of explorations into brain function. Oxford University Press, Oxford

Finger S, Eling P (2019) Franz Joseph Gall. Oxford University Press, New York. https://doi.org/10.1093/oso/9780190464622.001.0001

Fountoulakis K (2015a) Historical perspective. In: Fountoulakis K (ed) Bipolar disorder: an evidence-based guide to manic depression. Springer-Verlag, Berlin, pp 1–25. https://doi.org/10.1007/978-3-642-37216-2

Fountoulakis K (2015b) The role of temperament. In: Fountoulakis K (ed) Bipolar disorder: an evidence-based guide to manic depression. Springer-Verlag, Berlin, pp 163–179. https://doi.org/10.1007/978-3-642-37216-2

Fountoulakis KN (2019) Ethology, evolutionary psychology, sociobiology, and evolutionary psychiatry. In: Fountoulakis KN, Nimatoudis I (eds) Psychobiology of behaviour. Springer International Publishing, Cham, pp 345–395. https://doi.org/10.1007/978-3-030-18323-3_12
George B, Carpentier A, Evans J, Apuzzo ML (2000) Rene Descartes: mind, reason, and challenging assumptions. Neurosurgery 47(1):244–249. https://doi.org/10.1097/00006123-200007000-00054
Haas LF (1992) Rene Descartes 1596-1650. J Neurol Neurosurg Psychiatry 55(3):176. https://doi.org/10.1136/jnnp.55.3.176
Janiak A (2015) Space and motion in nature and scripture: Galileo, Descartes, Newton. Stud Hist Phil Sci 51:89–99. https://doi.org/10.1016/j.shpsa.2015.02.004
Jung C (1916) Collected papers on analytical psychology. Bailliere, Tindall and Cox, London
Jung C (1960) Psychology and religion. The Vail-Ballou Press, Binghamton
Jung C, Pauli W (1955) The interpretation of nature and psyche. Pantheon Books, New York
Kandel E (2000) The brain and behavior. In: Kandel E, Schwarts J, Jessell T (eds) Principles of neural science. McGraw-Hill, New York, pp 5–18
Kogge W, Richter M (2013) Synthetic biology and its alternatives. Descartes, Kant and the idea of engineering biological machines. Stud Hist Phil Biol Biomed Sci 44(2):181–189. https://doi.org/10.1016/j.shpsc.2013.03.013
Lieberman JA (2016) Shrinks: the untold story of psychiatry. Back Bay Books, New York
Livianos-Aldana L, Rojo-Moreno L, Sierra-Sanmiguel P (2007) F.J. Gall and the phrenological movement. Am J Psychiatry 164(3):414. https://doi.org/10.1176/ajp.2007.164.3.414
Lokhorst GJ, Kaitaro TT (2001) The originality of Descartes' theory about the pineal gland. J Hist Neurosci 10(1):6–18. https://doi.org/10.1076/jhin.10.1.6.5629
Majno G (1975) The healing hand. Harvard University Press, Cambridge
Mendler R (2014) False truths: the error of relying on authority. Hamilton Books, Falls Village
Nierenberg AA (2009) Failure to integrate brain and mind (or blaming Descartes to understand a very modern dilemma about explanatory models, psychiatric treatment, and underfunding of clinical care and research). CNS Spectr 14(11):599–600. https://doi.org/10.1017/s109285290002383x
O'Callaghan M, Carroll D (1982) Psychosurgery: a scientific analysis. MTP Press Limited, Lancaster
O'Malley C (1964) Andreas Vesalius of Brussels, 1514–1564. University of California Press, Berkeley
Okasha A, Okasha T (2016) Notes on mental disorders in pharaonic Egypt. Hist Psychiatry 11(44):413–424. https://doi.org/10.1177/0957154x0001104406
Roccatagliata G (1973) Storia de la psichiatria antica. Hoepli, Milano
Rodis-Lewis G (1999) Descartes his life and thought. Cornell University Press, Ithaca
Seeskin K (2008) Plato and the origin of mental health. Int J Law Psychiatry 31(6):487–494. https://doi.org/10.1016/j.ijlp.2008.09.004
Sessa B (2009) Let's not give in to Descartes: we need pharmacology and psychotherapy. J Psychopharmacol 23(2):227–228. https://doi.org/10.1177/0269881108093884
Shampo MA, Kyle RA (1975) Rene Descartes (1596-1650). JAMA 234(12):1247. https://doi.org/10.1001/jama.234.12.1247
Shapiro L (1999) Princess Elizabeth and Descartes: the union of soul and body and the practice of philosophy. Br J Hist Philos 7(3):503–520. https://doi.org/10.1080/09608789908571042
Shiraev E (2010) A history of psychology: a global perspective. Sage Publications, Los Angeles
Sigerist H (1961) A history of medicine. In: Early Greek, Hindu, and Persian medicine, vol 2. Oxford University Press, New York
Smith CU (1998) Descartes' pineal neuropsychology. Brain Cogn 36(1):57–72. https://doi.org/10.1006/brcg.1997.0954
Temkin O (1947) Gall and the phrenological movement. Bull Hist Med 21(3):275–321
Torrey FE (1992) Freudian fraud: the malignant effect of Freud's theory on American thought and culture. Harper Collins, New York

Urban E (2018) On matters of mind and body: regarding Descartes. J Anal Psychol 63(2):228–240. https://doi.org/10.1111/1468-5922.12395
Van Gijn J (2005) Rene Descartes (1596-1650). J Neurol 252(2):241–242. https://doi.org/10.1007/s00415-005-0775-2
Van Wyhe J (2002) The authority of human nature: the Schadellehre of Franz Joseph Gall. Br J Hist Sci 35(124 Pt 1):17–42. https://doi.org/10.1017/s0007087401004599
Ventriglio A, Bhugra D (2015) Descartes' dogma and damage to Western psychiatry. Epidemiol Psychiatr Sci 24(5):368–370. https://doi.org/10.1017/S2045796015000608
Watson RA (1982) What moves the mind—an excursion in cartesian dualism. Am Philos Q 19(1):73–81
Wickkiser B (2008) Asklepios, medicine, and the politics of healing in fifth-century Greece: between craft and cult. Johns Hopkins Press, Baltimore
Woollam D (1958) Concepts of the brain and its functions in classical antiquity. In: Perrin M (ed) The brain and its functions. Blackwell, Wellcome Foundation, Oxford, pp 5–18

Gross Anatomy of the Brain

2

Contents

2.1 Gross Functional Anatomy

The *central nervous system* (*CNS*), which is the brain, constitutes the seat of mental functions. A basic picture according to frontal, sagittal, and transverse planes is shown in Fig. 2.1.

At the macroscopic level, the central nervous system includes the *spinal cord*, the *medulla oblongata* (responsible for vital autonomic functions), the *pons*, the *cerebellum* (which is responsible for motor function), the *midbrain* (which coordinates visual and auditory reflexes and controls sensory and motor functions), the *diencephalon* (which includes the thalamus and the hypothalamus), and the *brain hemispheres*. The brain hemispheres are connected and communicate through the *corpus callosum*.

Another description includes three parts of the human brain: the hindbrain or posterior brain (medulla, pons, and cerebellum), the midbrain, and the forebrain (thalamus, hypothalamus, and hemispheres). The midbrain and hindbrain are also called *brain stem* (Fig. 2.2).

K. N. Fountoulakis, *The Human Connectome*,
https://doi.org/10.1007/978-3-031-10351-3_2

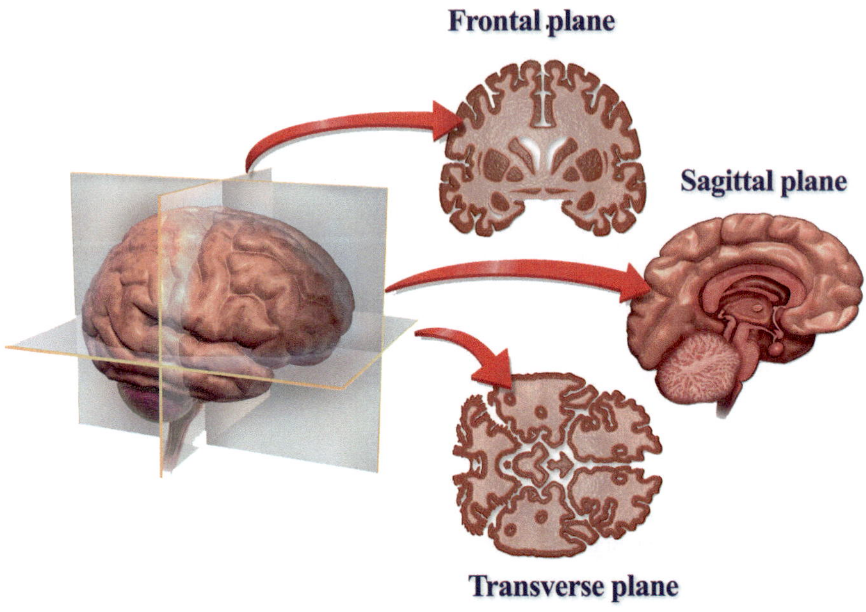

Fig. 2.1 A basic picture of the brain according to a frontal, sagittal, and transverse planes. (From Blausen.com staff (2014). "Medical gallery of Blausen Medical 2014." WikiJournal of Medicine 1 (2))

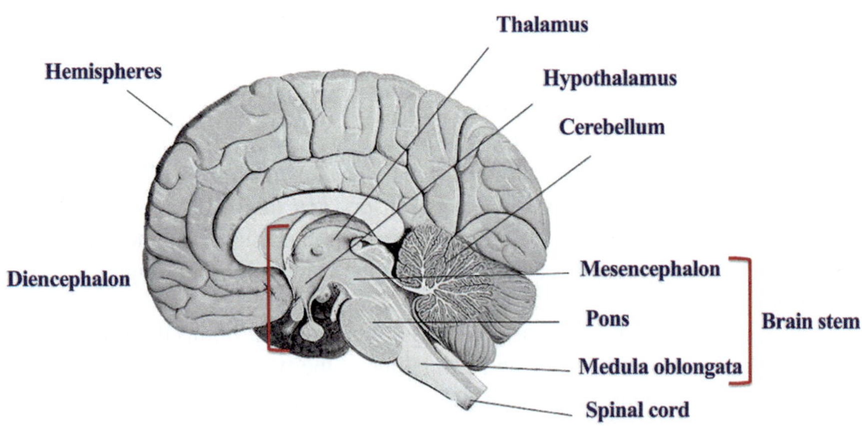

Fig. 2.2 The human brain and its parts: the hindbrain or posterior brain (medulla, pons, and cerebellum), the midbrain, and the forebrain (thalamus, hypothalamus, and hemispheres)

Brain
- Brain hemispheres
- Corpus callosum
- Diencephalon
- Brain stem
  - Midbrain
  - Pons
  - Medulla oblongata
  - Cerebellum

Spinal cord
- Dermatomes (neurotomes)

The *spinal cord* starts from the end of the medulla and it does not belong to the brain. It receives bodily sensory information and contains motor neurons responsible for both voluntary and reflex movements of the trunk and limbs. There are 31 spinal nerves serving these functions which concern the rest of the body which they innervate according to a specific topographic distribution, creating the *dermatomes* or *neurotomes* (Fig. 2.3).

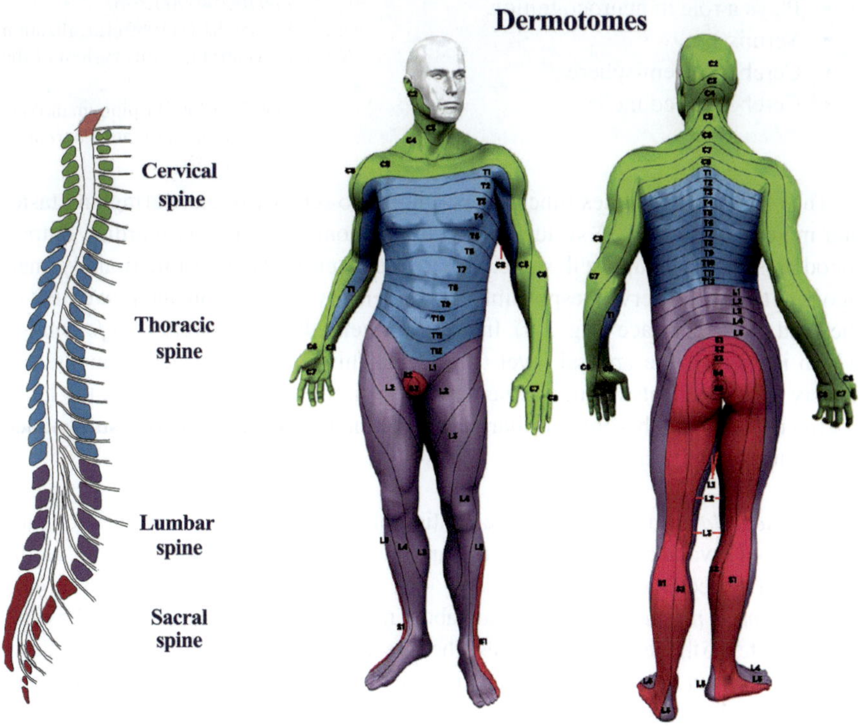

Fig. 2.3 The spinal cord and the spinal nerves. (From https://paininjuryrelief.com)

Spinal Cord and Medulla Oblongata

- They both receive body sensory information
- They host motor neurons
- From them they originate the following:
  - Twelve cranial nerves for the innervation of the head, face, and neck
  - Thirty-one spinal nerves for the innervation of the rest of the body

The *cerebellum* is the part of the brain which contains a far greater number of neurons and synapses than any other part. It includes the *vermis* and the *cerebellar hemispheres*, and it is attached to the brain stem through the *cerebellar peduncles*. It receives somatosensory and motor information, and it is important for maintaining posture, coordinating head and eye movements, and fine-tuning the movements of the muscle. It is important in learning motor skills but also in language and other neurocognitive functions (Fig. 2.4).

Cerebellum

- The highest neuronal and synaptic density in the CNS
- Both somatosensory and motor processing
- Coordinates body movements
- Plays a role in neurocognition
- Vermis
- Cerebellar hemispheres
- Cerebellar peduncles

The *brain stem* includes functions pertaining to sensory (e.g., hearing and taste) and motor (e.g., balance) systems as well as autonomic function (cardiac output, blood pressure, eye pupil reflex, etc.). There are *12 cranial nerves* functionally analogous to the spinal nerves responsible for the sensory information and movement of the head, neck, and face (Fig. 2.5). In the brain stem, there is the *reticular formation* which influences the arousal level (Fig. 2.6). This particular structure includes a variety of components including the raphe nuclei, the ventral tegmentum, the locus coeruleus, and the tuberomammillary nuclei of the hypothalamus and is surrounded by the nuclei of the cranial nerves:

- The *medulla* participates in the regulation of blood pressure and respiration, and it is also involved in taste, hearing, and maintenance of balance as well as the control of neck and facial muscles.
- The *pontine nuclei* relay information about movement and sensation from the cerebral cortex to the cerebellum, and also they are involved in respiration, taste, and sleep.
- The *midbrain* provides important linkages between cerebellum, the basal ganglia, and the cerebral hemispheres.

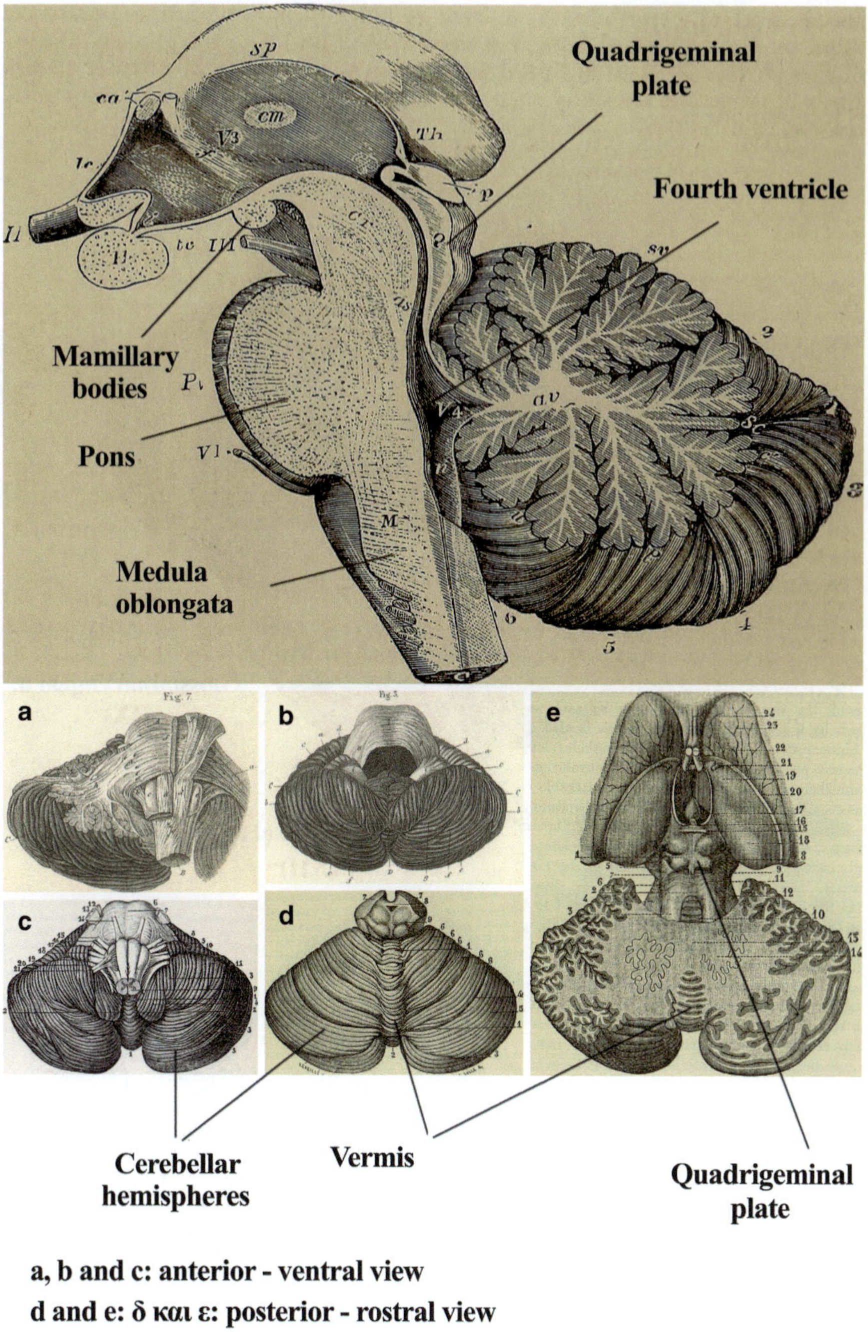

Fig. 2.4 The cerebellum in successive sections and revealing of nearby structure from ventral and rostral views. (Pictures from *Handbuch der Anatomie des Menschen* by Carl Ernest Bock (1841) Leipzig, Germany, *Human physiology* by Dunglison and Robley (1856), and *The brain as an organ of mind* by Bastian, H. Charlton (1896), New York: D. Appleton and Company)

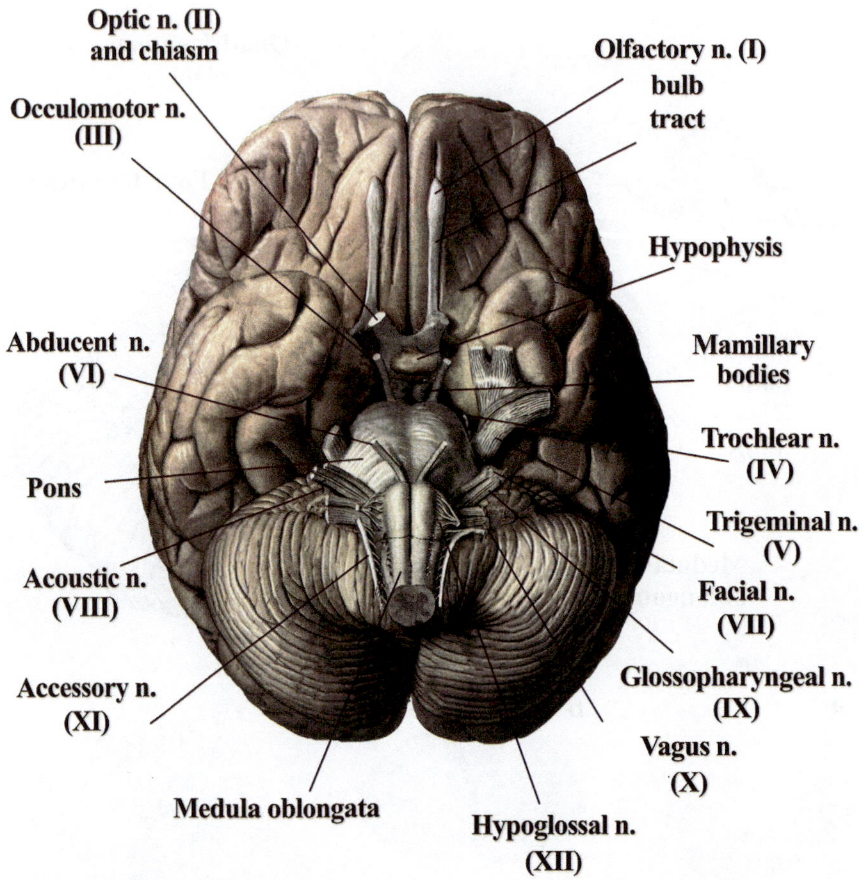

Fig. 2.5 Brain stem and cranial nerves

Fig. 2.6 The brain stem and the reticular formation

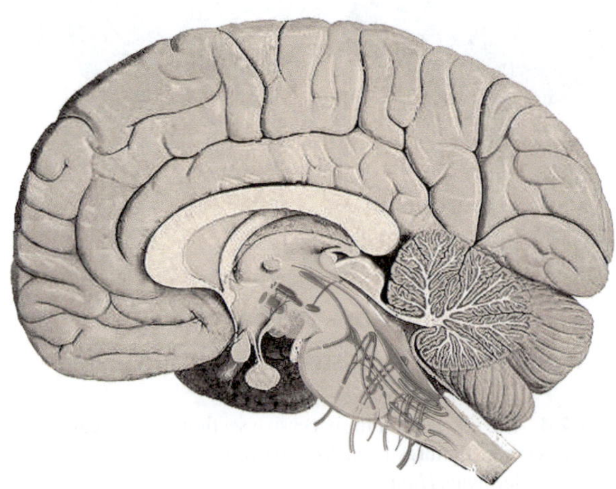

Brain Stem
- Sensory functions
- Motor functions
- Autonomic functions
- Regulation of the level of arousal

The *diencephalon* contains the thalamus and hypothalamus:

- The *thalamus* determines whether sensory information will reach conscious awareness in the neocortex. It participates in the integration of motor information from the cerebellum and the basal ganglia and subsequently transmits the processed result in the cortex. It includes a significant number of nuclei which are topographically classified into anterior, median, rostral, and ventral lateral, reticular, intralaminar, and midline nuclei. The right and the left thalami are connected with the massa intermedia (Fig. 2.7).
- The *hypothalamus* regulates several bodily vital behaviors that are essential for homeostasis and reproduction including eating, drinking, maternal behavior, and circadian rhythms (Fig. 2.8).

Diencephalon
- Control of the information flow to the cortex.
- Regulation of vital bodily functions.

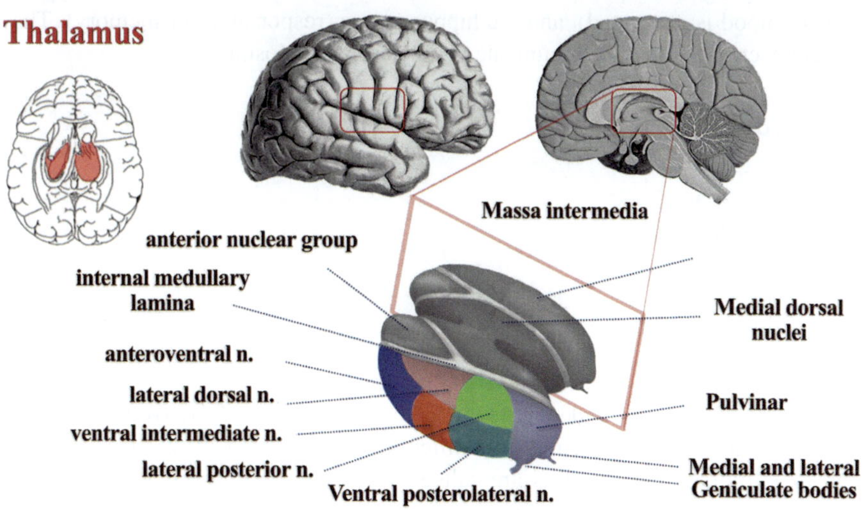

Fig. 2.7 Thalamus and its nuclei

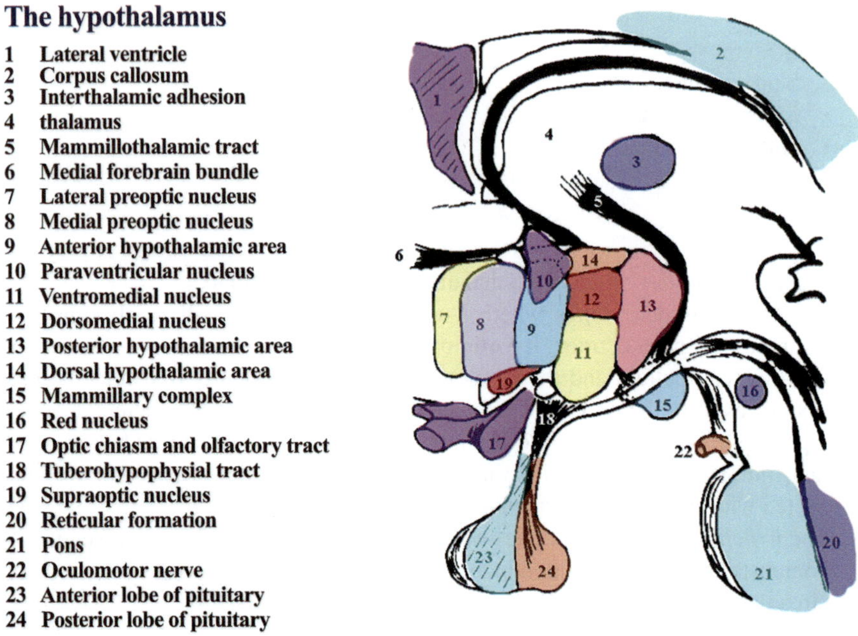

Fig. 2.8 Hypothalamus and its nuclei

The *cerebral hemispheres* constitute the largest region of the human brain and are concerned with perceptual, motor, and higher cognitive functions, including memory, emotion, and decision-making (Ungerleider 1995).

They include the cerebral cortex (2–4 mm thick), the underlying white matter, the basal ganglia (responsible for the control of fine movement), the amygdala (where mood is generated), and the hippocampus (responsible for memory). They are connected and they communicate via the corpus callosum.

Brain Hemispheres (a)

- Cortex
- White matter
- Basal ganglia
- Amygdala
- Hippocampus
- Corpus callosum

Each hemisphere is divided into four anatomically distinct *lobes* (frontal, parietal, temporal, and occipital) each hosting different functions (Fig. 2.9). Their surface is full of infoldings whose crests are called *gyri* while the grooves are called *sulci* (Fig. 2.10). These infoldings serve the purpose to increase the surface of the brain (and the gray matter) while keeping the overall volume as small as possible. The major gyri and sulci are stable across persons and they hold specific names (Fig. 2.11).

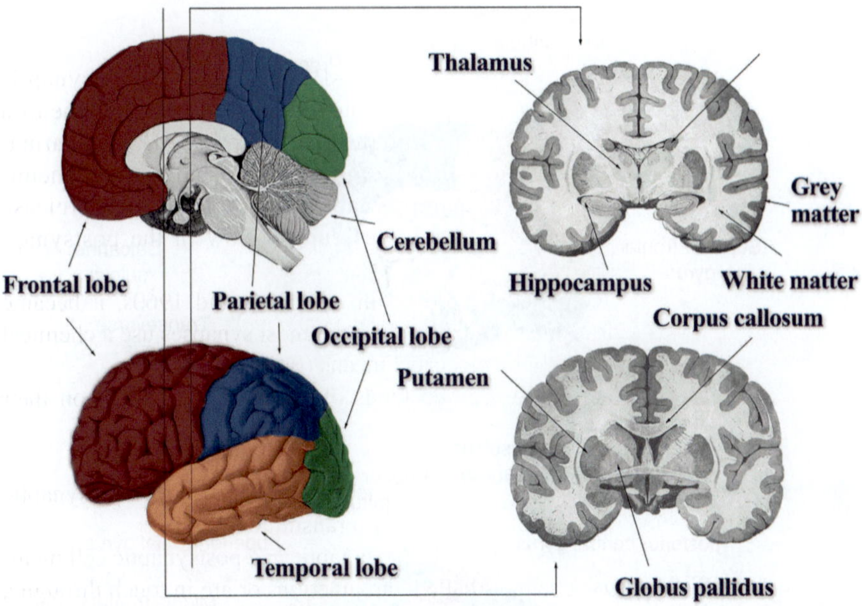

Fig. 2.9 Basic macro-anatomy of the human brain

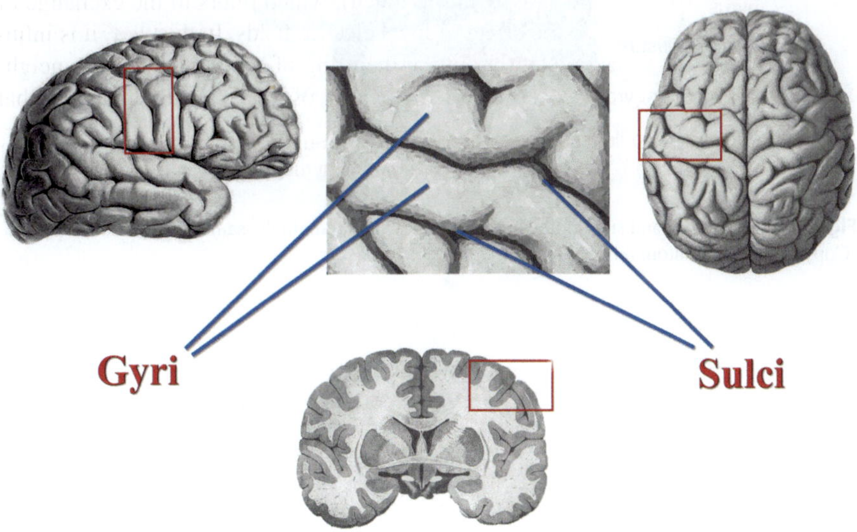

Fig. 2.10 The crests of brain infoldings are called gyri while the grooves are called sulci

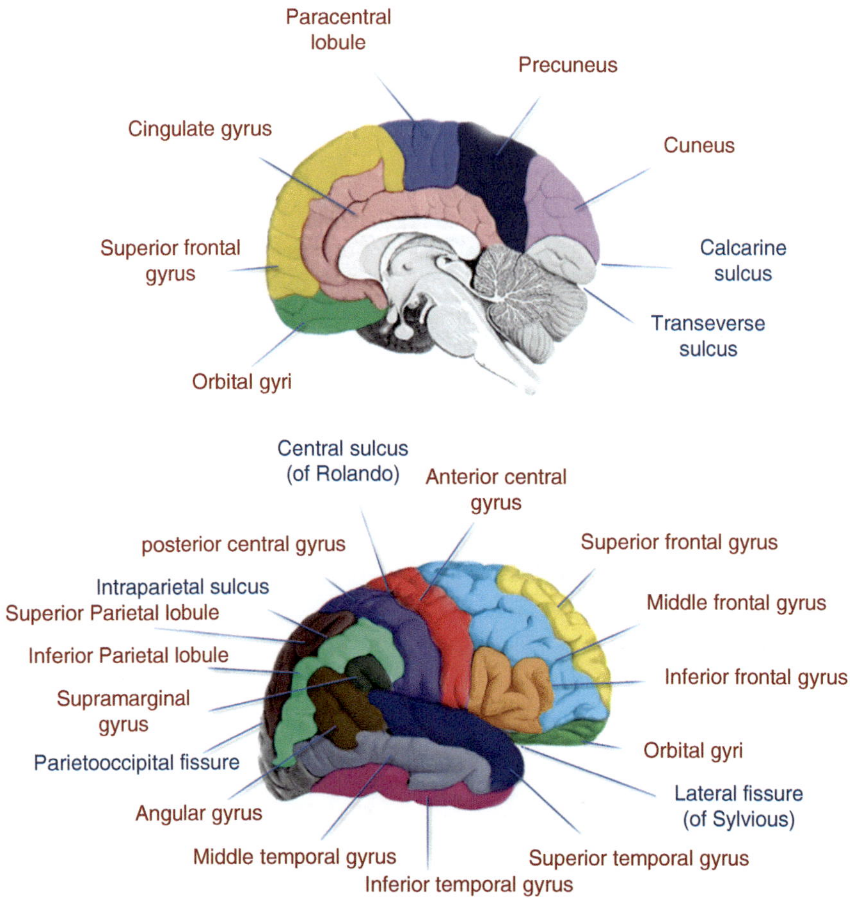

Fig. 2.11 Major gyri and sulci of the human brain, which remain the same from person to person. (Copyright KN Fountoulakis 2021)

Brain Hemispheres (b)

- Frontal lobe
- Parietal lobe
- Temporal lobe
- Occipital lobe
- Cortical sulci
- Cortical gyri

An important characteristic of the functioning of the hemispheres is that each hemisphere's function is primarily related to sensory and motor function of the opposite (*contralateral*) half of the body and hemispheres; although they are similar in appearance, they are not similar or symmetrical in function (Kandel 2000).

The two hemispheres are interconnected by the *corpus callosum*, which is essentially a thick set of fibers (the largest of the commissures) that connect symmetrical and similar regions of the left and right sides of the brain. The two brain hemispheres are connected also with the anterior and posterior commissures (Figs. 2.9 and 2.12).

The *basal ganglia* are the nuclei of the telencephalon and are consisted by groups of neurons (gray matter) at the base of each hemisphere. Their role concerns mainly the control of posture and voluntary movements. They include the following:

- The corpus striatum (nucleus caudatus and lentiform nucleus which is consisted by the putamen and the globus pallidus)
- The claustrum
- The amygdala (amygdaloid body)

Different classifications are the neostriatum (head of the nucleus caudatus and putamen), the paleostriatum (globus pallidus), and the archistriatum (amygdala or amygdaloid body).

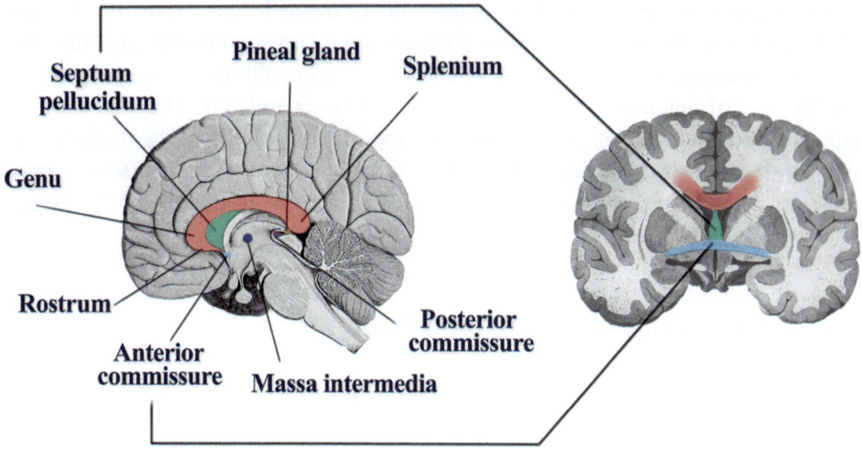

Fig. 2.12 Corpus callosum and the anterior and posterior commissure of the brain. (Copyright KN Fountoulakis 2021)

Basal Ganglia

- Corpus striatum
  - Nucleus caudatus
  - Lentiform nucleus
    - Putamen
    - Globus pallidus
- Claustrum
- Amygdala (amygdaloid body)
- Neostriatum
  - Head of the nucleus caudatus
  - Putamen
- Paleostriatum
  - Globus pallidus
- Archistriatum
  - Amygdala or amygdaloid body

The hippocampus lies deep inside the temporal lobe, and it is a small structure which together with the fornix, the mammillary bodies and other structures form the limbic system. A more detailed description of these structures will follow in Chap. 10.

The functional specialization of different brain regions and the localization of function are more pronounced in adults and to a much less extent in children. However, these specialization and localization are relative rather than absolute, and in case of damage in a certain area, other brain areas could take over and compensate the loss of function to a significant degree if not completely. The degree of this compensation depends on the specific function, the magnitude of the damage, and the age. It is much more difficult in adults to fully restore a function whose brain area has been damaged by mobilizing other regions and resources. In children, a similar damage could be completely restored functionally.

2.2 Connections Among Brain Areas

The *connections* between the brain areas and structures are many and complex but not infinite. Some are but others are not reciprocal, and some are uni- and others are bidirectional; some are convergent (which gather information from many sources to a single point), while others are divergent (which distribute information from a single source to many recipients); some are hierarchical (they transfer the information from a to b and then to c following by d), while others are parallel (they transfer information from a1 to b1 and then to c1, and the same information travels from a2 to b2 and then to c2) (Fig. 2.13).

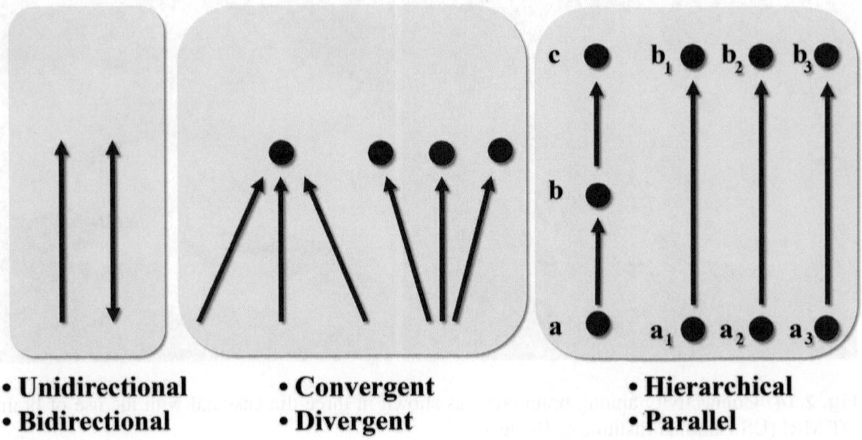

Fig. 2.13 Ways of connection among brain areas. Unidirectional–bidirectional, convergent–divergent, hierarchical–parallel

Connections Among Brain Areas
- Unidirectional
- Bidirectional
- Convergent
- Divergent
- Hierarchical
- Parallel

Connectivity among the brain areas was recently possible to image in three-dimensional with the use of diffuse-tensor MRI or DT-MRI (Fig. 2.14).

Since the localization of brain functions exists but it is not absolute and the functioning of one area seems all too often to be supported or controlled by the functions of another area, this makes the brain mainly a neural network on top of locally embedded functions, rather than the mere sum of relatively independent and localized functions.

Damage of any kind in the white matter (stroke, trauma, etc.), which disrupts connectivity and communication among brain structures, causes variable symptomatology depending on the location and the magnitude of the lesion.

Topographically, the position of the *internal capsule* is extremely important, since it concerns the passage of white matter fibers between the thalamus, the nucleus caudatus, and the lentiform nucleus (Fig. 2.15). From this passage, fibers pass with high density. They concern the whole body, and as a consequence, lesions in the area of the internal capsule result in the manifestation of severe symptoms and signs.

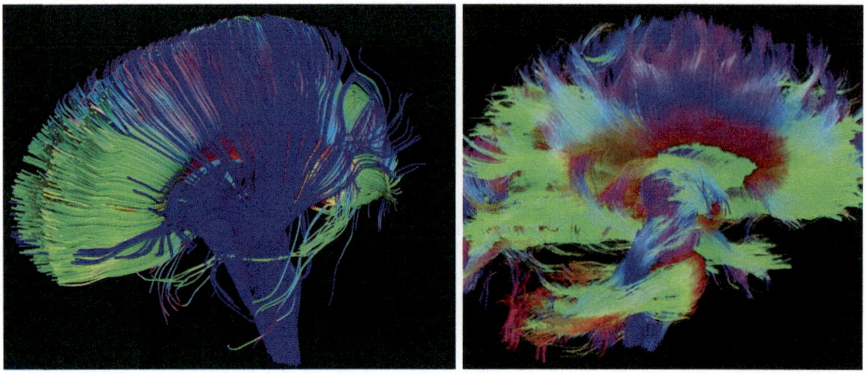

Fig. 2.14 Connectivity among brain areas as shown in three-dimensional with the use of brain DT-MRI (US National Institute of Health)

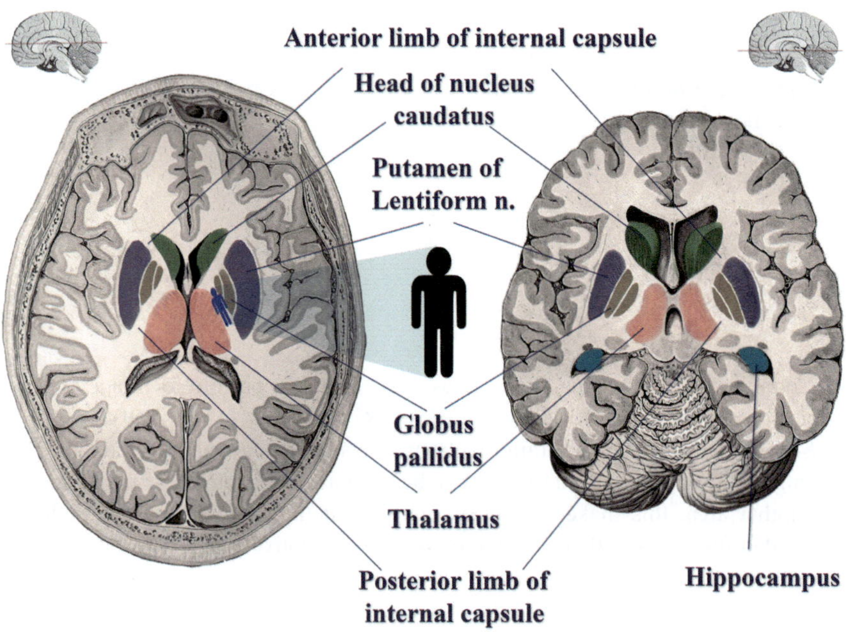

Fig. 2.15 Internal capsule in two transverse brain sections. (Copyright KN Fountoulakis 2021)

2.3 Brain Blood Flow

The detailed description of brain blood circulation is beyond the aims of the current book. However, an elementary description is necessary, since the way of the blood irrigation of the brain and its disorders are crucial for the normal as abnormal brain states.

Blood-Brain Irrigation

- Four main extracerebral arteries
  - Two internal carotid arteries
  - Two vertebral arteries
- Two-thirds of the blood supply to the brain come from the vertebrobasilar system and only one-third from the carotid arteries.
- Willis circle
- Brain arteries
  - Have thinner muscle tissue
  - Do not have elastic fibers
  - Are surrounded by a thinner outer layer
- Blood-brain barrier
- Brain corresponds to 2% of total body weight
- Receives 30% of total blood supply
- Brain veins
  - Two networks
    Superficial veins
    Deep cerebral veins
  - Venus sinuses

The blood supply to the brain (Figs. 2.16 and 2.17) is based on four main extracerebral arteries, the *two internal carotid arteries* and the *two vertebral arteries*. Two-thirds of the blood supply to the brain come from the vertebrobasilar system and only one-third from the carotid arteries.

These four basic vessels join the two anterior and the two posterior cerebral arteries which in turn join the anterior and the two posterior communicating arteries, to form the *Willis circle*, which was named after Thomas Willis (1621–1675), who described it for the first time in 1664. This circle serves the communication between the right and left carotid arteries as well as between the carotid and the vertebrobasilar systems and constitutes the main option of securing brain irrigation in cases of an obstruction in a major artery branch. Similar communications and arterial circles exist mainly through small communicating arterioles of the pia matter.

In comparison to the arteries of the rest of the body, brain arteries are characterized by less muscle tissue, lack of elastic fibers, and are surrounded by thinner outer layer. The main disadvantage of the thinner muscle layer is the development of vulnerable points at the sites where major branches arise from the big arteries of the

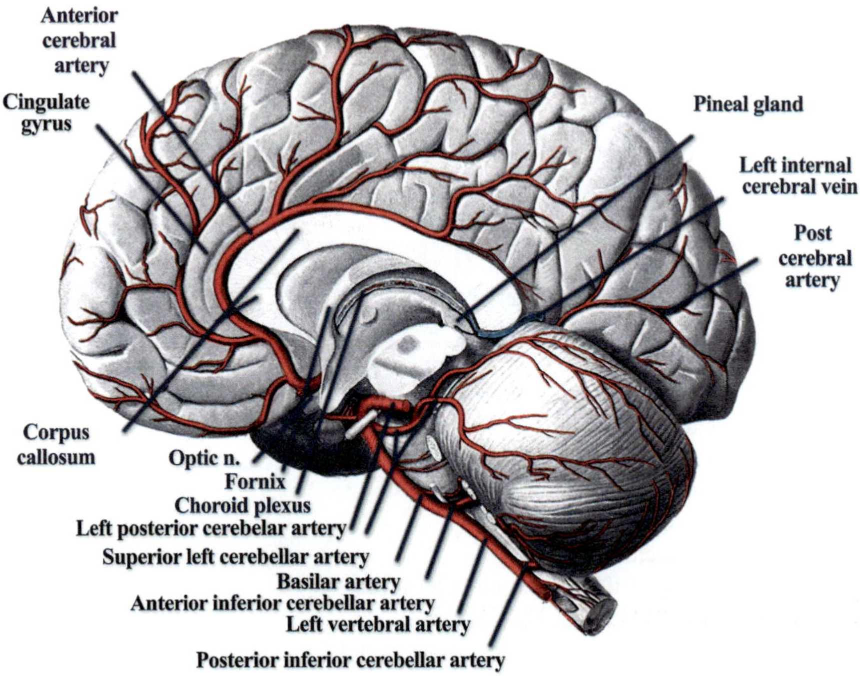

Fig. 2.16 The main routes of brain blood flow. (From Sobotta 1909)

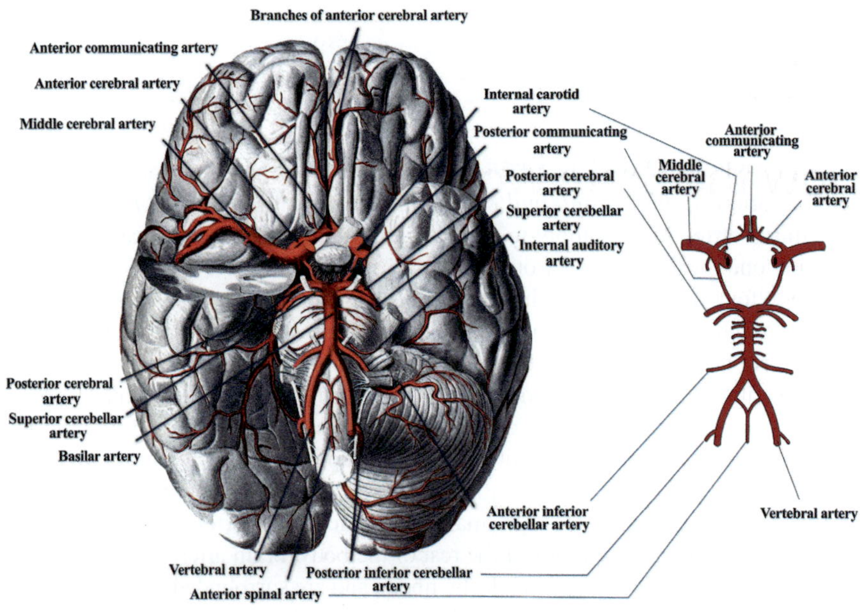

Fig. 2.17 The base of the brain and Willis' circle. (From Sobotta 1909)

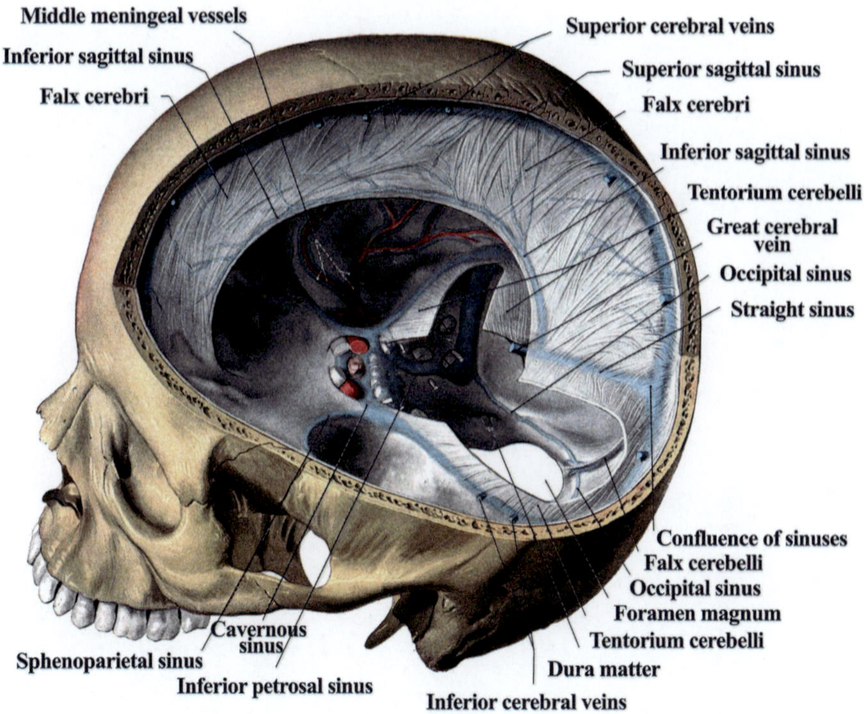

Fig. 2.18 The system of cerebral veins. (From Sobotta 1909)

base of the brain. As a result, aneurysms might develop and their rupture could lead even to death.

An important characteristic of brain arteries is the presence of the *blood-brain barrier*, which isolates the brain and the cerebrospinal fluid (CSF) from the rest of the body and precludes various substances that circulate in the peripheral body from entering into the brain. For details, see Chap. 3 and Fig. 3.2.

Although brain weights not more than 2% of total body weight, at baseline state, it receives 30% of blood supply and consumes the respected energy and percentage of metabolism.

The *cerebral vein system* consists of two networks, the *superficial* and the *deep cerebral veins*. These two systems drain into the *venous sinuses* which are large venous cavities which in turn drain into the two spaghetti veins (Fig. 2.18).

References

Kandel E (2000) The brain and behavior. In: Kandel E, Schwarts J, Jessell T (eds) Principles of neural science. McGraw-Hill, New York, pp 5–18

Ungerleider LG (1995) Functional brain imaging studies of cortical mechanisms for memory. Science 270(5237):769–775. https://doi.org/10.1126/science.270.5237.769

Cellular Structure and Histological Architecture of the Brain

3

Contents

3.1 Structural Components of the Brain

The brain includes an extraordinary number of nerve cells at the order of 100 billions (10^{11}). These cells can be classified in more than 1000 different types, but they are mainly divided into 2 broad categories: the *nerve cells* (*neurons*) and the *glial cells* (*glia*) (Kuehnel 2003; Erlandsen and Magney 1985). Glial cells outnumber neurons 10–50:1, and their name comes from the Greek word *γλοια* (meaning glue). In terms of information theory, neurons are the "processors," and glia is the supporting environment which provides the supporting skeleton, routes for feeding and removing residual substances, etc. The axons of neurons correspond to cables.

The *glial cells* are further divided into microglia and macroglia:

- *Microglia* are phagocytes mobilized in case of injury and infection of certain diseases. They are physiologically and embryologically unrelated to the other brain cells.
- *Macroglia* include oligodendrocytes which are located in the brain, Schwann cells which are located in the peripheral nervous system, and astrocytes. Oligodendrocytes provide the brain with structure and support and separate and also insulate neurons with a myelin sheath by wrapping their processes around the axons (Fig. 3.1) (Bunge 1968). Each oligodendrocyte insulates more than 15 axons.

K. N. Fountoulakis, *The Human Connectome*,
https://doi.org/10.1007/978-3-031-10351-3_3

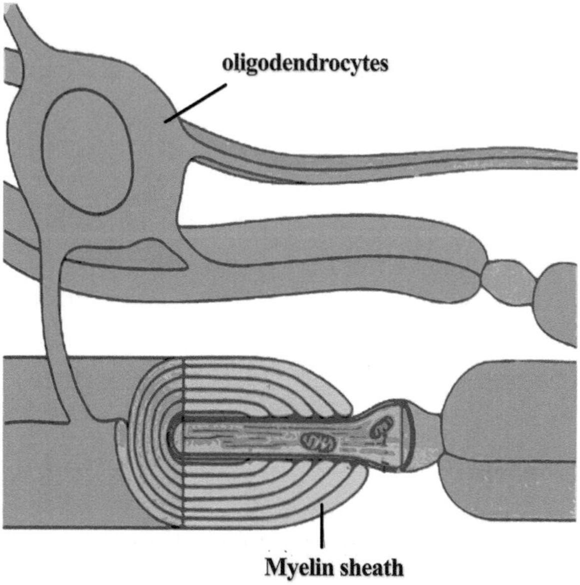

Fig. 3.1 Oligodendrocytes with myelin sheath which wraps nearby neurons. (Copyright KN Fountoulakis 2021)

Neural Cells
- Neurons
- Glial cells
  - Microglia
- Phagocytes
  - Macroglia
    Oligodendrocytes
    Astrocytes
      - Blood-brain barrier

Some of glia cells act as scavengers and regulate the environment where neurons operate, with the secretion of nutritional and growth agents. Astrocytes have processes, some of which terminate in end feet. Some of them place these end feet on the walls of blood vessels where they cause the formation of tight junctions between endothelial cells. In this way, they create the *blood-brain barrier* which isolates blood vessels from the brain and prevents various substances to enter the brain (Fig. 3.2). Astrocytes also take up the excess of potassium and neurotransmitters, and in this way, they isolate and protect neighboring neurons. The functions of glia include its mobilization in case of damage of brain tissue. In that case, microglial cells migrate to the damaged area, fill the gaps, and create a "scar" which is called *gliosis*.

Fig. 3.2 The blood-brain barrier. (After permission by Ben Brahim Mohammed)

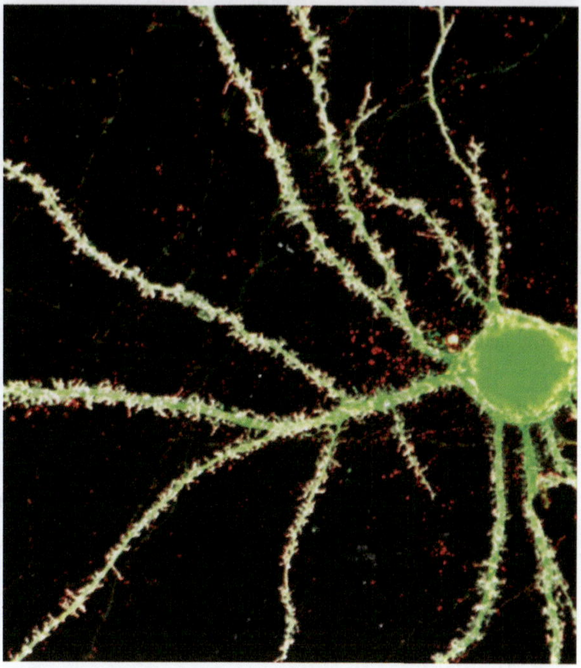

Fig. 3.3 Neuron (National Institute of Neurological Disorders and Stroke, NIH)

The *neurons* have a similar shape and structure throughout the entire brain (Figs. 3.3 and 3.4), and their structure resembles that of all the other cells of the human body. On the basis of shape, neurons are classified into *unipolar*, *bipolar*, and *multipolar*. Multipolar neurons are those who predominate in the brain. They are also classified into *sensory*, *motor*, and *interneurons*. Interneurons are the largest class and are divided into *relay or projection interneurons* which have long axons and transmit information over long distances and *local interneurons* with short axons which transmit information within local neural circuits. Neurons also differ concerning the neurotransmitters they contain but also in their function (e.g., some do not produce an action potential).

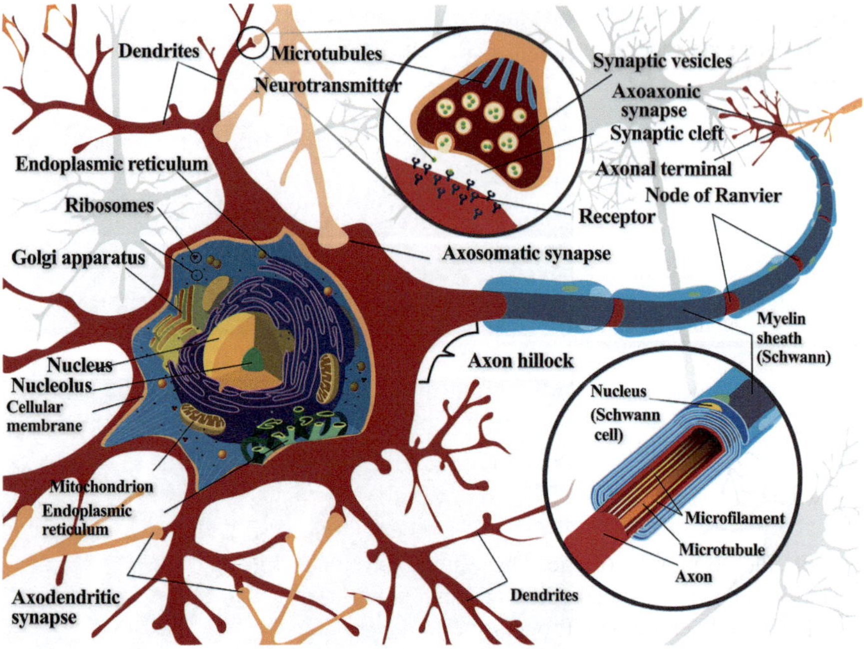

Fig. 3.4 Detailed representation of the neuron and its parts as well as its connections with nearby neurons

There is striking similarity between the proteins in the brain and those found in the rest of the body. Also, the nerve cells share more or less a similar structure throughout the brain. They serve different functions not because of their different structures but because they participate in distinct neuronal networks and circuits.

Neurons

- Unipolar
- Bipolar
- Multipolar
- Cell body
  - Nucleus (genes and DNA)
  - Endoplasmic reticulum (protein synthesis)
  - Mitochondria (energy production, mt-DNA)
- Dendrites
- Axon

Their structure includes four morphologically defined regions (Figs. 3.5, 3.6, and 3.7):

(a) The *cell body* contains the nucleus (with the genes and the DNA), the endoplasmic reticulum where proteins are synthesized, and the mitochondria which constitute the power stations of the cell, the places where energy is produced and manipulated. The function of mitochondria produces a significant quantity of garbage material which is toxic for the cell and it is the equivalent of having a nuclear plant at the center of a middle size city. Therefore, mitochondria are isolated from the cytoplasm by a membrane which is similar to the cellular membrane. The communication between the interior of the mitochondria and the cell body is kept at a minimum and this is why mitochondria have their own DNA (mt-DNA) which is inherited directly from the mother.

(b) The *dendrites* are short processes which arise from the cell body and serve the connection with neighboring cells.

(c) The *axon* is a single and very long dendrite which extends far away from the cell body. It is the main conducting unit and serves communication with distant neurons. The end-branches are called telodendria. It can carry signals up to 3 m away. The axons are wrapped with a myelin sheath stemming from oligodendrocytes. This sheath isolates them from one another. In this way, the electrical phenomena occurring in one axon do not spread to neighboring axons and cells (Fig. 3.8).

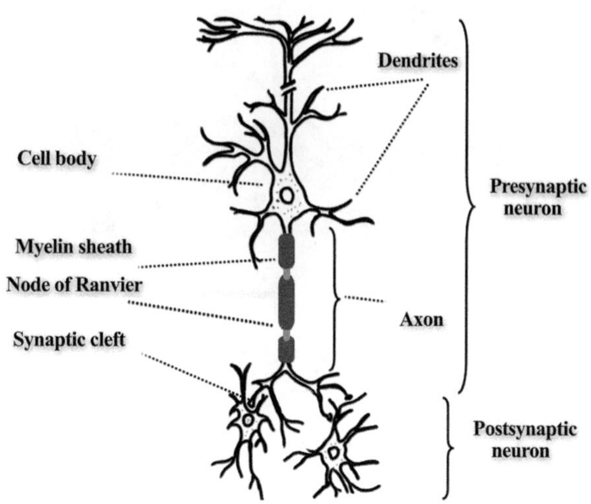

Fig. 3.5 The neuron and its main components. (Copyright KN Fountoulakis 2021)

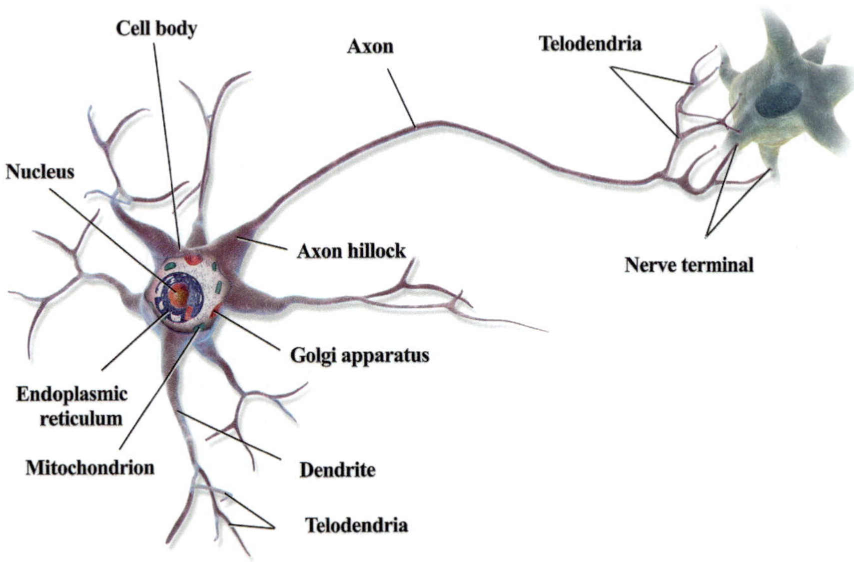

Fig. 3.6 The neuron and its component in space and in relationship to other neurons

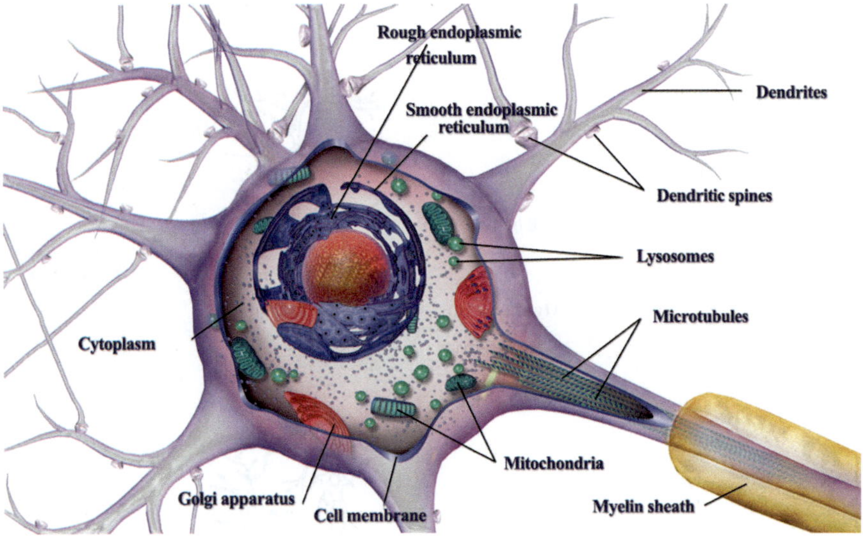

Fig. 3.7 The cell body of the neuron and its functioning parts

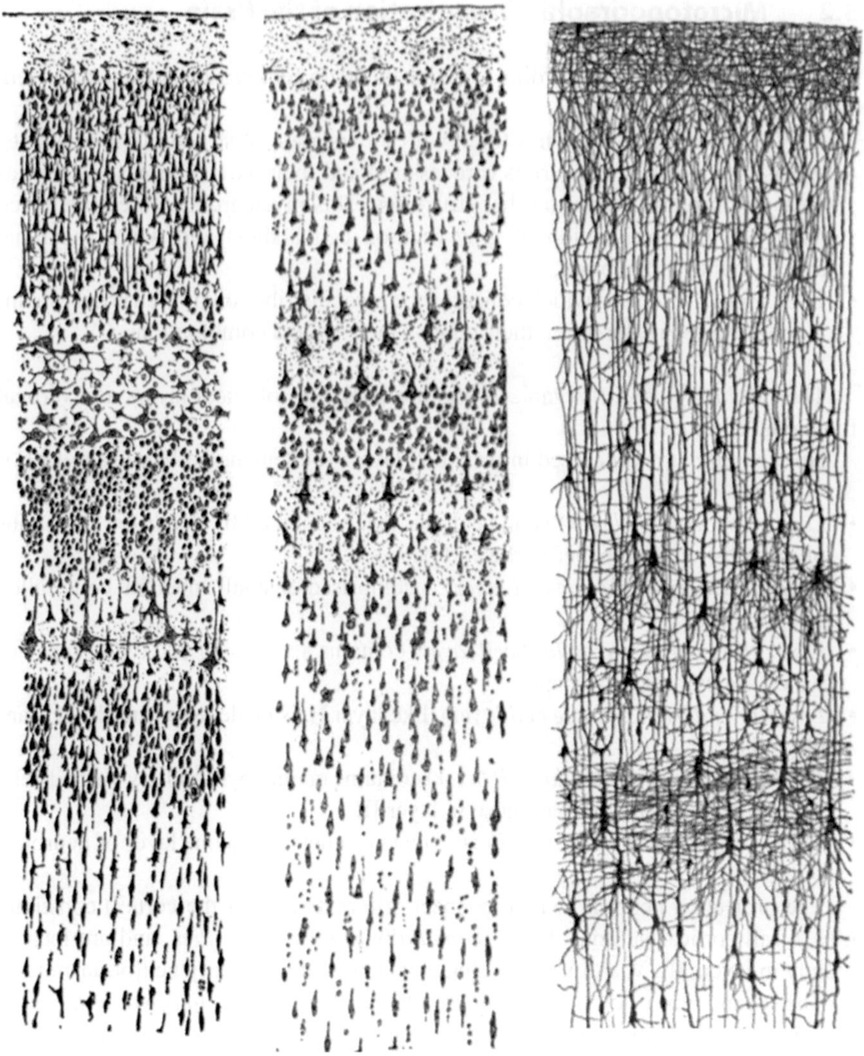

Fig. 3.8 Layers of the brain cortex, according to three designs by Santiago Ramon y Cajal. The sections are vertical with the outside of the brain on top. Left: visual cortex of an adult person (Nissl stain). Middle: motor cortex of an adult person (Nissl stain). Right: cortex of a 45-day-old infant. Nissl stain marks the cell bodies and Golgi stain the axons and dendrites

3.2 Microtopographic Organization of the Brain

There are two principles, according to which the brain cells are located and arranged:

- *Myeloarchitecture*, which has to do with the location of structural components, that is, the location of neurons (processors) and their axons (cables) as well as their position in relation to other structures. This myeloarchitecture is reflected mainly in the localization of gray (neuron cell bodies) and white (neuronal axons) matter in the brain.
- *Chemical architecture* which concerns the local distribution of the various chemical substances in the brain, their relative density, and combinations.

The *cerebral cortex* is organized into horizontal histological layers and vertical functional columns.

Horizontally, it is organized into six cell layers (Felleman and Van Essen 1991):

- *Layer I* (*molecular layer*) is an acellular layer occupied by dendrites and axons of the cells located deeper in the cortex.
- *Layer II* (*external granule cell layer*) is comprised of small spherical cells called granule cells.
- *Layer III* (*external pyramidal cell layer*) contains a variety of cell types, many of which are pyramidally shaped.
- *Layer IV* (*internal granule cell layer*), like layer II, is made up primarily of granule cells.
- *Layer V* (*internal pyramidal cell layer*) contains mainly pyramidally shaped cells that are typically larger than those in layer III.
- *Layer VI* (*polymorphic or multiform layer*) is a heterogeneous layer of neurons.

The projection neurons (which connect different areas of the cortex) constitute 60% of all cortical neurons, have *pyramidally shaped cell bodies*, and are located mainly in layers III, V, and VI with glutamate being their primary neurotransmitter.

The second type of cortical neurons are the *stellar cells* (*or non-pyramidal cells*) which are small neurons that participate in local circuits. These local interneurons (which connect different neurons within the same area of the cortex) are located in all layers and use GABA as their primary neurotransmitter.

The vertical organization of the cortex in *functional columns* gives rise to complete neuronal circuits which are able to process the input of a stimuli in a comprehensive way. The detailed charting of these functional columns started with the works of Korbinian Brodmann (1868–1918) who described the structure and the histological composition of 52 distinct cortical areas (areas according to Brodmann), which are classified into 11 histological groups (Fig. 1.27).

Those areas of the cortex which manifest a definitely recognizable cell structure with six layers are called "the isocortex" while those with a different structure are called "the allocortex" (usually the greatest difference concerns that layer IV is either missing or much thicker).

- Myeloarchitecture
- Chemical architecture

Cortex organization into histological layers and functional columns

- Layer I (molecular layer: dendrites and axons neurons)
- Layer II (external granule cell layer: granule cells)
- Layer III (external pyramidal cell layer: many pyramidally shaped cells)
- Layer IV (internal granule cell layer: granule cells)
- Layer V (internal pyramidal cell layer: large pyramidally shaped cells)
- Layer VI (polymorphic or multiform layer: heterogeneous layer)
- Projection pyramidal neurons: in layers III, V, and VI and glutamate
- Stellar cells: local interneurons, in all layers, local neural circuits, GABA
- Ομοτυπικός φλοιός
- Αλλοτυπικός φλοιός

References

Bunge RP (1968) Glial cells and the central myelin sheath. Physiol Rev 48(1):197–251. https://doi.org/10.1152/physrev.1968.48.1.197

Erlandsen S, Magney J (1985) Human histology: a microfiche atlas. University of Minnesota Press, Minneapolis

Felleman DJ, Van Essen DC (1991) Distributed hierarchical processing in the primate cerebral cortex. Cereb Cortex 1(1):1–47. https://doi.org/10.1093/cercor/1.1.1-a

Kuehnel W (2003) Color atlas of cytology, histology, and microscopic anatomy, 4th edn. Thieme, Stuttgart

The Cell Membrane

4

Contents

4.1 The Membrane

The cell membrane is an extremely important structural part of the cell, since it separates and isolates the cell and its internal space from the environment, and thus it permits the unopposed and specialized functioning of cellular organelles. It consists of 40% lipids mainly in the form of phospholipids and cholesterol and 60% proteins, while carbohydrates also exist at varying proportions (1–10%). Phospholipids form a membrane with two layers with cholesterol contributing to its stability and endurance. Proteins are necessary for most of membrane functions like receptors, ion channels, transmembrane passage of substances, and catalysis of chemical reactions (Fig. 4.1).

As *the membrane isolates* the cell from the other cells and at the same time isolates the internal of the cell from the outer surrounding, there is the need for communication and exchange of matter, that is, the need for feeding and excretion of the byproducts of metabolism. These two needs are satisfied because of a basic property of the membrane, to be selectively permeable either in a passive way or with active transportation, and this permeability could be modified according to the needs of the moment. Some molecules like water or oxygen seem to pass through it almost

K. N. Fountoulakis, *The Human Connectome*,
https://doi.org/10.1007/978-3-031-10351-3_4

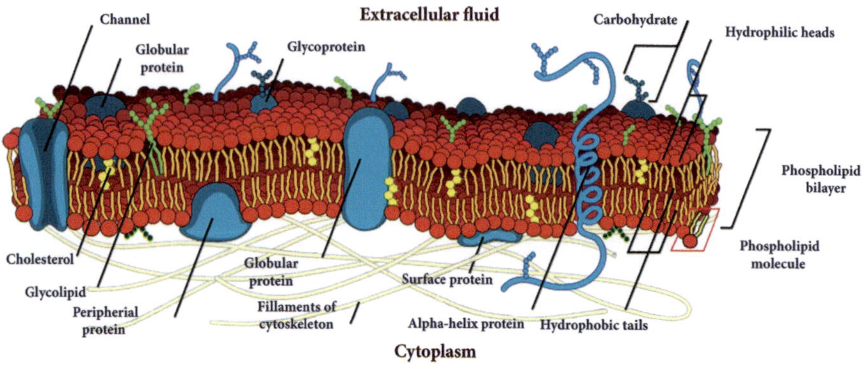

Fig. 4.1 The cell membrane and its structural components

freely, while others like big proteins seem to be completely unable to pass. The *ion channels* play a decisive role in the permeability of the membrane concerning various ions. These channels regulate not only the rate but also specifically which ions are allowed to cross.

4.2 The Resting Potential

Communication among neurons is based in principle on the production of electric signals. This production relies on the existence of the membrane resting potential, that is, a difference in electric potential between the intra- and the extracellular spaces. As a consequence, neurons, when at rest, sustain a difference in electric potential between the two surfaces of the cell membrane. This difference is on average equal to −65 mV and is called *resting potential*. It depends on the unequal distribution of ions and more specifically of sodium (Na+) and potassium (K+) ions as well as on the selective permeability of the membrane by potassium ions alone. The potential is maintained by the Na^+–K^+ *pump* which pumps Na^+ ions out of the cell and K^+ ions into the cell at a ratio of 3:2, which results in a negative potential for the internal of the cell. This pump is extremely energy-demanding, as it consumes up to 40% of the total energy usage by the neuron.

The phenomenon of increase of the resting potential is called *hyperpolarization*. This means that the potential of the neuron, which is negative, manifests an "increase" in its absolute number and becomes more negative. Hyperpolarization is an inhibitory phenomenon since the ability to produce an action potential by the neuron is reduced. On the contrary, the decrease of resting potential, which is of the decrease in absolute number which makes the difference less negative, is called *depolarization*, and since it enhances the ability of the neuron to produce an action potential, it is considered to be excitatory.

Resting Potential

- It constitutes a difference in electric potential between the intra- and the extracellular spaces.
- On average it is −65 mV.
- It depends on the unequal distribution of sodium (Na^+) and potassium (K^+) ions.
- The role of Na^+–K^+ pump.
- Hyperpolarization: The increase of resting potential (more negative)—inhibitory phenomenon.
- Depolarization: The decrease of resting potential (less negative).
- The resting potential plays the role of readiness potential.

The above mechanism aims to sustain an expensive energy-consuming difference in electric potential between the two sides of the cell membrane and between the intra- and the extracellular spaces. This significant difference in potential, in case the ion channels open abruptly, leads to a great and fast entering of Na^+ into the cell, resulting in a very fast response of the neuron to external stimuli if needed. Therefore, in essence, the resting potential plays the role of a *readiness potential*.

4.3 The Action Potential

On the basis of this energy-consuming resting potential, in case needed, the neuron can produce electric potentials which are called *action potentials*, and they are of about 100 mV amplitude and 1 ms duration and a speed of up to 100 m/s (Fig. 4.2). These potentials constitute an *all or none* phenomenon which does not decay and travels to distant targets, periodically regenerated. They are initiated at the *hillock* which is the specific region of the cell body where the axon originates and where ion channels are denser.

However, while the amplitude of the action potential is an 'all or none' phenomenon, the duration is not, and this duration of the action potential determines the amount of Ca^{2+} that flows into the terminal, with more Ca^{2+} flowing into the cell after prolonged action potentials (Smith and Augustine 1988). In this frame, the steady-state Ca^{2+} influx is enhanced by depolarization and decreased by hyperpolarization, while vice versa the action potential is modulated by *voltage-sensitive Ca^{2+} channels* (Baker et al. 1971) (Fig. 4.3).

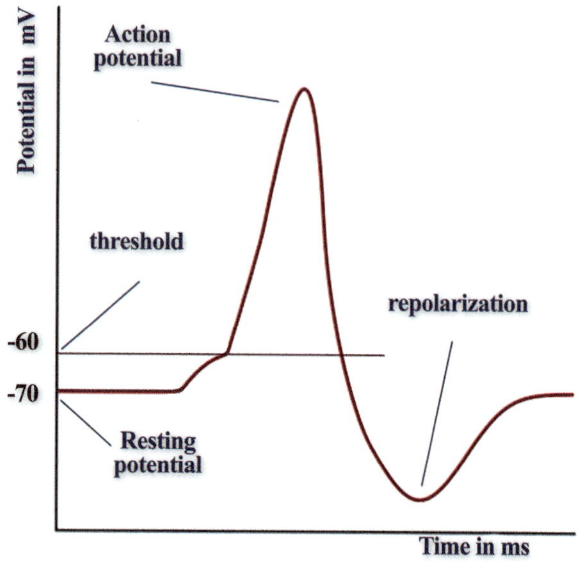

Fig. 4.2 Typical form of action potential

Action Potentials

- "All or none" phenomena
- Resting potential
- Na^+–K^+ pump
- Hyperpolarization—inhibitory phenomenon
- Depolarization—excitatory phenomenon
- Refractory period
- Calcium ions (Ca^{2+})
  - Intracellular messengers
- Voltage-sensitive Ca^{2+} channels
  - Regulation of action potentials
- Tetanic stimulation
  - Potentiation
  - Post-tetanic potentiation
- Effect of neuron on another neuron
  - Postsynaptic inhibition
  - Presynaptic inhibition
  - Presynaptic facilitation
- Neuronal plasticity (activity records)
  - Residual Ca^{2+} in terminals
  - Free Ca^{2+} concentration in the presynaptic terminal
- Long-term potentiation (LTP)
- Local action potentials
  - Graded

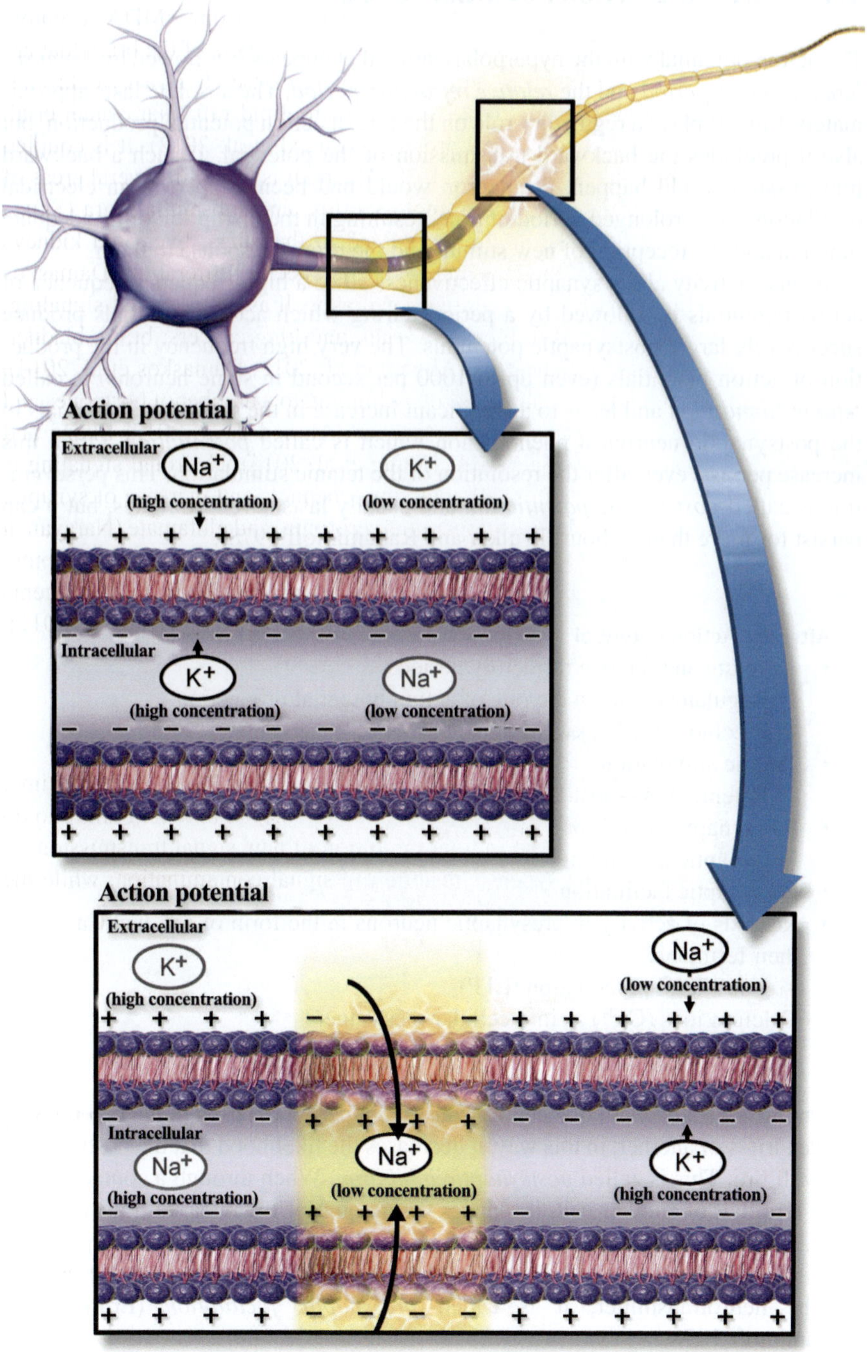

Fig. 4.3 Graphic representation of the underlying processes that produce the action potential. (From Blausen.com staff (2014). "Medical gallery of Blausen Medical 2014." *WikiJournal of Medicine* 1 (2))

4.4 After the Action Potential Period

The action potential with the hyperpolarization that appears is followed by the *absolute refractory period* and the *relative refractory period*. The absolute lasts approximately 1 ms. It plays a regulatory role on the rate of action potential production, but also it precludes the backward transmission of the potential. If such a backward transmission would happen, the neuron would had been trapped in an electrical oscillation for a prolonged period of time, resulting in the disruption of its adequate function and the reception of new stimuli and the proper response to them.

Intense activity alters synaptic effectiveness since a high-frequency sequence of action potentials is followed by a period during which action potentials produce successively larger postsynaptic potentials. The very high frequency in the production of action potentials (even up to 1000 per second in some neurons) is called *tetanic stimulation* and leads to a significant increase in the size of the potentials in the postsynaptic neuron, a phenomenon which is called *potentiation*. Often this increase persists even after the resolution of the tetanic stimulation. This perseveration is called *post-tetanic potentiation* and usually lasts several minutes, but it can persist for more than an hour (Erulkar and Rahamimoff 1978).

After the Action Potential

- Absolute and relative refractory periods
  - Regulatory role on the rate of action potential production
  - Precludes the backward transmission of the potential
- Tetanic stimulation
  - Potentiation—post-tetanic potentiation
- Postsynaptic inhibition
- Presynaptic inhibition
- Presynaptic facilitation
- Records of activity in presynaptic neurons in the form of residual Ca^{2+} in their terminals
  - Long-term potentiation (LTP)
- Calcium ions (Ca^{2+}) as intracellular messengers

On the contrary, when one neuron causes a hyperpolarization in the cell body or the dendrites of another, in this way, it decreases the likelihood that the postsynaptic cell will fire. This is called *postsynaptic inhibition*. When through a contact on the axon terminal of another cell, it reduces the amount of the released transmitter by the second cell (presynaptic neuron) with a third cell as a target (postsynaptic neuron), this action is called *presynaptic inhibition*, while when it increases the amount of the neurotransmitter, it is called *presynaptic facilitation* (Erulkar and Rahamimoff 1978).

One very interesting feature is that presynaptic neurons keep *records of their activity* in the form of residual Ca^{2+} in their terminals. This eventually leads to *long-term potentiation (LTP)* which could last from hours to days (Kandel 1981). Similar mechanisms which also involve the free Ca^{2+} concentration in the presynaptic terminal govern the plastic capabilities of the neuron (Dunlap et al. 1995; Klein et al. 1980).

Essentially the *calcium ions* (Ca^{2+}) are intracellular messengers, which translate the electrical activity of depolarization into all the nonelectrical activities inside the neuron. Thus, these ions serve the output of the system, and any delay in the initiation of these nonelectrical activities reflects the time it takes for Ca^{2+} channels to open and for Ca^{2+} to trigger the process.

4.5 Local Potentials

While the action potential is a long-distance signal, the neuron produces also a number of local signals in the form of receptor or synaptic potentials, which typically spread only for a few millimeters. While action potential is an "all or none" phenomenon, the *local potentials* could be graded, based on the amount of chemical neurotransmitter released and the duration these molecules are active in the synaptic cleft. On the contrary, the action potential carries information only in terms of frequency of potential generation and duration.

References

Baker PF, Hodgkin AL, Ridgway EB (1971) Depolarization and calcium entry in squid giant axons. J Physiol 218(3):709–755. https://doi.org/10.1113/jphysiol.1971.sp009641

Dunlap K, Luebke JI, Turner TJ (1995) Exocytotic Ca2+ channels in mammalian central neurons. Trends Neurosci 18(2):89–98

Erulkar SD, Rahamimoff R (1978) The role of calcium ions in tetanic and post-tetanic increase of miniature end-plate potential frequency. J Physiol 278:501–511. https://doi.org/10.1113/jphysiol.1978.sp012320

Kandel ER (1981) Calcium and the control of synaptic strength by learning. Nature 293(5835):697–700. https://doi.org/10.1038/293697a0

Klein M, Shapiro E, Kandel ER (1980) Synaptic plasticity and the modulation of the Ca2+ current. J Exp Biol 89:117–157

Smith SJ, Augustine GJ (1988) Calcium ions, active zones and synaptic transmitter release. Trends Neurosci 11(10):458–464. https://doi.org/10.1016/0166-2236(88)90199-3

Neurotransmitters

5

Contents

5.1 General Concepts

The term *neurotransmitter* refers to a group of substances with diverse chemical structure but with the common function to serve the synaptic transmission. The term was introduced for the first time by Henry Dale (1875–1968) and George Barger (1878–1939), ironically in an article attacking the concept (Barger and Dale 1910). The concept became popular in the 1930s after the Nobel Laureate Otto Loewi (1873–1961) showed the release of acetylcholine from the vagus nerve terminals into the heart of a frog and later when Henry Dale himself published his work on cholinergic and adrenergic neurotransmission (Fig. 5.1). Both received the Nobel Prize. In the same line of thinking, another Nobel Laureate Hermann Emil Louis Fischer (1852–1919) created the metaphor of the *lock and the key* concerning the reaction between messenger substances and receptors, in 1890 (Lemieux and Spohr 1994).

K. N. Fountoulakis, *The Human Connectome*,
https://doi.org/10.1007/978-3-031-10351-3_5

Fig. 5.1 Left: Henry Dale (1875–1968). Right: Otto Loewi (1873–1961). Both received the Nobel Prize in 1936 for their work on neurotransmission

However, the concept that receptors were behind neurotransmission and the effects of medication on the human body was a matter of debate for long. At the root of this debate was a resistance to accept a materialistic view of the human body since electrophysiology was leaving space for spirituality, but this was not the case with chemistry. There are authors suggesting that the biological aspect of psychiatry has finally been accepted in part because of the "spiritual" and cultural effect of psychedelic drugs (Gerard 1949, 1955a, b; Gordon 1948).

The essential function and characteristic of a neurotransmitter is that it is released in the synaptic cleft by the presynaptic neuron and affects the condition of the post-synaptic neuron (see also Chap. 6). In contrast to hormones, neurotransmitters are never released in the bloodstream (to act as neurotransmitters), but instead their target neuron is in contact or very close to the secreting neuron. While their *direct effects are short-lived* and last from milliseconds to minutes, they might result in long-term changes in the postsynaptic neuron.

Although an important characteristic of neurotransmitters is that their effects are transient, lasting from milliseconds to minutes, neurotransmitter action can result in long-term changes in target cells lasting hours or days. In principle, neurotransmission has an immediate and short-lasting effect on the postsynaptic neuron. Another type of effect is called *neuromodulation* and concerns longer-term effects.

Neurotransmitters' Role

- They serve the synaptic neurotransmission
- They are stored in the synaptic vesicles
- They are released in the synaptic cleft from the presynaptic neuron
- They are stored and released in fixed and stable quantities (quanta)
- They attach to receptors on the postsynaptic membrane
- They modify the condition of the postsynaptic neuron:
  - Immediate results
  - Short-term results
  - Neuromodulation

5.2 Definition of a Neurotransmitter

The modern *definition of a neurotransmitter* demands the following four criteria to be met:

1. It is synthesized in the presynaptic neuron.
2. It is located in the presynaptic terminal (stored in vesicles) and is released in amounts sufficient to exert a specific action on the postsynaptic neuron.
3. When administered exogenously, it mimics the action of the endogenously released transmitter exactly.
4. A specific mechanism exists for removing it from its site of action in the synaptic cleft.

Neurotransmitters: Definition

- Is synthesized presynaptically
- Is released in sufficient quantity
- Exogenous administration mimics endogenous action
- A special mechanism exists to remove it from the area of action

All neurotransmitters are synthesized in the cell body and are subsequently stored in areas called *synaptic vesicles*, which are located in the presynaptic terminal (Fig. 5.2). Each vesicle contains several thousand molecules of a single specific neurotransmitter (Kelly 1993). The synaptic vesicles tend to cluster at regions of the membrane which are specialized for releasing the neurotransmitter, and they are called *active zones* (Unwin and Zampighi 1980) (Fig. 5.3).

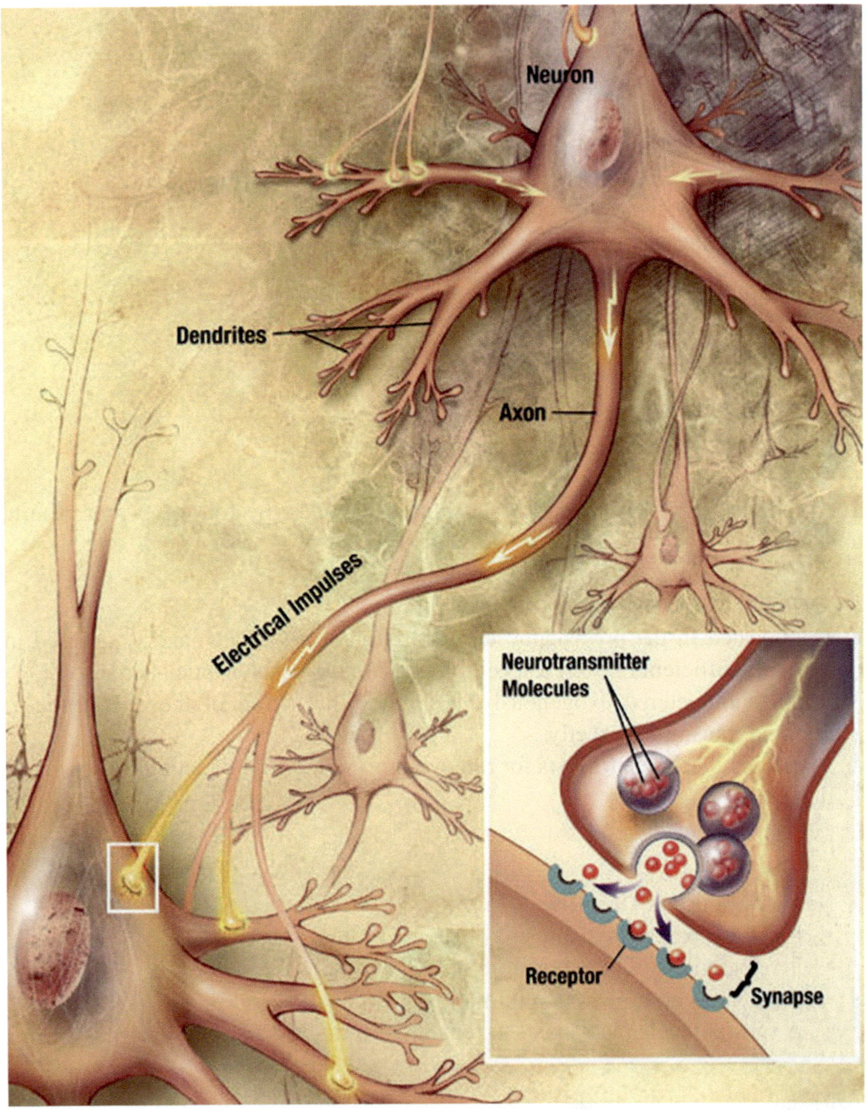

Fig. 5.2 Synaptic vesicles containing molecules or neurotransmitters

It is important to note that the storage and subsequently the release of neurotransmitters are done in standardized and fixed amounts (*quanta*) which are specific for each neurotransmitter and correspond to several thousand molecules. Following this, each vesicle affords to store only one quantum of the specific neurotransmitter and releases its entire contents into the synaptic cleft during exocytosis (Lindau and Almers 1995; Matthews 1996). The only exception identified so far to the *quantal mode of neurotransmitter release* is the retina. Each action potential in the brain releases only 1–10 quanta in comparison to an average of 150 quanta released at the

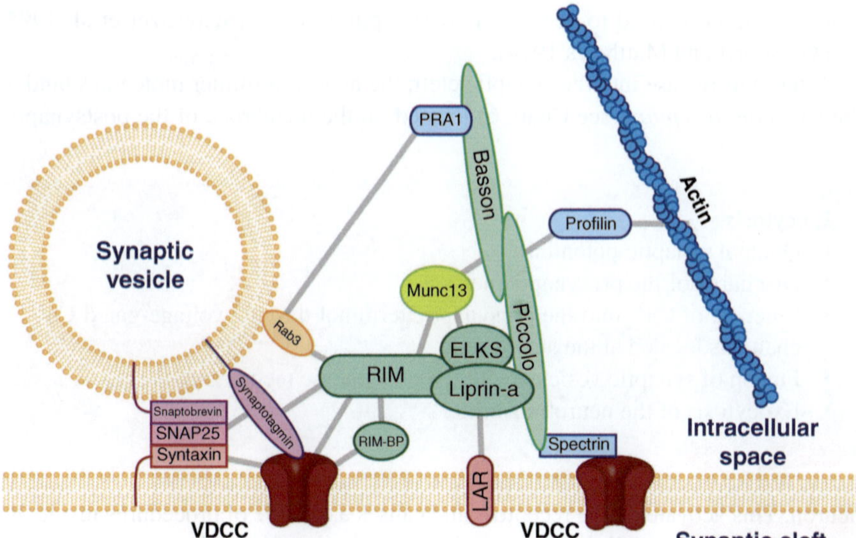

Fig. 5.3 Simplified design of the active zone: With gray color the interactions among proteins. With green color the Scaffold proteins. (Modified from Mittelstaedt, Alvarez-Baron, and Schoch. RIM proteins and their role in synapse function. Biol. Chem., Vol. 391, 599–606)

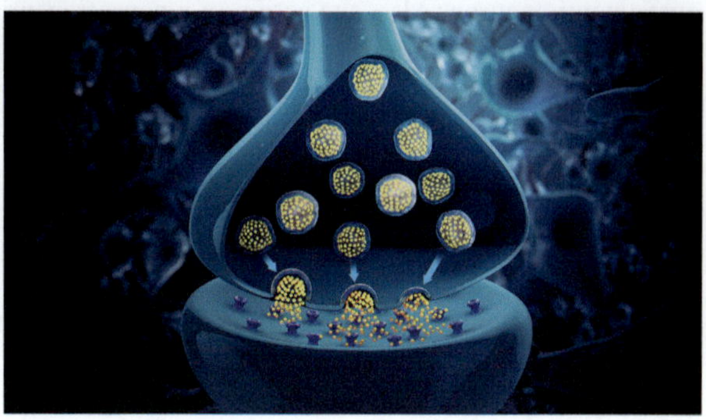

Fig. 5.4 Exocytosis of neurotransmitters from the synaptic vesicles to the synaptic cleft. (From https://www.scientificanimations.com)

neuromuscular synapse. Each quantum of transmitter produces a postsynaptic potential of fixed size, called the *quantal synaptic potential* (Liley 1956).

The release of the neurotransmitter into the synaptic cleft follows a complex cascade of events, but the principal role is played by Ca^{2+}, which during the discharge of a presynaptic action potential, they enter the presynaptic terminal through voltage-gated Ca^{2+} channels located at the active zone. This abrupt rise in the concentration of intracellular Ca^{2+} causes the vesicles to fuse with the cell membrane (Bahler and Greengard 1987; Hanson et al. 1997). In this way, a process called *exocytosis* (Fig. 5.4) permits the release of the neurotransmitter molecules into the synaptic cleft (Lawson et al. 1977). After exocytosis, the vesicle membrane which had been fused with the cellular membrane is retrieved rapidly and is recycled so that its material is used to generate new synaptic vesicles (Schweizer et al. 1995; von Gersdorff and Matthews 1994).

After their release into the synaptic cleft, the neurotransmitter molecules bind to their *specific receptors* (see Chap. 6) located on the membrane of the postsynaptic

Exocytosis

- Quantal synaptic potential
- Discharge of the presynaptic neuron
- Entering of Ca^{2+} into the presynaptic terminal through voltage-gated Ca^{2+} channels located at the active zone
- Fusion of synaptic vesicles with the presynaptic membrane
- Exocytosis of the neurotransmitter

neuron. This activates these receptors and causes a cascade of procedures inside the postsynaptic neuron. All these steps take time, varying from 0.3 ms to several milliseconds or even longer, and this is why chemical synapses are not as fast as electrical synapses. However, chemical transmission has the important property of amplification, since a small presynaptic electrical current can result in the release of thousands of neurotransmitter molecules in the cleft and in turn to the activation of a large number of postsynaptic receptors with a profound effect even on a large postsynaptic neuron. Also, the action of a neurotransmitter in the postsynaptic neuron *depends on the properties of the postsynaptic receptors* that recognize and bind the specific neurotransmitter rather than on the chemical properties of the transmitter itself.

5.3 Classification of Neurotransmitters

The neurotransmitters can be roughly classified into two big groups:

5.3.1 Small-Molecule Neurotransmitters or Biogenic Amines

This second term is rather imprecise in chemical terms but is being used for decades to label a group of neurotransmitters, including the *catecholamines* (*dopamine, epinephrine, and norepinephrine/noradrenaline*), *acetylcholine, and serotonin* but often also *histamine* which is chemically remote from both the catecholamines and the indolamines.

The catecholamines are synthesized from the essential amino acid tyrosine through a common biosynthetic pathway. During the first step of this pathway, tyrosine hydroxylase converts tyrosine to L-dihydroxyphenylalanine (L-DOPA). At the next step, L-DOPA is decarboxylated by a decarboxylase giving dopamine and CO_2. Then at the third step, dopamine β-hydroxylase converts dopamine to norepinephrine.

Serotonin (5-hydroxytryptamine or 5-HT) is synthesized from the essential amino acid tryptophan. They both are indoles with an aromatic structure. Histamine has been identified as important during inflammatory reactions in the body. It is synthesized from histidine by decarboxylation.

Acetylcholine is synthesized from choline with choline acetyltransferase as a catalyst and acetyl-coenzyme A as a co-substrate.

5.3.2 Excitatory Amino Acids and Neuroactive Peptides

This is a heterogenous group of substances with a neurotransmitter profile and includes molecules with a diverse biological action on the human body, hormones, etc.

All types of neurotransmitters and neuroactive molecules can coexist in the same neuron. The combination usually consists of one of the small-molecule transmitters and one or more peptides derived from one kind of polyprotein (Kupfermann 1991).

Small-Molecule Neurotransmitters or Biogenic Amines

- Acetylcholine
- Serotonin
- Catecholamines
  - Epinephrine (adrenaline)
  - Norepinephrine (noradrenaline)
  - Dopamine
- Histamine

Excitatory amino acids

- Glutamate
- Gamma-aminobutyric acid
- Glycine

Neuroactive peptides

- Vasopressin
- Gastrine
- Oxytocin
- Somatostatin
- Thyrotropin-releasing hormone (TRH)
- Purines, etc.

5.3.3 Small-Molecule Neurotransmitters

5.3.3.1 Acetylcholine (Ach)

Acetylcholine (*Ach*) is not an amino acid or its direct derivative, but still it is the only low-molecular-weight amine considered to be a neurotransmitter (Fig. 5.5). It is synthesized from choline (which derives from diet) in a reaction which includes a single step, catalyzed by choline-acetyltransferase with acetyl-coenzyme A (acetyl-CoA) as co-substrate. Acetylcholine is catabolized by *acetylcholinesterase* (*AchE*) and *butyrylcholinesterase* (*BCHE*).

Ach is widely distributed in the mammalian and the human brain (Fig. 5.6). It is produced in the cell bodies of neurons located mainly in the Meynert nucleus (nucleus basalis magnocellularis). According to the literature, there are eight regions where the cell bodies of cholinergic neurons are located (Ch1–Ch8) (Mesulam 1994; Mesulam et al. 1983). The Ch1 group concerns those cholinergic neurons located in the medial septal nucleus (Fujishiro et al. 2006), the Ch2 those in the nucleus of the vertical limb of the diagonal band (Fujishiro et al. 2006), the Ch3 in the lateral portion of the horizontal limb nucleus of the diagonal band (Mesulam et al. 1983), the Ch4 in the Meynert nucleus (Liu et al. 2015), the Ch5 in the pars compacta and pars dissipata of the pedunculopontine nucleus (Manaye et al. 1999), the Ch6 is centered around the laterodorsal tegmental nucleus and spreads into the central gray and medial longitudinal fasciculus (Mesulam et al. 1989), the Ch7 in the medial habenula (Dautan et al. 2016), and the Ch8 in the parabigeminal nucleus (Mufson et al. 1986). The Ch1 and Ch2 groups provide the main cholinergic

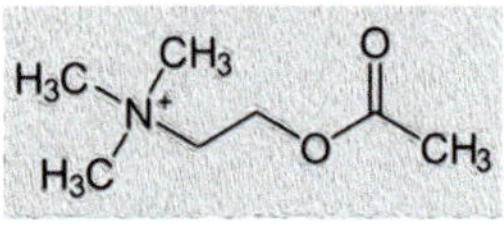

Fig. 5.5 The acetylcholine molecule (Ach)

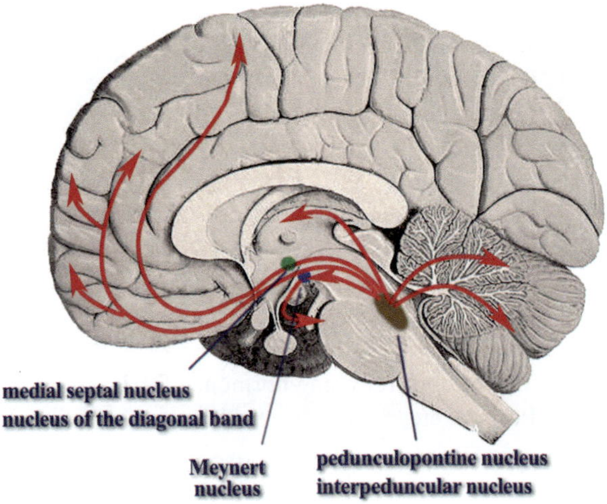

Fig. 5.6 The acetylcholine pathways in the human brain. (Copyright KN Fountoulakis 2021)

innervation to the hippocampus, the Ch3 to the olfactory bulb, the Ch4 to the hemisphere cortex and the amygdala, the Ch5 and Ch6 to the thalamus, the Ch7 to the interpeduncular nucleus, and the Ch8 to the superior colliculus. There are also projections of lesser importance from the Ch1–Ch4 and Ch8 to the thalamus and from the Ch5–Ch6 to the cortex (Geula and Mesulam 2011).

Acetylcholine
- Low-molecular-weight amine
- Synthesized from choline
- Synthesized in areas Ch1–Ch8 (mainly in Ch4 or Meynert nucleus)
- Innervates
  - Hippocampus
  - Olfactory bulb
  - Hemisphere cortex
  - Amygdala
  - Thalamus
  - Interpeduncular nucleus
  - Superior colliculus
- Excitatory action
- Expression of learned behavior

The main function of cholinergic system (see also Chap. 9) is to turn non-cholinergic neurons more sensitive to other excitatory stimuli. There is a bulk of pharmacological data suggesting that the central cholinergic system plays an important role in the acquisition as well as the expression of learned behaviors since it seems that scopolamine and atropine exert a negative effect on this kind of behaviors (Reiner and Fibiger 1994).

5.3.3.2 Serotonin (5-HT)

Serotonin (Fig. 5.7) was discovered in 1900 (Brodie 1900) and received its name in 1948 (Rapport et al. 1948), while its presence in the mammalian brain was verified in 1953 (Twarog and Page 1953) and its role as a neurotransmitter was established in 1956 (Bogdanski et al. 1956; Brodie and Shore 1957). Of the total circulating body serotonin, only 1–2% is found in the brain (Cooper and Fornal 1995). It is also called *5-hydroxytryptamine* (*5-HT*) (Marsden 1996), and it is an indoleamine which can be found not only in humans but also in most animals and even in plants since

Fig. 5.7 The serotonin molecule (5-HT)

it is one of the phylogenetically oldest neurotransmitters (Umbriaco et al. 1990). In mammals, it can be found in the platelets, the mast cells, and the chromaffin cells of the intestine.

5-HT is synthesized in a variety of cell types but not in the platelets. It does not cross the blood-brain barrier (BBB), and therefore in the brain, only the locally synthesized neurotransmitter in the raphe nucleus of the brain stem exists. Initially, the cells of the raphe nuclei were classified in the groups B9–B1 on the basis of their location from rostral to caudal (Dahlstroem and Fuxe 1964), but today, they are classified into two big subgroups, the superior, which includes the *caudal linear nucleus* (B8), the *median raphe nucleus* (B5 and B8), and the *dorsal raphe nucleus* (B6 and B7) and has more than half of serotonergic neurons, and the inferior which includes the *nucleus raphe obscurus* (B2), the *nucleus raphe magnus* (B3), the *nucleus raphe pallidus* (B1 and B4), the *ventrolateral medulla* (*lateral paragigantocellular nucleus* and the *intermediate reticular nuclei*, Bl/B3), and the *area postrema* (Fig. 5.8) (Jacobs and Azmitia 1992).

It is synthesized from the essential amino acid *tryptophan* (*TRP;* both are indoles with aromatic structure). Synthesis is catalyzed by the enzyme tryptophan hydroxylase which hydroxylates TRP to 5-hydroxytryptophan (5-HTP). The aromatic-L-amino acid decarboxylase (AADC), along with pyridoxal-5-phosphate (vitamin B6) as coenzyme, decarboxylates 5-HTP, thus leading to the creation of 5-HT. In this pathway, the rate-limiting step is the availability of 5-HTP, and for this, the availability of TRP is essential (Kema et al. 2000). TRP is carried through the blood-brain barrier by the same transporter other amino acids (isoleucine, leucine, phenylalanine, tyrosine, and valine) use. This is called the *large neutral amino acid* (*LNAA*) *transporter*, and TRP competes with other amino acids for transportation with the LNAA. This is the mechanism behind the rapid tryptophan depletion (RTD) test (Bell et al. 2001; Delgado et al. 1999; Moore et al. 2000; Reilly et al. 1997), which decreases brain 5-HT synthesis and release,

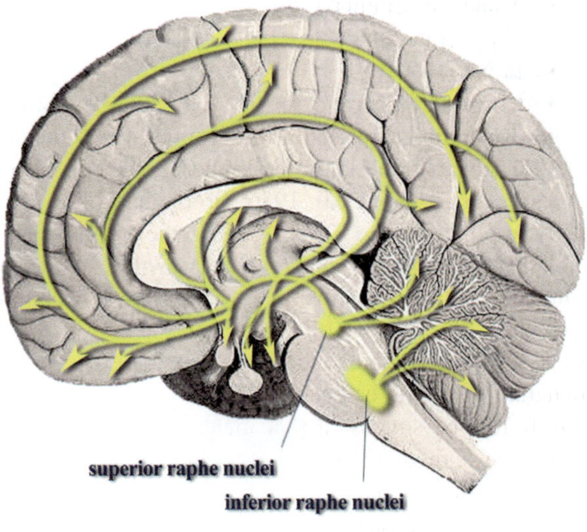

Fig. 5.8 The serotonin pathways in the human brain. (Copyright KN Fountoulakis 2021)

by temporarily reducing or even eliminating the availability of TRP. This is achieved through a dietary modification which concerns the administration of a load of LNAA not containing TRP.

Serotonin is catabolized mainly by *mono-amin-oxidase-A* (*MAO-A*) to 5-hydroxyindoleacetaldehyde, which is turned to 5-hydroxyindoleacetic acid (5-HIAA) by aldehyde dehydrogenase or to 5-hydroxytryptophol (5-HTOL) by alcohol dehydrogenase. 5-HIAA can be found in the cerebrospinal fluid and in brain tissue and constitutes that main metabolic product of 5-HT. Other pathways for the catabolizing of 5-HT include N-acetylation by serotonin-*N*-acetyltransferase and 5-O-methylation by hydroxyindole-*O*-methyltransferase which leads to the synthesis of melatonin (Cooper and Fornal 1995).

Serotonin
- Indolamine
- Synthesized from tryptophan
- Synthesized in the raphe nuclei of the brain stem
- Innervates
  - Cortex
  - Thalamus
  - Hippocampus
  - Basal ganglia
  - Amygdala
  - Nucleus accumbens
- Process of sensory stimuli
- Regulates motor activity
- Motive—reward processes
- Approach and avoidance behaviors
- Affect

Serotonin plays a role in the process of sensory stimuli, the regulation of motor activity, motives and reward, approach and avoidance behaviors, and the regulation of affect (see also Chap. 7).

5.3.3.3 Norepinephrine (NE)

Norepinephrine (*or noradrenaline—NA*) is *the main catecholamine in the brain* and one of the most important neurotransmitters (Fig. 5.9). For the first time, it was identified in the autonomic nervous system (Cannon and Uridil 1921) and later was also found in the brain (Holtz 1950; Vogt 1954). Initially, it was called "sympathin." Its current name was given by Von Euler (1946a, b), and it comes from the Greek name of the adrenal glands which is *επινεφρίδια* (*epinefridia*, meaning "on top of the kidney").

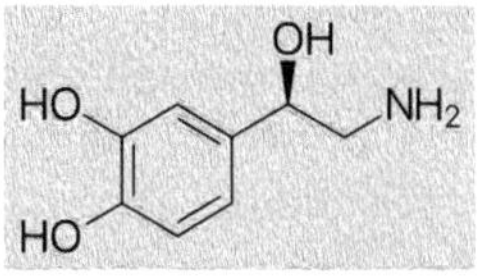

Fig. 5.9 The norepinephrine molecule (NE)

Norepinephrine

- The main catecholamine in the brain
  - Synthesized mainly in the locus coeruleus (for the brain)
  - Synthesized from phenylalanine and tyrosine
  - Metabolized by MAO-A and MAO-B and COMT
- Main pathways
  - Ventral norepinephric bundle
  - Dorsal norepinephric bundle
- Projects to
  - Thalamus
  - Hypothalamus
  - Hippocampus
  - Cortex
  - Limbic system
- Fight-or-flight reaction
- Volition
- Motor activity
- Reward mechanisms
- REM sleep

Norepinephrine is abundant in the human brain (Fig. 5.10) (Dahlstroem and Fuxe 1964; Lindvall and Bjorklund 1983). It is synthesized via the catecholamine synthesis pathway from *phenylalanine* and *tyrosine*. More specifically, L-tyrosine, which is an aromatic amino acid derived from food proteins, serves as the precursor for the catecholamines. Tyrosine hydroxylase (TH) catalyzes its hydroxylation resulting in the production of L-3,4-dihydroxyphenylalanine (L-DOPA). In turn, aromatic L-amino acid decarboxylase turns L-DOPA into dopamine, and finally dopamine-β-hydroxylase (DBH) turns dopamine into norepinephrine. Epinephrine (adrenaline) is produced by methylation of norepinephrine by phenylethanolamine *N*-methyltransferase (PNMT).

In this cascade of events, *the level of TH is the rate-limiting step*. The activity of TH is under acute inhibitory control (negative feedback) by the already synthesized catecholamines (end-product inhibition), but it is also under gene expression control and is significantly influenced by overall neuronal activity (Masserano et al. 1989). Also, electrical stimulation of neurons (Roth et al. 1975) and specific protein kinases (e.g., A and C) (Masserano et al. 1989; Edelman et al. 1978) could enhance norepinephrine synthesis.

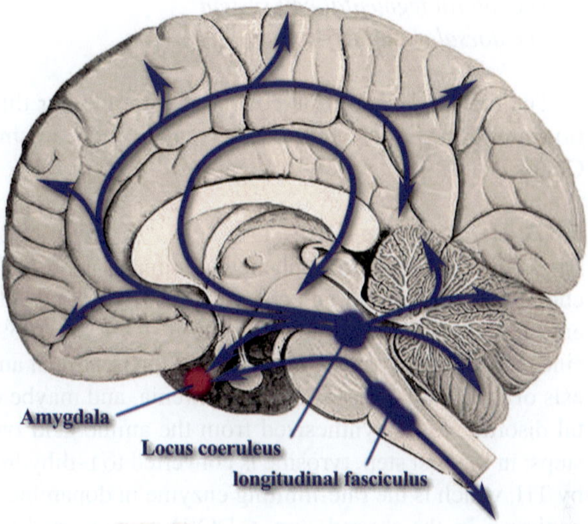

Fig. 5.10 The norepinephrine pathways in the human brain. (Copyright KN Fountoulakis 2021)

Norepinephrine is metabolized by two enzymes, the *monoamine oxidase* (*MAO*) *type A* and *B* and the *catechol-O-methyltransferase* (*COMT*). Both types of MAO exist in the neurons but also in the glial cells, while MAO-A is predominant in the locus coeruleus (LC) where norepinephrine is synthesized.

- MAO causes deamination which leads to 3,4-dihydroxyphenylglycolaldehyde (DHPGA). The next step includes two alternatives, conversion to 3,4-dihydroxyphenylglycol (DHPG) and then to 3-metoxy-4-hydroxyglycol (MHPG) or to 3,4-dihydromandelic acid (DHMA) and then to vanillylmandelic acid (VMA).
- COMT turns norepinephrine to normetanephrine (NMN), which afterward is converted to 3-metoxy-4-hydroxyphenylglycoaldehyde (MHPGA) which in turn is converted to either VMA or MHPG.

Eventually, in humans, MHPG is the main metabolite product which can be found in the cerebrospinal fluid, while on the contrary, VMA is predominant in the plasma and urine (Berry et al. 1994; Butcher et al. 1990; Costa et al. 1972; Elsworth et al. 1983; Kopin 1985; Mardh et al. 1981; Widerlov and Lewander 1978).

The specific localization of norepinephrine includes three main cell groups (see also Chap. 7):

- The *locus coeruleus* (*LC*), which constitutes the site where *the main norepinephrine pathway* originates from and projects to the thalamus, the dorsal hypothalamus, the hippocampus, and the cortex.
  - The *ventral norepinephric bundle* emerges ventral/rostral to the LC and reaches the hypothalamus and the subcortical parts of the limbic system.
  - The *dorsal norepinephric bundle* emerges from the LC and projects to the cortex.

- The *lateral tegmental cell system*
- The *dorsal medullary group*

The role of epinephrine concerns among other things the fight-or-flight reaction, volition, motor activity, the reward mechanism, and REM sleep (see also Chap. 7).

5.3.3.4 Dopamine (DA)

Dopamine (3,4-dihydroxyphenethylamine; Fig. 5.11) is the most abundant catecholamine in the brain and constitutes 80% of total catecholamines. It was discovered in the brain during the late 1950s (Carlsson 1959; Carlsson et al. 1957), and since then, it has a proven major role in the treatment and possibly the etiopathogenesis of Parkinson's disease, schizophrenia, and maybe other neurological and mental disorders. It is synthesized from the amino acid *tyrosine*. The process has two steps; in the first step, tyrosine is converted to L-dihydroxyphenylalanine (L-DOPA) by TH, which is the rate-limiting enzyme in dopamine synthesis, as it is for norepinephrine. In the second step, L-DOPA is converted to dopamine by the aromatic L-amino acid decarboxylase.

Dopamine is metabolized intracellularly by MAO to dihydroxyphenylacetic acid (DOPAC) and extracellularly by *COMT* and *MAO* to 3-methoxytyramine (3-MT) and *homovanillic acid* (*HVA*). In the human brain, HVA appears to be the main metabolite (Cooper et al. 1996).

The dopaminergic system is significantly more complex in comparison to the norepinephric. Below its classical structure is described, which is considered to be important for educational purposes, although it is currently disputed due to the accumulation of more data (Fig. 5.12).

Traditionally, four main dopaminergic pathways are mentioned as follows:

- The *nigrostriatal pathway* whose fibers originate from the midbrain tegmentum and more specifically from the substantia nigra zona compacta (SN-ZC or A9) and the ventral tegmental area of Tsai (VTA or A10) (Dahlstroem and Fuxe 1964; Anden et al. 1964) and project to the nucleus caudatus and the putamen of the lentiform nucleus as well as to the nucleus accumbens (NAC), the amygdala, the hippocampus, the olfactory tubercle, the medial prefrontal, the cingulate and entorhinal cortices, and other cortical areas (Anden et al. 1966; Ungerstedt 1971).

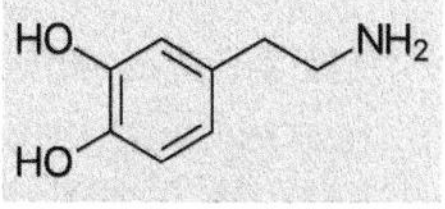

Fig. 5.11 The dopamine molecule (DA)

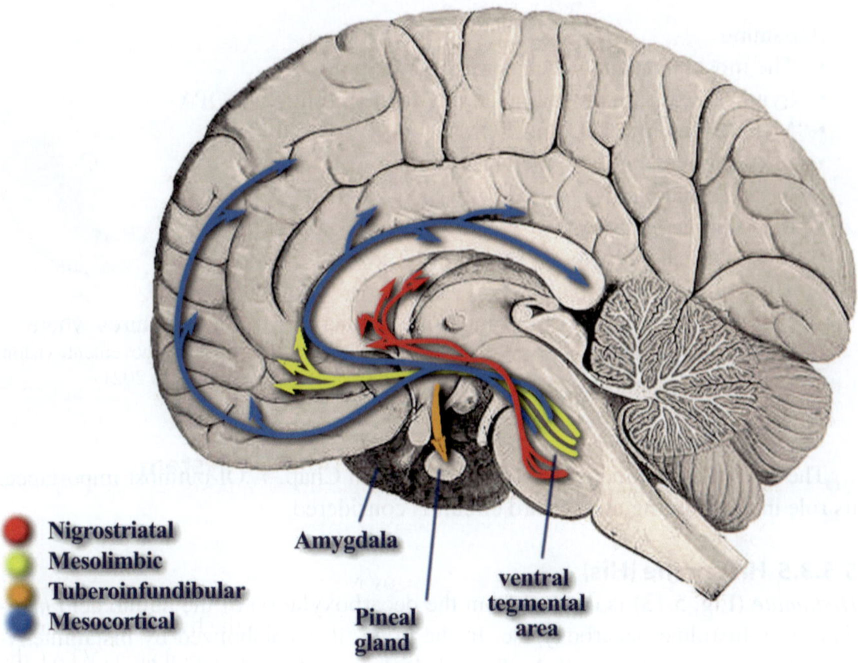

Fig. 5.12 The dopamine pathways in the human brain. (Copyright KN Fountoulakis 2021)

- The *mesolimbic pathway* whose fibers originate from the VTA (mainly from the ventrally located paranigral nucleus) and project to the amygdaloid nucleus, the dorsolateral septal nucleus, and the nucleus accumbens. The experience of stress activates this pathway (Fallon 1981; Swanson 1982).
- The *mesocortical pathway* whose fibers originate from the VTA (mainly from the parabrachial nucleus) and project to the frontal cortex and areas of the septum and the hippocampus (Fallon 1981; Swanson 1982; Deniau et al. 1980; Oades and Halliday 1987; Phillipson 1979; Simon et al. 1979). Again the experience of stress by the person activates this pathway.
- The *tuberoinfundibular pathway* (A12 region) (Moore and Lookingland 1994), whose fibers originate from the arcuate nucleus in the hypothalamus and project to the median eminence. Essentially, they constitute a small minority of dopaminergic neurons; however, they release dopamine in the stream of the hypophyseal portal veins, which eventually reaches the anterior pituitary gland where it inhibits the secretion of prolactin. Experiencing stress by the individual could lead to the inhibition of this pathway resulting in an increase in prolactin secretion and subsequently in sexual disorders in males and abnormal lactation and amenorrhea in females.

Dopamine

- The most abundant catecholamine in the brain
- Synthesized from the amino acid tyrosine through L-DOPA
- Metabolized by MAO and COMT
- Its major metabolite in the brain is homovanillic acid
- Nigrostriatal pathway
- Mesolimbic pathway
- Mesocortical pathway
- Tuberoinfundibular pathway
- Regulates and permits the synthesis of functions in the structures where dopaminergic neurons project
- Dopaminergic theory of reward

The role of dopamine is described in detail in Chap. 7. Of outmost importance, its role in the learning and reward circuit is considered.

5.3.3.5 Histamine (His)

Histamine (Fig. 5.13) is derived from the decarboxylation of the amino acid *histidine* by L-histidine decarboxylase. In the brain, it is catabolized by histamine-*N*-methyltransferase. The cell bodies of histamine neurons are located in the *tuberomammillary nuclei of the thalamus*, and they project throughout the brain, including the cortex, through the medial forebrain bundle (Fig. 5.14).

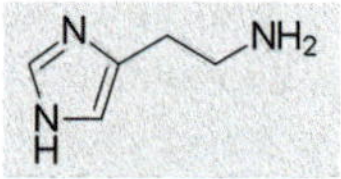

Fig. 5.13 The histamine molecule (His)

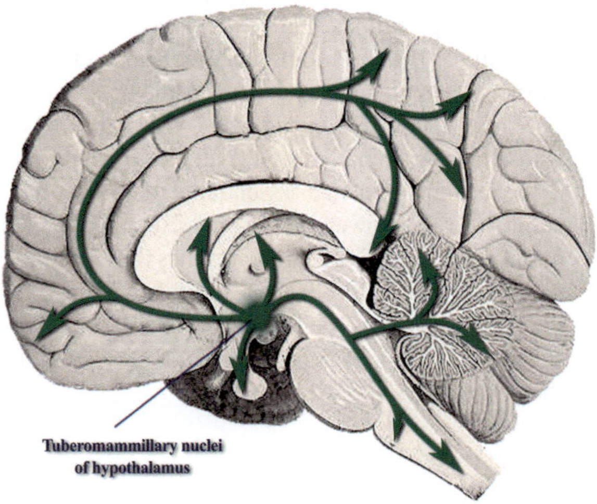

Fig. 5.14 The histamine pathways in the human brain. (Copyright KN Fountoulakis 2021)

Histamine

- Is synthesized in the tuberomammillary nuclei of the thalamus
- Histaminergic neurons project to the whole brain
- They are related to the sleep-wakefulness cycle
- The histaminergic neuron pattern of firing:
  - Low in periods of relaxation or tiredness
  - High during wakefulness
  - Complete absence of firing during sleep

5.3.4 Excitatory Amino Acids and Neuroactive Peptides

5.3.4.1 Glutamate (Glu)

Glutamate (Fig. 5.15) is the most important and frequently used excitatory amino acid throughout the brain, with glutamatergic neurons utilizing between 60 and 80% of total brain metabolic activity (Fig. 5.16) (Rothman et al. 2003). It is also major part of a wide variety of proteins, and therefore, it is one of the most abundant amino acids in the human body (Meldrum 2000).

Normally, it is *obtained from the diet* and there is no need to be synthesized. However, it can also be synthesized from alpha-ketoglutaric acid (part of the citric acid cycle which has citrate as starting point). Glu is actively transported through the blood-brain barrier (Shigeri et al. 2004; Vandenberg 1998). As a neurotransmitter, it is dominant throughout the cortex and in most subcortical pathways (Stahl 2008).

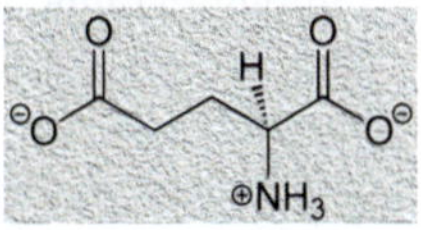

Fig. 5.15 The glutamate molecule (Glu)

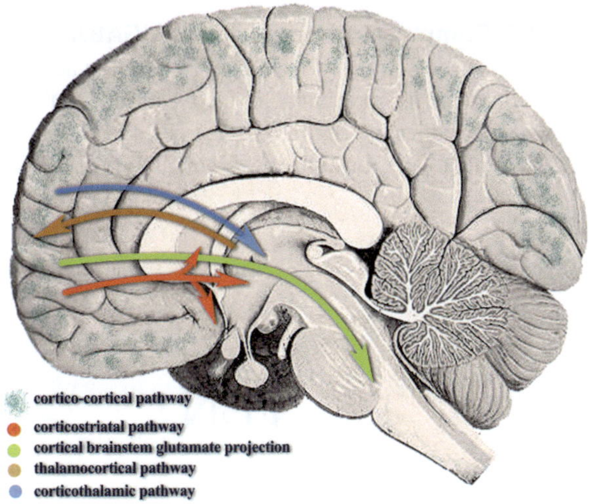

Fig. 5.16 The glutamate pathways in the human brain. (Copyright KN Fountoulakis 2021)

Glutamate

- Obtained from the diet
- Dominant throughout the cortex
- Major role in most subcortical pathways
- Excitatory role
- Brain plasticity
- Higher neurocognitive functions
- The most important function of all other neurotransmitters is to regulate Glu activity

There are five important glutamatergic pathways (Stahl 2008) (Fig. 5.16) as follows:

- The *cortical brain stem glutamate projection* from the prefrontal cortex to the brain stem (raphe nuclei, locus coeruleus, ventral tegmental area, substantia nigra).
- The *corticostriatal glutamate pathway* from the prefrontal cortex to the striatum and the *cortico-accumbens glutamate pathway* to the nucleus accumbens.
- The *thalamocortical glutamate pathway* from the thalamus to the pyramidal neurons in the cortex.
- The *corticothalamic glutamate pathway* from the prefrontal cortex to the thalamus.
- The *cortico-cortical glutamatergic pathway* which concerns the communication among pyramidal neurons within the cortex.

It plays a role in the mechanisms concerning brain *plasticity* and *higher neurocognitive functions.* It could be said that the most important function of all other neurotransmitters is to regulate Glu activity.

5.3.4.2 Gamma-Aminobutyric Acid (GABA)

GABA (Fig. 5.17) is *produced from Glu* via the action of the enzyme glutamic acid decarboxylase. It is present at high concentrations throughout the brain and can also be found in other tissues (e.g., pancreas and the adrenal gland). It is secreted by 15–20% of all neurons. In the brain, it is the major *inhibitory* neurotransmitter (Paul 1994).

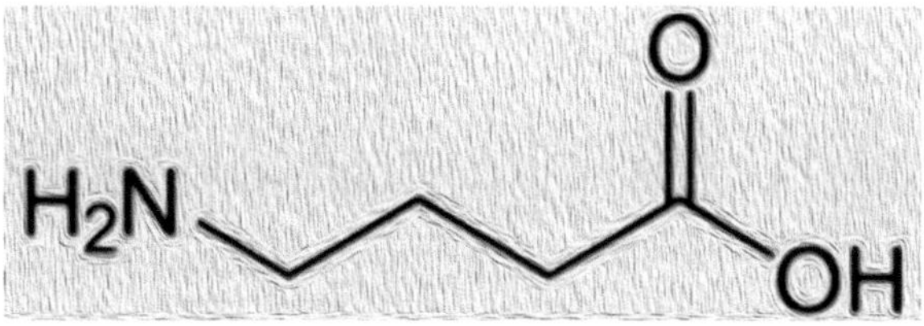

Fig. 5.17 The gamma-aminobutyric acid molecule (GABA)

5.3.4.3 Glycine (Gly)

Gly is the major transmitter *in inhibitory interneurons of the spinal cord.* It is synthesized from serine, and it is the smallest of the 20 amino acids commonly found in proteins. It bounds to *ionotropic receptors*, and it is a required *co-agonist* along with glutamate for NMDA receptors.

5.3.4.4 Neuroactive Peptides

There have been discovered over 100 distinct neurobiologically active neuropeptides in the mammalian CNS, and it is estimated that there exist at least several hundreds. All are defined by convention as encompassing approximately 90 or fewer amino acids. Their molecular weight is less than 10,000 Da and they are phylogenetically ancient molecules. Their structure has been conserved across phyla to a remarkable degree. Our information concerns mainly studies in the rat brain, and the extent of this knowledge reflects also their existence, distribution, and function in the human brain which remains to be proven beyond doubt. Abnormal peptidergic functioning may involve neurotransmitter, neuromodulator, and neurohormone effects.

Some of these peptides had been previously identified as hormones (e.g., angiotensin and gastrin) or as products of neuroendocrine secretion (e.g., oxytocin, vasopressin, somatostatin, and thyrotropin-releasing hormone—TRH) (A decade of neuropeptides: past, present, and future. Tenth Annual Winter Neuropeptide Conference. Breckenridge, Colorado, January 16–20, 1989. Proceedings 1990; Hokfelt et al. 1986). Adenosine triphosphate (ATP) and its products (e.g., adenosine) act as transmitters in some synapses. Adenine and guanine and their products are called purines, and they also serve as nucleotide bases in DNA and RNA.

Most of the peptides play a significant role in the mechanisms of (see also Chap. 7):

- Thermoregulation
- Food and fluid intake
- Memory
- Learning
- The reaction to stress and pain

Most of them do not satisfy all criteria to be considered a neurotransmitter but several do (Myers 1994). Neurons could contain and release several peptides which come from the same polyprotein, and their release leads to potentially different postsynaptic actions (Fisher et al. 1988).

Peptides can be grouped into the following:

- *Hypothalamic*
- *Brain-borne* (pituitary hormones, pro-opiomelanocortin peptides, opioid peptides, brain-borne gastrointestinal or brain-gut neuropeptides)
- *Other* neuropeptides
- *Neuroactive steroids*

Neurohormones are released into the bloodstream, and then they cross the blood-brain barrier and they get into the brain.

Neuroactive Peptides

- There have been discovered over 100 distinct neurobiologically active neuropeptides
- Some of these peptides had been previously identified as hormones
- Role in the mechanisms of
  - Thermoregulation
  - Food and fluid intake
  - Memory
  - Learning
  - The reaction to stress
  - Pain
- Most of them do not satisfy all criteria to be considered a neurotransmitter
- Neurons contain and release several peptides which come from the same polyprotein
- Peptides can be grouped into
  - Hypothalamic
  - Brain-borne
  - Other neuropeptides
  - Neuroactive steroids
- Following the binding on the receptor
  - Incorporation inside the postsynaptic cell
  - Release back into the synaptic cleft
- Fragments of neuropeptides with the active area are active on the receptor
- Receptors are connected to G-protein-coupled second messengers
- The neuropeptide-receptor complex is translocated to the cell nucleus and influences gene expression
- The duration of the physiological effects of neuropeptides lasts for minutes to hours
- The concentrations of neuropeptides range from the picomolar to femtomolar
- Low concentration and prolonged action

Fragments of neuropeptides containing the active area could be active on the receptor in a similar way the complete molecule is. All these receptors are connected to G-protein-coupled second messengers. It is interesting that neuropeptides (e.g., neurotensin or TRH), after binding on a postsynaptic receptor, create a complex which is then translocated from the neuronal membrane to the *cell nucleus*, where it directly regulates specific gene expression mechanisms.

The *duration* of the physiological effects of neuropeptides lasts for minutes to hours while that of classical neurotransmitters lasts for seconds or minutes. Also the concentrations of neuropeptides range from the picomolar to femtomolar, relative to the nanomolar to micromolar concentrations of the classical chemical neurotransmitters or the millimolar concentrations of certain amino acid neurotransmitters, such as glutamic acid. This means that in comparison to biogenic amines and excitatory amino acids, neuroactive peptides appear in much lower concentrations, but their action is much prolonged.

Hypothalamic Neuropeptides

This group of peptides are synthesized mainly in the hypothalamus. They mainly act as neurohormones at the pituitary level or in the periphery, and they exist at high concentrations in the median eminence.

Hypothalamic Neuropeptides

- Luteinizing hormone-releasing hormone or gonadotropin-releasing hormone (LHRH or GnRH)
- Thyrotropin-releasing hormone (TRH)
- Corticotropin-releasing hormone (CRH)
- The growth hormone-releasing hormone (GHRH)
- Somatostatin
- Cortistatin
- Vasopressin
- Oxytocin
- Neuromodulation
- Reproduction
- *Αναπαραγωγή*
- Arousal
- Response to stress
- Social behavior
- Hormone regulation
- Gastrointestinal system regulation
- Autonomic function
- Immune response

- The *luteinizing hormone-releasing hormone or gonadotropin-releasing hormone* (*LHRH or GnRH*) is synthesized mainly in the arcuate nucleus, is secreted into portal vessels, and is transported to the anterior pituitary. Its main function is reproduction in all vertebrates by stimulating the secretion of luteinizing hormone (LH) and follicle-stimulating hormone (FSH) (Jew et al. 1984; Palkovits 1984).
- The *thyrotropin-releasing hormone* (*TRH*) is synthesized in the hypothalamus and more specifically in the paraventricular nucleus (PVN), the preoptic supra-

chiasmatic nucleus, the dorsomedial nucleus, and the perifornical nucleus of the hypothalamus and the premamillary region. It can be found at high concentration in the amygdala, the lateral septal nucleus, and the nucleus tractus solitarius (NTS) of the medulla oblongata. Its main function is to stimulate the production and release of thyrotrophin hormone in the anterior pituitary. It also functions as a neurotransmitter and neuromodulator in the brain. Its secretion is under the control of noradrenaline, somatostatin, and opioids, and it is influenced by external stressful stimuli.

- The *corticotropin-releasing hormone* (*CRH*) is widely distributed mainly in the hypothalamus but also in other areas of the brain. Its function is to stimulate the release of adrenocorticotropic hormone (ACTH) in the anterior pituitary. The main CRH projections emerge from the central nucleus of the amygdala, the paraventricular nucleus (PVN), and bed nucleus of the stria terminalis (BNST) and terminate in the locus coeruleus (LC) (Van Bockstaele 1998). Their function serves mainly the enhanced arousal and focused attention to stressors with an emotional component. In the raphe nuclei, CRH interacts with serotonin (Price et al. 1998) while in the substantia nigra interacts with dopamine (Steckler and Holsboer 1999). There are two receptors identified thus far, the CRH-1 and CRH-2, and both are coupled with a G-protein (Hauger et al. 2003). CRH-1 receptors are found in the pituitary, cerebellum, and frontal cortex and play a role in neurocognition. It is interesting that urocortin which is a neuropeptide that is 45% homologous to CRH binds more potently to the CRH-2-alpha receptor, than CRH itself (Donaldson et al. 1996). CRH constitutes the main hormone of the reaction to stress mechanisms (Liu et al. 1997; McEwen 2000; Schulkin et al. 1998; Thrivikraman et al. 2000), while urocortin does not seem to possess similar properties (Turnbull et al. 1999). There is negative feedback at the level of the PVN, at the anterior pituitary, and through the ventral hippocampus to the BNST (Herman et al. 1992; Swanson and Simmons 1989), while positive feedback happens at the level of the central nucleus of the amygdala (CeA) and the dorsal PVN (Swanson and Simmons 1989).
- The growth hormone-releasing hormone (GHRH) exists in two forms, the GRH-40 and the GRH-44. They are synthesized in the medial basal hypothalamus (in the arcuate and ventromedial nuclei), and both stimulate the release of growth hormone from the anterior pituitary.
- *Somatostatin* is produced in various areas of the brain and in the periphery. It is highly concentrated in the hypothalamus. It exerts an inhibitory effect on GH and TSH in the anterior pituitary and also acts as a neurotransmitter in the brain. In the periphery, it regulates the activity of pancreatic islet and gastrointestinal and immune cell functions.
- *Cortistatin* is a somatostatin-like peptide which can be found almost exclusively in the cerebral cortex and the hippocampus (Spier and de Lecea 2000).
- *Vasopressin* (also named antidiuretic hormone (ADH), arginine vasopressin (AVP), and argipressin) is synthesized in the hypothalamus where it can be found at high concentrations. Its fibers in addition to oxytocin fibers constitute the hypothalamo-hypophyseal neurosecretory tract. In the brain, it binds to AVP-1A

and AVP-1B receptors and plays a role in stress-related and social behaviors (Insel 2010; Bielsky et al. 2004; Wersinger et al. 2004, 2007; Thibonnier et al. 1998, 2001).

- *Oxytocin* is synthesized only in the magnocellular hypothalamic nuclei. There, it is coexpressed with cholecystokinin (CCK) and enkephalins. Its fibers, together with vasopressin fibers, constitute the hypothalamo-hypophyseal neurosecretory tract (Insel 2010).

Brain-Borne Neuropeptides

Pituitary Hormones

The pituitary peptides are of pituitary origin, but they are also synthesized elsewhere and especially in the hypothalamus. They include, among others, the *luteinizing hormone* (*LH*), *thyrotropin* (*TSH*), *growth hormone* (*GH*), and *prolactin.*

Pituitary Hormones
- Luteinizing hormone (LH)
- Thyrotropin (TSH)
- Growth hormone (GH)
- Prolactin

Pro-opiomelanocortin (POMC) Peptides

Their common characteristic is that they arise from a common 31-kDa precursor molecule, and they colocalize in the same neurons of the medial basal hypothalamus, almost exclusively in the arcuate and ventral premamillary nuclei.

Pro-opiomelanocortin (POMC) Peptides
- Adrenocorticotropic hormone (ACTH)
- β-Endorphin
- Melanocyte-stimulating hormones (MSH)

- The *adrenocorticotropic hormone* (*ACTH*), which is also found in the NTS and in several hypothalamic (mainly in the PVN) and extrahypothalamic nuclei.
- The *β-endorphin* which is one of the opioid peptides, which is also found in the NTS, several hypothalamic (mainly in the PVN), and extrahypothalamic nuclei, the amygdala, the thalamic PVN, the midbrain central gray, the raphe nuclei, and the lateral parabrachial nucleus.
- The *melanocyte-stimulating hormones* (*MSH*) which include the α-, β-, and γ-MSH. They are present in the hypothalamus (arcuate nucleus, dorsal part of the lateral hypothalamic area, the perifornical nucleus, and the ventral portion of the

zona incerta) but also in the dorsolateral tegmentum of the pons and the NTS. Their fibers project exclusively to the cerebral cortex, hippocampus, olfactory bulb, and striatum. It seems that α-MSH exerts an anorexigenic action.

Brain-Borne Opioid Peptides

Brain-Borne Opioid Peptides
- Met- and Leu-enkephalins
- β-Endorphin
- Dynorphin-related peptides
- Endomorphins
- Nociceptin (N) or orphanin FQ (OFQ)

- The *Met-* and *Leu-enkephalins* are widely distributed in the CNS, but they are most abundant in the hypothalamic ventromedial nucleus, the globus pallidus, the nucleus accumbens, the central gray, the NTS, the spinal trigeminal nucleus, and the dorsal horn.
- The *β-endorphin* is described above.
- There are five *dynorphin-related peptides* with high analgesic activities at both hypothalamic (action on β-endorphin neurons) and spinal cord levels. Their highest concentration is found in the hypothalamus.
- The *endomorphins* include endomorphin 1 (EM1) and endomorphin 2 (EM2), and both exhibit high affinity and selectivity for the μ-opiate receptor. They have potent analgesic activities (Martin-Schild et al. 1999).
- *Nociceptin* (*N*) or *orphanin FQ* (OFQ) is an opioid-like endogenous agonist of the opioid receptor homologue, N/OFQ receptor. It has a role in many behavioral, autonomic, and physiological functions (Neal et al. 1999; Houtani et al. 2000).

Brain-Borne Gastrointestinal or Brain-Gut Neuropeptides

Brain-Borne Gastrointestinal or Brain-Gut Neuropeptides
- Substance P
- Neurokinins A and B
- Neuropeptide Y (NPY)
- Vasoactive intestinal polypeptide (VIP)
- Peptide histidine-isoleucine with 27 amino acids (PHI-27)
- Pituitary adenylate cyclase-activating peptide (PACAP)
- Cholecystokinin (CCK)
- Gastrin-releasing peptide (GRP)
- Galanin

- *Substance P* and *neurokinins A* and *B* are present in the central and peripheral nervous systems and in gut endocrine cells. Substance P is present in the hypothalamus but not in the cerebral cortex, constitutes the major transmitter in the striatonigral connections, and is strongly involved in pain mechanisms.
- *Neuropeptide Y* (*NPY*) is one of the most abundant neuropeptides in the brain. The receptor protein that NPY operates on is a G-protein-coupled receptor in the rhodopsin like 7-transmembrane G-protein-coupled receptors family. There are five subtypes of the NPY receptor (Y1–Y5), but only four of them exist in humans (Michel et al. 1998). NPY stimulates feeding behavior, influences the activity of the gastrointestinal and cardiovascular systems, and also has a role in body temperature regulation.
- *Vasoactive intestinal polypeptide* (*VIP*) has a wide range of effects on the cardiovascular, respiratory, and gastrointestinal systems.
- The *peptide histidine-isoleucine with 27 amino acids* (*PHI-27*) has considerable structural similarities to VIP, and it is highly concentrated in the suprachiasmatic nucleus. It plays a role in the control of prolactin, ACTH, and growth hormone secretion in the anterior pituitary (Hokfelt et al. 1983).
- The *pituitary adenylate cyclase-activating peptide* (*PACAP*) is a potent stimulator of pituitary adenylate cyclase and has its highest density in hypothalamic nuclei (Arimura and Shioda 1995).
- *Cholecystokinin* (*CCK*). Its name comes from Greek *χολή* ("chole" that is bile), *κύστη* ("cystis" meaning sac), and *κίνησις* ("kinisis" meaning the move). The complete name refers to the substance that promotes the moves of the gallbladder. Previously it was called pancreozymin. It is synthesized and secreted by enteroendocrine cells in the duodenum. It is present in different forms with the CCK-8 being the main representative in the brain. It plays a role in pain control (strong analgesic), and its effect takes place in the brain stem, the thalamus, and the cerebral cortex through binding to opioid receptors. Its receptors are also named CCK-A and CCK-B. While CCK-A receptors can be found mainly on pancreatic acinar cells, CCK-B are located mostly in the brain and stomach. CCK has been shown to interact with orexin neurons, which control appetite and sleep and reduce food intake (Tsujino et al. 2005).
- The *gastrin-releasing peptide* (*GRP*) can be found at high density in the hypothalamus and especially in the arcuate nucleus. It is released by G-cells in the pyloric antrum of the stomach, the duodenum, and the pancreas. It plays an important role in thermoregulation and regulation of blood pressure, it inhibits food intake, and also it is antidiuretic. It binds on CCK-B receptors.
- *Galanin* is widespread in the brain, but it can also be found with a high density in the hypothalamus, especially in the median eminence. It stimulates the release of GH in the plasma, increases prolactin levels, inhibits insulin secretion, and also plays a role in food intake.

Other Neuropeptides

Other Neuropeptides
- Neurotensin
- Peptides of the renin-angiotensin system
- Apelin
- Atrial natriuretic peptides (ANP)
- Bradykinin
- Calcitonin gene-related peptide (CGRP)
- Cocaine- and amphetamine-regulated transcript (CART)
- Orexin (hypocretin) peptides
- Melanin-concentrating hormone (MCH)
- Agouti-related protein (AgRP)
- Tuberoinfundibular peptide of 39 residues (TIP39)

This "residual" group includes a number of neuropeptides which all participate in central baroreceptor mechanisms.

- *Neurotensin* is present in the caudate nucleus, in nucleus accumbens, in the globus pallidus and mammillary body, in the arcuate, periventricular, and medial preoptic nuclei, in the periventricular subdivision of the PVN, and in the lateral hypothalamic area as well as in the endocrine cells of the gastrointestinal mucosa. Its action reduces body temperature, locomotor activity, and food intake (Sicard et al. 2005; Rostene and Alexander 1997).
- All the *peptides of the renin-angiotensin system*, including renin; angiotensinogen; angiotensin-converting enzyme (ACE); angiotensins I, II, III, and IV; angiotensins I–VII; and angiotensin receptors can also be found in the brain. Angiotensin II and III increase blood pressure and enhance drinking behavior and salt appetite, and also they increase the secretion of pituitary hormones.
- *Apelin* is located at high density in the arcuate nucleus, but it can also be found in the preoptic region, supraoptic and paraventricular nuclei, the pons, and the medulla oblongata (Reaux et al. 2002). It probably plays a role in the regulation of feeding behaviors, pituitary hormone release, and circadian rhythms.
- The *atrial natriuretic peptides* (*ANP*) include the ANP, the cANP, and the brain natriuretic peptide (BNP) and exist in high concentrations in the hypothalamus, the limbic system, and the brain stem. Their function includes the regulation of thirst and drinking behavior, fluid balance, and cardiovascular function. They also inhibit the secretion of vasopressin.
- *Bradykinin* plays an important role in the regulation of blood pressure and also acts as a potent pain transmitter in the brain.
- *Calcitonin gene-related peptide* (*CGRP*) is widely distributed in the brain. It inhibits growth hormone release and feeding and gastric acid secretion. It also plays a role in pain mechanisms.
- *Cocaine- and amphetamine-regulated transcript* (*CART*) is located in the paraventricular, arcuate, and dorsomedial hypothalamic nuclei, the lateral hypotha-

lamic area, the olfactory bulb, the lateral septal nucleus, and the medial posterodorsal amygdaloid nucleus (Koylu et al. 1998).

- There are two *orexin* (*hypocretin*) *peptides*, that is, orexin-33 (hypocretin-1) and orexin-28 (hypocretin-2). They are located exclusively in the lateral hypothalamus. They play a major role in the regulation of sleep-wake cycle (Broberger et al. 1998).
- *Melanin-concentrating hormone* (*MCH*) exists in the dorsal part of the caudal hypothalamus, and it is involved in the mechanisms which control food intake (Broberger et al. 1998).
- *Agouti-related protein* (*AgRP*) is a melanocortin receptor antagonist and a potent orexigenic peptide (Broberger et al. 1998).
- *Tuberoinfundibular peptide of 39 residues* (*TIP39*) can be found in the caudal hypothalamus and more specifically in the subparafascicular area as well as in the caudal paralemniscal nucleus in the pons. It is involved with mechanisms of pain perception (Usdin et al. 2000).

Neuroactive Steroids

Neuroactive Steroids

- Inhibitory neurosteroids
  - Allopregnanolone
  - Pregnanolone
  - Tetrahydrodeoxycorticosterone
  - 3α-Androstanediol
  - Cholesterol
- Excitatory neurosteroids
  - Pregnenolone sulfate
  - Epipregnanolone
  - Isopregnanolone
  - Dehydroepiandrosterone
  - 24(*S*)-hydroxycholesterol
- Pheromones
  - Androstadienol
  - Androstadienone
  - Androstanol
  - Androstenone
  - Estratetraenol
- Others
  - Pregnenolone
  - Progesterone
  - Estradiol
  - Corticosterone
  - Products of the metabolism of neurosteroids might be physiologically active
- Interact mainly with ligand-gated ion channels
- Receptors inside the cell nucleus
- Synthesized from cholesterol

Neuroactive steroids (neurosteroids) are endogenous or exogenous steroids that interact mainly with *ligand-gated ion channels* and other cell surface receptors (Paul and Purdy 1992; Lan and Gee 1994; Reddy 2010; Srivastava et al. 2011). Additionally, they regulate gene expression via *nuclear steroid hormone receptors*. Most of them are synthesized from cholesterol (Mellon 2007; Mellon and Deschepper 1993) and are classified (Reddy 2010) into the following:

- *Inhibitory neurosteroids* which act as positive allosteric modulators of GABA-A receptors. They include allopregnanolone (Pinna et al. 2000), pregnanolone (Carl et al. 1994), tetrahydrodeoxycorticosterone (Reddy 2003), 3α-androstanediol (Reddy and Jian 2010), and cholesterol (Henin et al. 2014; Rosenhouse-Dantsker 2017; Levitan et al. 2014).
- *Excitatory neurosteroids* which act as negative allosteric modulators of GABA-A receptors and positive allosteric modulators of NMDA receptor, and they also act as agonists of sigma-1 receptors. They include pregnenolone sulfate (Harteneck 2013), epipregnanolone (O'Dell et al. 2005), isopregnanolone (Hedstrom et al. 2009), dehydroepiandrosterone (Friess et al. 2000), and 24(*S*)-hydroxycholesterol (Paul et al. 2013).
- *Pheromones*, which include androstadienol (Cooke and Gower 1981), androstadienone (Hummer and McClintock 2009), androstenol (Kaminski et al. 2006), androstenone (Pierce et al. 2004), and estratetraenol (Jacob et al. 2001). They exert their effect mainly on the hypothalamus by activating the vomeronasal receptor cells (Monti-Bloch et al. 1994, 1998).
- There are a number of *other endogenous steroids*, including pregnenolone (Marx et al. 2009, 2014), progesterone (Baulieu and Schumacher 2000), estradiol (Srivastava et al. 2011), and corticosterone (Albrecht et al. 2013). Also some products of the metabolism of other neurosteroids are physiologically active. This is a heterogeneous group of substances acting on a variety of different cells and receptors.

5.3.5 Gliotransmitters

Gliotransmitters are molecules that are secreted from glial cells and regulate the communication among neurons and the other glial cells (Koizumi et al. 2005), possibly in collaboration with Ca^{2+} ions. However, neither their role nor the mechanisms they are involved are clear (Halassa et al. 2007; Agulhon et al. 2010), as it is also unclear whether they are secreted mainly by astrocytes.

A characteristic structure they participate is the so-called tripartite synapse, which includes the pre- and postsynaptic neuron together with a glial cell (Araque et al. 1999; Perea and Araque 2005).

Astrocytes seem to secrete mainly glutamate and ATP (Martineau et al. 2006; Zhang and Haydon 2005). While glutamate exerts its well-known actions on its receptors, ATP exerts an inhibitory action on neuronal activity through P2X, P2Y, and A1 receptors (Halassa et al. 2007). Its actions include the surfacing of AMPA receptors from the postsynaptic membrane and remyelination through its effect on purine receptors (Walter et al. 2004; Koizumi et al. 2005; Zhang and Haydon 2005). Other gliotransmitters include homocysteic acid, taurine, tumor necrosis factor-alpha (TNF-alpha), and even GABA.

It is believed that glial cells respond to intracellular waves of Ca^{2+} ions when ATP exists in the extracellular space. Mechanical stimulation seems to lead to an ATP release of this kind (Koizumi et al. 2005), and this mechanism seems to be part of the glia response to exogenous damage.

5.4 Cotransmission

Cotransmission refers to the release of more than one neurotransmitters from the same presynaptic neuron. This seems to be the rule rather than the exception in the brain and permits the manifestation of higher complexity in neuronal communication (Trudeau and Gutierrez 2007).

Cotransmission could take many forms, and neurotransmitters could share the same synaptic vesicle or occupy separate vesicles with similar or different rate and pattern of release into the same or different synapses of the same neuron. The pairs of neurotransmitters which are co-secreted are called *cotransmitters*, and cotransmission usually serves the stabilization together with the development of negative feedback in the synaptic cleft when the neuronal circuit lacks inhibitory interneurons (Thomas and Bornstein 2003). Commonly, pairs of cotransmitters are GABA together with glycine and glutamate with dopamine or acetylcholine. The cotransmission of glutamate together with GABA is somewhat unusual, but it constitutes a particularly interesting case as it is observed in the ventral tegmental area, the globus pallidus, and the supramammillary nuclei, having as final target the habenular nuclei and the dentate gyrus in the hippocampus (Root et al. 2018).

References

A decade of neuropeptides: past, present, and future. Tenth Annual Winter Neuropeptide Conference. Breckenridge, Colorado, January 16-20, 1989. Proceedings (1990). Ann N Y Acad Sci 579:1–280

Agulhon C, Fiacco TA, McCarthy KD (2010) Hippocampal short- and long-term plasticity are not modulated by astrocyte Ca2+ signaling. Science 327(5970):1250–1254. https://doi.org/10.1126/science.1184821

Albrecht A, Caliskan G, Oitzl MS, Heinemann U, Stork O (2013) Long-lasting increase of corticosterone after fear memory reactivation: anxiolytic effects and network activity modulation in the ventral hippocampus. Neuropsychopharmacology 38(3):386–394. https://doi.org/10.1038/npp.2012.192

Anden NE, Carlsson A, Dahlstroem A, Fuxe K, Hillarp NA, Larsson K (1964) Demonstration and mapping out of nigro-neostriatal dopamine neurons. Life Sci (1962) 3:523–530. https://doi.org/10.1016/0024-3205(64)90161-4

Anden NE, Dahlstrom A, Fuxe K, Olson L, Ungerstedt U (1966) Ascending noradrenaline neurons from the pons and the medulla oblongata. Experientia 22(1):44–45. https://doi.org/10.1007/BF01897761

Araque A, Parpura V, Sanzgiri RP, Haydon PG (1999) Tripartite synapses: glia, the unacknowledged partner. Trends Neurosci 22(5):208–215. https://doi.org/10.1016/s0166-2236(98)01349-6

Arimura A, Shioda S (1995) Pituitary adenylate cyclase activating polypeptide (PACAP) and its receptors: neuroendocrine and endocrine interaction. Front Neuroendocrinol 16(1):53–88. https://doi.org/10.1006/frne.1995.1003

Bahler M, Greengard P (1987) Synapsin I bundles F-actin in a phosphorylation-dependent manner. Nature 326(6114):704–707. https://doi.org/10.1038/326704a0

Barger G, Dale H (1910) β-imidazolylethylamine, a depressor constituent of intestinal mucosa. J Physiol 41:499–503

Baulieu EE, Schumacher M (2000) Progesterone as a neuroactive neurosteroid, with special reference to the effect of progesterone on myelination. Hum Reprod 15(Suppl 1):1–13. https://doi.org/10.1093/humrep/15.suppl_1.1

Bell C, Abrams J, Nutt D (2001) Tryptophan depletion and its implications for psychiatry. Br J Psychiatry 178:399–405. https://doi.org/10.1192/bjp.178.5.399

Berry MD, Juorio AV, Paterson IA (1994) The functional role of monoamine oxidases A and B in the mammalian central nervous system. Prog Neurobiol 42(3):375–391. https://doi.org/10.1016/0301-0082(94)90081-7

Bielsky IF, Hu SB, Szegda KL, Westphal H, Young LJ (2004) Profound impairment in social recognition and reduction in anxiety-like behavior in vasopressin V1a receptor knockout mice. Neuropsychopharmacology 29(3):483–493. https://doi.org/10.1038/sj.npp.1300360

Bogdanski DF, Pletscher A, Brodie BB, Undenfriend S (1956) Identification and assay of serotonin in brain. J Pharmacol Exp Ther 117(1):82–88

Broberger C, De Lecea L, Sutcliffe JG, Hokfelt T (1998) Hypocretin/orexin- and melanin-concentrating hormone-expressing cells form distinct populations in the rodent lateral hypothalamus: relationship to the neuropeptide Y and agouti gene-related protein systems. J Comp Neurol 402(4):460–474

Brodie T (1900) The immediate action of an intravenous injection of bloodserum. J Physiol 26:48–71

Brodie BB, Shore PA (1957) A concept for a role of serotonin and norepinephrine as chemical mediators in the brain. Ann N Y Acad Sci 66(3):631–642. https://doi.org/10.1111/j.1749-6632.1957.tb40753.x

Butcher SP, Fairbrother IS, Kelly JS, Arbuthnott GW (1990) Effects of selective monoamine oxidase inhibitors on the in vivo release and metabolism of dopamine in the rat striatum. J Neurochem 55(3):981–988. https://doi.org/10.1111/j.1471-4159.1990.tb04587.x

Cannon WB, Uridil JE (1921) Studies on the conditions of activity in endocrine glands. VIII. Some effects on the denervated heart of stimulating the nerves of the liver. Am J Physiol 58:353–354

Carl P, Hogskilde S, Lang-Jensen T, Bach V, Jacobsen J, Sorensen MB, Gralls M, Widlund L (1994) Pharmacokinetics and pharmacodynamics of eltanolone (pregnanolone), a new steroid intravenous anaesthetic, in humans. Acta Anaesthesiol Scand 38(7):734–741. https://doi.org/10.1111/j.1399-6576.1994.tb03987.x

Carlsson A (1959) The occurrence, distribution and physiological role of catecholamines in the nervous system. Pharmacol Rev 11(2, Part 2):490–493

Carlsson A, Lindqvist M, Magnusson T (1957) 3,4-Dihydroxyphenylalanine and 5-hydroxytryptophan as reserpine antagonists. Nature 180(4596):1200. https://doi.org/10.1038/1801200a0

Cooke GM, Gower DB (1981) Investigations into the possible effects of naturally occurring steroids on biosynthesis of 16-androstenes and androgens in microsomes of boar testis. J Endocrinol 88(3):409–418. https://doi.org/10.1677/joe.0.0880409

Cooper BL, Fornal CA (1995) Serotonin and behavior: a general hypothesis. In: Bloom FE, Kupfer DJ (eds) Psychopharmacology: the fourth generation of progress. Raven Press, New York, pp 461–469

Cooper JR, Bloom FE, Roth RH (1996) The biochemical basis of neuropharmacology, 7th edn. Oxford University Press, Oxford

Costa E, Groppetti A, Naimzada MK (1972) Effects of amphetamine on the turnover rate of brain catecholamines and motor activity. Br J Pharmacol 44(4):742–751. https://doi.org/10.1111/j.1476-5381.1972.tb07312.x

Dahlstroem A, Fuxe K (1964) Evidence for the existence of monoamine-containing neurons in the central nervous system. I. Demonstration of monoamines in the cell bodies of brain stem neurons. Acta Physiol Scand Suppl 232:231–255

Dautan D, Hacioglu Bay H, Bolam JP, Gerdjikov TV, Mena-Segovia J (2016) Extrinsic sources of cholinergic innervation of the striatal complex: a whole-brain mapping analysis. Front Neuroanat 10:1. https://doi.org/10.3389/fnana.2016.00001

Delgado PL, Miller HL, Salomon RM, Licinio J, Krystal JH, Moreno FA, Heninger GR, Charney DS (1999) Tryptophan-depletion challenge in depressed patients treated with desipramine or fluoxetine: implications for the role of serotonin in the mechanism of antidepressant action. Biol Psychiatry 46(2):212–220. https://doi.org/10.1016/s0006-3223(99)00014-1

Deniau JM, Thierry AM, Feger J (1980) Electrophysiological identification of mesencephalic ventromedial tegmental (VMT) neurons projecting to the frontal cortex, septum and nucleus accumbens. Brain Res 189(2):315–326. https://doi.org/10.1016/0006-8993(80)90093-1

Donaldson CJ, Sutton SW, Perrin MH, Corrigan AZ, Lewis KA, Rivier JE, Vaughan JM, Vale WW (1996) Cloning and characterization of human urocortin. Endocrinology 137(9):3896. https://doi.org/10.1210/endo.137.9.8756563

Edelman AM, Raese JD, Lazar MA, Barchas JD (1978) In vitro phosphorylation of a purified preparation of bovine corpus striatal tyrosine hydroxylase. Commun Psychopharmacol 2(6):461–465

Elsworth JD, Roth RH, Redmond DE Jr (1983) Relative importance of 3-methoxy-4-hydroxyphenylglycol and 3,4-dihydroxyphenylglycol as norepinephrine metabolites in rat, monkey, and humans. J Neurochem 41(3):786–793. https://doi.org/10.1111/j.1471-4159.1983.tb04809.x

Fallon JH (1981) Collateralization of monoamine neurons: mesotelencephalic dopamine projections to caudate, septum, and frontal cortex. J Neurosci 1(12):1361–1368

Fisher JM, Sossin W, Newcomb R, Scheller RH (1988) Multiple neuropeptides derived from a common precursor are differentially packaged and transported. Cell 54(6):813–822. https://doi.org/10.1016/s0092-8674(88)91131-2

Friess E, Schiffelholz T, Steckler T, Steiger A (2000) Dehydroepiandrosterone--a neurosteroid. Eur J Clin Investig 30(Suppl 3):46–50. https://doi.org/10.1046/j.1365-2362.2000.0300s3046.x

Fujishiro H, Umegaki H, Isojima D, Akatsu H, Iguchi A, Kosaka K (2006) Depletion of cholinergic neurons in the nucleus of the medial septum and the vertical limb of the diagonal band in dementia with Lewy bodies. Acta Neuropathol 111(2):109–114. https://doi.org/10.1007/s00401-005-0004-1

Gerard RW (1949) Physiology and psychiatry. Am J Psychiatry 106(3):161–173. https://doi.org/10.1176/ajp.106.3.161

Gerard RW (1955a) The biological roots of psychiatry. Am J Psychiatry 112(2):81–90. https://doi.org/10.1176/ajp.112.2.81

Gerard RW (1955b) Biological roots of psychiatry. Science 122(3162):225–230. https://doi.org/10.1126/science.122.3162.225

Geula C, Mesulam MM (2011) Brainstem cholinergic systems. In: Mai J, Paxinos G (eds) The human nervous system. Academic Press, New York, pp 456–470

Gordon WW (1948) Cerebral physiology and psychiatry. J Ment Sci 94(394):118–132. https://doi.org/10.1192/bjp.94.394.118

Halassa MM, Fellin T, Haydon PG (2007) The tripartite synapse: roles for gliotransmission in health and disease. Trends Mol Med 13(2):54–63. https://doi.org/10.1016/j.molmed.2006.12.005

Hanson PI, Heuser JE, Jahn R (1997) Neurotransmitter release - four years of SNARE complexes. Curr Opin Neurobiol 7(3):310–315. https://doi.org/10.1016/s0959-4388(97)80057-8

Harteneck C (2013) Pregnenolone sulfate: from steroid metabolite to TRP channel ligand. Molecules 18(10):12012–12028. https://doi.org/10.3390/molecules181012012

Hauger RL, Grigoriadis DE, Dallman MF, Plotsky PM, Vale WW, Dautzenberg FM (2003) International Union of Pharmacology. XXXVI. Current status of the nomenclature for receptors for corticotropin-releasing factor and their ligands. Pharmacol Rev 55(1):21–26. https://doi.org/10.1124/pr.55.1.3

Hedstrom H, Bixo M, Nyberg S, Spigset O, Zingmark E, Backstrom T (2009) Studies of pharmacokinetic and pharmacodynamic properties of isoallopregnanolone in healthy women. Psychopharmacology 203(1):85–98. https://doi.org/10.1007/s00213-008-1372-8

Henin J, Salari R, Murlidaran S, Brannigan G (2014) A predicted binding site for cholesterol on the GABAA receptor. Biophys J 106(9):1938–1949. https://doi.org/10.1016/j.bpj.2014.03.024

Herman JP, Cullinan WE, Young EA, Akil H, Watson SJ (1992) Selective forebrain fiber tract lesions implicate ventral hippocampal structures in tonic regulation of paraventricular nucleus corticotropin-releasing hormone (CRH) and arginine vasopressin (AVP) mRNA expression. Brain Res 592(1–2):228–238. https://doi.org/10.1016/0006-8993(92)91680-d

Hokfelt T, Fahrenkrug J, Tatemoto K, Mutt V, Werner S, Hulting AL, Terenius L, Chang KJ (1983) The PHI (PHI-27)/corticotropin-releasing factor/enkephalin immunoreactive hypothalamic neuron: possible morphological basis for integrated control of prolactin, corticotropin, and growth hormone secretion. Proc Natl Acad Sci U S A 80(3):895–898. https://doi.org/10.1073/pnas.80.3.895

Hokfelt T, Everitt B, Meister B, Melander T, Schalling M, Johansson O, Lundberg JM, Hulting AL, Werner S, Cuello C et al (1986) Neurons with multiple messengers with special reference in neuroendocrine systems. Recent Prog Horm Res 42:1–70. https://doi.org/10.1016/b978-0-12-571142-5.50005-7

Holtz P (1950) Ueber die sumpahticomimetische Wirksamkeit von Gehirnextgrackten. Acta Physiol Scand 20:354–362

Houtani T, Nishi M, Takeshima H, Sato K, Sakuma S, Kakimoto S, Ueyama T, Noda T, Sugimoto T (2000) Distribution of nociceptin/orphanin FQ precursor protein and receptor in brain and spinal cord: a study using in situ hybridization and X-gal histochemistry in receptor-deficient mice. J Comp Neurol 424(3):489–508

Hummer TA, McClintock MK (2009) Putative human pheromone androstadienone attunes the mind specifically to emotional information. Horm Behav 55(4):548–559. https://doi.org/10.1016/j.yhbeh.2009.01.002

Insel TR (2010) The challenge of translation in social neuroscience: a review of oxytocin, vasopressin, and affiliative behavior. Neuron 65(6):768–779. https://doi.org/10.1016/j.neuron.2010.03.005

Jacob S, Hayreh DJ, McClintock MK (2001) Context-dependent effects of steroid chemosignals on human physiology and mood. Physiol Behav 74(1–2):15–27. https://doi.org/10.1016/s0031-9384(01)00537-6

Jacobs BL, Azmitia EC (1992) Structure and function of the brain serotonin system. Physiol Rev 72(1):165–229. https://doi.org/10.1152/physrev.1992.72.1.165

Jew JY, Leranth C, Arimura A, Palkovits M (1984) Preoptic LH-RH and somatostatin in the rat median eminence. An experimental light and electron microscopic immunocytochemical study. Neuroendocrinology 38(3):169–175. https://doi.org/10.1159/000123886

Kaminski RM, Marini H, Ortinski PI, Vicini S, Rogawski MA (2006) The pheromone androstenol (5 alpha-androst-16-en-3 alpha-ol) is a neurosteroid positive modulator of GABAA receptors. J Pharmacol Exp Ther 317(2):694–703. https://doi.org/10.1124/jpet.105.098319

Kelly RB (1993) Storage and release of neurotransmitters. Cell 72(Suppl):43–53. https://doi.org/10.1016/s0092-8674(05)80027-3

Kema IP, de Vries EG, Muskiet FA (2000) Clinical chemistry of serotonin and metabolites. J Chromatogr B Biomed Sci Appl 747(1–2):33–48. https://doi.org/10.1016/s0378-4347(00)00341-8

Koizumi S, Fujishita K, Inoue K (2005) Regulation of cell-to-cell communication mediated by astrocytic ATP in the CNS. Purinergic Signal 1(3):211–217. https://doi.org/10.1007/s11302-005-6321-y

Kopin IJ (1985) Catecholamine metabolism: basic aspects and clinical significance. Pharmacol Rev 37(4):333–364

Koylu EO, Couceyro PR, Lambert PD, Kuhar MJ (1998) Cocaine- and amphetamine-regulated transcript peptide immunohistochemical localization in the rat brain. J Comp Neurol 391(1):115–132

Kupfermann I (1991) Functional studies of cotransmission. Physiol Rev 71(3):683–732. https://doi.org/10.1152/physrev.1991.71.3.683

Lan NC, Gee KW (1994) Neuroactive steroid actions at the GABAA receptor. Horm Behav 28(4):537–544. https://doi.org/10.1006/hbeh.1994.1052

Lawson D, Raff MC, Gomperts B, Fewtrell C, Gilula NB (1977) Molecular events during membrane fusion. A study of exocytosis in rat peritoneal mast cells. J Cell Biol 72(2):242–259. https://doi.org/10.1083/jcb.72.2.242

Lemieux RU, Spohr U (1994) How Emil Fischer was led to the lock and key concept for enzyme specificity. Adv Carbohydr Chem Biochem 50:1–20

Levitan I, Singh DK, Rosenhouse-Dantsker A (2014) Cholesterol binding to ion channels. Front Physiol 5:65. https://doi.org/10.3389/fphys.2014.00065

Liley AW (1956) The quantal components of the mammalian end-plate potential. J Physiol 133(3):571–587. https://doi.org/10.1113/jphysiol.1956.sp005610

Lindau M, Almers W (1995) Structure and function of fusion pores in exocytosis and ectoplasmic membrane fusion. Curr Opin Cell Biol 7(4):509–517. https://doi.org/10.1016/0955-0674(95)80007-7

Lindvall O, Bjorklund A (1983) Dopamine and norepinephrine containing neuron systems: their anatomy in the rat brain. In: Emson PC (ed) Chemical neuroanatomy. Raven Press, New York, pp 229–255

Liu D, Diorio J, Tannenbaum B, Caldji C, Francis D, Freedman A, Sharma S, Pearson D, Plotsky PM, Meaney MJ (1997) Maternal care, hippocampal glucocorticoid receptors, and hypothalamic-pituitary-adrenal responses to stress. Science 277(5332):1659–1662. https://doi.org/10.1126/science.277.5332.1659

Liu AK, Chang RC, Pearce RK, Gentleman SM (2015) Nucleus basalis of Meynert revisited: anatomy, history and differential involvement in Alzheimer's and Parkinson's disease. Acta Neuropathol 129(4):527–540. https://doi.org/10.1007/s00401-015-1392-5

Manaye KF, Zweig R, Wu D, Hersh LB, De Lacalle S, Saper CB, German DC (1999) Quantification of cholinergic and select non-cholinergic mesopontine neuronal populations in the human brain. Neuroscience 89(3):759–770. https://doi.org/10.1016/s0306-4522(98)00380-7

Mardh G, Sjoquist B, Anggard E (1981) Norepinephrine metabolism in man using deuterium labelling: the conversion of 4-hydroxy-3-methoxyphenylglycol to 4-hydroxy-3-methoxymandelic acid. J Neurochem 36(3):1181–1185. https://doi.org/10.1111/j.1471-4159.1981.tb01716.x

Marsden C (1996) The neuropharmacology of serotonin in the central nervous system. In: Feighner J, Boyer W (eds) Selective serotonin re-uptake inhibitors. John Wiley and Sons, Chichester, pp 1–35

Martineau M, Baux G, Mothet JP (2006) Gliotransmission at central glutamatergic synapses: D-serine on stage. J Physiol Paris 99(2–3):103–110. https://doi.org/10.1016/j.jphysparis.2005.12.011

Martin-Schild S, Gerall AA, Kastin AJ, Zadina JE (1999) Differential distribution of endomorphin 1- and endomorphin 2-like immunoreactivities in the CNS of the rodent. J Comp Neurol 405(4):450–471

Marx CE, Keefe RS, Buchanan RW, Hamer RM, Kilts JD, Bradford DW, Strauss JL, Naylor JC, Payne VM, Lieberman JA, Savitz AJ, Leimone LA, Dunn L, Porcu P, Morrow AL, Shampine LJ (2009) Proof-of-concept trial with the neurosteroid pregnenolone targeting cognitive and negative symptoms in schizophrenia. Neuropsychopharmacology 34(8):1885–1903. https://doi.org/10.1038/npp.2009.26

Marx CE, Lee J, Subramaniam M, Rapisarda A, Bautista DC, Chan E, Kilts JD, Buchanan RW, Wai EP, Verma S, Sim K, Hariram J, Jacob R, Keefe RS, Chong SA (2014) Proof-of-

concept randomized controlled trial of pregnenolone in schizophrenia. Psychopharmacology 231(17):3647–3662. https://doi.org/10.1007/s00213-014-3673-4
Masserano JM, Vuilliet PR, Tank AW, Weiner N (1989) The role of tyrosine hydroxylase in the regulation of catecholamine synthesis. In: Trendelenburg U, Weiner N (eds) Handbook of experimental pharmacology. Springer-Verlag, Berlin, pp 427–469
Matthews G (1996) Synaptic exocytosis and endocytosis: capacitance measurements. Curr Opin Neurobiol 6(3):358–364. https://doi.org/10.1016/s0959-4388(96)80120-6
McEwen BS (2000) Effects of adverse experiences for brain structure and function. Biol Psychiatry 48(8):721–731. https://doi.org/10.1016/s0006-3223(00)00964-1
Meldrum BS (2000) Glutamate as a neurotransmitter in the brain: review of physiology and pathology. J Nutr 130(4S Suppl):1007S–1015S. https://doi.org/10.1093/jn/130.4.1007S
Mellon SH (2007) Neurosteroid regulation of central nervous system development. Pharmacol Ther 116(1):107–124. https://doi.org/10.1016/j.pharmthera.2007.04.011
Mellon SH, Deschepper CF (1993) Neurosteroid biosynthesis: genes for adrenal steroidogenic enzymes are expressed in the brain. Brain Res 629(2):283–292. https://doi.org/10.1016/0006-8993(93)91332-m
Mesulam M (1994) Structure and function of cholinergic pathways in the cerebral cortex, limbic system, basal ganglia and thalamus of the human brain. In: Bloom F, Kupfer D (eds) Psychopharmacology: the fourth generation of progress. Raven Press, New York, pp 135–146
Mesulam MM, Mufson EJ, Wainer BH, Levey AI (1983) Central cholinergic pathways in the rat: an overview based on an alternative nomenclature (Ch1-Ch6). Neuroscience 10(4):1185–1201. https://doi.org/10.1016/0306-4522(83)90108-2
Mesulam MM, Geula C, Bothwell MA, Hersh LB (1989) Human reticular formation: cholinergic neurons of the pedunculopontine and laterodorsal tegmental nuclei and some cytochemical comparisons to forebrain cholinergic neurons. J Comp Neurol 283(4):611–633. https://doi.org/10.1002/cne.902830414
Michel MC, Beck-Sickinger A, Cox H, Doods HN, Herzog H, Larhammar D, Quirion R, Schwartz T, Westfall T (1998) XVI. International Union of Pharmacology recommendations for the nomenclature of neuropeptide Y, peptide YY, and pancreatic polypeptide receptors. Pharmacol Rev 50(1):143–150
Monti-Bloch L, Jennings-White C, Dolberg DS, Berliner DL (1994) The human vomeronasal system. Psychoneuroendocrinology 19(5–7):673–686. https://doi.org/10.1016/0306-4530(94)90049-3
Monti-Bloch L, Jennings-White C, Berliner DL (1998) The human vomeronasal system. A review. Ann N Y Acad Sci 855:373–389. https://doi.org/10.1111/j.1749-6632.1998.tb10595.x
Moore K, Lookingland K (1994) Dopaminergic neuronal systems in the hypothalamus. In: Bloom F, Kupfer D (eds) Psychopharmacology: the fourth generation of progress. Raven Press, New York, pp 245–256
Moore P, Landolt HP, Seifritz E, Clark C, Bhatti T, Kelsoe J, Rapaport M, Gillin JC (2000) Clinical and physiological consequences of rapid tryptophan depletion. Neuropsychopharmacology 23(6):601–622. https://doi.org/10.1016/S0893-133X(00)00161-5
Mufson EJ, Martin TL, Mash DC, Wainer BH, Mesulam MM (1986) Cholinergic projections from the parabigeminal nucleus (Ch8) to the superior colliculus in the mouse: a combined analysis of horseradish peroxidase transport and choline acetyltransferase immunohistochemistry. Brain Res 370(1):144–148. https://doi.org/10.1016/0006-8993(86)91114-5
Myers RD (1994) Neuroactive peptides: unique phases in research on mammalian brain over three decades. Peptides 15(2):367–381. https://doi.org/10.1016/0196-9781(94)90025-6
Neal CR Jr, Mansour A, Reinscheid R, Nothacker HP, Civelli O, Watson SJ Jr (1999) Localization of orphanin FQ (nociceptin) peptide and messenger RNA in the central nervous system of the rat. J Comp Neurol 406(4):503–547
O'Dell LE, Purdy RH, Covey DF, Richardson HN, Roberto M, Koob GF (2005) Epipregnanolone and a novel synthetic neuroactive steroid reduce alcohol self-administration in rats. Pharmacol Biochem Behav 81(3):543–550. https://doi.org/10.1016/j.pbb.2005.03.020

Oades RD, Halliday GM (1987) Ventral tegmental (A10) system: neurobiology. 1. Anatomy and connectivity. Brain Res 434(2):117–165. https://doi.org/10.1016/0165-0173(87)90011-7

Palkovits M (1984) Neuropeptides in the hypothalamo-hypophyseal system: lateral retrochiasmatic area as a common gate for neuronal fibers towards the median eminence. Peptides 5(Suppl 1):35–39. https://doi.org/10.1016/0196-9781(84)90262-6

Paul S (1994) GABA and glycine. In: Bloom F, Kupfer D (eds) Psychopharmacology: the fourth generation of progress. Raven Press, New York, pp 87–94

Paul SM, Purdy RH (1992) Neuroactive steroids. FASEB J 6(6):2311–2322

Paul SM, Doherty JJ, Robichaud AJ, Belfort GM, Chow BY, Hammond RS, Crawford DC, Linsenbardt AJ, Shu HJ, Izumi Y, Mennerick SJ, Zorumski CF (2013) The major brain cholesterol metabolite 24(S)-hydroxycholesterol is a potent allosteric modulator of N-methyl-D-aspartate receptors. J Neurosci 33(44):17290–17300. https://doi.org/10.1523/JNEUROSCI.2619-13.2013

Perea G, Araque A (2005) Synaptic regulation of the astrocyte calcium signal. J Neural Transm (Vienna) 112(1):127–135. https://doi.org/10.1007/s00702-004-0170-7

Phillipson OT (1979) The cytoarchitecture of the interfascicular nucleus and ventral tegmental area of Tsai in the rat. J Comp Neurol 187(1):85–98. https://doi.org/10.1002/cne.901870106

Pierce JD Jr, Cohen AB, Ulrich PM (2004) Responsivity to two odorants, androstenone and amyl acetate, and the affective impact of odors on interpersonal relationships. J Comp Psychol 118(1):14–19. https://doi.org/10.1037/0735-7036.118.1.14

Pinna G, Uzunova V, Matsumoto K, Puia G, Mienville JM, Costa E, Guidotti A (2000) Brain allopregnanolone regulates the potency of the GABA(A) receptor agonist muscimol. Neuropharmacology 39(3):440–448. https://doi.org/10.1016/s0028-3908(99)00149-5

Price LH, Malison RT, McDougle CJ, Pelton GH, Heninger GR (1998) The neurobiology of tryptophan depletion in depression: effects of intravenous tryptophan infusion. Biol Psychiatry 43(5):339–347. https://doi.org/10.1016/s0006-3223(97)00284-9

Rapport MM, Green AA, Page IH (1948) Serum vasoconstrictor (serotonin) IV: isolation and characterization. J Biol Chem 176:1248–1251

Reaux A, Gallatz K, Palkovits M, Llorens-Cortes C (2002) Distribution of apelin-synthesizing neurons in the adult rat brain. Neuroscience 113(3):653–662. https://doi.org/10.1016/s0306-4522(02)00192-6

Reddy DS (2003) Is there a physiological role for the neurosteroid THDOC in stress-sensitive conditions? Trends Pharmacol Sci 24(3):103–106. https://doi.org/10.1016/S0165-6147(03)00023-3

Reddy DS (2010) Neurosteroids: endogenous role in the human brain and therapeutic potentials. Prog Brain Res 186:113–137. https://doi.org/10.1016/B978-0-444-53630-3.00008-7

Reddy DS, Jian K (2010) The testosterone-derived neurosteroid androstanediol is a positive allosteric modulator of GABAA receptors. J Pharmacol Exp Ther 334(3):1031–1041. https://doi.org/10.1124/jpet.110.169854

Reilly JG, McTavish SF, Young AH (1997) Rapid depletion of plasma tryptophan: a review of studies and experimental methodology. J Psychopharmacol 11(4):381–392. https://doi.org/10.1177/026988119701100416

Reiner P, Fibiger C (1994) Functional heterogeneity of central cholinergic systems. In: Bloom F, Kupfer D (eds) Psychopharmacology: the fourth generation of progress. Raven Press, New York, pp 147–153

Root DH, Zhang S, Barker DJ, Miranda-Barrientos J, Liu B, Wang HL, Morales M (2018) Selective brain distribution and distinctive synaptic architecture of dual glutamatergic-GABAergic neurons. Cell Rep 23(12):3465–3479. https://doi.org/10.1016/j.celrep.2018.05.063

Rosenhouse-Dantsker A (2017) Insights into the molecular requirements for cholesterol binding to ion channels. Curr Top Membr 80:187–208. https://doi.org/10.1016/bs.ctm.2017.05.003

Rostene WH, Alexander MJ (1997) Neurotensin and neuroendocrine regulation. Front Neuroendocrinol 18(2):115–173. https://doi.org/10.1006/frne.1996.0146

Roth RH, Morgenroth VH 3rd, Salzman PM (1975) Tyrosine hydroxylase: allosteric activation induced by stimulation of central noradrenergic neurons. Naunyn Schmiedebergs Arch Pharmacol 289(4):327–343. https://doi.org/10.1007/BF00508408

Rothman DL, Behar KL, Hyder F, Shulman RG (2003) In vivo NMR studies of the glutamate neurotransmitter flux and neuroenergetics: implications for brain function. Annu Rev Physiol 65:401–427. https://doi.org/10.1146/annurev.physiol.65.092101.142131

Schulkin J, Gold PW, McEwen BS (1998) Induction of corticotropin-releasing hormone gene expression by glucocorticoids: implication for understanding the states of fear and anxiety and allostatic load. Psychoneuroendocrinology 23(3):219–243. https://doi.org/10.1016/s0306-4530(97)00099-1

Schweizer FE, Betz H, Augustine GJ (1995) From vesicle docking to endocytosis: intermediate reactions of exocytosis. Neuron 14(4):689–696. https://doi.org/10.1016/0896-6273(95)90213-9

Shigeri Y, Seal RP, Shimamoto K (2004) Molecular pharmacology of glutamate transporters, EAATs and VGLUTs. Brain Res Brain Res Rev 45(3):250–265. https://doi.org/10.1016/j.brainresrev.2004.04.004

Sicard F, De Groot D, Gras M, Leprince J, Conlon JM, Roubos EW, Vaudry H, Delarue C (2005) Neuroendocrine regulation of frog adrenocortical cells by neurotensin. Ann N Y Acad Sci 1040:200–205. https://doi.org/10.1196/annals.1327.025

Simon H, Le Moal M, Calas A (1979) Efferents and afferents of the ventral tegmental-A10 region studied after local injection of [3H]leucine and horseradish peroxidase. Brain Res 178(1):17–40. https://doi.org/10.1016/0006-8993(79)90085-4

Spier AD, de Lecea L (2000) Cortistatin: a member of the somatostatin neuropeptide family with distinct physiological functions. Brain Res Brain Res Rev 33(2–3):228–241. https://doi.org/10.1016/s0165-0173(00)00031-x

Srivastava DP, Waters EM, Mermelstein PG, Kramar EA, Shors TJ, Liu F (2011) Rapid estrogen signaling in the brain: implications for the fine-tuning of neuronal circuitry. J Neurosci 31(45):16056–16063. https://doi.org/10.1523/JNEUROSCI.4097-11.2011

Stahl SM (2008) Stahl's essential psychopharmacology: neuroscientific basis and practical applications, 3rd edn. Cambridge University Press, Cambridge

Steckler T, Holsboer F (1999) Corticotropin-releasing hormone receptor subtypes and emotion. Biol Psychiatry 46(11):1480–1508. https://doi.org/10.1016/s0006-3223(99)00170-5

Swanson LW (1982) The projections of the ventral tegmental area and adjacent regions: a combined fluorescent retrograde tracer and immunofluorescence study in the rat. Brain Res Bull 9(1–6):321–353. https://doi.org/10.1016/0361-9230(82)90145-9

Swanson LW, Simmons DM (1989) Differential steroid hormone and neural influences on peptide mRNA levels in CRH cells of the paraventricular nucleus: a hybridization histochemical study in the rat. J Comp Neurol 285(4):413–435. https://doi.org/10.1002/cne.902850402

Thibonnier M, Conarty DM, Preston JA, Wilkins PL, Berti-Mattera LN, Mattera R (1998) Molecular pharmacology of human vasopressin receptors. Adv Exp Med Biol 449:251–276. https://doi.org/10.1007/978-1-4615-4871-3_34

Thibonnier M, Coles P, Thibonnier A, Shoham M (2001) The basic and clinical pharmacology of nonpeptide vasopressin receptor antagonists. Annu Rev Pharmacol Toxicol 41(1):175–202. https://doi.org/10.1146/annurev.pharmtox.41.1.175

Thomas EA, Bornstein JC (2003) Inhibitory cotransmission or after-hyperpolarizing potentials can regulate firing in recurrent networks with excitatory metabotropic transmission. Neuroscience 120(2):333–351. https://doi.org/10.1016/s0306-4522(03)00039-3

Thrivikraman KV, Nemeroff CB, Plotsky PM (2000) Sensitivity to glucocorticoid-mediated fast-feedback regulation of the hypothalamic–pituitary–adrenal axis is dependent upon stressor specific neurocircuitry. Brain Res 870(1–2):87–101. https://doi.org/10.1016/s0006-8993(00)02405-7

Trudeau LE, Gutierrez R (2007) On cotransmission & neurotransmitter phenotype plasticity. Mol Interv 7(3):138–146. https://doi.org/10.1124/mi.7.3.5

Tsujino N, Yamanaka A, Ichiki K, Muraki Y, Kilduff TS, Yagami K, Takahashi S, Goto K, Sakurai T (2005) Cholecystokinin activates orexin/hypocretin neurons through the cholecystokinin A receptor. J Neurosci 25(32):7459–7469. https://doi.org/10.1523/JNEUROSCI.1193-05.2005

Turnbull AV, Vaughan J, Rivier JE, Vale WW, Rivier C (1999) Urocortin is not a significant regulator of intermittent electrofootshock-induced adrenocorticotropin secretion in the intact male rat. Endocrinology 140(1):71–78. https://doi.org/10.1210/endo.140.1.6419

Twarog BM, Page IH (1953) Serotonin content of some mammalian tissues and urine and a method for its determination. Am J Phys 175(1):157–161. https://doi.org/10.1152/ajplegacy.1953.175.1.157

Umbriaco D, Anctil M, Descarries L (1990) Serotonin-immunoreactive neurons in the cnidarian Renilla koellikeri. J Comp Neurol 291(2):167–178. https://doi.org/10.1002/cne.902910202

Ungerstedt U (1971) Stereotaxic mapping of the monoamine pathways in the rat brain. Acta Physiol Scand Suppl 367:1–48. https://doi.org/10.1111/j.1365-201x.1971.tb10998.x

Unwin PN, Zampighi G (1980) Structure of the junction between communicating cells. Nature 283(5747):545–549. https://doi.org/10.1038/283545a0

Usdin TB, Wang T, Hoare SR, Mezey E, Palkovits M (2000) New members of the parathyroid hormone/parathyroid hormone receptor family: the parathyroid hormone 2 receptor and tuberoinfundibular peptide of 39 residues. Front Neuroendocrinol 21(4):349–383. https://doi.org/10.1006/frne.2000.0203

Van Bockstaele EJ (1998) Morphological substrates underlying opioid, epinephrine and γ-aminobutyric acid inhibitory actions in the rat locus coeruleus. Brain Res Bull 47(1):1–15. https://doi.org/10.1016/s0361-9230(98)00062-8

Vandenberg RJ (1998) Molecular pharmacology and physiology of glutamate transporters in the central nervous system. Clin Exp Pharmacol Physiol 25(6):393–400. https://doi.org/10.1111/j.1440-1681.1998.tb02221.x

Vogt M (1954) The concentration of sympathin in different parts of the central nervous system under normal conditions and after the administration of drugs. J Physiol 123(3):451–481. https://doi.org/10.1113/jphysiol.1954.sp005064

von Euler US (1946a) The presence of a substance with sympathin properties in spleen extracts. Acta Physiol Scand 11:168–186

von Euler US (1946b) A specific sympathomimetic ergone in adrenergic nerve fibers (sympathin) and its relations to adrenaline and noradrenaline. Acta Physiol Scand 12:73–97

von Gersdorff H, Matthews G (1994) Dynamics of synaptic vesicle fusion and membrane retrieval in synaptic terminals. Nature 367(6465):735–739. https://doi.org/10.1038/367735a0

Walter L, Dinh T, Stella N (2004) ATP induces a rapid and pronounced increase in 2-arachidonoylglycerol production by astrocytes, a response limited by monoacylglycerol lipase. J Neurosci 24(37):8068–8074. https://doi.org/10.1523/JNEUROSCI.2419-04.2004

Wersinger SR, Kelliher KR, Zufall F, Lolait SJ, O'Carroll AM, Young WS III (2004) Social motivation is reduced in vasopressin 1b receptor null mice despite normal performance in an olfactory discrimination task. Horm Behav 46(5):638–645. https://doi.org/10.1016/j.yhbeh.2004.07.004

Wersinger SR, Caldwell HK, Martinez L, Gold P, Hu SB, Young WS III (2007) Vasopressin 1a receptor knockout mice have a subtle olfactory deficit but normal aggression. Genes Brain Behav 6(6):540–551. https://doi.org/10.1111/j.1601-183X.2006.00281.x

Widerlov E, Lewander T (1978) Inhibition of the in vivo biosynthesis and changes of catecholamine levels in rat brain after alpha-methyl-p-tyrosine; time- and dose-response relationships. Naunyn Schmiedebergs Arch Pharmacol 304(2):111–123. https://doi.org/10.1007/BF00495547

Zhang Q, Haydon PG (2005) Roles for gliotransmission in the nervous system. J Neural Transm (Vienna) 112(1):121–125. https://doi.org/10.1007/s00702-004-0119-x

The Neural Synapse

6

Contents

6.1 Background

The point where two neurons communicate is called the *synapse* (Jessell and Kandel 1993). The word "synapse" comes from the Greek verb *συνάπτω* (bring together, join), and it was introduced in 1897 by Michael Foster (1836–1907) at the suggestion of Arthur Woollgar Verrall (1851–1912) which was an English classical scholar (Tansey 1997). The attribution of the first usage of the term to the Nobel Laureate Charles Scott Sherrington (1857–1952) is probably wrong.

The neuron that transmits the signal is called *presynaptic* and the neuron which receives it is *postsynaptic*. Depending on the parts of the neurons which participate in their formation, synapses are classified into (Fig. 6.1) the following:

- Axosecretory
- Axoaxonic
- Axodendritic
- Axoextracellular
- Axosomatic
- Axosynaptic

K. N. Fountoulakis, *The Human Connectome*,
https://doi.org/10.1007/978-3-031-10351-3_6

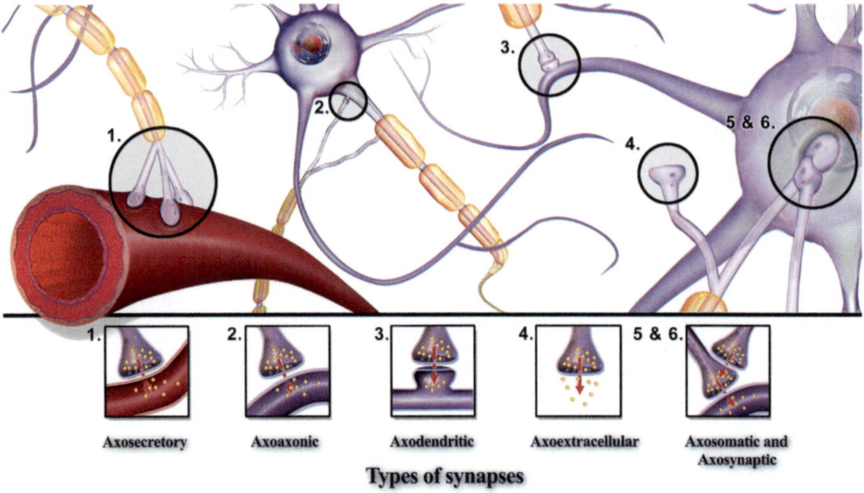

Fig. 6.1 Types of synapses. (From Blausen.com staff (2014). Medical gallery of Blausen Medical 2014. *WikiJournal of Medicine* 1 (2))

The physiologists, led by John Eccles (1903–1997), argued that all synaptic transmission is electrical and that the action potential in the presynaptic neuron generates a current that flows passively into the postsynaptic cell. The pharmacologists, led by Henry Dale (1875–1968), argued that transmission is chemical and that the action potential in the presynaptic neuron leads to the release of a chemical substance that in turn initiates current flow in the postsynaptic cell.

When physiological techniques improved in the 1950s and 1960s, it became clear that both forms of transmission exist. Although most synapses use a chemical transmitter, some operate purely by electrical means (Eccles 1976).

So today, we know that there are two kinds of synapses, depending on their mechanism of function:

- *Chemical synapses* (Fig. 6.2), where the electrical activity in the presynaptic neuron is converted into the release of a neurotransmitter.
- *Electrical synapses* (Fig. 6.3), where the presynaptic and postsynaptic cell membranes are connected by channels called gap junctions or are in touch through a synaptic cleft that is capable of passing electric current.

It is to be noted that synaptic communication is distinct from *ephaptic coupling* (from Greek *εφάπτομαι* meaning being in touch), which refers to the exchange of ions between the cells, or to the effect of local electric fields. In this way, it is influenced by the synchronization and timing of the firing of action potentials in neighboring neurons. It is exactly the phenomenon of ephaptic interactions that myelination prohibits (Arvanitaki 1942).

Types of Synapses Depending on Their Mechanism of Function

- Chemical synapses
  - Synaptic cleft
  - Neurotransmitters
  - Neurotransmitter reuptake
  - Receptors
  - Slow neurotransmission with the ability to amplify and variable signaling
- Electric synapses
  - Gap junctions
  - Fast neurotransmission in the form "all or nothing"
- Receptors
- Ephaptic coupling
- Principle of dynamic polarization—unidirectional transmission
- Principle of connectional specificity—neuronal networks

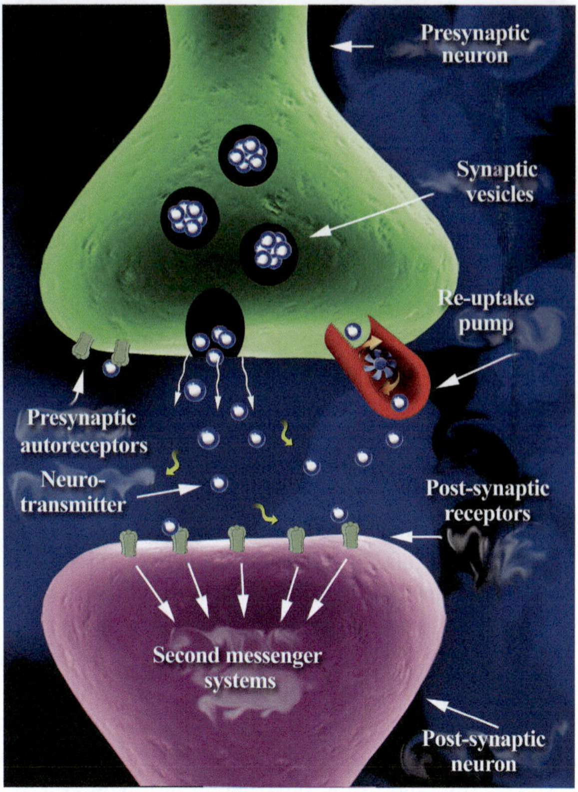

Fig. 6.2 The chemical synapse and its parts. (Copyright KN Fountoulakis 2021)

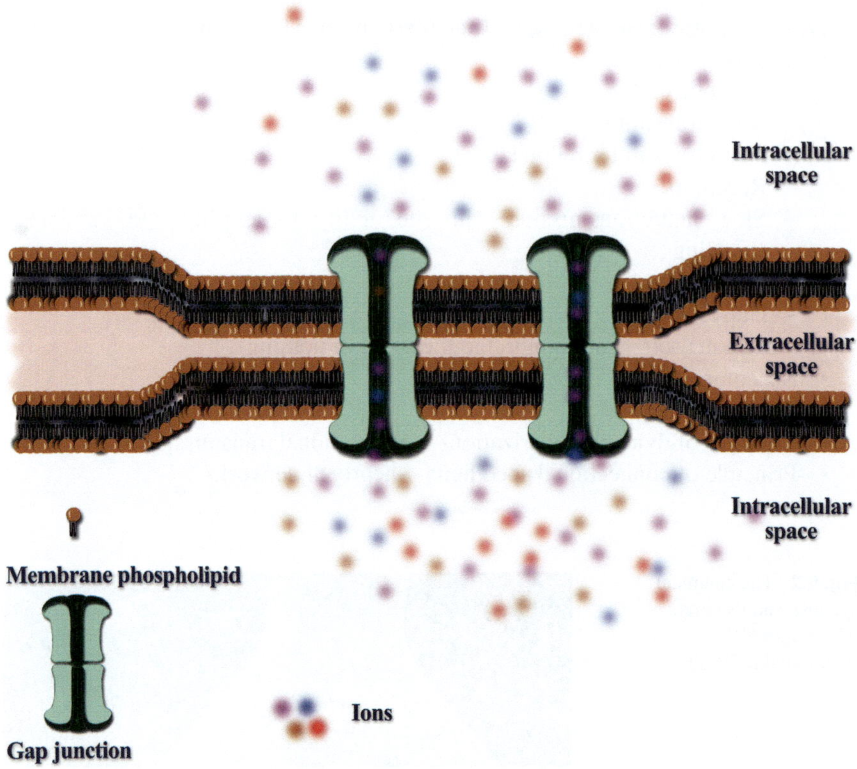

Fig. 6.3 The electrical synapse and its parts. (Copyright KN Fountoulakis 2021)

In the brain, electrical synaptic transmission is very rapid and rather stereotyped, and it is triggered explosively in an all-or-none manner. In contrast, chemical synapses are capable of more variable signaling and thus can produce more complex behaviors. They can mediate either excitatory or inhibitory actions in postsynaptic cells and produce electrical changes in the postsynaptic cell that last from milliseconds to many minutes. This comes at the price of longer latency, that is, longer time between the presynaptic spike and the postsynaptic potential, because chemical transmission demands several biochemical steps. On the contrary, at electrical synapses, the gap junction channels connect the cytoplasm of the two cells, and in this way, they allow a direct ion flow between them.

One important characteristic of the communication between neurons is that its strength can be enhanced or diminished by cellular activity, a phenomenon which is called *plasticity*, and it is crucial to memory and other higher brain functions.

One important principle in the function of neurons and synapses is that they transmit a signal always in one direction (*principle of dynamic polarization*) from the receiving site to the presynaptic terminal. A second principle is that neurons connect to each other in a way which leads to the formation of neuronal networks which serve some specific function (*principle of connectional specificity*). However, the complexity of these networks is extremely high. On average, a spinal motor neuron receives 10,000 contacts from other neurons, of which 2000 on the cell body and the rest on its dendrites, while a Purkinje neuron in the cerebellum could receive up to 150,000 contacts.

6.2 Chemical Synapses

In chemical synapses (Fig. 6.2), the two neurons are not in physical contact; instead there is an empty space between them, of the size of approximately 20–40 nm, called the *synaptic cleft*. The end of the presynaptic neuron is called *presynaptic terminal*, and it can end almost anywhere on the postsynaptic neuron but preferentially on dendrites (Fig. 6.1).

6.2.1 Receptors

In chemical synapses and in the membrane of the pre- as well as the postsynaptic neuron, and oriented toward the synaptic cleft, there are molecules called *receptors*. The notion of a receptor was introduced in the late nineteenth century by the German bacteriologist Paul Ehrlich (1854–1915) to explain the selective action of toxins and other pharmacological agents and the great specificity of immunological reactions, and as a metaphor, the *lock and key* relationship was used.

These receptors serve as docking sites for neurotransmitters. The *postsynaptic receptors* are a sort of "receivers" of the signal which is transmitted by the presynaptic neuron, while the *presynaptic receptors* are a kind of safety switch which reduces or even stops the secretion when the quantity of the neurotransmitter in the synaptic cleft is sufficient by regulating the opening of ion channels in the presynaptic neuron either directly or indirectly.

The change in postsynaptic receptor sensitivity, either toward an increase (*upregulation*) or a decrease (*downregulation*), constitutes an additional safety mechanism as well as an adaptation of the system to prolonged high or low secretion of a neurotransmitter in the synaptic cleft, with the return of final total postsynaptic response to previous levels before the occurrence of this chronic change.

After their release into the synaptic cleft, the neurotransmitter molecules bind to *specific for that neurotransmitter receptors* which are located on the postsynaptic membrane. This, in turn, activates the receptors and causes a cascade of events inside the postsynaptic neuron. The time needed for all these steps to take place varies from 0.3 ms to a few milliseconds or more, and this is the reason why chemical synapses are not as fast as electric ones. However, chemical neurotransmission has the important property of amplification. This means that a small presynaptic electric current could cause the release of thousands of neurotransmitter molecules inside the cleft, and this in turn could cause the activation of a large number of postsynaptic receptors, with a significant effect even on a huge postsynaptic neuron.

One important characteristic of transmission is that *it is the receptor that determines* whether a particular synapse is excitatory or inhibitory and whether an ion channel will be activated directly by the transmitter or indirectly through a second messenger. The action of a particular neurotransmitter on the postsynaptic neuron depends more on the properties of the postsynaptic receptors which recognize and bind this specific neurotransmitter, rather than on the chemical properties of the neurotransmitter itself. It should be noted that the end result is the sum of the neurotransmitter effect on a number of different receptors, which manifest different sensitivity concerning their binding to the neurotransmitter, different threshold for activation, different signaling inside the postsynaptic neuron, and also different magnitude of this signaling.

Receptors

- "Lock and key" metaphor
- Presynaptic receptors
- Postsynaptic receptors
- Ionotropic receptors
  - Fast and short-lived actions
  - Only increase of ion passage through the channel
- Metabotropic receptors
  - Slow and prolonged actions
  - Wider and more complex results
  - Either increase or decrease of ion passage through a channel
  - Modification of status and properties of the postsynaptic neuron
- The result of neurotransmission is determined by the combination of the binding of the neurotransmitter on different receptors with different properties each

There are *two main types of receptors* as follows:

- *Ionotropic receptors*: They exist throughout the brain. Their structure includes an ion channel, which opens as a response to the binding of the neurotransmitter (ligand-gated; Fig. 6.4). This allows the flux of Na^+, K^+, and sometimes Ca^{++} ions, thus causing depolarization (Unwin 1993).
- *Metabotropic receptors*: They act by stimulation of the production of second messengers (Fig. 6.5), which are small freely diffusible intracellular metabolites such as cAMP and diacylglycerol. Subsequently, these second messengers activate protein kinases that phosphorylate a variety of proteins, or they mobilize Ca^{2+} ions from intracellular stores (Tanaka and Nishizuka 1994). Often metabotropic receptors gate ion channels, but this happens only indirectly through second messengers. There are two types of metabotropic receptors: the G-protein-coupled receptors and the receptor tyrosine kinases. The G-protein

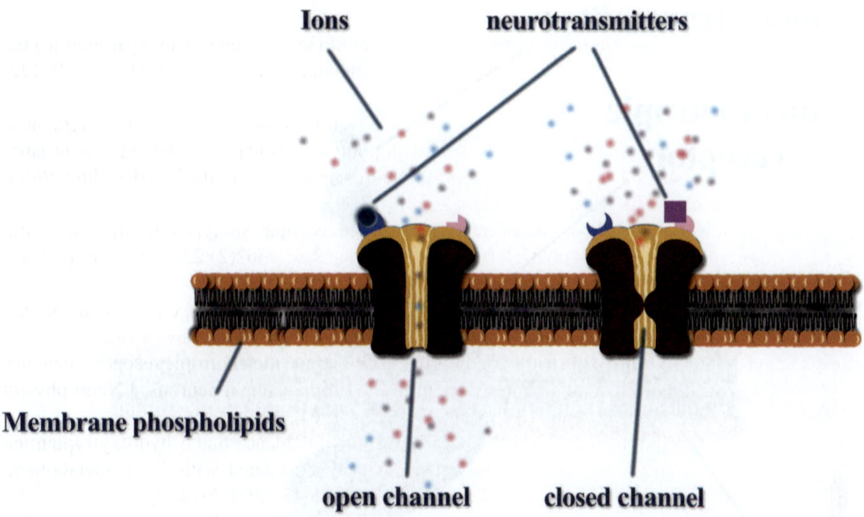

Fig. 6.4 Ionotropic receptors: they include an ion channel which opens or closes as a response to the binding of a neurotransmitter (ligand-gated). (Copyright KN Fountoulakis 2021)

family contains the α- and β-adrenergic receptors, the muscarinic Ach receptors, the $GABA_B$, and others (Gilman 1989, 1990, 1995). Receptor tyrosine kinases are activated by hormones, growth factors, and neuropeptides.

Ionotropic receptors	Metabotropic receptors
• Fast action	• Slower action
• Duration in milliseconds	• Duration in seconds or minutes
• They mediate fast simple behaviors	• Complex behaviors/regulations
• Function of the on-off switch type	• Either increase or decrease of ion passage
• "All or nothing"	• Regulate the electrophysiological properties of the cell
• Only increase of ion passage	• Usually they are unable to trigger an action potential
• They trigger an action potential	• Regulatory effect on the postsynaptic neuron

Ionotropic and metabotropic receptors have different functions as the following:

The ionotropic receptors produce relatively *fast synaptic actions* lasting only milliseconds. These are commonly found in neural circuits that mediate rapid behaviors, such as the stretch receptor reflex.

The metabotropic receptors produce *slower synaptic actions* lasting seconds to minutes. These slower actions can modulate behavior by altering the excitability of neurons and the strength of the synaptic connections of the neural circuitry mediating behavior. Such modulatory synaptic pathways often act as crucial reinforcing pathways in the process of learning.

Additionally, ligand-gated channels function as *simple on-off switches* triggering an *all-or-nothing effect*, while on the contrary, the metabotropic receptors manifest

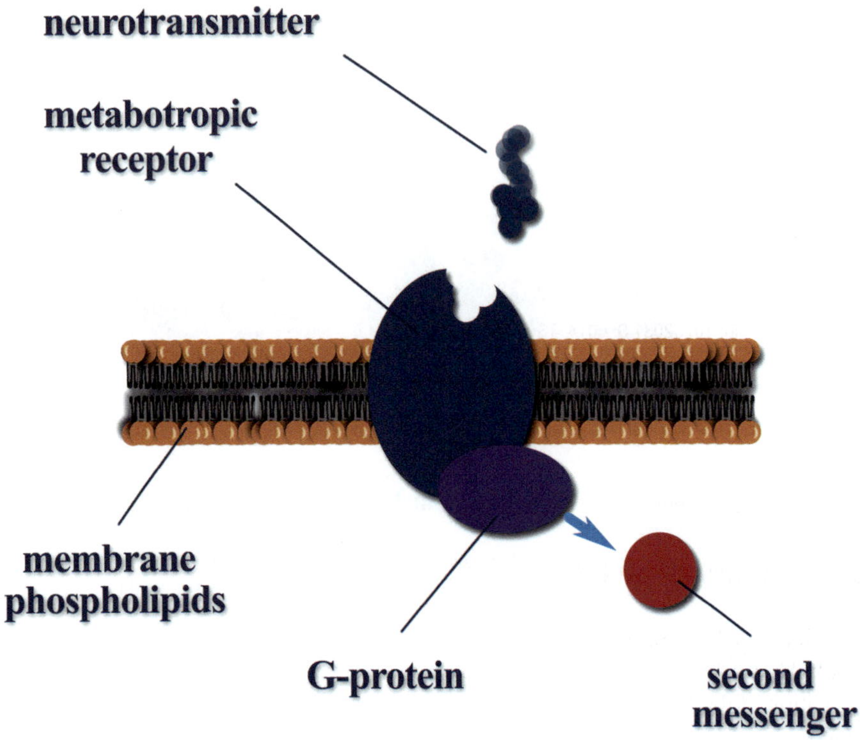

Fig. 6.5 Metabotropic receptors: regulation of ion channels and link to second messenger systems. (Copyright KN Fountoulakis 2021)

a complex function with the ability to vary in space and time as well as in the number of intracellular systems affected. While ionotropic excitation can only increase the ion flux through the channel, metabotropic activity can either increase or decrease (Hille 1994).

Essentially, metabotropic receptors exert a *modulatory effect* on the postsynaptic neuron since they are normally not capable of firing an action potential. Their role is mainly to regulate the electrophysiological properties of a cell (e.g., resting potential, action potential duration, repetitive firing characteristics, etc.).

6.2.1.1 Acetylcholine Receptors

Cholinergic receptors are classified into two main families, the *muscarinic* and the *nicotinic*, on the basis of muscarine and nicotine to mimic the effects of Ach.

Acetylcholine Receptors

- Nicotinic receptors
  - Two alpha (α1–α7) subunits and one beta (β1–β4)
  - Retina
  - Hippocampus
  - Respiratory center of the brain stem
  - Cerebral cortex
  - Thalamus
  - Hypothalamus
  - Substantia nigra
  - Striatum
  - Locus coeruleus
- Muscarinic receptors
  - **M1** (excitatory)
    Forebrain
    Sympathetic ganglia
    Hippocampus
    Cortex
    Striatum
    Olfactory bulb
    Thalamus
    Higher density in the limbic system and the association cortices
  - **M2** (inhibitory)
    Low density in the brain but homogenous distribution
    Relatively higher density in the primary sensory and motor areas
  - **M3** (excitatory)
    Low density throughout the brain
  - **M4** (inhibitory)
    Telencephalon
    Striatum
    Olfactory bulb
  - **M5** (excitatory)
    Less than 2% of total muscarinic receptors
- Attention
- Concentration
- Memory
- Learning
- Affect
- Role in body posture

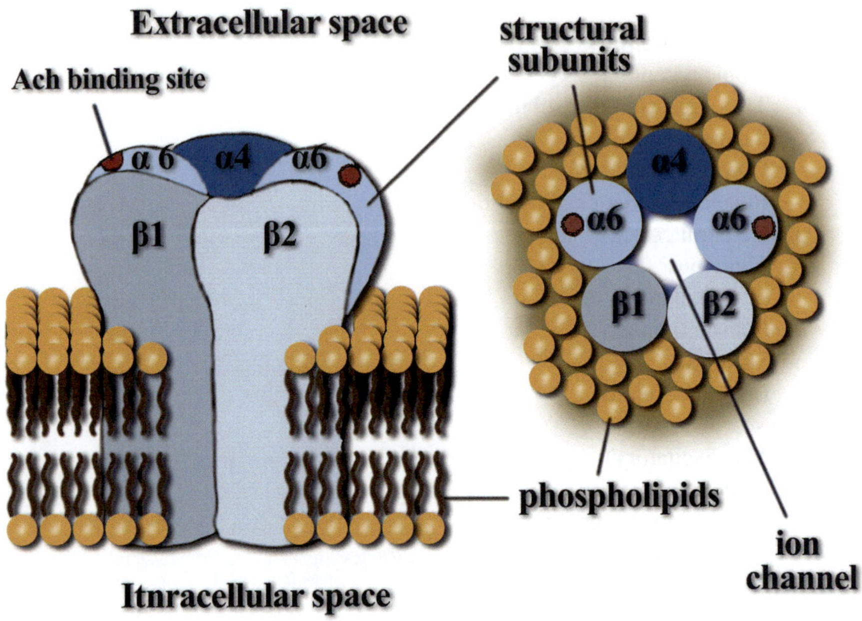

Fig. 6.6 Nicotinic cholinergic receptors. (Copyright KN Fountoulakis 2021)

- *Nicotinic receptors* (Arneric et al. 1994) in the brain consist of alpha (α1–α7) and beta (β1–β4) subunits (Fig. 6.6). In the muscles, there are also other subunits that participate in the nicotinic structure (γ, δ, and ε). Different combinations of these subunits result in different properties of the receptor. The only documented nicotinic neurotransmission in the brain concerns the communication between the motor neuron and a Renshaw cell in the spinal cord. However, a nicotinic type of activity has been recorded in the retina, the hippocampus, the respiratory center in the brain stem, the cortex, the thalamus, the hypothalamus, the substantia nigra, the striatum, and the locus coeruleus. Nicotine interacts with nicotinic receptors and increases the secretion of various neurotransmitters, while the cholinergic system in general regulates and governs a number of important functions including attention and concentration, brain regional blood flow, glucose consumption, and the electrical activity of the cortex. Each one of these functions is enhanced by nicotine and attenuated by its antagonists. Additionally, nicotine could improve memory and learning, and it has an anxiolytic effect probably through the release of endogenous substances acting on GABA receptors. Nicotine-related disorders are related to depression, but the nature of this relationship is unknown (Edwards et al. 2011; Lyons et al. 2008; Fu et al. 2007; Killen et al. 2003; Cardenas et al. 2002; Breslau et al. 1993).
- On the other hand, *muscarinic receptors* (Ehlert et al. 1994) are widely distributed in the human body. Their activation in the periphery causes drop of heart

frequency, vasodilation, constriction of airways, increase of gastric tube mobility, and miosis in the eye pupil, among others. In the brain, they are implicated in function including memory, learning, and the control of posture.

There have been identified four types of muscarinic receptors with the use of pharmacological methods and five (M1–M5) with molecular biology methods. There is almost complete correspondence between the types identified with these two methods, and all receptors seem to act through their coupling with a second messenger system of a G-protein:

- The *M1* type is abundant in the forebrain and the sympathetic ganglia while its density is lower in the hindbrain. Its highest concentration is in the hippocampus, followed by the cortex, the striatum, the olfactory bulb, and the thalamus, while the concentration is far lower in the hypothalamus. This receptor is denser in the limbic system and the association cortices. Its activation results in an increase in the hydrolysis of phosphatidylinositol.
- The *M2* type is widespread in the body, including the brain, but its density is rather low. It is the only muscarinic receptor located in the myocardium. It is the main muscarinic receptor of smooth muscles, excluding the intestine ones. Its density in the brain is low but the distribution is rather homogenous. It is relatively denser in the cerebellum, the medulla oblongata, and the mesencephalon, followed by the hypothalamus and the thalamus. Its highest density is in the primary sensory and motor areas. Its activation leads to the inhibition of adenyl cyclase.
- The *M3* type is the main muscarinic receptor of exocrine glands. It also triggers the constriction of smooth muscle fibers in spite of the fact that it constitutes a minority among the muscarinic receptors in these fibers. Probably, it constitutes the main receptor governing the intestine mobility. It exists in low density throughout the brain, and it is slightly denser in the forebrain. Its activation results in an increase in the hydrolysis of phosphatidylinositol.
- The *M4* type is abundant in the telencephalon, the striatum, and the olfactory bulb. It is denser in the forebrain and its activation leads to the inhibition of adenyl cyclase.
- The *M5* type constitutes less than 2% of total muscarinic receptors, and it is found exclusively in the brain. Its activation results in an increase in the hydrolysis of phosphatidylinositol.

6.2.1.2 Serotonin Receptors

A significant number of 5-HT receptors (most recent number mounts to 14) have been identified so far, but for the role of most of them, we know little (Peroutka and Snyder 1979; Barnes and Sharp 1999; Hoyer and Martin 1996, 1997; Hoyer et al. 1994); however, we do know that they are widely distributed in the brain, mainly localized in parts of the cortex, the limbic system, the hippocampus, the hypothalamus, the septum pellucidum, the striatum, and the amygdala (Blier 2001; Kilpatrick et al. 1996; Pineyro and Blier 1999).

The main are the following:

- *5-HT₁ (A, B, D, E, and F)*: They are negatively linked to the adenyl cyclase system. They inhibit the transformation of ATP into c-AMP (in the hippocampus, substantia nigra, gray matter around the Sylvius aqueduct). Increase in the 5-HT_1 activity in the raphe nucleus leads to increased appetite as it reduces serotonergic activity. Increase of the activity of the rest 5-HT_1 receptors leads to reduced appetite (Montgomery and Fineberg 1989). The 5-HT_{1A}, 5-HT_{1B}, and 5-HT_{1D} are both pre- and postsynaptic. 5-HT_{1E} and 5-HT_{1F} are exclusively postsynaptic. The density of 5-HT_{1A} receptors is very high in the raphe and parts of the hippocampus, relatively high in the hypothalamus, insula, temporal, cingulate, and ventral prefrontal cortices, moderately high in the occipital and parietal cortices, and very low in the cerebellum, striatum, and thalamus and also in the white matter. They are expressed presynaptically in the raphe (somatodendritic autoreceptors) and postsynaptically in other brain regions (Azmitia et al. 1996). There are abundant postsynaptic 5-HT_{1A} in the astrocytes and many in the glial cells in the frontal and limbic cortices but not in the striatum, the thalamus, or the cerebellum (Azmitia et al. 1996; Whitaker-Azmitia et al. 1993). In the cortex, the hippocampus, and the amygdala, the stimulation of postsynaptic 5-HT_{1A} receptors inhibits glutamate-related excitability (Azmitia 1999; Azmitia et al. 1996; Sprouse and Aghajanian 1988). Stimulation of astrocyte-based 5-HT_{1A} results in neurotrophic effects, while their antagonism results in losses of dendrites, spines, and synapses in adult and developing animals (Whitaker-Azmitia and Azmitia 1989, 1991). Stimulation of presynaptic 5-HT1A autoreceptors inhibits the firing of raphe neurons and also reduces serotonin synthesis and release (Briley and Moret 1993; Chaput et al. 1986, 1990; Sprouse and Aghajanian 1988). Additionally, the stimulation of 5-HT_{1A} located on neuronal bodies is essential for the maintenance of the cytoskeleton (Drevets 2000; McEwen 1999).
- *5-HT_2 (A, B, and C)*: They are positively linked to the phosphatidylinositol system and lead to an increase of intracellular Ca or to protein kinase activation. Animal research has shown that social isolation causes hypersensitivity of 5-HT_{2C}. These same receptors of the amygdala-hippocampus circuit probably lead to the manifestation of aggressive behaviors when danger is confronted, by overcoming the inhibitory effects of 5-HT_1 and 5-HT_{2A} located in the gray matter around the aqueduct. The increase in the activity of these receptors again could lead to a reduction in appetite, especially in bulimic patients which also manifest affective disorders and disordered saturation feeling mediated by cholecystokinin (CCK). In general, 5-HT_2 exert a tonic inhibition on noradrenalin in the hippocampus. All 5-HT_2 receptors are postsynaptic but 5-HT_{2C} may also be presynaptic.
- *5-HT_3*: They have positive link with fast ion channels and increase their permeability (Hoyer et al. 1994; Hoyer and Martin 1996). They are both pre- and postsynaptic. They inhibit Ach activity in the cortex but augment dopaminergic in the corpus striatum and the limbic system.

- *5-HT$_4$*: They have a positive link with the adenyl cyclase system. They increase the transformation of ATP into c-AMP. Their relationship to behavior is unknown. They are postsynaptic receptors.
- *5-HT$_5$ (A* and *B)*: It is unknown the exact second messenger system they relate to and equally unknown is their relationship to behavior. They are postsynaptic receptors.
- *5-HT$_6$*: They have a positive link with the adenyl cyclase system. They increase the transformation of ATP into c-AMP. Their relationship to behavior is unknown. They are postsynaptic receptors.
- *5-HT$_7$*: They have a positive link with the adenyl cyclase system. They increase the transformation of ATP into c-AMP. Their relationship to behavior is unknown. Recently, it has been suggested that they mediate the antidepressant effect of lurasidone (Cates et al. 2013). They are postsynaptic receptors.

Serotonin Receptors

- 5-HT$_1$ (A, B, D, E, and F)
  - Negative link with the adenyl cyclase system
  - Hippocampus
  - Substantia nigra
  - Gray matter around the Sylvius aqueduct
  - Appetite regulation
- 5-HT$_2$ (A, B, and C)
  - Positive link with the phosphatidylinositol system
  - Amygdala
  - Hippocampus
  - Gray matter around the Sylvius aqueduct
  - Sociality
  - Aggressiveness
  - Appetite regulation
- 5-HT$_3$
  - Positive link with fast ion channels
  - They inhibit Ach activity in the cortex
  - They increase dopaminergic activity in the striatum and the limbic system
- 5-HT$_4$
  - Positive link with the adenyl cyclase system
- 5-HT$_5$ (A and B)
- 5-HT$_6$
  - Positive link with the adenyl cyclase system
- 5-HT$_7$
  - Positive link with the adenyl cyclase system
- Ot 5-HT$_1$, 5HT$_2$, and 5-HT$_3$ can be found either pre- or postsynaptically
- All the rest 5-HT receptors can be found exclusively postsynaptically

6.2.1.3 Norepinephrine Receptors

Norepinephrine (adrenergic) receptors belong to the G-protein-coupled superfamily. They can be found pre- and postsynaptically on neurons but also on glial cells (Salm and McCarthy 1992; Stone and Ariano 1989).

They have been originally classified into α (alpha)- and β (beta)-receptors (Ahlquist 1948). This classification is based on the relative preference of norepinephrine for the alpha- and isoprenaline for the beta-receptors. Isoproterenol and salbutamol are beta-receptor agonists, while propranolol and pindolol are nonselective antagonists (Hieble and Ruffolo 1991). They are further subdivided into subgroups on the basis of pharmacological properties and molecular biology (Bylund et al. 1994; Lands et al. 1967; Frielle et al. 1988; Shorr et al. 1981), for example, selective antagonist for beta-1 is atenolol and ICI-118, 551 is for beta-2 (Hieble and Ruffolo 1991).

- *Alpa-1 (α1; a, b, c, and d)*: Postsynaptic excitatory receptors linked to the phosphatidylinositol system (Hieble et al. 1995; Bylund et al. 1994). They can also be found in the muscles of the vessel wall and the smooth muscles of the intestine and the heart.
- *Alpha-2 (α2; a, b, c, and d)*: They can be found presynaptically, and they are inhibitory receptors linked to the adenyl cyclase system (Patel et al. 1981; Langer 1974; Aghajanian and VanderMaelen 1982; Anden et al. 1970; Cedarbaum and Aghajanian 1977; Langer and Arbilla 1990; Starke 1971; Svensson et al. 1975; Bylund et al. 1994; O'Dowd et al. 1989; Venter et al. 1989). They mediate their functions through the Gi class of G-proteins (Limbird 2004), and they can be found both post- and presynaptically (Cedarbaum and Aghajanian 1977; Langer 1997; Miller 1998; Stjärne 1989). Apart from the brain, they can be found in the muscle of the vessel wall, the smooth muscles of the intestine, and the platelet membrane. Subtypes are unevenly distributed in the brain, with high densities of α2a and α2c in the LC, amygdala, and hippocampus, while α2b is found mainly in the thalamus (Scheinin et al. 1994; Rosin et al. 1996; Talley et al. 1996; Wamsley et al. 1992; Nicholas et al. 1993a).
- *Beta-1 (β1)*: They are excitatory receptors linked to the adenyl cyclase system (Frielle et al. 1987; Shorr et al. 1982; Nicoll et al. 1990). They predominate in the cerebral cortex, the dentate gyrus, the CA1, and the medial dorsal hypothalamic nuclei (Nicholas et al. 1993b; Palacios and Kuhar 1982; Rainbow et al. 1984). Apart from the brain, they can be found in the heart.
- *Beta-2 (β2)*: They are excitatory receptors linked to the adenyl cyclase system (Tholanikunnel et al. 1999; Ostrowski et al. 1992; O'Dowd et al. 1988; Nicoll et al. 1990). They are more abundant in the cerebellum and reticular, paraventricular, and central thalamic nuclei (Nicholas et al. 1993b; Palacios and Kuhar 1982; Rainbow et al. 1984), but they can also be found both in the smooth and skeletal muscles and the liver. They are also present in the membrane of the lymphocytes where their function is unknown.
- *Beta-3 (β3)*: They can be found in the adipose tissue and they might be related to lipolysis. Overall little is known about them (Emorine et al. 1989).

Norepinephrine Receptors

- Alpha receptors
  - Alpha-1 (α1: a, b, c, and d)
    Postsynaptic
    Excitatory
    They are linked to the phosphatidylinositol system
    In the smooth muscle mainly of the blood vessel wall
  - Alpha-2 (α2: a, b, c, and d)
    Presynaptic
    Inhibitory
    They are linked to the adenyl cyclase system
    Locus coeruleus
    Amygdala
    Hippocampus
    Thalamus
- Beta receptors
  - Beta-1 (β1)
    Excitatory
    They are linked to the adenyl cyclase system
    Dominant in the cerebral cortex
    Dentate gyrus
    CA1 area
    Dorsal hypothalamic nuclei
  - Beta-2 (β2)
    Excitatory
    They are linked to the adenyl cyclase system
    Cerebellum
    Reticular, paraventricular, and central thalamic nuclei
  - Beta-3 (β3)
    Adipose tissue

6.2.1.4 Dopamine Receptors

The following dopaminergic receptors have been identified so far (Civelli 1994; Mansur and Watson 1994). They can grossly classified on the basis of their sequence homology, into D1-like (D1 and D5) which activate adenylyl cyclase and D2-like (D2, D3, and D4) which inhibit adenylyl cyclase (Bunzow et al. 1988; Sokoloff et al. 1990; Van Tol et al. 1991). Both the D1 and D2 families can be found postsynaptically, but only D2-like can be found presynaptically:

- *D1:* It is a postsynaptic excitatory receptor, linked to the adenyl cyclase system through a G-protein. It is abundant in the anterior cingulate gyrus, the cortex of the frontal pole, the insular (Reil cortex) and the olfactory cortex, the nucleus caudatus, the putamen, the nucleus accumbens, and the olfactory bulb and to lower density in the septum, the hypothalamus, and the rest of the cortex. It is absent from the substantia nigra, the VTA and the hippocampus. It is dominant in the amygdaloid nucleus where D2 is absent. D1 is present solely in the brain and not in the rest of the body (Boyson et al. 1986).

- *D2*: It is a presynaptic autoreceptor and inhibits the adenyl cyclase system through a G-protein; however, there is evidence that at least a subgroup of these receptors acts independently of this second messenger system (Roth and Elsworth 1994). It is the dominant dopamine receptor in the substantia nigra, the VTA, the globus palidus, the NAC, and the hippocampus where D1 is absent. It is also abundant in the olfactory cortex, the nucleus caudatus and the putamen, the nucleus accumbens, the olfactory bulb, and the hypothalamus and in lower density in the septum and the rest of the cortex (Bouthenet et al. 1991; Gehlert et al. 1992; Gehlert and Wamsley 1985; Jackson and Westlind-Danielsson 1994; Meador-Woodruff et al. 1989; Weiner and Brann 1989; Yung et al. 1995; Levey et al. 1993). It is absent from the amygdaloid nucleus where D1 is dominant. Outside the brain, it can be found in the cortex of the adrenal glands, the pituitary, and the retina. It is possible that D2 plays a role in pain regulation though receptors are located in the reticular nucleus of thalamus and the gigantocellular reticular nucleus in the hindbrain.
- *D3*: It is a presynaptic autoreceptor and probably it is linked to the same G-protein D2 is linked but this is not solidly proven. It is located in the same areas with D1 and D2 but with a much lower density. It is expressed with high density in forebrain and limbic areas (Sokoloff et al. 1990), preferentially in the islands of Calleja, in the granule cells and in the NAC. It can be found in lower density in the dorsal striatum, in the substantia nigra pars compacta, in the VTA, in the Purkinje cells, in the hippocampus, in the septal area, and in various cortical layers of the medial-temporal lobe (Bouthenet et al. 1991; Levesque et al. 1992). It is not found outside the brain.
- *D4*: Probably it is linked to the same G-protein D2 is linked but this is not solidly proven. It is located in the same areas with D1 and D2 but with a much lower density. It also manifests some selectivity for the limbic system. Probably, it is also located in the kidneys and the heart (O'Malley et al. 1992; Mrzljak et al. 1996).
- *D5:* It is an excitatory receptor linked to the adenyl cyclase system through a G-protein. It is located solely in the hippocampus, the hypothalamus, and the parafascicular nucleus in the thalamus, and it is implicated in functions concerning affect, pain, and neuroendocrine secretion. It is probably located also in the kidneys and the heart.

Dopamine Receptors

- They are linked to the adenyl cyclase system through a G-protein
- D1
  - Postsynaptic
  - Excitatory
  - Anterior cingulate gyrus
  - Frontal pole cortex
  - Reil cortex
  - Olfactory cortex
  - Nucleus caudatus
  - Putamen
  - Nucleus accumbens
  - Septum
  - Hypothalamus
- D2
  - Presynaptic autoreceptor
  - Inhibitory
  - Substantia nigra
  - Ventral tegmental area
  - Globus pallidus
  - Hippocampus
  - Olfactory cortex
  - Nucleus caudatus
  - Putamen
  - Nucleus accumbens
  - Hypothalamus
- D3
  - Presynaptic autoreceptor
- D4
- D5
  - Excitatory
  - Hippocampus
  - Hypothalamus
  - Parafascicular nucleus in the thalamus
  - Affect
  - Pain
  - Neuroendocrine secretion

There seems to be a pattern in the localization of dopamine receptors in the brain with D3 and D4 restricted to the mesocortical and the mesolimbic systems which are areas associated with cognition and emotion and D2 expressed mainly in areas of subcortical regions associated with motor control (Ross and Pearlson 1996; Schwartz et al. 2000).

6.2.1.5 Histamine Receptors

Histamine receptors are classified as H1–H4 and are all bound to a G-protein. It is also believed that histamine activates ligand-gated chloride channels in the brain. These are located in the thalamus and hypothalamus and produce fast inhibitory postsynaptic potentials (Panula et al. 2015):

- *H1*: It is produced in the tuberomammillary nucleus, and fibers project to the dorsal raphe and the locus coeruleus among other areas (Panula et al. 2015; Blandina et al. 2012).
- *H2*: Their receptors are located exclusively in the gastrointestinal system and in vascular smooth muscle cells (Wouters et al. 2016; Panula et al. 2015).
- *H3*: It manifests both autoreceptor and heteroreceptor functions and exerts an inhibitory effect on the secretion of histamine, Ach, norepinephrine, and 5-HT (Gilman 1990).
- *H4*: It is found primarily in the basophils, the bone marrow, the thymus, the small intestine, the spleen, and the colon (Panula et al. 2015).

6.2.1.6 Glutamate Receptors

Glu is excitatory at ionotropic receptors and modulatory at metabotropic receptors (Kew and Kemp 2005; Laezza et al. 1999).

Glutamate Receptors

- Ionotropic Glu receptors
  - *N*-Methyl-D-aspartate (NMDA)
    Part of large macromolecular synaptic signaling complexes
    They are not activated by single synaptic events
    Glycine binding as co-agonist
    Multiple structural subunits (NR1, NR2, and NR3)
    Synaptic plasticity
    Synaptic development
    Attention
    Learning
    Memory
    General knowledge
    Mood anxiety
    Excitotoxicity-neurodegeneration
  - α-Amino-3-hydroxy-5-methyl-4-isoxazolepropionic acid receptor (AMPA)
    Four subunits (GluA1–GluA4)
  - Kainic acid (KA)
    Receptor subunits GluR 5–7 and KA1 and KA2
- Metabotropic Glu receptors
  - Linked with a G-protein system
  - Three groups and eight subtypes (mGluR1–mGluR8)

- The *ionotropic receptors of Glu* are classified on the basis of the agonist they bind to, into the following types (Tsumoto 1990; Macdermott 1987; Watkins 1981; Watkins and Evans 1981):
  - *N-Methyl-*d*-aspartate (NMDA)*: These receptors (Fig. 6.7) are essentially a ligand-gated rapidly transmitting calcium ion channel (Flatman et al. 1983; MacDonald et al. 1982). Glutamate binding causes the channel to open and the neuron to be excited. They are composed of multiple subunits which may belong to three related families: NR1, NR2, and NR3 (Nakanishi 1992). Both the NR1 and NR2 subunits are required for the receptor to function, while a third unit, NR3, plays a modulatory role. NR1 subunits contain a glycine binding site, while NR2 subunits contain a glutamate binding site (Hirai et al. 1996; Laube et al. 1997, 1998). Every NMDA receptor is composed of two NR1 and two NR2 subunits, and this is why it is called a "tetramer" (Laube et al. 1998). NMDA receptors with different subunit compositions show distinct distributions in the brain and different properties and regulations (Yamakura and Shimoji 1999). It is important to note that NR1 receptor subunits are distributed ubiquitously in the brain, and NR1 subunit mRNA are expressed throughout the different stages of neurodevelopment (Mony et al. 2009; Yamakura and Shimoji 1999; Laurie et al. 1997).

    The glycine binding site is located on the NR1 subunit while the Glu site is located on the NR2 subunit (Laube et al. 1998). There are two NR1 subunits and two NR2 subunits in a single NMDA receptor complex, which is often called

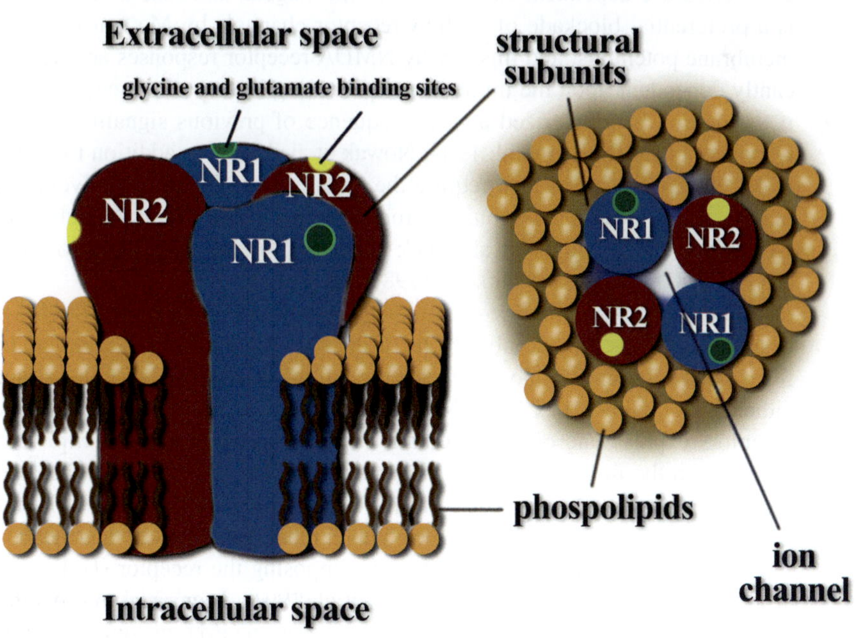

Fig. 6.7 *N*-methyl-D-aspartate (NMDA) receptors. (Copyright KN Fountoulakis 2021)

tetramer (Laube et al. 1998). There is a single NR1 gene which results in eight distinct NR1 isoforms (Hollmann et al. 1993; Sugihara et al. 1992). On the contrary, there are four genes encoding the four different members of the NR2 subunit family (NR2A–NR2D), and thus it is evident that the properties of NMDA receptors depend on the specific NR2 subunit which is present in the heteromeric complex (Buller et al. 1994; Monyer et al. 1992, 1994; Williams 1993).

NMDA receptors are located in synaptic, presynaptic, and extrasynaptic sites, but in most neurons, NMDA receptor density is higher in dendritic spines within the postsynaptic density (Fig. 6.8) which is located in the postsynaptic membrane and close to the active area of the postsynaptic neuron (Mony et al. 2009). They are often found co-localized with AMPA receptors and are not activated by single synaptic events like the Glu binding alone (Huang and Bergles 2004). NMDA receptors possess unique characteristics that allow them to be part of large macromolecular synaptic signaling complexes. The simple activation of the NMDA receptor produces only a very small and difficult-to-detect result (Collingridge et al. 1988). Two simultaneously occurring events are necessary for their activation: presynaptic release and binding of glutamate to NMDA receptors and sufficiently strong postsynaptic membrane depolarization leading to the removal of Mg^{++} blockade which is normally present (Wu and Zhuo 2009). The NMDA receptors become fully activated upon extensive stimulation of the synapse, because the repetitive activation of AMPA receptors results in sufficient depolarization of the postsynaptic membrane and thus the lifting of the Mg^{++} block, during a process called use-dependent calcium influx (Monaghan and Jane 2009). There is a preferential blockade of NMDA receptor channels by Mg^{++} at negative membrane potentials, and this is why NMDA receptor responses are significantly dependent upon the immediately preceding history of the neuron, and if the neuron is depolarized as a consequence of previous signaling, larger responses occur (Mayer et al. 1984; Nowak et al. 1984). In addition to glutamate, NMDA receptors also require the simultaneous binding of glycine or maybe D-serine as a co-agonist (Monaghan and Jane 2009; Asztely and Gustafsson 1996; Mayer et al. 1984; Monaghan et al. 1989; Johnson and Ascher 1987; Danysz and Parsons 1998).

– *α-Amino-3-hydroxy-5-methyl-4-isoxazolepropionic acid receptors (AMPA)*: They are composed by combinations of the four subunits which combine to form hetero-tetrameric blocks consisting of two subblocks of GluA2 and either GluR1, GluR3, or GluR4 (Mayer 2005; Greger et al. 2007). Also AMPA receptors express one of two alternative splice cassettes, termed "flip" and "flop," with the flop form manifesting more rapid desensitization. When the receptor channels are composed of GluR-B (or GluR2), the channel is not Ca^{++} permeable. On the contrary, the channel manifests high Ca^{++} permeability when the other types of subunits are composing the receptor (Hollmann et al. 1991; Hume et al. 1991; Verdoorn et al. 1991). In general, the AMPA receptors represent the predominant fast excitatory neurotransmitter receptor in the CNS, and their distribution in the brain is similar to that of the NMDA.

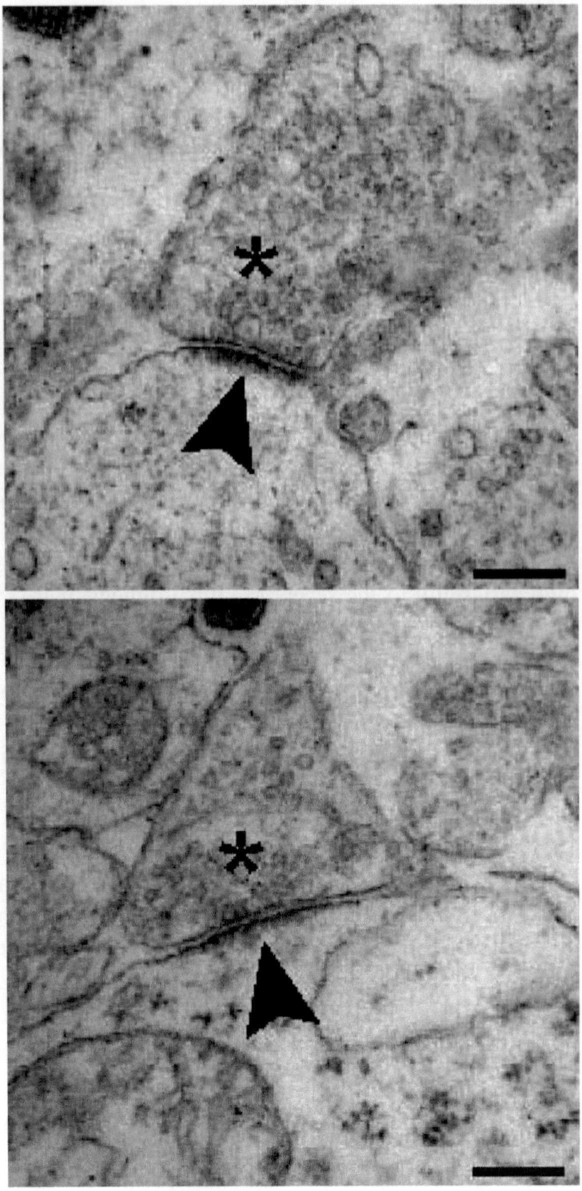

Fig. 6.8 The area of the protein of postsynaptic density, which is located on the postsynaptic membrane close to the active area of the postsynaptic neuron. Electron microscope image of the synapses in the brain stem of rats, aged 20 days. The scale is 250 nm. (από Heupel et al. Neural Development 2008 3:25)

- *Kainate (KA) receptors*: They are composed by heteromeric combinations of two related subfamilies of proteins which comprise the receptor subunits, GluR 5–7 and KA1 and KA2 and their variants (Bettler et al. 1990, 1992; Egebjerg et al. 1991; Herb et al. 1992; Sakimura et al. 1992; Werner et al. 1991). They have both presynaptic and postsynaptic actions, they are fewer in comparison to the other ionotropic receptors, and their function is not well understood (Contractor et al. 2011).

- The *metabotropic Glu receptors* are coupled with a G-protein system and are divided into three groups based on their transduction mechanisms, sequence homology, and pharmacology (Nakanishi 1992) and eight subtypes, *mGluR1–mGluR8*, which are coded by eight different genes (Pin and Acher 2002; Anwyl 1992, 1999; Cartmell and Schoepp 2000; Conn and Pin 1997; De Blasi et al. 2001; Pin and Duvoisin 1995):
  - *Group I* includes mGlu1 and mGlu5. They are both linked positively to phospholipase C (Conn and Pin 1997).
  - *Group II* includes mGlu2 and mGlu3 which are coupled negatively with adenylyl cyclase (Tanabe et al. 1993).
  - *Group III* includes mGlu4 and mGlu6–mGlu8 which are coupled negatively with adenylyl cyclase (Tanabe et al. 1993).

All mGlu receptors include a large extracellular N-terminal (which is a free amine group at the end of the polypeptide) and a cytoplasmic carboxy-terminal domain which varies in length. There are a significant number of splice variants of mGlu receptors, and in combination with the variable length of the cytoplasmic domain and the existence of cysteine residues in the extracellular part, they give high variability to the receptors' three-dimensional structure and subsequently to their properties (Tanabe et al. 1993).

6.2.1.7 Gamma-Aminobutyric Acid (GABA) Receptors

There are two main GABA receptors, named A and B (Zorumski and Isenberg 1991), while there is also a third group, C, which constitutes a subcategory of A. In most brain regions, both GABA subtypes coexist:

- The *$GABA_A$* is a fast ionotropic receptor linked to an ion channel (Cossart et al. 2005). It is made of at least five subunits which are transmembrane and are arranged in a ring creating a pore in the center, which constitutes the ion channel (Fig. 6.9). There are three major types of subunits named α, β, and γ (but also the

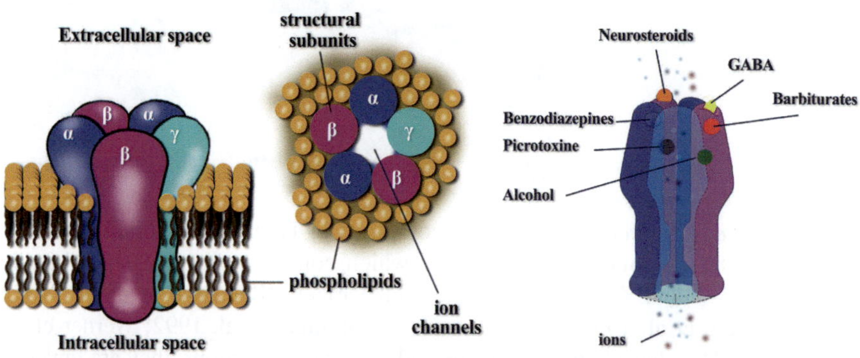

Fig. 6.9 GABA receptor. The depiction of binding sites is indicative and not exhaustive. (Copyright KN Fountoulakis 2021)

existence of δ, ε, π, and θ has been reported). The usual combination is two α-, two β-, and one γ-subunits. These three subunits are allosterically connected, which means that binding to any of them modifies the binding ability of the others (Olsen and Sieghart 2009).

- The *$GABA_B$* is a metabotropic receptor and is linked to a G-protein. It is located presynaptically in the neurons containing GABA or Glu.
- The presence of a distinct type of subunit, named ρ, determines a subgroup of $GABA_A$ receptors named *$GABA_C$* which do not manifest allosteric properties (Enz and Cutting 1998).

Most areas of the brain contain both A and B GABA receptors.

Gamma-Aminobutyric Acid (GABA) Receptors

- $GABA_A$
  - Ionotropic receptor
  - Is made of at least five transmembrane subunits (two α, two β, and one γ)
  - The three subunits are allosterically connected
- $GABA_B$
  - Metabotropic receptor
  - Linked to a G-protein
  - Presynaptic receptor in neurons containing GABA or Glu
- $GABA_C$
  - Subcategory of A with the presence of the ρ-subunit
  - Does not manifest allosteric properties

6.2.1.8 Opioid Receptors

Opioid receptors are a group of inhibitory G-protein-coupled receptors with opioids as ligands (Bjorklund et al. 1990, 1992; Janecka et al. 2004; Takemori and Portoghese 1985; Waldhoer et al. 2004; Dhawan et al. 1996). They are approximately 40% identical to somatostatin receptors.

Opioid Receptors

- Inhibitory receptors
- Linked to a G-protein
- Delta receptors (δ)
  - δ-1
  - δ-2

    Pontine nuclei
    Amygdala
    Olfactory bulb
    Deep cortex layers
    Peripheral sensory neurons

- Kappa receptors (κ)
  - κ1
  - κ2
  - κ3
    Hypothalamus
    Periaqueductal gray
    Claustrum
    Spinal cord
    Substantia gelatinosa
    Peripheral sensory neurons
- Mu receptors (μ)
  - μ1
  - μ2
  - μ3
    Laminae III and IV of the cortex
    Thalamus
    Striosomes
    Periaqueductal gray
    Rostral ventromedial medulla
    Spinal cord
    Substantia gelatinosa
    Peripheral sensory neurons
    Intestinal tract
- Opioid growth factor receptor (OGFr or ζ-opioid receptor)
  - Heart
  - Liver
  - Skeletal muscle
  - Kidney
  - Brain
  - Pancreas
  - Fetal tissue

- The *delta (δ) receptors* include δ-1 and δ-2 subtypes and can be found in the pontine nuclei, the amygdala, the olfactory bulbs, and the deep cortex layers and in some peripheral sensory neurons. Their functions include analgesia, antidepressant, and convulsant effects and physical dependence. They also could modulate μ-opioid receptor-mediated respiratory depression (Holden et al. 2005).
- The *kappa (κ) receptors* include κ1, κ2, and κ3 subtypes and can be found in the hypothalamus, the periaqueductal gray, the claustrum, the spinal cord, and the substantia gelatinosa and in some peripheral sensory neurons. Their function includes analgesia, anticonvulsant effects, induction of depression, dissociative and hallucinogenic effects, diuresis, dysphoria, miosis, neuroprotection, sedation, and induction of stress (Holden et al. 2005).
- The *mu (μ) receptors* include the μ1, μ2, and μ3 subtypes. They can be found in the laminae III and IV of the cortex, the thalamus, the striosomes, the periaque-

ductal gray, the rostral ventromedial medulla, the spinal cord, and the substantia gelatinosa and in some peripheral sensory neurons and in the intestinal tract. The functions of the μ1 include analgesia and physical dependence; those of μ2 include respiratory depression, miosis, euphoria, reduced gastrointestinal motility, and physical dependence, while those of the μ3 are unknown and possibly include vasodilation (Holden et al. 2005).

- The *opioid growth factor receptor (OGFr or the ζ-opioid receptor)* is a receptor for opioid growth factor (OGF), also known as [Met(5)]-enkephalin (Zagon et al. 2000). It was originally named as opioid receptor zeta (ζ); however, it was subsequently found that it shares little structure similarity with the other opioid receptors. It can be found in the heart, the liver, the skeletal muscle, the kidney, the brain, the pancreas, and the fetal tissue. Its functions include regulation of tissue growth (Cheng et al. 2009; Malendowicz et al. 2005; Wu et al. 1998; Zagon et al. 2002), embryonic development (Zagon et al. 1999b), and wound repair (Sassani et al. 2003), and it plays a role in certain types of cancer (Cheng et al. 2007; Donahue et al. 2009; McLaughlin et al. 1999; Zagon et al. 1999a).

6.2.1.9 Sigma Receptors

The *sigma (σ)-receptors* were initially thought to be a subtype of opioid receptors, but eventually it was proven that their function was unrelated to that of the opioid receptors. Although these functions are poorly understood (Leonard 2004), it seems that their activation may induce hallucinogenic effects, and they might mediate opioid overdose convulsions. Other actions include hypertonia, tachycardia, tachypnea, antitussive effects, and mydriasis. Endogenous ligands include DHEA and pregnenolone, and endogenous antagonist is progesterone. Substances which act as agonists on σ-receptors include cocaine, morphine, opipramol, PCP, fluvoxamine, methamphetamine, dextromethorphan, and the herbal antidepressant berberine (Yang et al. 2007; Fontanilla et al. 2009; Skuza and Rogoz 2006). In terms of behavioral effects, σ-receptor agonists exert heterogeneous effects; some subjects experience euphoric or significant antidepressive effects, while others experience dysphoria and anxiety (Wang et al. 2007; Guitart et al. 2004; Zhang and Cuevas 2005).

Sigma Receptors (σ)

- Agonists
  - Cocaine
  - Morphine
  - Οπιπραμόλη
  - Φαινκυκλιδίνη (PCP)
  - Φλουβοξαμίνη
  - Μεθαμφεταμίνη
  - Δεξτρομεθορφάνη
  - Βερβερίνη
- Sigma-1 receptor (σ1)
- Sigma-2 receptor (σ2)

- The *sigma-1 receptor (σ1)* probably modulates calcium signaling through the inositol trisphosphate (IP3) receptor (Hayashi and Su 2001, 2004, 2007; Monassier and Bousquet 2002). It is unclear whether it is coupled with a G-protein (Brimson et al. 2011; Hong and Werling 2000), but probably it exists in a complex with voltage-gated K+ channels, leading to the idea that σ1-receptors are auxiliary subunits (Monassier and Bousquet 2002; Lupardus et al. 2000). Some authors suggest that σ1-receptors are intracellular signal transduction amplifiers (Su and Hayashi 2003). While an endogenous ligand has yet to be conclusively identified, a number of endogenous substances including tryptaminergic trace amines and neuroactive steroids all activate the receptor (Fontanilla et al. 2009; Guitart et al. 2004). In terms of exogenous ligands, they have affinity for various structurally and pharmacologically distinct substances including haloperidol and cocaine (Su and Hayashi 2003; Lee et al. 2008). They are expressed in certain regions of the central nervous system (Weissman et al. 1988), and they have been implicated in several phenomena, including cardiovascular function, schizophrenia, clinical depression, the effects of cocaine abuse, and cancer (Guitart et al. 2004; Zhang and Cuevas 2005).
- The *sigma-2 receptor (σ2)* is highly expressed in malignant cancer cells (van Waarde et al. 2015), and it was difficult to distinguish from NMDA receptors (Skuza and Rogoz 2002). It is more densely located in parts of the brain that are responsible for motor function and emotional response. It is a cytochrome-related protein (Ahmed et al. 2012) located in the lipid raft (Narayanan et al. 2011) and is associated with P450 proteins (Ahmed et al. 2012). It is coupled with various ion channels (Huang et al. 2014), and it is found in several areas of the brain, including the cerebellum, motor cortex, and substantia nigra (Skuza and Rogoz 2002). It is also highly expressed in the lungs, liver, and kidneys (Ahmed et al. 2012). Its functions include a role in cell proliferation (Damaskos et al. 2014; Huang et al. 2014) and neuronal as well as nonneuronal signaling. They are highly expressed in breast, ovarian, and lung cancers; brain, bladder, and colon cancers; and melanoma (Ahmed et al. 2012; Damaskos et al. 2014). Nonneuronal signaling is achieved through binding to a number of hormones and steroids via a calcium secondary messenger (Ahmed et al. 2012; Skuza and Rogoz 2002; Narayanan et al. 2011; Huang et al. 2014). Neuronal signaling is done by regulation of calcium and potassium channels and increase of synaptic vesicular transport and release of dopamine, serotonin, and glutamate (Narayanan et al. 2011; Skuza 2012; Skuza and Rogoz 2002). Finally, the sigma-2 receptors are the locus of binding for several antipsychotics and antidepressant agents (Ahmed et al. 2012; Skuza 2012; Skuza and Rogoz 2002; Narayanan et al. 2011).

6.2.2 Neurotransmitter Reuptake

The neurotransmitter effect in the synaptic cleft lasts only for short periods of time. After this, the neurotransmitter molecules *should be removed in time*. Failure to do so could result in "noise" or even in total inhibition of new signal transmission, in the diffusion to neighboring synapsis resulting in signal contamination, while the signal itself these molecules carry will tend to degrade. Therefore, the timely removal of transmitters from the synaptic cleft is critical for the quality of synaptic transmission. Neuropeptides are removed more slowly than small-molecule transmitters, and this probably contributes to their more lasting effects.

Neurotransmitter Reuptake

- Reuptake pump
- Material recycling
- Fast total flow of information
- Fast reaction to environmental stimuli
- High dynamism of the system of information transmission within the brain
- Focusing of the signal
- The signal is clean and of high quality
- Maximum control on the temporal characteristics of the signal
- Easily and quickly replaceable by another signal

Transmitter molecules are removed from the cleft by three mechanisms:

- Diffusion
- Enzymatic degradation
- Reuptake

While diffusion removes some fraction of all chemical messengers, it cannot remove the complete quantity. Enzymatic degradation of transmitter molecules is used primarily by the cholinergic synapses.

The most important mechanism of removal is *reuptake*, and it is important also because it "salvages" the neurotransmitter and its fragments for recycling. It is mediated by transporter molecules (*reuptake pumps*; Fig. 6.10) which are classically proteins which use transmembrane ion gradients and electrical potential to transport neurotransmitter across the membrane. There are several transporters for each transmitter (Nelson and Lill 1994; Amara and Arriza 1993; Ciliax et al. 1995).

The fast reuptake and the quick recycling of the secreted neurotransmitters make possible both the quick overall flow of information and the fast reaction to environmental stimuli, thus contributing to the high dynamism of the system of information transmission in the brain. Thanks to these mechanisms, the signal focuses, it is clean, of high quality, and there is a high control of its temporal characteristics, while at the same time, it is easy and fast to replace it with a new one.

Concerning neuropeptides specifically, after the binding on the receptor, two alternative events could follow:

- The first concerns the *incorporation* of the neuropeptide or of its active part into the postsynaptic neuron.
- The second concerns its detachment from the receptor and its *release* back into the synaptic cleft.

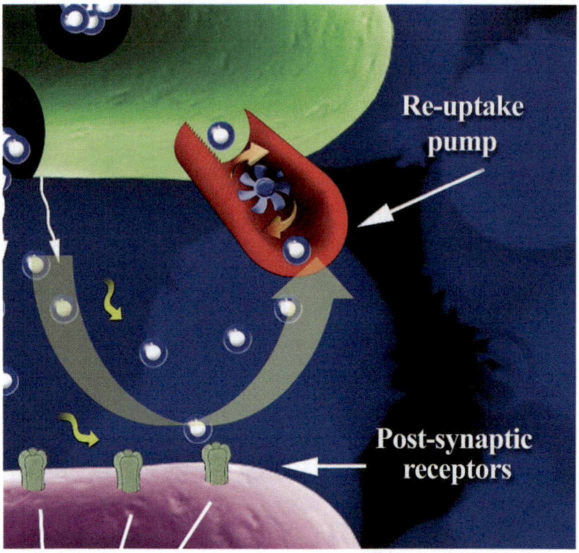

Fig. 6.10 The neurotransmitters' reuptake pump. (Copyright KN Fountoulakis 2021)

There are no data suggestive of the existence of some kind of reuptake event of an entire neuropeptide molecule into the presynaptic neuron, similar to what happens with biogenic amines. Special enzyme peptidases break the active peptide into fragments, and eventually into single amino acids, which then could be reuptaken and be part of the general metabolic storage of amino acid within the presynaptic neuron.

6.3 Electrical Synapses

They can be found almost everywhere in the brain (Eugenin et al. 2012; Pereda et al. 2013). In this kind of synapses, the pre- and the postsynaptic cell membranes are connected with channels which are called *gap junctions* (Gibson et al. 2005; Bennett and Zukin 2004), or they are in touch through a synaptic cleft which allows the passage of electric current in the form of ions. In these connections, the distance between the membranes of the two neurons could be approximately 3.8 nm, that is, 5–10 times smaller in comparison to that of the chemical synapse (Hormuzdi et al. 2004) (Fig. 6.3).

Electrical Synapses

- Gap junctions
- Fast
- "All or nothing"
- They serve simple functions
- Bidirectional transmission of the signal
- Plasticity

The electrical synapses transmit the signal much faster in comparison to the chemical synapses; however, the electrical synaptic transmission is stereotypical and it is activated in an "all-or-nothing" fashion. In these synapses, the gap junctions connect the cytoplasms of the two neurons, and in this way, the fast and direct flow of ions is made possible and to both directions (Purves et al. 2018). However, only relatively simple functions could be served, while the phenomenon of plasticity also appears (Haas et al. 2011; Palacios-Prado et al. 2013). One significant advantage is the easier synchronization in the triggering of neurons (Gibson et al. 2005; Bennett and Zukin 2004).

6.4 Autapses

Autapses are called chemical or electrical synapses which are formed by a neuron *between its own axis and its own dendrites* (Yilmaz et al. 2016; Seung et al. 2000).

They could be *excitatory* and their activity to be related to glutamate or *inhibitory* and to be related to GABA (Ikeda and Bekkers 2006).

They are more frequently found in the neocortex, the substantia nigra, and the hippocampus (Van der Loos and Glaser 1972; Park et al. 1980; Karabelas and Purpura 1980; Cobb et al. 1997). In the neocortex, almost 80% of the pyramidal neurons of layer V has autapses (Lubke et al. 1996).

Their functional significance is not clear; however, they seem to regulate the pattern of neuron discharge. Neurons without autapses seem to manifest variable periods of chaotic and rhythmic discharge as well as periods during which they remain silent. On the contrary, neurons with autapses manifest chaotic behavior (Wang et al. 2014). It seems that the inhibitory excite the silent neurons and the excitatory activate the silent neurons (Qin et al. 2015; Saada et al. 2009), although their capacity to trigger the neuron by themselves alone is limited (Bekkers 2009).

References

Aghajanian GK, VanderMaelen CP (1982) alpha 2-adrenoceptor-mediated hyperpolarization of locus coeruleus neurons: intracellular studies in vivo. Science 215(4538):1394–1396. https://doi.org/10.1126/science.6278591

Ahlquist RP (1948) A study of the adrenotropic receptors. Am J Physiol 153(3):586–600. https://doi.org/10.1152/ajplegacy.1948.153.3.586

Ahmed IS, Chamberlain C, Craven RJ (2012) S2R(Pgrmc1): the cytochrome-related sigma-2 receptor that regulates lipid and drug metabolism and hormone signaling. Expert Opin Drug Metab Toxicol 8(3):361–370. https://doi.org/10.1517/17425255.2012.658367

Amara SG, Arriza JL (1993) Neurotransmitter transporters: three distinct gene families. Curr Opin Neurobiol 3(3):337–344. https://doi.org/10.1016/0959-4388(93)90126-j

Anden NE, Corrodi H, Fuxe K, Hokfelt B, Hokfelt T, Rydin C, Svensson T (1970) Evidence for a central noradrenaline receptor stimulation by clonidine. Life Sci 9(9):513–523. https://doi.org/10.1016/0024-3205(70)90207-9

Anwyl R (1992) Metabotropic glutamate receptors: electrophysiological properties and role in plasticity. Rev Neurosci 3(3):217–231. https://doi.org/10.1515/REVNEURO.1992.3.3.217

Anwyl R (1999) Metabotropic glutamate receptors: electrophysiological properties and role in plasticity. Brain Res Brain Res Rev 29(1):83–120. https://doi.org/10.1016/s0165-0173(98)00050-2

Arneric S, Sullivan J, Decker MW (1994) Neuronal nicotinic acetylcholine receptors: novel targets for CNS therapeutics. In: Bloom F, Kupfer D (eds) Psychopharmacology: the fourth generation of progress. Raven Press, New York, pp 95–110

Arvanitaki A (1942) Effects evoked in an axon by the activity of a contiguous one. J Neurophysiol 5(2):89–108

Asztely F, Gustafsson B (1996) Ionotropic glutamate receptors. Their possible role in the expression of hippocampal synaptic plasticity. Mol Neurobiol 12(1):1–11. https://doi.org/10.1007/BF02740744

Azmitia EC (1999) Serotonin neurons, neuroplasticity, and homeostasis of neural tissue. Neuropsychopharmacology 21(2 Suppl):33S–45S. https://doi.org/10.1016/S0893-133X(99)00022-6

Azmitia E, Gannon P, Kheck N, Whitakerazinitia P (1996) Cellular localization of the 5-HT receptor in primate brain neurons and glial cells. Neuropsychopharmacology 14(1):35–46. https://doi.org/10.1016/s0893-133x(96)80057-1

Barnes NM, Sharp T (1999) A review of central 5-HT receptors and their function. Neuropharmacology 38(8):1083–1152. https://doi.org/10.1016/s0028-3908(99)00010-6

Bekkers JM (2009) Synaptic transmission: excitatory autapses find a function? Curr Biol 19(7):R296–R298. https://doi.org/10.1016/j.cub.2009.02.010

Bennett MV, Zukin RS (2004) Electrical coupling and neuronal synchronization in the mammalian brain. Neuron 41(4):495–511. https://doi.org/10.1016/s0896-6273(04)00043-1

Bettler B, Boulter J, Hermans-Borgmeyer I, O'Shea-Greenfield A, Deneris ES, Moll C, Borgmeyer U, Hollmann M, Heinemann S (1990) Cloning of a novel glutamate receptor subunit, GluR5: expression in the nervous system during development. Neuron 5(5):583–595. https://doi.org/10.1016/0896-6273(90)90213-y

Bettler B, Egebjerg J, Sharma G, Pecht G, Hermans-Borgmeyer I, Moll C, Stevens CF, Heinemann S (1992) Cloning of a putative glutamate receptor: a low affinity kainate-binding subunit. Neuron 8(2):257–265. https://doi.org/10.1016/0896-6273(92)90292-l

Bjorklund A, Hokfelt T, Kuhar MJ (1990) Neuropeptide receptors in the CNS. In: Bjorklund A, Hokfelt T (eds) Handbook of chemical neuroanatomy, vol 9. Elsevier, Amsterdam

Bjorklund A, Hokfelt T, Kuhar MJ (1992) Neuropeptide receptors in the CNS, part II. In: Bjorklund A, Hokfelt T (eds) Handbook of chemical neuroanatomy, vol 11. Elsevier, Amsterdam

Blandina P, Munari L, Provensi G, Passani MB (2012) Histamine neurons in the tuberomammillary nucleus: a whole center or distinct subpopulations? Front Syst Neurosci 6:33. https://doi.org/10.3389/fnsys.2012.00033

Blier P (2001) Pharmacology of rapid-onset antidepressant treatment strategies. J Clin Psychiatry 62(Suppl 15):12–17

Bouthenet ML, Souil E, Martres MP, Sokoloff P, Giros B, Schwartz JC (1991) Localization of dopamine D3 receptor mRNA in the rat brain using in situ hybridization histochemistry: comparison with dopamine D2 receptor mRNA. Brain Res 564(2):203–219. https://doi.org/10.1016/0006-8993(91)91456-b

Boyson SJ, McGonigle P, Molinoff PB (1986) Quantitative autoradiographic localization of the D1 and D2 subtypes of dopamine receptors in rat brain. J Neurosci 6(11):3177–3188

Breslau N, Kilbey MM, Andreski P (1993) Nicotine dependence and major depression. New evidence from a prospective investigation. Arch Gen Psychiatry 50(1):31–35. https://doi.org/10.1001/archpsyc.1993.01820130033006

Briley M, Moret C (1993) Neurobiological mechanisms involved in antidepressant therapies. Clin Neuropharmacol 16(5):387–400. https://doi.org/10.1097/00002826-199310000-00002

Brimson JM, Brown CA, Safrany ST (2011) Antagonists show GTP-sensitive high-affinity binding to the sigma-1 receptor. Br J Pharmacol 164(2b):772–780. https://doi.org/10.1111/j.1476-5381.2011.01417.x

Buller AL, Larson HC, Schneider BE, Beaton JA, Morrisett RA, Monaghan DT (1994) The molecular basis of NMDA receptor subtypes: native receptor diversity is predicted by subunit composition. J Neurosci 14(9):5471–5484

Bunzow JR, Van Tol HH, Grandy DK, Albert P, Salon J, Christie M, Machida CA, Neve KA, Civelli O (1988) Cloning and expression of a rat D2 dopamine receptor cDNA. Nature 336(6201):783–787. https://doi.org/10.1038/336783a0

Bylund DB, Eikenberg DC, Hieble JP, Langer SZ, Lefkowitz RJ, Minneman KP, Molinoff PB, Ruffolo RR Jr, Trendelenburg U (1994) International Union of Pharmacology nomenclature of adrenoceptors. Pharmacol Rev 46(2):121–136

Cardenas L, Tremblay LK, Naranjo CA, Herrmann N, Zack M, Busto UE (2002) Brain reward system activity in major depression and comorbid nicotine dependence. J Pharmacol Exp Ther 302(3):1265–1271. https://doi.org/10.1124/jpet.302.3.1265

Cartmell J, Schoepp DD (2000) Regulation of neurotransmitter release by metabotropic glutamate receptors. J Neurochem 75(3):889–907. https://doi.org/10.1046/j.1471-4159.2000.0750889.x

Cates LN, Roberts AJ, Huitron-Resendiz S, Hedlund PB (2013) Effects of lurasidone in behavioral models of depression. Role of the 5-HT(7) receptor subtype. Neuropharmacology 70:211–217. https://doi.org/10.1016/j.neuropharm.2013.01.023

Cedarbaum JM, Aghajanian GK (1977) Catecholamine receptors on locus coeruleus neurons: pharmacological characterization. Eur J Pharmacol 44(4):375–385. https://doi.org/10.1016/0014-2999(77)90312-0

Chaput Y, de Montigny C, Blier P (1986) Effects of a selective 5-HT reuptake blocker, citalopram, on the sensitivity of 5-HT autoreceptors: electrophysiological studies in the rat brain. Naunyn Schmiedebergs Arch Pharmacol 333(4):342–348. https://doi.org/10.1007/BF00500007

Chaput Y, Lesieur P, de Montigny C (1990) Effects of short-term serotonin depletion on the efficacy of serotonin neurotransmission: electrophysiological studies in the rat central nervous system. Synapse 6(4):328–337. https://doi.org/10.1002/syn.890060404

Cheng F, Zagon IS, Verderame MF, McLaughlin PJ (2007) The opioid growth factor (OGF)-OGF receptor axis uses the p16 pathway to inhibit head and neck cancer. Cancer Res 67(21):10511–10518. https://doi.org/10.1158/0008-5472.CAN-07-1922

Cheng F, McLaughlin PJ, Verderame MF, Zagon IS (2009) The OGF-OGFr axis utilizes the p16INK4a and p21WAF1/CIP1 pathways to restrict normal cell proliferation. Mol Biol Cell 20(1):319–327. https://doi.org/10.1091/mbc.E08-07-0681

Ciliax BJ, Heilman C, Demchyshyn LL, Pristupa ZB, Ince E, Hersch SM, Niznik HB, Levey AI (1995) The dopamine transporter: immunochemical characterization and localization in brain. J Neurosci 15(3 Pt 1):1714–1723

Civelli O (1994) Molecular biology of the dopamine receptor subtypes. In: Bloom F, Kupfer D (eds) Psychopharmacology: the fourth generation of progress. Raven Press, New York, pp 155–161

Cobb SR, Halasy K, Vida I, Nyiri G, Tamas G, Buhl EH, Somogyi P (1997) Synaptic effects of identified interneurons innervating both interneurons and pyramidal cells in the rat hippocampus. Neuroscience 79(3):629–648. https://doi.org/10.1016/s0306-4522(97)00055-9

Collingridge GL, Herron CE, Lester RA (1988) Frequency-dependent N-methyl-D-aspartate receptor-mediated synaptic transmission in rat hippocampus. J Physiol 399:301–312. https://doi.org/10.1113/jphysiol.1988.sp017081

Conn PJ, Pin JP (1997) Pharmacology and functions of metabotropic glutamate receptors. Annu Rev Pharmacol Toxicol 37:205–237. https://doi.org/10.1146/annurev.pharmtox.37.1.205

Contractor A, Mulle C, Swanson GT (2011) Kainate receptors coming of age: milestones of two decades of research. Trends Neurosci 34(3):154–163. https://doi.org/10.1016/j.tins.2010.12.002

Cossart R, Bernard C, Ben-Ari Y (2005) Multiple facets of GABAergic neurons and synapses: multiple fates of GABA signalling in epilepsies. Trends Neurosci 28(2):108–115. https://doi.org/10.1016/j.tins.2004.11.011

Damaskos C, Garmpis N, Karatzas T, Kostakis ID, Nikolidakis L, Kostakis A, Kouraklis G (2014) Nuclear receptors in pancreatic tumor cells. Anticancer Res 34(12):6897–6911

Danysz W, Parsons CG (1998) Glycine and N-methyl-D-aspartate receptors: physiological significance and possible therapeutic applications. Pharmacol Rev 50(4):597–664

De Blasi A, Conn PJ, Pin J, Nicoletti F (2001) Molecular determinants of metabotropic glutamate receptor signaling. Trends Pharmacol Sci 22(3):114–120. https://doi.org/10.1016/s0165-6147(00)01635-7

Dhawan BN, Cesselin F, Raghubir R, Reisine T, Bradley PB, Portoghese PS, Hamon M (1996) International Union of Pharmacology. XII. Classification of opioid receptors. Pharmacol Rev 48(4):567–592

Donahue RN, McLaughlin PJ, Zagon IS (2009) Cell proliferation of human ovarian cancer is regulated by the opioid growth factor-opioid growth factor receptor axis. Am J Physiol Regul Integr Comp Physiol 296(6):R1716–R1725. https://doi.org/10.1152/ajpregu.00075.2009

Drevets WC (2000) Neuroimaging studies of mood disorders. Biol Psychiatry 48(8):813–829. https://doi.org/10.1016/s0006-3223(00)01020-9

Eccles J (1976) From electrical to chemical transmission in the central nervous system. Notes Rec R Soc Lond 30(2):219–230. https://doi.org/10.1098/rsnr.1976.0015

Edwards AC, Maes HH, Pedersen NL, Kendler KS (2011) A population-based twin study of the genetic and environmental relationship of major depression, regular tobacco use and nicotine dependence. Psychol Med 41(2):395–405. https://doi.org/10.1017/S0033291710000589

Egebjerg J, Bettler B, Hermans-Borgmeyer I, Heinemann S (1991) Cloning of a cDNA for a glutamate receptor subunit activated by kainate but not AMPA. Nature 351(6329):745–748. https://doi.org/10.1038/351745a0

Ehlert F, Roeske W, Yamamura H (1994) Molecular biology, pharmacology and brain distribution of subtypes of the muscarinic receptor. In: Bloom F, Kupfer D (eds) Psychopharmacology: the fourth generation of progress. Raven Press, New York, pp 111–124

Emorine LJ, Marullo S, Briend-Sutren MM, Patey G, Tate K, Delavier-Klutchko C, Strosberg AD (1989) Molecular characterization of the human beta 3-adrenergic receptor. Science 245(4922):1118–1121. https://doi.org/10.1126/science.2570461

Enz R, Cutting GR (1998) Molecular composition of GABAC receptors. Vision Res 38(10):1431–1441. https://doi.org/10.1016/s0042-6989(97)00277-0

Eugenin EA, Basilio D, Saez JC, Orellana JA, Raine CS, Bukauskas F, Bennett MV, Berman JW (2012) The role of gap junction channels during physiologic and pathologic conditions of the human central nervous system. J Neuroimmune Pharmacol 7(3):499–518. https://doi.org/10.1007/s11481-012-9352-5

Flatman JA, Schwindt PC, Crill WE, Stafstrom CE (1983) Multiple actions of N-methyl-D-aspartate on cat neocortical neurons in vitro. Brain Res 266(1):169–173. https://doi.org/10.1016/0006-8993(83)91323-9

Fontanilla D, Johannessen M, Hajipour AR, Cozzi NV, Jackson MB, Ruoho AE (2009) The hallucinogen N,N-dimethyltryptamine (DMT) is an endogenous sigma-1 receptor regulator. Science 323(5916):934–937. https://doi.org/10.1126/science.1166127

Frielle T, Collins S, Daniel KW, Caron MG, Lefkowitz RJ, Kobilka BK (1987) Cloning of the cDNA for the human beta 1-adrenergic receptor. Proc Natl Acad Sci U S A 84(22):7920–7924. https://doi.org/10.1073/pnas.84.22.7920

Frielle T, Kobilka B, Lefkowitz RJ, Caron MG (1988) Human beta 1- and beta 2-adrenergic receptors: structurally and functionally related receptors derived from distinct genes. Trends Neurosci 11(7):321–324. https://doi.org/10.1016/0166-2236(88)90095-1

Fu Q, Heath AC, Bucholz KK, Lyons MJ, Tsuang MT, True WR, Eisen SA (2007) Common genetic risk of major depression and nicotine dependence: the contribution of antisocial traits in a United States veteran male twin cohort. Twin Res Hum Genet 10(3):470–478. https://doi.org/10.1375/twin.10.3.470

Gehlert DR, Wamsley JK (1985) Dopamine receptors in the rat brain: quantitative autoradiographic localization using [(3)H]sulpiride. Neurochem Int 7(4):717–723. https://doi.org/10.1016/0197-0186(85)90070-1

Gehlert DR, Gackenheimer SL, Seeman P, Schaus J (1992) Autoradiographic localization of [3H]quinpirole binding to dopamine D2 and D3 receptors in rat brain. Eur J Pharmacol 211(2):189–194. https://doi.org/10.1016/0014-2999(92)90528-c

Gibson JR, Beierlein M, Connors BW (2005) Functional properties of electrical synapses between inhibitory interneurons of neocortical layer 4. J Neurophysiol 93(1):467–480. https://doi.org/10.1152/jn.00520.2004

Gilman AG (1989) The Albert Lasker Medical Awards. G proteins and regulation of adenylyl cyclase. JAMA 262(13):1819–1825

Gilman AG (1990) Regulation of adenylyl cyclase by G proteins. Adv Second Messenger Phosphoprotein Res 24:51–57

Gilman AG (1995) Nobel Lecture. G proteins and regulation of adenylyl cyclase. Biosci Rep 15(2):65–97. https://doi.org/10.1007/BF01200143

Greger IH, Ziff EB, Penn AC (2007) Molecular determinants of AMPA receptor subunit assembly. Trends Neurosci 30(8):407–416. https://doi.org/10.1016/j.tins.2007.06.005

Guitart X, Codony X, Monroy X (2004) Sigma receptors: biology and therapeutic potential. Psychopharmacology 174(3):301–319. https://doi.org/10.1007/s00213-004-1920-9

Haas JS, Zavala B, Landisman CE (2011) Activity-dependent long-term depression of electrical synapses. Science 334(6054):389–393. https://doi.org/10.1126/science.1207502

Hayashi T, Su TP (2001) Regulating ankyrin dynamics: roles of sigma-1 receptors. Proc Natl Acad Sci U S A 98(2):491–496. https://doi.org/10.1073/pnas.021413698

Hayashi T, Su TP (2004) Sigma-1 receptors at galactosylceramide-enriched lipid microdomains regulate oligodendrocyte differentiation. Proc Natl Acad Sci U S A 101(41):14949–14954. https://doi.org/10.1073/pnas.0402890101

Hayashi T, Su TP (2007) Sigma-1 receptor chaperones at the ER-mitochondrion interface regulate Ca(2+) signaling and cell survival. Cell 131(3):596–610. https://doi.org/10.1016/j.cell.2007.08.036

Herb A, Burnashev N, Werner P, Sakmann B, Wisden W, Seeburg PH (1992) The KA-2 subunit of excitatory amino acid receptors shows widespread expression in brain and forms ion channels with distantly related subunits. Neuron 8(4):775–785. https://doi.org/10.1016/0896-6273(92)90098-x

Hieble JP, Ruffolo RRJ (1991) Subclassification of β-adrenoceptors. In: Ruffolo RRJ (ed) Progress in basic and clinical pharmacology, B-adrenoceptors: molecular biology, biochemistry and pharmacology, vol 7. Karger, Basel, pp 1–25

Hieble JP, Bylund DB, Clarke DE, Eikenburg DC, Langer SZ, Lefkowitz RJ, Minneman KP, Ruffolo RR Jr (1995) International Union of Pharmacology. X. Recommendation for nomenclature of alpha 1-adrenoceptors: consensus update. Pharmacol Rev 47(2):267–270

Hille B (1994) Modulation of ion-channel function by G-protein-coupled receptors. Trends Neurosci 17(12):531–536. https://doi.org/10.1016/0166-2236(94)90157-0

Hirai H, Kirsch J, Laube B, Betz H, Kuhse J (1996) The glycine binding site of the N-methyl-D-aspartate receptor subunit NR1: identification of novel determinants of co-agonist potentiation in the extracellular M3-M4 loop region. Proc Natl Acad Sci U S A 93(12):6031–6036. https://doi.org/10.1073/pnas.93.12.6031

Holden JE, Jeong Y, Forrest JM (2005) The endogenous opioid system and clinical pain management. AACN Clin Issues 16(3):291–301. https://doi.org/10.1097/00044067-200507000-00003

Hollmann M, Hartley M, Heinemann S (1991) Ca2+ permeability of KA-AMPA--gated glutamate receptor channels depends on subunit composition. Science 252(5007):851–853. https://doi.org/10.1126/science.1709304

Hollmann M, Boulter J, Maron C, Beasley L, Sullivan J, Pecht G, Heinemann S (1993) Zinc potentiates agonist-induced currents at certain splice variants of the NMDA receptor. Neuron 10(5):943–954. https://doi.org/10.1016/0896-6273(93)90209-a

Hong W, Werling LL (2000) Evidence that the σ1 receptor is not directly coupled to G proteins. Eur J Pharmacol 408(2):117–125. https://doi.org/10.1016/s0014-2999(00)00774-3

Hormuzdi SG, Filippov MA, Mitropoulou G, Monyer H, Bruzzone R (2004) Electrical synapses: a dynamic signaling system that shapes the activity of neuronal networks. Biochim Biophys Acta 1662(1–2):113–137. https://doi.org/10.1016/j.bbamem.2003.10.023

Hoyer D, Martin GR (1996) Classification and nomenclature of 5-HT receptors: a comment on current issues. Behav Brain Res 73(1–2):263–268. https://doi.org/10.1016/0166-4328(96)00109-x
Hoyer D, Martin G (1997) 5-HT receptor classification and nomenclature: towards a harmonization with the human genome. Neuropharmacology 36(4–5):419–428. https://doi.org/10.1016/s0028-3908(97)00036-1
Hoyer D, Clarke DE, Fozard JR, Hartig PR, Martin GR, Mylecharane EJ, Saxena PR, Humphrey PP (1994) International Union of Pharmacology classification of receptors for 5-hydroxytryptamine (serotonin). Pharmacol Rev 46(2):157–203
Huang YH, Bergles DE (2004) Glutamate transporters bring competition to the synapse. Curr Opin Neurobiol 14(3):346–352. https://doi.org/10.1016/j.conb.2004.05.007
Huang YS, Lu HL, Zhang LJ, Wu Z (2014) Sigma-2 receptor ligands and their perspectives in cancer diagnosis and therapy. Med Res Rev 34(3):532–566. https://doi.org/10.1002/med.21297
Hume RI, Dingledine R, Heinemann SF (1991) Identification of a site in glutamate receptor subunits that controls calcium permeability. Science 253(5023):1028–1031. https://doi.org/10.1126/science.1653450
Ikeda K, Bekkers JM (2006) Autapses. Curr Biol 16(9):R308. https://doi.org/10.1016/j.cub.2006.03.085
Jackson DM, Westlind-Danielsson A (1994) Dopamine receptors: molecular biology, biochemistry and behavioural aspects. Pharmacol Ther 64(2):291–370. https://doi.org/10.1016/0163-7258(94)90041-8
Janecka A, Fichna J, Janecki T (2004) Opioid receptors and their ligands. Curr Top Med Chem 4(1):1–17. https://doi.org/10.2174/1568026043451618
Jessell TM, Kandel ER (1993) Synaptic transmission: a bidirectional and self-modifiable form of cell-cell communication. Cell 72(Suppl):1–30. https://doi.org/10.1016/s0092-8674(05)80025-x
Johnson JW, Ascher P (1987) Glycine potentiates the NMDA response in cultured mouse brain neurons. Nature 325(6104):529–531. https://doi.org/10.1038/325529a0
Karabelas AB, Purpura DP (1980) Evidence for autapses in the substantia nigra. Brain Res 200(2):467–473. https://doi.org/10.1016/0006-8993(80)90935-x
Kew JN, Kemp JA (2005) Ionotropic and metabotropic glutamate receptor structure and pharmacology. Psychopharmacology 179(1):4–29. https://doi.org/10.1007/s00213-005-2200-z
Killen JD, Fortmann SP, Schatzberg A, Hayward C, Varady A (2003) Onset of major depression during treatment for nicotine dependence. Addict Behav 28(3):461–470. https://doi.org/10.1016/s0306-4603(01)00266-0
Kilpatrick GJ, Hagan RM, Gale JD (1996) 5-HT3 and 5-HT4 receptors in terminal regions of the mesolimbic system. Behav Brain Res 73(1–2):11–13. https://doi.org/10.1016/0166-4328(96)00063-0
Laezza F, Doherty JJ, Dingledine R (1999) Long-term depression in hippocampal interneurons: joint requirement for pre- and postsynaptic events. Science 285(5432):1411–1414. https://doi.org/10.1126/science.285.5432.1411
Lands AM, Arnold A, McAuliff JP, Luduena FP, Brown TG Jr (1967) Differentiation of receptor systems activated by sympathomimetic amines. Nature 214(5088):597–598. https://doi.org/10.1038/214597a0
Langer SZ (1974) Presynaptic regulation of catecholamine release. Biochem Pharmacol 23(13):1793–1800. https://doi.org/10.1016/0006-2952(74)90187-7
Langer SZ (1997) 25 years since the discovery of presynaptic receptors: present knowledge and future perspectives. Trends Pharmacol Sci 18(3):95–99. https://doi.org/10.1016/s0165-6147(96)01034-6
Langer SZ, Arbilla S (1990) Presynaptic receptors on peripheral noradrenergic neurons. Ann N Y Acad Sci 604:7–16. https://doi.org/10.1111/j.1749-6632.1990.tb31978.x
Laube B, Hirai H, Sturgess M, Betz H, Kuhse J (1997) Molecular determinants of agonist discrimination by NMDA receptor subunits: analysis of the glutamate binding site on the NR2B subunit. Neuron 18(3):493–503. https://doi.org/10.1016/s0896-6273(00)81249-0
Laube B, Kuhse J, Betz H (1998) Evidence for a tetrameric structure of recombinant NMDA receptors. J Neurosci 18(8):2954–2961

Laurie DJ, Bartke I, Schoepfer R, Naujoks K, Seeburg PH (1997) Regional, developmental and interspecies expression of the four NMDAR2 subunits, examined using monoclonal antibodies. Brain Res Mol Brain Res 51(1–2):23–32. https://doi.org/10.1016/s0169-328x(97)00206-4

Lee IT, Chen S, Schetz JA (2008) An unambiguous assay for the cloned human sigma1 receptor reveals high affinity interactions with dopamine D4 receptor selective compounds and a distinct structure-affinity relationship for butyrophenones. Eur J Pharmacol 578(2–3):123–136. https://doi.org/10.1016/j.ejphar.2007.09.020

Leonard BE (2004) Sigma receptors and sigma ligands: background to a pharmacological enigma. Pharmacopsychiatry 37(Suppl 3 (S 3)):S166–S170. https://doi.org/10.1055/s-2004-832674

Levesque D, Diaz J, Pilon C, Martres MP, Giros B, Souil E, Schott D, Morgat JL, Schwartz JC, Sokoloff P (1992) Identification, characterization, and localization of the dopamine D3 receptor in rat brain using 7-[3H]hydroxy-N,N-di-n-propyl-2-aminotetralin. Proc Natl Acad Sci U S A 89(17):8155–8159. https://doi.org/10.1073/pnas.89.17.8155

Levey AI, Hersch SM, Rye DB, Sunahara RK, Niznik HB, Kitt CA, Price DL, Maggio R, Brann MR, Ciliax BJ (1993) Localization of D1 and D2 dopamine receptors in brain with subtype-specific antibodies. Proc Natl Acad Sci U S A 90(19):8861–8865. https://doi.org/10.1073/pnas.90.19.8861

Limbird LE (2004) α2-adrenergic receptors. Primer on the autonomic nervous system. Elsevier, Amsterdam. https://doi.org/10.1016/b978-012589762-4/50013-x

Lubke J, Markram H, Frotscher M, Sakmann B (1996) Frequency and dendritic distribution of autapses established by layer 5 pyramidal neurons in the developing rat neocortex: comparison with synaptic innervation of adjacent neurons of the same class. J Neurosci 16(10):3209–3218. https://doi.org/10.1523/jneurosci.16-10-03209.1996

Lupardus PJ, Wilke RA, Aydar E, Palmer CP, Chen Y, Ruoho AE, Jackson MB (2000) Membrane-delimited coupling between sigma receptors and K+ channels in rat neurohypophysial terminals requires neither G-protein nor ATP. J Physiology 526 Pt 3(3):527–539. https://doi.org/10.1111/j.1469-7793.2000.00527.x

Lyons M, Hitsman B, Xian H, Panizzon MS, Jerskey BA, Santangelo S, Grant MD, Rende R, Eisen S, Eaves L, Tsuang MT (2008) A twin study of smoking, nicotine dependence, and major depression in men. Nicotine Tob Res 10(1):97–108. https://doi.org/10.1080/14622200701705332

Macdermott A (1987) Transmitters and receptors: excitatory amino acid transmission. Science 237(4821):1517–1518. https://doi.org/10.1126/science.237.4821.1517-a

MacDonald JF, Porietis AV, Wojtowicz JM (1982) L-Aspartic acid induces a region of negative slope conductance in the current-voltage relationship of cultured spinal cord neurons. Brain Res 237(1):248–253. https://doi.org/10.1016/0006-8993(82)90575-3

Malendowicz LK, Rebuffat P, Tortorella C, Nussdorfer GG, Ziolkowska A, Hochol A (2005) Effects of met-enkephalin on cell proliferation in different models of adrenocortical-cell growth. Int J Mol Med 15(5):841–845. https://doi.org/10.3892/ijmm.15.5.841

Mansur A, Watson S (1994) Dopamine receptor expression in the central nervous system. In: Bloom F, Kupfer D (eds) Psychopharmacology: the fourth generation of progress. Raven Press, New York, pp 207–220

Mayer ML (2005) Glutamate receptor ion channels. Curr Opin Neurobiol 15(3):282–288. https://doi.org/10.1016/j.conb.2005.05.004

Mayer ML, Westbrook GL, Guthrie PB (1984) Voltage-dependent block by Mg2+ of NMDA responses in spinal cord neurons. Nature 309(5965):261–263. https://doi.org/10.1038/309261a0

McEwen BS (1999) Stress and hippocampal plasticity. Annu Rev Neurosci 22(1):105–122. https://doi.org/10.1146/annurev.neuro.22.1.105

McLaughlin PJ, Levin RJ, Zagon IS (1999) Regulation of human head and neck squamous cell carcinoma growth in tissue culture by opioid growth factor. Int J Oncol 14(5):991–998. https://doi.org/10.3892/ijo.14.5.991

Meador-Woodruff JH, Mansour A, Bunzow JR, Van Tol HH, Watson SJ Jr, Civelli O (1989) Distribution of D2 dopamine receptor mRNA in rat brain. Proc Natl Acad Sci U S A 86(19):7625–7628. https://doi.org/10.1073/pnas.86.19.7625

Miller RJ (1998) Presynaptic receptors. Annu Rev Pharmacol Toxicol 38(1):201–227. https://doi.org/10.1146/annurev.pharmtox.38.1.201

Monaghan DT, Jane DE (2009) Pharmacology of NMDA receptors. In: Van Dongen AM (ed) Biology of the NMDA receptor. Frontiers in neuroscience. CRC Press, Boca Raton, pp 257–283

Monaghan DT, Bridges RJ, Cotman CW (1989) The excitatory amino acid receptors: their classes, pharmacology, and distinct properties in the function of the central nervous system. Annu Rev Pharmacol Toxicol 29:365–402. https://doi.org/10.1146/annurev.pa.29.040189.002053

Monassier L, Bousquet P (2002) Sigma receptors: from discovery to highlights of their implications in the cardiovascular system. Fundam Clin Pharmacol 16(1):1–8. https://doi.org/10.1046/j.1472-8206.2002.00063.x

Montgomery SA, Fineberg N (1989) Is there a relationship between serotonin receptor subtypes and selectivity of response in specific psychiatric illnesses? Br J Psychiatry Suppl 8:63–69

Mony L, Kew JN, Gunthorpe MJ, Paoletti P (2009) Allosteric modulators of NR2B-containing NMDA receptors: molecular mechanisms and therapeutic potential. Br J Pharmacol 157(8):1301–1317. https://doi.org/10.1111/j.1476-5381.2009.00304.x

Monyer H, Sprengel R, Schoepfer R, Herb A, Higuchi M, Lomeli H, Burnashev N, Sakmann B, Seeburg PH (1992) Heteromeric NMDA receptors: molecular and functional distinction of subtypes. Science 256(5060):1217–1221. https://doi.org/10.1126/science.256.5060.1217

Monyer H, Burnashev N, Laurie DJ, Sakmann B, Seeburg PH (1994) Developmental and regional expression in the rat brain and functional properties of four NMDA receptors. Neuron 12(3):529–540. https://doi.org/10.1016/0896-6273(94)90210-0

Mrzljak L, Bergson C, Pappy M, Huff R, Levenson R, Goldman-Rakic PS (1996) Localization of dopamine D4 receptors in GABAergic neurons of the primate brain. Nature 381(6579):245–248. https://doi.org/10.1038/381245a0

Nakanishi S (1992) Molecular diversity of glutamate receptors and implications for brain function. Science 258(5082):597–603. https://doi.org/10.1126/science.1329206

Narayanan S, Bhat R, Mesangeau C, Poupaert JH, McCurdy CR (2011) Early development of sigma-receptor ligands. Future Med Chem 3(1):79–94. https://doi.org/10.4155/fmc.10.279

Nelson N, Lill H (1994) Porters and neurotransmitter transporters. J Exp Biol 196:213–228

Nicholas AP, Pieribone V, Hokfelt T (1993a) Distributions of mRNAs for alpha-2 adrenergic receptor subtypes in rat brain: an in situ hybridization study. J Comp Neurol 328(4):575–594. https://doi.org/10.1002/cne.903280409

Nicholas AP, Pieribone VA, Hokfelt T (1993b) Cellular localization of messenger RNA for beta-1 and beta-2 adrenergic receptors in rat brain: an in situ hybridization study. Neuroscience 56(4):1023–1039. https://doi.org/10.1016/0306-4522(93)90148-9

Nicoll RA, Malenka RC, Kauer JA (1990) Functional comparison of neurotransmitter receptor subtypes in mammalian central nervous system. Physiol Rev 70(2):513–565. https://doi.org/10.1152/physrev.1990.70.2.513

Nowak L, Bregestovski P, Ascher P, Herbet A, Prochiantz A (1984) Magnesium gates glutamate-activated channels in mouse central neurons. Nature 307(5950):462–465. https://doi.org/10.1038/307462a0

O'Dowd BF, Hnatowich M, Regan JW, Leader WM, Caron MG, Lefkowitz RJ (1988) Site-directed mutagenesis of the cytoplasmic domains of the human beta 2-adrenergic receptor. Localization of regions involved in G protein-receptor coupling. J Biol Chem 263(31):15985–15992

O'Dowd BF, Lefkowitz RJ, Caron MG (1989) Structure of the adrenergic and related receptors. Annu Rev Neurosci 12(1):67–83. https://doi.org/10.1146/annurev.ne.12.030189.000435

O'Malley KL, Harmon S, Tang L, Todd RD (1992) The rat dopamine D4 receptor: sequence, gene structure, and demonstration of expression in the cardiovascular system. New Biol 4(2):137–146

Olsen RW, Sieghart W (2009) GABA A receptors: subtypes provide diversity of function and pharmacology. Neuropharmacology 56(1):141–148. https://doi.org/10.1016/j.neuropharm.2008.07.045

Ostrowski J, Kjelsberg MA, Caron MG, Lefkowitz RJ (1992) Mutagenesis of the beta 2-adrenergic receptor: how structure elucidates function. Annu Rev Pharmacol Toxicol 32:167–183. https://doi.org/10.1146/annurev.pa.32.040192.001123

Palacios J, Kuhar MJ (1982) Beta adrenergic receptor localization in rat brain by light microscopic autoradiography. Neurochem Int 4(6):473–490. https://doi.org/10.1016/0197-0186(82)90036-5

Palacios-Prado N, Hoge G, Marandykina A, Rimkute L, Chapuis S, Paulauskas N, Skeberdis VA, O'Brien J, Pereda AE, Bennett MV, Bukauskas FF (2013) Intracellular magnesium-dependent modulation of gap junction channels formed by neuronal connexin36. J Neurosci 33(11):4741–4753. https://doi.org/10.1523/JNEUROSCI.2825-12.2013

Panula P, Chazot PL, Cowart M, Gutzmer R, Leurs R, Liu WL, Stark H, Thurmond RL, Haas HL (2015) International Union of Basic and Clinical Pharmacology. XCVIII. Histamine receptors. Pharmacol Rev 67(3):601–655. https://doi.org/10.1124/pr.114.010249

Park MR, Lighthall JW, Kitai ST (1980) Recurrent inhibition in the rat neostriatum. Brain Res 194(2):359–369. https://doi.org/10.1016/0006-8993(80)91217-2

Patel S, Patel U, Vithalani D, Verma SC (1981) Regulation of catecholamine release by presynaptic receptor system. Gen Pharmacol 12(6):405–422. https://doi.org/10.1016/0306-3623(81)90063-x

Pereda AE, Curti S, Hoge G, Cachope R, Flores CE, Rash JE (2013) Gap junction-mediated electrical transmission: regulatory mechanisms and plasticity. Biochim Biophys Acta 1828(1):134–146. https://doi.org/10.1016/j.bbamem.2012.05.026

Peroutka SJ, Snyder SH (1979) Multiple serotonin receptors: differential binding of [3H]5-hydroxytryptamine, [3H]lysergic acid diethylamide and [3H]spiroperidol. Mol Pharmacol 16(3):687–699

Pin JP, Acher F (2002) The metabotropic glutamate receptors: structure, activation mechanism and pharmacology. Curr Drug Targets CNS Neurol Disorders 1(3):297–317. https://doi.org/10.2174/1568007023339328

Pin JP, Duvoisin R (1995) The metabotropic glutamate receptors: structure and functions. Neuropharmacology 34(1):1–26. https://doi.org/10.1016/0028-3908(94)00129-g

Pineyro G, Blier P (1999) Autoregulation of serotonin neurons: role in antidepressant drug action. Pharmacol Rev 51(3):533–591

Purves D, Augustine G, Fitzpatrick D, Hall W, LaMantia A, Mooney R, White L, Platt M (2018) Neuroscience, 6th edn. Oxford University Press, Oxford

Qin H, Wu Y, Wang C, Ma J (2015) Emitting waves from defects in network with autapses. Commun Nonlinear Sci Numer Simul 23(1–3):164–174. https://doi.org/10.1016/j.cnsns.2014.11.008

Rainbow TC, Parsons B, Wolfe BB (1984) Quantitative autoradiography of beta 1- and beta 2-adrenergic receptors in rat brain. Proc Natl Acad Sci U S A 81(5):1585–1589. https://doi.org/10.1073/pnas.81.5.1585

Rosin DL, Talley EM, Lee A, Stornetta RL, Gaylinn BD, Guyenet PG, Lynch KR (1996) Distribution of alpha 2C-adrenergic receptor-like immunoreactivity in the rat central nervous system. J Comp Neurol 372(1):135–165. https://doi.org/10.1002/(SICI)1096-9861(19960812)372:1<135::AID-CNE9>3.0.CO;2-4

Ross CA, Pearlson GD (1996) Schizophrenia, the heteromodal association neocortex and development: potential for a neurogenetic approach. Trends Neurosci 19(5):171–176. https://doi.org/10.1016/s0166-2236(96)10022-9

Roth R, Elsworth J (1994) Biochemical pharmacology of midbrain dopamine neurons. In: Bloom F, Kupfer D (eds) Psychopharmacology: the fourth generation of progress. Raven Press, New York, pp 227–243

Saada R, Miller N, Hurwitz I, Susswein AJ (2009) Autaptic excitation elicits persistent activity and a plateau potential in a neuron of known behavioral function. Curr Biol 19(6):479–484. https://doi.org/10.1016/j.cub.2009.01.060

Sakimura K, Morita T, Kushiya E, Mishina M (1992) Primary structure and expression of the gamma 2 subunit of the glutamate receptor channel selective for kainate. Neuron 8(2):267–274. https://doi.org/10.1016/0896-6273(92)90293-m

Salm AK, McCarthy KD (1992) The evidence for astrocytes as a target for central noradrenergic activity: expression of adrenergic receptors. Brain Res Bull 29(3–4):265–275. https://doi.org/10.1016/0361-9230(92)90056-4

Sassani JW, Zagon IS, McLaughlin PJ (2003) Opioid growth factor modulation of corneal epithelium: uppers and downers. Curr Eye Res 26(5):249–262. https://doi.org/10.1076/ceyr.26.4.249.15427

Scheinin M, Lomasney JW, Hayden-Hixson DM, Schambra UB, Caron MG, Lefkowitz RJ, Fremeau RT Jr (1994) Distribution of alpha 2-adrenergic receptor subtype gene expression in rat brain. Brain Res Mol Brain Res 21(1–2):133–149. https://doi.org/10.1016/0169-328x(94)90386-7

Schwartz JC, Diaz J, Pilon C, Sokoloff P (2000) Possible implications of the dopamine D(3) receptor in schizophrenia and in antipsychotic drug actions. Brain Res Brain Res Rev 31(2–3):277–287. https://doi.org/10.1016/s0165-0173(99)00043-0

Seung HS, Lee DD, Reis BY, Tank DW (2000) The autapse: a simple illustration of short-term analog memory storage by tuned synaptic feedback. J Comput Neurosci 9(2):171–185. https://doi.org/10.1023/a:1008971908649

Shorr RG, Lefkowitz RJ, Caron MG (1981) Purification of the beta-adrenergic receptor. Identification of the hormone binding subunit. J Biol Chem 256(11):5820–5826

Shorr RG, Strohsacker MW, Lavin TN, Lefkowitz RJ, Caron MG (1982) The beta 1-adrenergic receptor of the turkey erythrocyte. Molecular heterogeneity revealed by purification and photoaffinity labeling. J Biol Chem 257(20):12341–12350

Skuza G (2012) Pharmacology of sigma (sigma) receptor ligands from a behavioral perspective. Curr Pharm Des 18(7):863–874. https://doi.org/10.2174/138161212799436458

Skuza G, Rogoz Z (2002) A potential antidepressant activity of SA4503, a selective sigma 1 receptor agonist. Behav Pharmacol 13(7):537–543. https://doi.org/10.1097/00008877-200211000-00003

Skuza G, Rogoz Z (2006) The synergistic effect of selective sigma receptor agonists and uncompetitive NMDA receptor antagonists in the forced swim test in rats. J Physiol Pharmacol 57(2):217–229

Sokoloff P, Giros B, Martres MP, Bouthenet ML, Schwartz JC (1990) Molecular cloning and characterization of a novel dopamine receptor (D3) as a target for neuroleptics. Nature 347(6289):146–151. https://doi.org/10.1038/347146a0

Sprouse JS, Aghajanian GK (1988) Responses of hippocampal pyramidal cells to putative serotonin 5-HT1A and 5-HT1B agonists: a comparative study with dorsal raphe neurons. Neuropharmacology 27(7):707–715. https://doi.org/10.1016/0028-3908(88)90079-2

Starke K (1971) Influence of -receptor stimulants on noradrenaline release. Die Naturwissenschaften 58(8):420. https://doi.org/10.1007/BF00591535

Stjärne L (1989) Basic mechanisms and local modulation of nerve impulse-induced secretion of neurotransmitters from individual sympathetic nerve varicosities. Rev Physiol Biochem Pharmacol 112:1–137. https://doi.org/10.1007/bfb0027496

Stone EA, Ariano MA (1989) Are glial cells targets of the central noradrenergic system? A review of the evidence. Brain Res Brain Res Rev 14(4):297–309. https://doi.org/10.1016/0165-0173(89)90015-5

Su TP, Hayashi T (2003) Understanding the molecular mechanism of sigma-1 receptors: towards a hypothesis that sigma-1 receptors are intracellular amplifiers for signal transduction. Curr Med Chem 10(20):2073–2080. https://doi.org/10.2174/0929867033456783

Sugihara H, Moriyoshi K, Ishii T, Masu M, Nakanishi S (1992) Structures and properties of seven isoforms of the NMDA receptor generated by alternative splicing. Biochem Biophys Res Commun 185(3):826–832. https://doi.org/10.1016/0006-291x(92)91701-q

Svensson TH, Bunney BS, Aghajanian GK (1975) Inhibition of both noradrenergic and serotonergic neurons in brain by the alpha-adrenergic agonist clonidine. Brain Res 92(2):291–306. https://doi.org/10.1016/0006-8993(75)90276-0

Takemori AE, Portoghese PS (1985) Affinity labels for opioid receptors. Annu Rev Pharmacol Toxicol 25(1):193–223. https://doi.org/10.1146/annurev.pa.25.040185.001205

Talley EM, Rosin DL, Lee A, Guyenet PG, Lynch KR (1996) Distribution of alpha 2A-adrenergic receptor-like immunoreactivity in the rat central nervous system. J Comp Neurol 372(1):111–134. https://doi.org/10.1002/(SICI)1096-9861(19960812)372:1<111::AID-CNE8>3.0.CO;2-6

Tanabe Y, Nomura A, Masu M, Shigemoto R, Mizuno N, Nakanishi S (1993) Signal transduction, pharmacological properties, and expression patterns of two rat metabotropic glutamate receptors, mGluR3 and mGluR4. J Neurosci 13(4):1372–1378

Tanaka C, Nishizuka Y (1994) The protein kinase C family for neuronal signaling. Annu Rev Neurosci 17:551–567. https://doi.org/10.1146/annurev.ne.17.030194.003003

Tansey EM (1997) Not committing barbarisms: Sherrington and the synapse, 1897. Brain Res Bull 44(3):211–212. https://doi.org/10.1016/s0361-9230(97)00312-2

Tholanikunnel BG, Raymond JR, Malbon CC (1999) Analysis of the AU-rich elements in the 3′-untranslated region of beta 2-adrenergic receptor mRNA by mutagenesis and identification of the homologous AU-rich region from different species. Biochemistry 38(47):15564–15572. https://doi.org/10.1021/bi9913348

Tsumoto T (1990) Excitatory amino acid transmitters and their receptors in neural circuits of the cerebral neocortex. Neurosci Res 9(2):79–102. https://doi.org/10.1016/0168-0102(90)90025-a

Unwin N (1993) Neurotransmitter action: opening of ligand-gated ion channels. Cell 72(Suppl):31–41. https://doi.org/10.1016/s0092-8674(05)80026-1

Van der Loos H, Glaser EM (1972) Autapses in neocortex cerebri: synapses between a pyramidal cell's axon and its own dendrites. Brain Res 48:355–360. https://doi.org/10.1016/0006-8993(72)90189-8

Van Tol HH, Bunzow JR, Guan HC, Sunahara RK, Seeman P, Niznik HB, Civelli O (1991) Cloning of the gene for a human dopamine D4 receptor with high affinity for the antipsychotic clozapine. Nature 350(6319):610–614. https://doi.org/10.1038/350610a0

van Waarde A, Rybczynska AA, Ramakrishnan NK, Ishiwata K, Elsinga PH, Dierckx RA (2015) Potential applications for sigma receptor ligands in cancer diagnosis and therapy. Biochim Biophys Acta 1848(10 Pt B):2703–2714. https://doi.org/10.1016/j.bbamem.2014.08.022

Venter JC, Fraser CM, Kerlavage AR, Buck MA (1989) Molecular biology of adrenergic and muscarinic cholinergic receptors. A perspective. Biochem Pharmacol 38(8):1197–1208. https://doi.org/10.1016/0006-2952(89)90325-0

Verdoorn TA, Burnashev N, Monyer H, Seeburg PH, Sakmann B (1991) Structural determinants of ion flow through recombinant glutamate receptor channels. Science 252(5013):1715–1718. https://doi.org/10.1126/science.1710829

Waldhoer M, Bartlett SE, Whistler JL (2004) Opioid receptors. Annu Rev Biochem 73(1):953–990. https://doi.org/10.1146/annurev.biochem.73.011303.073940

Wamsley JK, Alburges ME, Hunt MA, Bylund DB (1992) Differential localization of alpha 2-adrenergic receptor subtypes in brain. Pharmacol Biochem Behav 41(2):267–273. https://doi.org/10.1016/0091-3057(92)90097-y

Wang J, Mack AL, Coop A, Matsumoto RR (2007) Novel sigma (sigma) receptor agonists produce antidepressant-like effects in mice. Eur Neuropsychopharmacol 17(11):708–716. https://doi.org/10.1016/j.euroneuro.2007.02.007

Wang H, Ma J, Chen Y, Chen Y (2014) Effect of an autapse on the firing pattern transition in a bursting neuron. Commun Nonlinear Sci Numer Simul 19(9):3242–3254. https://doi.org/10.1016/j.cnsns.2014.02.018

Watkins JC (1981) Pharmacology of excitatory amino acid transmitters. Adv Biochem Psychopharmacol 29:205–212

Watkins JC, Evans RH (1981) Excitatory amino acid transmitters. Annu Rev Pharmacol Toxicol 21:165–204. https://doi.org/10.1146/annurev.pa.21.040181.001121

Weiner DM, Brann MR (1989) The distribution of a dopamine D2 receptor mRNA in rat brain. FEBS Lett 253(1–2):207–213. https://doi.org/10.1016/0014-5793(89)80960-3

Weissman AD, Su TP, Hedreen JC, London ED (1988) Sigma receptors in post-mortem human brains. J Pharmacol Exp Ther 247(1):29–33

Werner P, Voigt M, Keinanen K, Wisden W, Seeburg PH (1991) Cloning of a putative high-affinity kainate receptor expressed predominantly in hippocampal CA3 cells. Nature 351(6329):742–744. https://doi.org/10.1038/351742a0

Whitaker-Azmitia PM, Azmitia EC (1989) Stimulation of astroglial serotonin receptors produces culture media which regulates growth of serotonergic neurons. Brain Res 497(1):80–85. https://doi.org/10.1016/0006-8993(89)90972-4

Whitaker-Azmitia PM, Azmitia EC (1991) Serotonin trophic factors in development, plasticity and aging. serotonin: molecular biology, receptors and functional effects. Birkhäuser, Basel. https://doi.org/10.1007/978-3-0348-7259-1_5

Whitaker-Azmitia PM, Clarke C, Azmitia EC (1993) Localization of 5-HT1A receptors to astroglial cells in adult rats: implications for neuronal-glial interactions and psychoactive drug mechanism of action. Synapse 14(3):201–205. https://doi.org/10.1002/syn.890140303

Williams K (1993) Ifenprodil discriminates subtypes of the N-methyl-D-aspartate receptor: selectivity and mechanisms at recombinant heteromeric receptors. Mol Pharmacol 44(4):851–859

Wouters MM, Vicario M, Santos J (2016) The role of mast cells in functional GI disorders. Gut 65(1):155–168. https://doi.org/10.1136/gutjnl-2015-309151

Wu LJ, Zhuo M (2009) Targeting the NMDA receptor subunit NR2B for the treatment of neuropathic pain. Neurotherapeutics 6(4):693–702. https://doi.org/10.1016/j.nurt.2009.07.008

Wu Y, McLaughlin PJ, Zagon IS (1998) Ontogeny of the opioid growth factor, [Met5]-enkephalin, preproenkephalin gene expression, and the ζ opioid receptor in the developing and adult aorta of rat. Dev Dyn 211(4):327–337. https://doi.org/10.1002/(sici)1097-0177(199804)211:4<327::Aid-aja4>3.0.Co;2-j

Yamakura T, Shimoji K (1999) Subunit- and site-specific pharmacology of the NMDA receptor channel. Prog Neurobiol 59(3):279–298. https://doi.org/10.1016/s0301-0082(99)00007-6

Yang S, Bhardwaj A, Cheng J, Alkayed NJ, Hurn PD, Kirsch JR (2007) Sigma receptor agonists provide neuroprotection in vitro by preserving bcl-2. Anesth Analg 104(5):1179–1184. https://doi.org/10.1213/01.ane.0000260267.71185.73, tables of contents

Yilmaz E, Ozer M, Baysal V, Perc M (2016) Autapse-induced multiple coherence resonance in single neurons and neuronal networks. Sci Rep 6:30914. https://doi.org/10.1038/srep30914

Yung KK, Bolam JP, Smith AD, Hersch SM, Ciliax BJ, Levey AI (1995) Immunocytochemical localization of D1 and D2 dopamine receptors in the basal ganglia of the rat: light and electron microscopy. Neuroscience 65(3):709–730. https://doi.org/10.1016/0306-4522(94)00536-e

Zagon IS, Smith JP, McLaughlin PJ (1999a) Human pancreatic cancer cell proliferation in tissue culture is tonically inhibited by opioid growth factor. Int J Oncol 14(3):577–584. https://doi.org/10.3892/ijo.14.3.577

Zagon IS, Wu Y, McLaughlin PJ (1999b) Opioid growth factor and organ development in rat and human embryos. Brain Res 839(2):313–322. https://doi.org/10.1016/s0006-8993(99)01753-9

Zagon IS, Verderame MF, Allen SS, McLaughlin PJ (2000) Cloning, sequencing, chromosomal location, and function of cDNAs encoding an opioid growth factor receptor (OGFr) in humans. Brain Res 856(1–2):75–83. https://doi.org/10.1016/s0006-8993(99)02330-6

Zagon IS, Verderame MF, McLaughlin PJ (2002) The biology of the opioid growth factor receptor (OGFr). Brain Res Brain Res Rev 38(3):351–376. https://doi.org/10.1016/s0165-0173(01)00160-6

Zhang H, Cuevas J (2005) sigma Receptor activation blocks potassium channels and depresses neuroexcitability in rat intracardiac neurons. J Pharmacol Exp Ther 313(3):1387–1396. https://doi.org/10.1124/jpet.105.084152

Zorumski CF, Isenberg KE (1991) Insights into the structure and function of GABA-benzodiazepine receptors: ion channels and psychiatry. Am J Psychiatry 148(2):162–173. https://doi.org/10.1176/ajp.148.2.162

Functional Significance of Neurotransmitter Systems

7

Contents

K. N. Fountoulakis, *The Human Connectome*,
https://doi.org/10.1007/978-3-031-10351-3_7

The basic role of neurotransmitters is the communication among neurons, and each postsynaptic neuron "senses" the changes of the neurotransmitter concentration in the synaptic cleft, as a result of the changes in their release from the presynaptic neuron. The exposure of the postsynaptic neuron to prolonged stimulation leads to adaptation with desensitization and adjustment of sensitivity to recognize signals in more intense stimuli. The most frequent mechanisms of such an adaptation is *down-regulation* and *receptor sequestration* through endocytosis. On the contrary, the reduced exposure to stimuli leads to *upregulation* of the receptors and increase in the number of active receptors in the synaptic cleft. Overall, it seems that the most important function of all the major neurotransmitters is the regulation of glutaminergic activity.

7.1 The Functional Role of the Cholinergic System

The main function of the cholinergic system is to make non-cholinergic neurons more sensitive to excitatory stimuli. There is a large bulk of pharmacological data which suggest that the central cholinergic system plays a pivotal role in the acquisition as well as the manifestation of learned behaviors. In this frame, it seems that scopolamine and atropine exert a negative effect on this kind of behaviors (Reiner and Fibiger 1994).

The Functional Role of the Cholinergic System

- Makes non-cholinergic neurons more sensitive to excitatory stimuli.
- Acquisition and manifestation of learned behaviors.
- Nicotinic receptors.
  - Retina.
  - Hippocampus.
  - Respiratory center of the brain stem.
  - Cortex.
  - Thalamus.
  - Hypothalamus.
  - Substantia nigra.
  - Striatum.
  - Locus coeruleus.
  - There is no proven endogenous action.

- Muscarinic receptors.
  - Widely distributed in the human body.
  - Decrease of heart rate.
  - Vasodilation.
  - Airways contraction.
  - Increased mobility of gastroenterological tract.
  - Miosis.
  - Memory.
  - Learning.
  - Control of body posture.

Cholinergic action is exerted through two groups of cholinergic receptors, the nicotinic and the muscarinic receptors:

- Concerning *nicotinic receptors*, the only proven neurotransmitter-like activity in the brain concerns the communication between a motor neuron and a Renshaw cell in the spinal cord. However, a nicotinic type of activity has been found in the retina, the hippocampus, the respiratory center of the brain stem, the cortex, the thalamus, the substantia nigra, the striatum, and the locus coeruleus. Nicotine interacts with nicotinic receptors and increases the release of other neurotransmitters and in general regulates and controls various important functions, including attention and concentration, regional blood flow in the brain, glucose consumption, and electrical activity of the cortex. Each one of these functions is enhanced by nicotine and is reduced by its antagonists. Additionally, nicotine could improve memory and learning. It also possesses anxiolytic properties, probable through the release of endogenous agents which act on GABA receptors. Nicotine-related disorders are related to depression, but the nature of this relationship is unknown (Edwards et al. 2011; Lyons et al. 2008; Fu et al. 2007; Killen et al. 2003; Cardenas et al. 2002; Breslau et al. 1993).
- On the other hand, *muscarinic receptors* (Ehlert et al. 1994) are widely distributed in the human body (Fig. 7.1). Their activation in the peripheral body causes, among others, a decrease in heart rate, vasodilation, airway contraction, increase in the mobility of the gastrointestinal tract, and miosis of the eye pupil. In the brain, they are related to various functions, including memory, learning, and the control of body posture.

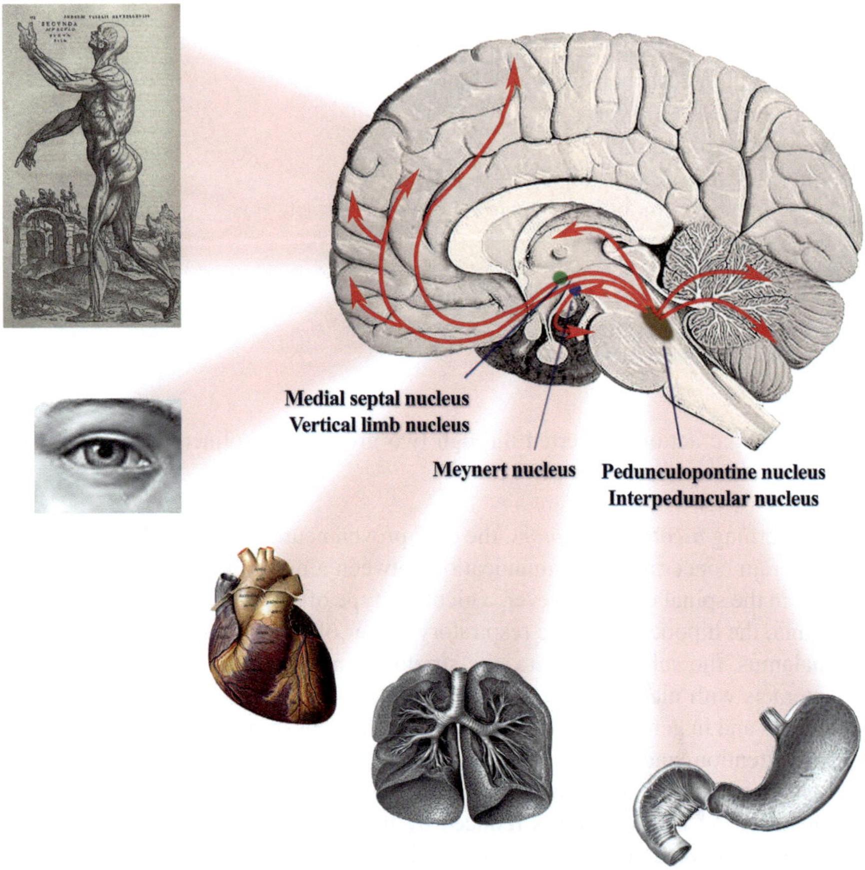

Fig. 7.1 Representation of muscarinic cholinergic receptors and their role in movement, vision, and the cardiovascular and gastrointestinal systems (copyright KN Fountoulakis 2021)

7.2 The Functional Role of the Serotonergic System

7.2.1 Pathways and Information Input into the Serotonin System

The *afferent input* into the raphe nuclei comes mainly from the hypothalamic areas, the substantia nigra, the locus coeruleus (LC), the ventral tegmental area (VTA), the prefrontal cortex (PFC), the amygdala, and the nucleus accumbens (NAC). The neurotransmitters involved include the GABA, excitatory amino acids, norepinephrine, substance P, and vasointestinal polypeptide. Of significant importance is the excitatory norepinephric input in the DRN which is mediated by postsynaptic alpha-1 adrenergic receptors (Svensson 2000; Svensson et al. 1975). This input is tonic, and when the noradrenergic input is inhibited after administration of either clonidine, which suppresses the activity of NA cells, or by blocking alpha-1 receptors, the activity of midbrain 5-HT neurons is inhibited (Aghajanian et al. 1968; Baraban and Aghajanian 1980; Svensson et al. 1975; Aghajanian and Vandermaelen 1982).

Afferent input into the serotonin system

Comes from

- Hypothalamus.
- Substantia nigra.
- Locus coeruleus.
- Ventral tegmental area.
- Prefrontal cortex.
- Amygdala.
- Nucleus accumbens.
- GABA.
- Excitatory amino acids.
- Norepinephrine.
- Substance P.
- Vasointestinal polypeptide.

7.2.2 Regulation of Serotonin Activity

- *Autoregulation* of serotonin release is achieved through somatodendritic 5-HT_{1A} and synaptic terminal 5-HT_{1B} and 5-HT_{1D} autoreceptors in the synaptic cleft (Crespi et al. 1990; Pineyro and Blier 1999).
- *Heteroregulation* of serotonin release in the DRN and the MRN occurs via inhibitory action of GABA and GABA agonists (Becquet et al. 1990). Heteroregulation in the terminal projection areas is mediated by alpha-2 receptors (Martire et al. 1989; Mongeau et al. 1993; Raiteri et al. 1990).
- The *facilitation of release* action is based on the activity of substance P and direct and indirect dopamine agonists (Ferre et al. 1994; Reisine et al. 1982), as well as with histaminergic H3 receptors (Fink et al. 1990; Summers and Giacobini 1995).

7.2.3 Projections of Serotonergic Pathways

Virtually all brain regions receive *serotonergic input* (Pineyro and Blier 1999) (Fig. 5.8), and this suggests a wide role for 5-HT in the brain (Heinz et al. 2001; Jacobs and Fornal 1995; Lanctot et al. 2001; Shiah and Yatham 2000; Rajkowska 2000). Particularly high concentrations of 5-HT can be found in the *limbic system* and imply its relationship with emotions. However, all the projections of serotonergic neurons stem from the dorsal and median raphe nuclei. The median provides with thick fibers while the dorsal with more thin, which are probably more sensitive to toxic agents:

- The cells in the *dorsal raphe nucleus (DRN)* project into the same regions of the brain in which the dopaminergic neurons of the *ventral tegmental area (VTA)* project and especially to the basal ganglia, the amygdala, and the nucleus accumbens. They are implicated mainly in the regulation of locomotor and incentive-motivation activity. Mainly, these neurons are related to the motor component of the mechanisms of response, and they function mainly through 5-HT2, 5-HT2A, and 5-HT2D receptors (Blue et al. 1988; Mamounas et al. 1991). The activity of 5-HT cells in the DRN is at its highest during active waking but becomes slower and more regular during quiet waking (Jacobs and Azmitia 1992). During slow-wave sleep (SWS), the 5-HT cell activity becomes even slower and also irregular. On the contrary, during REM sleep, the serotonergic neurons in the DRN are completely silent (Portas and McCarley 1994).
- On the contrary, the cells of the *median raphe nucleus (MRN)* project to the cortex, the thalamus, and the hippocampus, and they contribute to the process of sensory stimuli. Thus, the first group of neurons is related to the motor part of the response mechanisms and acts mainly through the 5-HT_2, 5-HT_{2A}, and 5-HT_{2D} receptors, while the second group is related to the sensory part and acts possibly through the 5-HT_{1A} receptors (Blue et al. 1988; Mamounas et al. 1991).

Projections of Serotonergic Pathways

- Virtually all brain regions receive serotonergic input.
  - Limbic system.
- The cells in the dorsal raphe nucleus (DRN) project into.
  - Basal ganglia.
  - Amygdala.
  - Nucleus accumbens.
  - Regulation of motor activity.
  - Incentive-motivation activity.
- *Τα κύτταρα του μέσου πυρήνα της ραφής, προβάλουν στον.*
  - Cortex.
  - Thalamus.
  - Hippocampus.
  - Process of sensory stimuli.
- Main action: To inhibit behaviors and to reduce impulsivity.
  - Sleep.
  - Food intake.
  - Thermoregulation.
  - Emotions.
  - Psychotic experiences.
  - Regulates approach-avoidance behaviors.
  - Attenuates the impact of stress.

It is widely accepted that these pathways also regulate the approach-avoidance behaviors (Graeff et al. 1993; Hodges et al. 1987; LeDoux et al. 1990). Another function of serotonin is regulation of sensory functions. This conclusion comes from the observation that there is high 5-HT activity in association with repetitive motor behavior and reduced 5-HT activity in association with orienting behavior (Jacobs and Azmitia 1992). Pharmacological studies with accompanying electrophysiological techniques also support this conclusion (Aston-Jones et al. 1991).

Overall, in the brain, 5-HT acts as a neurotransmitter and regulates a number of functions, including sleep, food intake, thermoregulation, emotions, and psychotic experiences. However, the main role of serotonin is *to inhibit behaviors* and *to reduce impulsivity*. The widespread and abundant expression of serotonin receptors in the limbic system, the raphe nuclei, and the mesolimbic pathways as well as in the choroid plexus suggest a role serotonin plays in the production and regulation of affect, modulation of dopamine function, and autoregulation of serotonin function and cerebrospinal fluid characteristics (Blier 2001; Kilpatrick et al. 1996; Pineyro and Blier 1999). Additionally, it seems that the mission of the serotonergic system in the forebrain is to attenuate the impact on human behavior stressful and adverse life events have. In other words, its mission is *prevention and adjustment to stress* (Deakin 2013; Deakin and Graeff 1991) (Fig. 7.2).

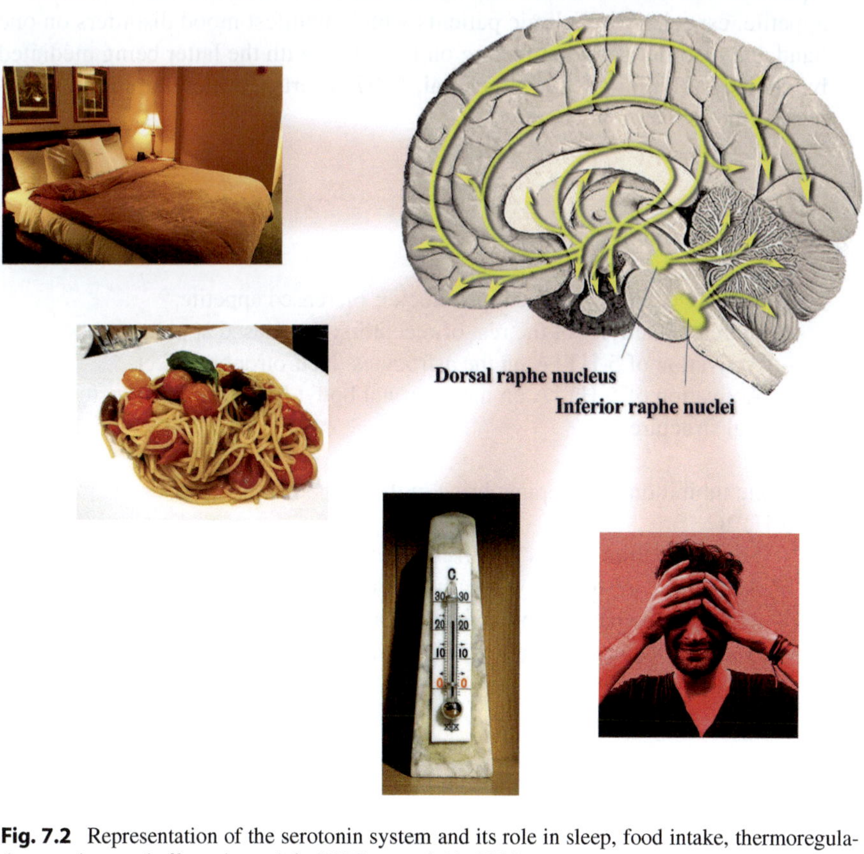

Fig. 7.2 Representation of the serotonin system and its role in sleep, food intake, thermoregulation, anxiety, and affects (copyright KN Fountoulakis 2021)

7.2.4 The Role of Serotonergic Receptors

- The activation of presynaptic 5-HT_{1A} autoreceptors inhibits the discharge of the raphe neurons, and additionally it reduces serotonin synthesis and release (Briley and Moret 1993; Chaput et al. 1986; Chaput et al. 1990; Sprouse and Aghajanian 1988). This leads to increased appetite as serotonin activity is reduced. The increase in the rest of 5-HT_1 receptors leads to reduction of appetite (Montgomery and Fineberg 1989). The excitation of astrocyte 5-HT_{1A} results in a neurotrophic effect, while on the contrary, their antagonism leads to loss of dendrites, axons, and synapses in adult but also in developing animals (Whitaker-Azmitia and Azmitia 1989; Whitaker-Azmitia and Azmitia 1991). Finally, the activation of 5-HT_{1A} receptors located in the neuronal body is essential for the maintenance of cytoskeletal structure (Drevets 2000; McEwen 1999).
- Animal studies have shown that social isolation causes hypersensitivity to 5-HT_{2C} receptors. The same receptors, in the amygdala-hippocampus circuit, are probably responsible for the manifestation of behaviors of aggression when danger is faced. This is accomplished by overcoming the inhibitory effect of 5-HT_1 and 5-HT_{2A} receptors located in the gray matter around the Sylvius aqueduct. The increase in the activity of these receptors could lead to decreased appetite, especially in bulimic patients which manifest mood disorders on one hand and disorder satiation feeling on the other, with the latter being mediated by cholecystokinin (CCK). In general, *5-HT_2* exerts a tonic inhibition on noradrenaline in the hippocampus.

The Role of Serotonergic Receptors

- 5-HT1.
  - Increased activity in the raphe nuclei: Increased appetite.
  - Increased activity of the rest of receptors: Decreased appetite.
  - Activation of 5-HT1A on the astrocytes: Neurotrophic effect.
  - Activation of 5-HT1A on the neuronal body: Maintenance of cytoskeletal structure.
- 5-HT2
- Tonic inhibition of norepinephrine in the hippocampus.
- 5-HT2C
  - Social isolation causes hypersensitivity.
  - Activation in the amygdala-hippocampus circuit: Manifestation of aggressive behaviors in the face of danger.
  - Appetite reduction especially in bulimic patients.

7.3 The Functional Role of the Norepinephrine System

7.3.1 Pathways and Information Input into the Norepinephrine System

The precise norepinephrine topography includes three major neuronal groups (Fig. 5.10):

- The *locus coeruleus (LC)* from where *the main norepinephrine pathway originates* in the brain and projects to the thalamus, the dorsal hypothalamus, the hippocampus, and the cortex:
  - The *ventral norepinephric bundle* emerges ventral/rostral to the LC and reaches the hypothalamus and the subcortical parts of the limbic system.
  - The *dorsal norepinephric bundle* emerges from the LC and projects to the cortex.

Pathways and Input to the Norepinephrine System

Τρείς κύριες ομάδες νευρικών κυττάρων:

- Locus coeruleus projects to.
  - Thalamus.
  - Dorsal hypothalamus.
  - Hippocampus.
  - Cortex.
  - Ventral norepinephric bundle.
  - Dorsal norepinephric bundle.
  - Volition.
  - Drive and motivation.
  - Reward mechanisms.
  - REM sleep.
  - Input to LC comes from.

    Feedback controls between the frontal cortex-nucleus accumbens-A9/A10 areas.

    Reciprocal regulation from the prefrontal cortex.

    Ventral tegmental area.

    Raphe nuclei.

    Hypothalamus.

    Paraventricular nucleus.

    Nucleus paragigantocellular.

    Nucleus prepositus hypoglossi.

- Waking-sleep cycle.
- Vigilance.
- Responses to environmental stimuli.
- Lateral tegmental system.
- Dorsal medullary group.
  - Regulation of blood pressure.
  - Fight-or-flight mechanisms.
  - Integration of communication between the central and the peripheral norepinephrine systems.
- Important role of norepinephrine in the response to both internal and external events including defense reactions.

Both bundles seem to relate with volition, drive, and motivation as well as with reward mechanisms and REM sleep. It is interesting that in spite of the fact that although there are less than 10,000 cells in the LC, each one has more than 10,000 synapses (Moore and Bloom 1979), and therefore the effect of LC innervation is huge in the human brain in general.

The *input to LC* comes from feedback controls between the frontal cortex-nucleus accumbens-A9/A10 areas, reciprocal regulation from the prefrontal cortex, and input from the ventral tegmental area (VTA), raphe nuclei, hypothalamus, paraventricular nucleus (PVN), nucleus paragigantocellular (Pgi; mediates the effect of environmental stimuli), and nucleus prepositus hypoglossi (PRH, mainly inhibitory action) (Jodo et al. 1998; Aston-Jones et al. 1986).

Norepinephrine activity particularly in the LC is strongly related to the waking-sleep cycle, vigilance, and responses to environmental stimuli (especially when novelty or fear are triggered). On the contrary, this activity is reduced with grooming, consumption of sweet water, and sleep (Aston-Jones and Bloom 1981; Foote and Aston-Jones 1995; Valentino and Aston-Jones 1995; Foote et al. 1983).

- The *lateral tegmental system.*
- The *dorsal medullary group.* In the brain stem and the dorsal nucleus of the vagus nerve, norepinephrine plays an essential role in the regulation of blood pressure and the fight-or-flight mechanisms. The lateral tegmental cell system and the dorsal medullary group serve the integration of communication between the central and the peripheral norepinephrine systems.

Overall these suggest an important role of norepinephrine in the response to both internal and external events including defense reactions (Svensson 1987; Aston-Jones et al. 1999).

7.3.2 Regulation of Norepinephrine Activity

There are *alpha-2 heteroreceptors*, which are located elsewhere than on noradrenergic neurons, and they are involved in release of serotonin (Gobert et al. 1998; Gobert et al. 1997; Mongeau et al. 1993; Raiteri et al. 1990), dopamine (Gobert et al. 1998; Gobert et al. 1997; Trendelenburg et al. 1994), histamine (Gulat-Marnay et al. 1989), and other neurotransmitters.

Apart from the rest, norepinephrine is synthesized in the adrenals and the post-ganglionic neurons of the sympathetic nervous system and circulates throughout the body, playing a role mainly in the function of the autonomic nervous system (Fig. 7.3).

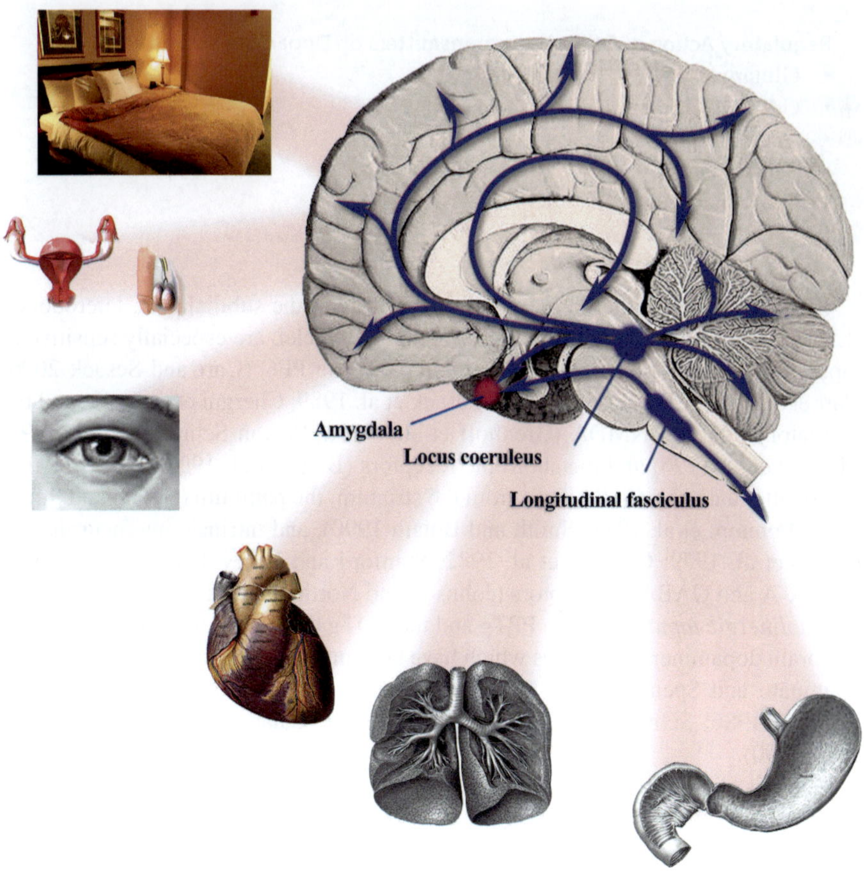

Fig. 7.3 Representation of the norepinephrine system and its role in sleep, reproduction, vision, and the cardiovascular and gastrointestinal systems (copyright KN Fountoulakis 2021)

7.4 The Functional Role of the Dopamine System

7.4.1 Pathways and Information Input into the Dopaminergic System

The *mesocortical* and *mesolimbic* dopaminergic pathways (Fig. 5.12) manifest several functional differences from each other, including differences in neuropeptides and autoreceptor and heteroreceptor regulation (Hokfelt et al. 1980; Schalling et al. 1990; Bean et al. 1990; Hertel et al. 1995; Lambert et al. 1995; Quirion et al. 1985). Also, the mesocortical neurons have a more variable firing pattern (Grenhoff et al. 1988b; Roth and Elsworth 1994).

Regulatory Action of Other Neurotransmitters on Dopamine

- Glutamate (+).
- GABA (−).
- Acetylcholine (+).
- Serotonin (−).
- Norepinephrine (+).

Some VTA dopaminergic neurons, among which the subthalamic, laterodorsal (LDTg), and pedunculopontine (PPTg) tegmental nuclei, are especially sensitive to input from excitatory *glutamatergic* input from the PFC (Carr and Sesack 2000; Carr et al. 1999; Christie et al. 1985; Sesack et al. 1989; Chergui et al. 1993) and the regulatory effect of NMDA (Grenhoff et al. 1988a; Servan-Schreiber et al. 1990; Mercuri et al. 1996) and metabotropic receptors (Bonci et al. 1997). VTA function is also attenuated by *GABA* input from the striatum, the pallidum (Bolam and Smith 1990; Fonnum et al. 1978; Smith and Bolam 1990), and intrinsic interneurons (Di Chiara et al. 1979; Grofova et al. 1982; Stanford and Lacey 1996), through both GABA-A and GABA-B receptors (Johnson and North 1992b; Erhardt et al. 2002).

Cholinergic inputs from the PPTg and the LDTg have an excitatory effect on the midbrain dopaminergic neurons which have both muscarinic and nicotinic receptors (Beninato and Spencer 1987; Bolam et al. 1991; Clarke et al. 1987; Cortes and Palacios 1986; Schwartz 1986; Wada et al. 1989; Calabresi et al. 1989; Lacey et al. 1990).

There is also an *inhibitory serotonergic input* (Ugedo et al. 1989) from the raphe nuclei to the substantia nigra (Dray et al. 1976), the nucleus caudatus and the putamen (Miller et al. 1975), and the VTA (Herve et al. 1987; Phillipson 1979; Van Bockstaele et al. 1994).

There is also excitatory *noradrenergic input* from the LC to the VTA (Grenhoff and Svensson 1989, 1993; Phillipson 1979; Grenhoff et al. 1995; Ranjbar-Slamloo and Fazlali 2019).

7.4.2 Regulation of Dopaminergic Activity

In some dopaminergic neurons, the presence of glutamate has been reported, thus forming glutamatergic synapses, called *autapses* (see also Chap. 6) (Sulzer et al. 1998). This serves the ability of these dopaminergic neurons to develop slow dopaminergic and fast glutamatergic potentials through different groups and types of synapses (Joyce and Rayport 2000).

Dopaminergic activity is also regulated by *autoreceptors* (Anden et al. 1973; Bunney et al. 1973; Kehr et al. 1972; Carlsson 1977), which can be activated by dopamine released by the neuron itself (autoinhibition) or from neighboring neurons (Innis and Aghajanian 1987; Lacey et al. 1987). These autoreceptors are present in the striatum (Kehr et al. 1972; Westerink and de Vries 1989; Zetterstrom and Ungerstedt 1984) but not in the mesocortical neurons (Chiodo et al. 1984).

7.4.3 Projections of Dopamine Pathways

Generally, the *dopaminergic input exerts an inhibitory effect*. Its inhibitory action on the PFC is mediated either by inhibition of the glutamatergic or by excitation of GABAergic interneurons (Zhou and Hablitz 1999; Lavin and Grace 2001). D2 receptor activation also plays a role in the control of thalamocortical circuits (Lavin and Grace 2001). The overall picture suggests that the dopaminergic system does not serve any specific function but instead it *regulates and permits the synthesis* of the functions which are hosted in the structures where this system projects (Le Moal 1994).

Dopaminergic Action

- General inhibitory effect.
  - Inhibition of glutamatergic activity.
  - Excitation of GABAergic activity.
- Regulates and permits the synthesis of functions.
  - Nonlinear action.
- Gating mechanism for the selection, acquisition, and maintenance of behavior.
- Pain.
- Attention.
- Affect.
- Speech.
- Working memory.
- Memory and learning.
- Executive functions.
- Motives and reward.

- The mesocortical and mesolimbic pathways participate mainly in high-level mental functions, including affect, speech, working memory, memory and learning, and executive functions.
- The midbrain systems participate mainly in body movements; however, the distinction of roles by no means can be considered to be absolute (Sawaguchi and Goldman-Rakic 1994; Koob et al. 1998; Le Moal and Simon 1991; Schultz 1997, 1998) (Fig. 7.4).

The effect of dopamine on these functions is not linear. For example, it is important to note that although low dopaminergic activity seems to relate to low ability to concentrate and reduced performance, also an excessive increase is related to deficits in attention and stereotypical behaviors (Murphy et al. 1996).

Dopamine neurons show two distinct patterns of firing:

- *Single spike firing* which is regular and of low frequency (e.g., <10 Hz).
- A second form which is characterized by burst firing which is a *sudden high-frequency* discharge of *multiple action potentials* (Grace and Bunney 1983, 1984).

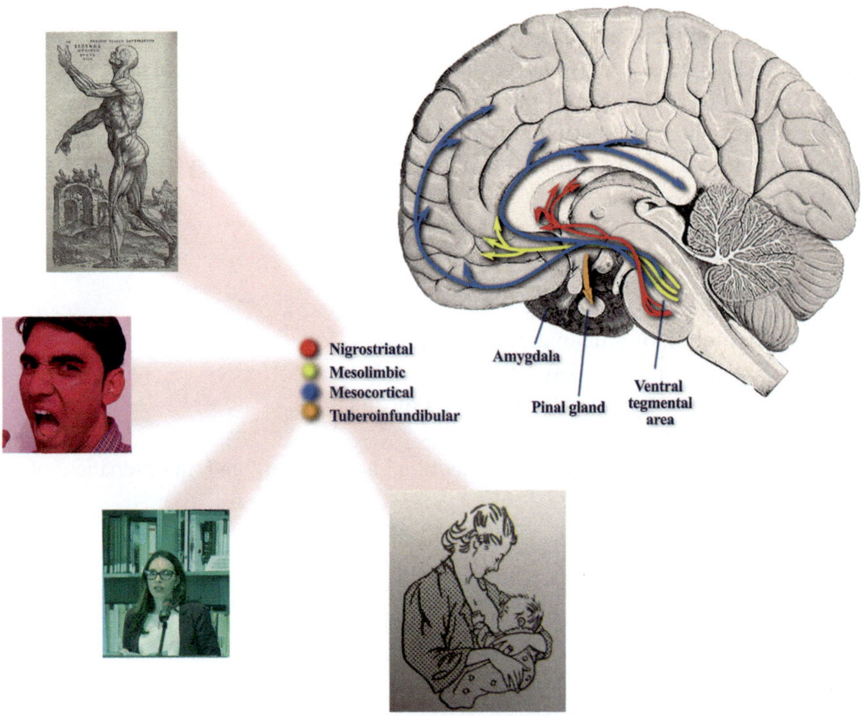

Fig. 7.4 Representation of the activity of the dopamine system and the role of the nigrostriatal pathway on movement; of the mesolimbic and mesocortical on anger, aggressiveness, speech, and social behavior; and of the tuberoinfundibular on reproduction and lactation (copyright KN Fountoulakis 2021)

Burst firing is more efficient in the release of dopamine in comparison with a regular firing pattern of the same average rate, and it subsequently causes more activation of postsynaptic neurons and more release of co-localized neurotransmitters (Bean and Roth 1991; Chergui et al. 1996; Chergui et al. 1997; Gonon 1988; Suaud-Chagny et al. 1992). These changes in firing activity occur usually in relation to *attentional, motivational, and reward processes in the frame of goal-oriented behaviors* (Nishino et al. 1987; Schultz 1986, 1998; Schultz et al. 1993). Probably one of the rare but most important specific dopaminergic effects is a dopaminergic involvement in reward systems (*dopaminergic theory of reward*) (Jacques 1979; Wise 2008).

It is of prime importance that *drugs of abuse*, including cocaine, amphetamine, opiates, ethanol, and nicotine, directly or indirectly cause an increased dopamine release (Di Chiara 1995; Yeomans et al. 1985; Yeomans and Baptista 1997; Yeomans et al. 1993; Calabresi et al. 1989; Grenhoff et al. 1986; Picciotto et al. 1998; Pidoplichko et al. 1997; Johnson and North 1992a). Dopamine neurons are activated by a primary reward, especially if unexpected. An interesting feature is that in case the reward is *coupled with a conditioning signal of any modality*, dopaminergic neurons will *switch the increasing of firing* from the rewarding stimulus to the conditioning stimulus. This means that the "rewarding" dopamine spike will happen not when the actual stimulus appears but on the contrary when the accompanying stimulus emerges. In turn, this means probably that the system is exploring the possibility of using surrounding stimuli to predict reward and to navigate behaviorally and with motivation cues through the environment (Redgrave et al. 1999; Schultz 1997, 1998).

Dopamine possibly plays a role in the *regulation of pain* through receptors located in the reticular nucleus of the thalamus and the gigantocellular reticular nucleus (magnocellularis) in the hindbrain.

It has been reported that intrapsychic conflict as a result of ambiguous situations with unclear solution leads to an increase in dopaminergic activity, and in this way, they trigger displacement activities which reduce the level of arousal and stabilize the system (Tazi et al. 1988; Tazi et al. 1986).

Dopaminergic input controls these mechanisms and functions by maintaining stability at baseline conditions and by enhancing or suppressing signals during stimulation and subsequently regulating learning and reward processes. Overall dopamine function seems to correspond to a *gating mechanism for selection, acquisition, and maintenance of appropriate behavioral programs.*

7.4.4 The Role of Dopamine Receptors

In general, there seems to be a *pattern of dopamine receptor localization* in the brain. D2 are found in the subcortical regions which are related to motor control, while with their presence in the reticular nucleus of the thalamus and the gigantocellular reticular nucleus in the hindbrain, they play a role in pain regulation. D3 and D4 are restricted to the mesocortical and the mesolimbic systems, that is, in areas related to mental functions and affectivity (Ross and Pearlson 1996; Schwartz et al. 2000).

The Role of Dopamine Receptors

- D1.
  - In the deep layers of the cerebral cortex, reducing excitability (in interaction with D2).
- D2.
  - Subcortical areas, motor control.
  - Reticular nucleus of the thalamus and the gigantocellular reticular nucleus in the hindbrain, pain regulation.
  - Fast attenuation of responses to rewarding stimuli.
  - In the deep layers of the cerebral cortex, reducing excitability (in interaction with D1).
- D3 and D4.
  - Mesocortical and mesolimbic system, mental functions, and affectivity.

In the deep layers of the cerebral cortex, dopaminergic neurons lack transporter systems (Sesack et al. 1998) which means that dopamine diffuses freely outside the synaptic cleft and is probably taken up in neighboring norepinephrine nerve terminals (Gresch et al. 1995), after activating extrasynaptic dopamine receptors (Smiley et al. 1994).

In both the dorsal striatum and the NAC, dopamine generally decreases the excitability of neurons through D1 receptor stimulation (Calabresi et al. 1987; O'Donnell and Grace 1996; White and Wang 1986) but also in a complex interplay with D2 activation (Onn et al. 2000), which among other functions determines also the fast attenuation of responses to rewarding stimuli.

7.5 The Functional Role of the Histamine System

7.5.1 The Role of Histamine in the Brain and the Body

Interestingly, histaminergic neurons in the tuberomammillary nuclei of the thalamus (Fig. 5.14) have a firing pattern which is *closely related to wakefulness*. During wakefulness, the fire frequency is high, while during periods of relaxation or tiredness, the frequency is significantly lower. It seems they completely stop firing during REM and non-REM sleep. When antihistamines (that is, older-type H1 receptor antagonists) cross the blood–brain barrier, they produce drowsiness. Similarly, destruction of histamine-releasing neurons or inhibition of histamine synthesis leads to an inability to maintain vigilance. On the contrary, H3 receptor antagonists increase wakefulness. Therefore, it is believed that histamine increases wakefulness and prevents sleep (Brown et al. 2001).

Histamine (Fig. 7.5) is also involved in *local immune responses*, during which it is secreted by basophils and mast cells and in the function of the gut. It increases the permeability of the capillaries to white blood cells and some proteins.

Additionally to its arousing and stimulatory effects, histaminergic activity also protects against the susceptibility to convulsion, drug sensitization, denervation supersensitivity, ischemic lesions, and stress, probably through a suppressive type

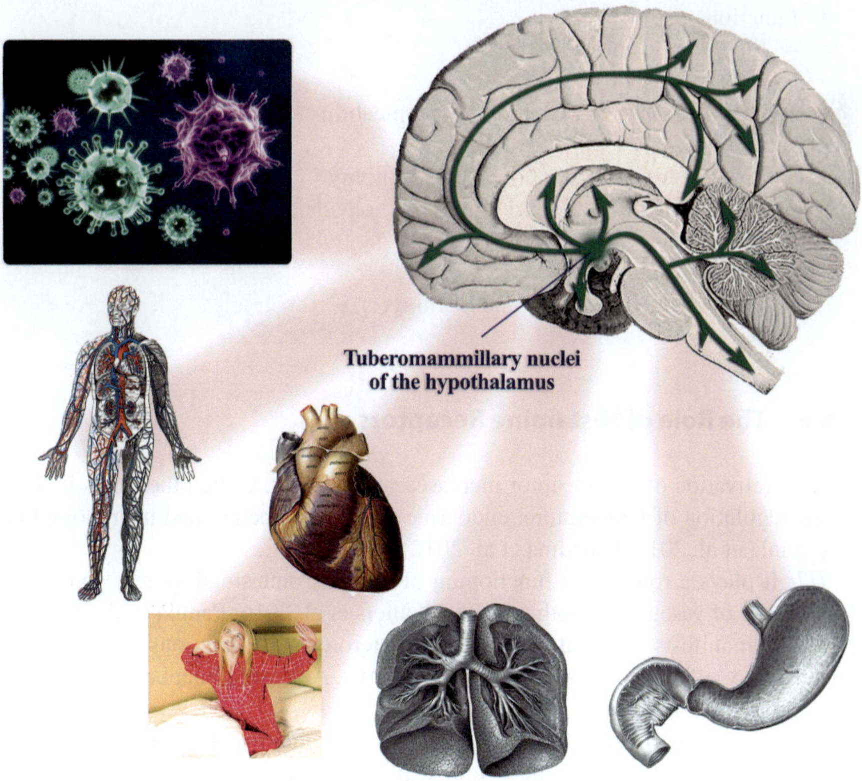

Fig. 7.5 Representation of the histaminergic system and its role on the immune, cardiovascular, respiratory, and gastroenterological systems as well as on wakefulness (copyright KN Fountoulakis 2021)

of effect (Yanai and Tashiro 2007). It has also been suggested that histamine controls the mechanisms by which memories and learning are forgotten (Alvarez 2009).

The Role of Histamine

- It is synthesized in the tuberomammillary nuclei of the thalamus.
- Histaminergic neurons project throughout the brain.
- Relationship with sleep-wakefulness phenomenon.
- Pattern of histaminergic neurons firing.
  - Low at relaxation or tiredness.
  - High during wakefulness.
  - Complete lack of firing during sleep.
- Functions.
  - Sleep-wakefulness.
  - Immune responses.
  - Regulation of the capillaries' permeability to white blood cells and proteins.
  - Protects from denervation supersensitivity.
  - Protects against the susceptibility to convulsion.
  - Protects from ischemic lesions.
  - Protects from stress.
  - Plays a role in memory and learning.

7.5.2 The Role of Histamine Receptors

- *H1:* Activation of this receptor increases wakefulness, while other roles include the regulation of temperature, endocrine function, appetite, and neurocognition (Panula et al. 2015; Blandina et al. 2012).
- *H2:* It plays a role in the functioning of the gastrointestinal system and in the mobility of vascular smooth muscle (Wouters et al. 2016; Panula et al. 2015).
- *H3:* It manifests both autoreceptor and heteroreceptor functions and exerts an inhibitory effect on the secretion of histamine, Ach, norepinephrine, and 5-HT (Gilman 1990).
- *H4:* It is found primarily in the basophils, the bone marrow, the thymus, the small intestine, the spleen, and the colon. It plays a role in various mechanisms of inflammation, etc. Its role on cognition is unclear (Panula et al. 2015).

The Role of Histamine Receptors

- H1.
  - Wakefulness.
  - Temperature regulation.
  - Endocrine function.
  - Appetite.
  - Neurocognitive function.

- H2.
  - Functioning of the gastrointestinal system.
  - Mobility of vascular smooth muscles.
- H3.
  - Autoreceptor and heteroreceptor functions.
  - Inhibitory effect on the secretion of.
    Histamine.
    Acetylcholine.
    Norepinephrine.
    Serotonin.
- H4.
  - Inflammation.
  - Neurocognitive function (?)

7.6 The Functional Role of the Glutamatergic System

Glutamate is *the major excitatory neurotransmitter in the brain*. It plays a role in the mechanisms concerning brain *plasticity* and *higher neurocognitive functions*. Dysfunction in glutamatergic activity with overexcitation leads to neuronal death on the basis of a phenomenon called *excitotoxicity*. It is particularly important to note that in a way, the most important function of all other neurotransmitters is to regulate glutamatergic activity.

There are five important glutamatergic pathways (Stahl 2008) (Fig. 5.16):

- The *cortical brain stem glutamate projection* from the prefrontal cortex to the brain stem (raphe nuclei, locus coeruleus, ventral tegmental area, substantia nigra).
- The *corticostriatal glutamate pathway* from the prefrontal cortex to the striatum and the *cortico-accumbens glutamate pathway* to the nucleus accumbens.
- The *thalamocortical glutamate pathway* from the thalamus to the pyramidal neurons in the cortex.
- The *corticothalamic glutamate pathway*, from the prefrontal cortex to the thalamus.
- The *cortico-cortical glutamatergic pathway* which concerns the communication among pyramidal neurons within the cortex.

7.6.1 The Role of Ionotropic Glutamate Receptors

7.6.1.1 The Role of the NMDA Receptors

Synaptic plasticity is characterized by long-term changes in the synaptic efficacy at glutamatergic synapses. Its most common forms are long-term potentiation (LTP) and long-term depression (LTD). LTP lasts for days or weeks and constitutes an enhancement of synaptic transmission. It begins with the activation of a sodium channel associated with an AMPA receptor. This leads to membrane depolarization and subsequently to the removal of the magnesium block of the NMDA receptor. Activation of NMDA receptors leads to calcium influx and activation of a variety of kinases and lipases. This last activation is responsible for the LTP that are part of a cascade that consolidates the LTP (Zheng and Gallagher 1992).

The Role of NMDA Receptors

- Synaptic plasticity.
- Synaptic development.
- Optimization of connectivity.
- Long-term potentiation (LTP).
- Long-term depression (LTD).
- Removal of the magnesium block of the NMDA receptor.
- Influx of ca ions.
- Memory formation and consolidation.
- Learning.
- General neurocognitive function.
- Mood.
- Anxiety.
- Excitotoxicity.
- Neurodegeneration.
- Excess excitatory activity of either NMDA or non-NMDA receptors.
- Entering of too much Ca^{2+} ions.

Therefore, they play a crucial role in learning and memory formation and consolidation and general cognition and attention but also in mood and anxiety (VanDongen 2009), that is, in neuronal function terms, they lead to *synaptic development* and *optimization of connectivity* in relationship with function and experience.

However, in contrast to the rest ion channels, NMDA receptors are permeable for Ca ions, which makes possible for these receptors to utilize Ca as a lead for the LTP and plasticity in the frame of experience-based phenomena.

However, they also mediate a phenomenon called *excitotoxicity* which leads to neurodegeneration (Olney 1989). The term refers to a paradoxical phenomenon, whereby the neuroexcitatory action of glutamate which is for normal brain functioning turns into reckless glutamate activity leading to the death of neurons (Stahl 2008). This could be the result of excess excitatory activity of either NMDA or non-NMDA receptors, causing a dangerous opening of the calcium channel allowing too much Ca^{2+} to enter. This in turn leads to the activation of intracellular enzymes and to the formation of free radicals, which exert a toxic action on cellular organelles and membranes. On the other hand, glutamate hypofunction could be also *autoexcitotoxic* (Choi et al. 1988; Rothman and Olney 1987). An additional glutamatergic mechanism leading to neuronal death could be the chronic potentiation of pathological intracellular primary oxidative processes and the diminishing of intracellular antioxidative protection. These processes are characterized by cell loss but not by remarkable gliosis, since many times an *apoptotic mechanism* is involved (Coyle and Puttfarcken 1993).

There are data suggesting that the excessive activation of excitatory amino acid receptors constitutes a basic mechanism of neuropathology in acute brain insults (e.g., stroke or trauma) and in mental and organic neurodegenerative disorders (e.g., schizophrenia, Huntington's disease, Alzheimer's disease, etc.) (Choi 1994; Meldrum 2000). Therefore, except from glutamate receptors, a very important role rests with the excitatory amino acid reuptake and the transporters responsible for it. These transporters do not only serve the recycling of the neurotransmitter and the shaping of the postsynaptic signal as any other transporter; additionally, they protect neurons from glutamate excitotoxicity (McBean and Roberts 1985; Robinson et al. 1993; Lievens et al. 1997; Rothstein 1995). However, on the contrary, under certain abnormal conditions, they constitute sites of cellular L-glutamate efflux leading to neurotoxicity (Takahashi et al. 1997; Koch et al. 1999; Longuemare and Swanson 1995).

7.6.1.2 The Role of the Kainate Receptors

Kainate receptors are abundant in the hippocampal mossy fiber synapse (Foster et al. 1981; Castillo et al. 1997; Vignes et al. 1997). Kainate receptors also exert presynaptic inhibition on GABA, which might be the explanation of the epileptogenic action of kainic acid and several similar substances (Rodriguez-Moreno et al. 1997).

7.6.2 The Role of the Metabotropic Glutamate Receptors

Metabotropic Glu receptors are also widespread in the brain. They can be found in the hippocampus, the cerebellum, and the spinal cord and on glial cells. Outside the brain, they can be found in osteoblasts, ileum, heart tissue, and elsewhere in the body (Skerry and Genever 2001).

The Role of the Metabotropic Glutamate Receptors

- Hippocampus.
- Cerebellum.
- Spinal cord.
- Glia.
- Group I mGlu receptors.
  - Pre- and postsynaptically.
  - Away from active zones.
- Group II mGlu receptors.
  - Presynaptic in the hippocampus.
  - Postsynaptically in the rest of the brain.
- Group III mGlu receptors.
  - Presynaptically or inside or close the active zones.

- Presynaptic autoreceptors.
  - Open ion channels and especially for potassium.
  - They inhibit calcium channels.
- Function of motor circuits.
  - Spinal cord.
  - Basal ganglia.
  - Cerebellum.
- Pain.
- Synaptic plasticity.
- Stroke.
- Epilepsy.
- Neurodegenerative disorders.

- The *group I mGlu* receptors are located postsynaptically, away from the active zones, but also they can be found presynaptically (Cartmell and Schoepp 2000; Baude et al. 1993; Lujan et al. 1997).
- The *group II mGlu* receptors are located presynaptically in the hippocampus and postsynaptically elsewhere in the brain (Petralia et al. 1996b; Petralia et al. 1996a; Shigemoto et al. 1997) with the mGlu3 subtype being abundant in glial cells (Mineff and Valtschanoff 1999).
- The *group III mGlu* receptors are located presynaptically and in or near the active zones (Shigemoto et al. 1997), with the exception of mGlu6, which can be found only in the bipolar cells of the retina.

The main function of mGlu receptors is acting as *presynaptic autoreceptors* for the regulation of the function of glutamatergic synapses (Cartmell and Schoepp 2000). The activation mainly of group II and group III mGlu receptors results in a reduction of glutamate release (Gereau and Conn 1995). They also regulate the release of other neurotransmitters (Cartmell and Schoepp 2000).

Metabotropic Glu receptors exert a *direct inhibitory effect* on neurons by opening potassium as well as other ion channels (Charpak et al. 1990; Desai and Conn 1991; Guerineau et al. 1994; Luthi et al. 1996; Linden et al. 1994; Crepel et al. 1994; Pozzo Miller et al. 1995), usually inhibiting calcium channels (Choi and Lovinger 1996; Chavis et al. 1995; Rothe et al. 1994; Sahara and Westbrook 1993; Trombley and Westbrook 1992) although activation of calcium channels has also been reported (Chavis et al. 1995; Rothe et al. 1994).

Their function is important in various motor circuits in the spinal cord, the basal ganglia, and the cerebellum. They are also involved in mechanisms of pain and synaptic plasticity, stroke, epilepsy, and neurodegenerative disorders (Anwyl 1999, 1992; Salt and Eaton 1994; Young et al. 1995; Bruno et al. 1995; Buisson et al. 1996).

7.7 The Functional Role of the GABAergic System

GABA is the *main inhibitory neurotransmitter in the brain*. As it is widely distributed in the brain, essentially its activity reflects the activity of the system it specifically inhibits.

The Functional Role of the GABAergic System

- GABA is the main inhibitory neurotransmitter in the brain.
- Its activity reflects the activity of the system it specifically inhibits.
- Has an effect on the secretion of.
  - Growth hormone (GH).
  - Melatonin.
- Sleep regulation.
- Reproduction.
- Inhibits motor signals in the basal ganglia.
- Role in the spinal cord in the formation of smoothness of movements.
- Reduces the respiratory rate acting at the level of the medulla oblongata.
- Role in pancreas function.
- GABAergic medication manifests the following effects.
  - Calming.
  - Anxiolytic.
  - Antiepileptic.
  - As adverse effect, anterograde and retrograde amnesia.

It does not cross the blood-brain barrier (Kuriyama and Sze 1971). In the peripheral body, it increases the secretion of growth hormone (GH) (Powers et al. 2008), while in the brain, the state of brain physiology determines whether this secretion will increase or decrease (Muller et al. 1999). It also plays a role in melatonin synthesis and the regulation of sleep as well as of reproductive functions (Sato et al. 2008; Balemans et al. 1983). It inhibits motor signals in the basal ganglia, plays a role in the spinal cord in the formation of the smoothness of movements, and reduces the respiratory rate by acting at the level of the medulla oblongata (Wang et al. 2019; Tong 2014; Raymond 2017). It is produced by the beta cells in the pancreas and inhibits the function of the alpha cells and stimulates the development of the beta and the transformation of alpha into beta (Khanna et al. 2013).

Medications with GABAergic action (with benzodiazepines being the prime example) manifest *calming, anxiolytic, and antiepileptic effects* (Foster and Kemp 2006; Chapouthier and Venault 2001), and as an adverse effect, they cause *anterograde and retrograde amnesia* (Campagna et al. 2003).

An abstracted representation of the interaction of basic neurotransmitters and the overlapping of their actions is shown in Fig. 7.6.

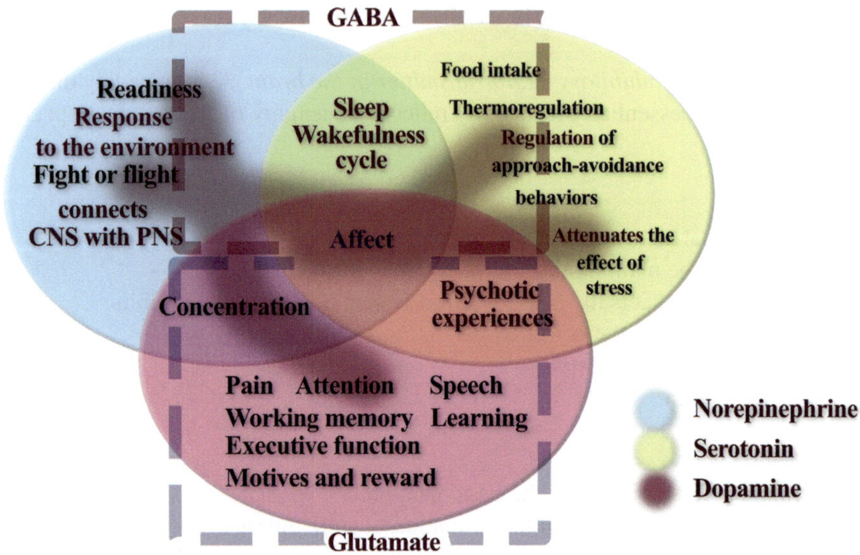

Fig. 7.6 Abstracted representation of the interaction among the basic neurotransmitters and of the overlapping of their actions (copyright KN Fountoulakis 2021)

7.8 The Functional Role of the Neuropeptide System

7.8.1 The General Functional Role of Neuropeptides

Most neuropeptides are synthesized in the hypothalamus but are present in almost all major brain areas. However, their density is highest in the median eminence. Most descending hypothalamic fibers are peptidergic (Sawchenko and Swanson 1982; Cechetto and Saper 1988; Moga et al. 1990). Neurons could contain and release several peptides which come from the same polyprotein, and their release leads to potentially different postsynaptic actions (Fisher et al. 1988). They coexist in the same neuron with classical neurotransmitters, but no pattern has been identified so far concerning this coexistence. When different cells synthesize different peptides but in the same nucleus, the term is *co-localization*, while when the same cell synthesizes more than one, the term is *coexpression*. Depending on the site of action, neuropeptides can act as neurotransmitters, neurohormones, or both.

While most of physiological functions of neurosteroids are unknown, the most important known include their participation in the following mechanisms:

- *Neural plasticity* (Benarroch 2007).
- *Higher neurocognitive functions* (Vallee et al. 2001).
- Various aspects of *mood* and *behavior* (King 2008; Miczek et al. 2003; Engel and Grant 2001; Kaur and Kulkarni 2001; Frye 2009).
- Vulnerability to manifest *epileptic seizures* (Joshi et al. 2013).

- They also seem to play an important role in a number of *sexually dimorphic behaviors* (King 2008).

The following mechanisms are influenced by neuropeptides:

- **Food intake.**
  - Neuropeptide Y (NPY).
  - Galanin.
  - Apelin.
  - Calcitonin gene-related peptide (CGRP).
  - Melanin-concentrating hormone (MCH).
  - Agouti-related protein (AgRP).
  - Gastrin-releasing peptide (GRP).
  - Neurotensin.
  - α-Melanocyte-stimulating hormones (α-MSH).
  - Cholecystokinin (CCK).
- **Fluid intake.**
  - Angiotensins II and III.
  - Atrial natriuretic peptides (ANP).
- **Thermoregulation.**
  - Neuropeptide Y (NPY).
  - Neurotensin.
  - Gastrin-releasing peptide (GRP).
- **Cardiovascular function and blood pressure.**
  - Neuropeptide Y (NPY).
  - Gastrin-releasing peptide (GRP).
  - Angiotensins I and II.
  - Atrial natriuretic peptides (ANP).
  - Bradykinin.
  - Vasoactive intestinal polypeptide (VIP).
- **Reproduction.**
  - Luteinizing hormone-releasing hormone (LHRH).
- **Hormonal regulation.**
  - Luteinizing hormone-releasing hormone (LHRH).
  - Thyrotropin-releasing hormone (TRH).
  - Corticotropin-releasing hormone (CRH).
  - Growth hormone (GH).
  - Somatostatin.
  - Adrenocorticotropic hormone (ACTH).
  - Melanocyte-stimulating hormones (MSH).
- Peptide histidine-isoleucine with 27 amino acids (PHI-27).
  - Galanin.
  - Angiotensins I and II.
  - Apelin.

- **Stress.**
  - Thyrotropin-releasing hormone (TRH).
  - Adrenocorticotropic hormone (ACTH).
  - Corticotropin-releasing hormone (CRH).
- **Pain.**
  - Endomorphins EM1 and EM2.
  - Substance P.
  - Neurokinins A and B.
  - Calcitonin gene-related peptide (CGRP).
  - Tuberoinfundibular peptide of 39 residues (TIP39).
- **Sleep and circadian rhythms.**
  - Cholecystokinin (CCK).
  - Apelin.
  - Orexin-33 (hypocretin-1) and orexin-28 (hypocretin-2).

7.8.2 The Role of Neuropeptide Receptors

The functional significance of neuropeptide receptors is unknown for most of them, as is for neuropeptides themselves. More and relatively certain knowledge exists for the following receptors:

- The *kappa (κ) opioid receptors*: Their function includes analgesia, anticonvulsant effects, induction of depression, dissociative and hallucinogenic effects, diuresis, dysphoria, miosis, neuroprotection, sedation, and induction of stress (Holden et al. 2005).
- The functions of *μ1 opioid receptors* include analgesia and somatic dependence on substances.
- The functions of *μ2 opioid receptors* include respiratory depression, miosis, euphoria, reduced gastroenterological mobility, and somatic dependence.
- Those of *μ3 receptors* are unknown although they most likely include vasodilation (Holden et al. 2005).
- The *opioid growth factor receptor (OGFr or ζ-opioid receptor)* plays a role in tissue growth (Cheng et al. 2009; Malendowicz et al. 2005; Wu et al. 1998; Zagon et al. 2002), embryonal development (Zagon et al. 1999b), and healing of wounds (Sassani et al. 2003), while it also plays an important role in certain types of cancer (Cheng et al. 2007; Donahue et al. 2009; McLaughlin et al. 1999; Zagon et al. 1999a).
- The *sigma (σ) receptors* had initially been considered to constitute a subclass of opioid receptors, but eventually it has been proven that their function was different and unrelated to opioids. The functions they serve are essentially unknown (Leonard 2004); however, it is known that their activation has a hallucinatory

effect, and they might mediate the triggering of epileptic seizures in case of opioid overdose. Other effects include hypertonia, tachycardia, increased respiratory rate, antitussive action, and mydriasis. The substances that have an agonist effect on these receptors include cocaine, morphine, opipramol, phencyclidine (PCP), fluvoxamine, methamphetamine, dextromethorphan, and the natural antidepressant berberine (Yang et al. 2007; Fontanilla et al. 2009; Skuza and Rogoz 2006). Concerning their effect on behavior, the sigma receptors manifest multiple actions. Some persons experience elevated mood or an antidepressant effect while others exactly the opposite, that is, dysphoria and anxiety (Wang et al. 2007; Guitart et al. 2004; Zhang and Cuevas 2005):

- The *sigma-1 receptors (σ1)* can be found in specific areas of the brain (Weissman et al. 1988) and play a role in a number of mechanisms and conditions, including cardiovascular function, schizophrenia, clinical depression, the effects of cocaine, and cancer (Guitart et al. 2004; Zhang and Cuevas 2005).
- The *sigma-2 receptor (σ2)* is expressed at high concentration on the surface of malignant cancer cells (van Waarde et al. 2015). It can be found a higher density in those brain areas responsible for motor function and emotional responses. It is linked to various ion channels (Huang et al. 2014) and can be found in the cerebellum, the motor cortex, and the substantia nigra (Skuza and Rogoz 2002). It is expressed also in the lungs, the liver, and the kidneys (Ahmed et al. 2012). Their functions include a role in cell proliferation (Damaskos et al. 2014; Huang et al. 2014) and neuronal as well as nonneuronal signal transduction. It can be found at a high density in breast, ovary, lung, brain, urinary bladder, and large intestine cancers as well as in melanoma (Ahmed et al. 2012; Damaskos et al. 2014).

The Functional Role of Neuropeptide Receptors

- Unknown for most of them.
- Kappa (κ) receptors.
  - Analgesia.
  - Anticonvulsant effects.
  - Induction of depression.
  - Dissociative and hallucinogenic effects.
  - Diuresis.
  - Dysphoria.
  - Miosis.
  - Neuroprotection.
  - Sedation.
  - Induction of stress.

- μ1-receptors,
  - Analgesia.
  - Somatic dependence on substances.
- μ2-receptors,
  - Respiratory depression.
  - Miosis.
  - Euphoria.
  - Reduced gastroenterological mobility.
  - Somatic dependence.
- μ3-receptors,
  - Unknown function.
  - Maybe vasodilation.
- Opioid growth factor receptor (OGFr).
  - Tissue growth.
  - Embryonal development.
  - Wound healing.
  - Important role in types of cancer.
- σ1-receptors,
  - Cardiovascular function.
  - Schizophrenia.
  - Clinical depression.
  - Cocaine.
  - Cancer.
- σ2-receptors.
  - Malignant cancer cells.
  - Motor function.
  - Emotional responses.
  - Role in cell proliferation.

7.9 General Principles of Neurotransmitter Systems Self-Regulation

Neurotransmitter systems possess a number of autoregulatory mechanisms, so that on one hand, the intensity of the stimuli does not cross a certain threshold and on the other hand, it is easier to regulate function. It should be noted that in case these systems are left without regulatory mechanisms, dysfunction is certain and neuronal death through apoptosis is highly possible.

Autoregulatory Mechanisms of Neurotransmitter Systems

- Changes in sensitivity (hypersensitivity-desensitization).
  - Masking/unmasking of receptor.
  - Accentuation of signal amplification.
  - Receptor sequestration.
- Synergy/counteraction.

Change of sensitivity leads to

- Tachyphylaxis.
- Tolerance.

The most important mechanisms are the following (Westerfield and Barrera 2020; Longo et al. 1992; Monshipouri et al. 2000; Qin et al. 2011; Stephenson 1956) (Fig. 7.7):

- *Changes in sensitivity (hypersensitivity-desensitization).* This mechanism causes a change in sensitivity of the whole system either toward an enhanced final signal or toward an attenuated, taken into consideration the specific amount of the neurotransmitter available in the synaptic cleft. The end result is either up- or down-regulation of receptors. The ways this change in sensitivity can be achieved are the following:
  - *Masking/unmasking of receptor:* Attenuation of the signal and desensitization could be achieved by masking the receptors with a protein which engulfs the intracellular end of the receptor, and in this way, it prevents the transmission of the signal to the second messenger. Removal of the protein and unmasking of the receptor leads to the increase in sensitivity of the system and the amplification of the signal. The mechanism of masking/unmasking makes particularly problematic the interpretation of receptor imaging studies with the use of radioligands, because these ligands bind to all receptors, both masked and unmasked, while on the contrary, neurotransmitters manifest a preference for unmasked receptors. Medications have properties similar to radioligands, and this constitutes a significant difference concerning their behavior on the receptor in comparison to the endogenous substance they antagonize.
  - *Accentuation of signal amplification:* This is achieved through the phosphorylation of receptors or the increase in enzyme activity as this is described in the chapter that follows (Chap. 8) which concerns second messengers.
  - *Receptor sequestration:* It refers to the complete incorporation of a receptor through endocytosis which subsequently is either carried to a lysosome for decomposition or remains in reserve and could emerge again later, on the outer surface of the cellular membrane.

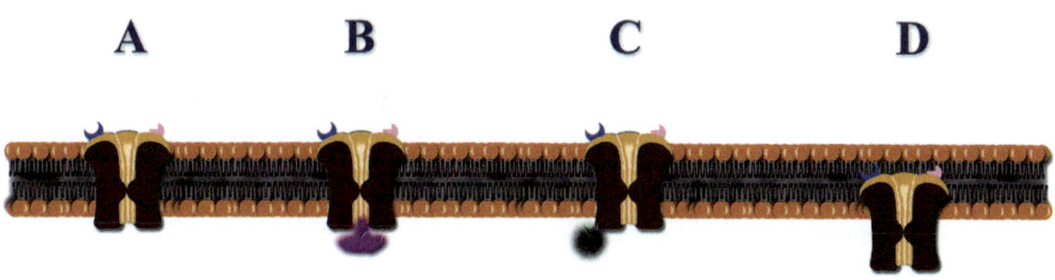

Fig. 7.7 Abstracted representation of the neurotransmission autoregulatory mechanisms at the receptor level. (**a**) Activated receptor. (**b**) Masking of the receptor by a protein at its intracellular end. (**c**) Accentuation of signal amplification (e.g., through phosphorylation). (**d**) Endocytosis of the receptor (copyright KN Fountoulakis 2021)

- *Synergy/counteraction:* Synergy occurs when the simultaneous activation of two receptors produces an effect which is stronger than the sum of the effects their separate activation would produce. Counteraction leads to a result which depends on the properties of the receptors. One such result are muscarinic cholinergic receptors, of whom, type M1 is excitatory and M2 is inhibitory (see also Chap. 6). Type M1 manifests a lower activation threshold, which means it can be activated with lower concentrations of the neurotransmitter in the synaptic cleft. In turn, this means it will be activated earlier than M2; however, its effect is weaker in comparison to that of the M2. Eventually, in a theoretical situation where the postsynaptic membrane would have only M1 and M2 receptors, after acetylcholine secretion in the cleft, initially M1 would be activated, and as a result, there would be an increase in cholinergic activity. However, if acetylcholine continues to be released and its concentration in the cleft increases above the M2 activation threshold, this will result in the stabilization or even the attenuation in cholinergic activity, since the full activation of all M2 receptors will lead to the complete dominance of their strong inhibitory effect versus the weak excitatory effect of the M1 (Fig. 7.8).

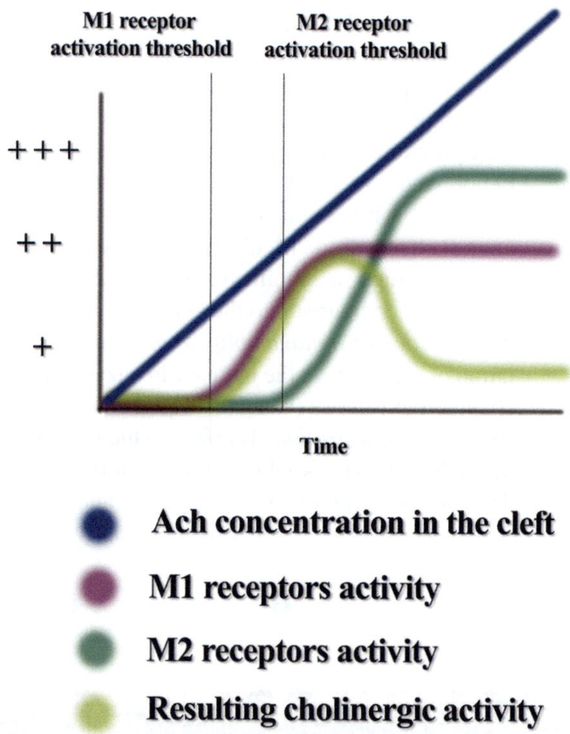

Fig. 7.8 Schematic representation of the interaction between the M1 and M2 cholinergic receptor actions in the frame of a theoretical and simplified model (copyright KN Fountoulakis 2021)

Change in sensitivity could lead to the manifestation of tachyphylaxis or tolerance.

- *Tachyphylaxis*: It refers to the fast reduction of the system's response to an agonist as a result of its repeated effect but over brief periods of time (Zuo et al. 2011; Swanson et al. 1999). In fact, it reflects a form of fast desensitization and appears without any change in the number of active receptors or in the receptor's ability to transduce the signal to the second messenger. What is important in this mechanism is that the increase in the dosage of the agonist does not always lead to the restoration of the response. In clinical trials of medications, the effect of tachyphylaxis is rather difficult to distinguish from the placebo effect, since both lead to relapse after a brief period of response.
- *Tolerance*: It concerns the gradual attenuation of response to an agonist after its prolonged administration (Miller et al. 1987; Weiner et al. 1980; Bespalov et al. 2016). Increasing the dosage almost always restores the signal, although this might not be the case after repeated increases which stretch the system to its limits.

References

Aghajanian GK, Foote WE, Sheard MH (1968) Lysergic acid diethylamide: sensitive neuronal units in the midbrain raphe. Science 161(3842):706–708. https://doi.org/10.1126/science.161.3842.706

Aghajanian GK, Vandermaelen CP (1982) Intracellular identification of central noradrenergic and serotonergic neurons by a new double labeling procedure. J Neurosci 2(12):1786–1792

Ahmed IS, Chamberlain C, Craven RJ (2012) S2R(Pgrmc1): the cytochrome-related sigma-2 receptor that regulates lipid and drug metabolism and hormone signaling. Expert Opin Drug Metab Toxicol 8(3):361–370. https://doi.org/10.1517/17425255.2012.658367

Alvarez EO (2009) The role of histamine on cognition. Behav Brain Res 199(2):183–189. https://doi.org/10.1016/j.bbr.2008.12.010

Anden NE, Magnusson T, Stock G (1973) Effects of drugs influencing monoamine mechanisms on the increase in brain dopamine produced by axotomy or treatment with gammahydroxybutyric acid. Naunyn Schmiedeberg's Arch Pharmacol 278(4):363–372. https://doi.org/10.1007/BF00501480

Anwyl R (1992) Metabotropic glutamate receptors: electrophysiological properties and role in plasticity. Rev Neurosci 3(3):217–231. https://doi.org/10.1515/REVNEURO.1992.3.3.217

Anwyl R (1999) Metabotropic glutamate receptors: electrophysiological properties and role in plasticity. Brain Res Brain Res Rev 29(1):83–120. https://doi.org/10.1016/s0165-0173(98)00050-2

Aston-Jones G, Akaoka H, Charlety P, Chouvet G (1991) Serotonin selectively attenuates glutamate-evoked activation of noradrenergic locus coeruleus neurons. J Neurosci 11(3):760–769

Aston-Jones G, Bloom FE (1981) Activity of norepinephrine-containing locus coeruleus neurons in behaving rats anticipates fluctuations in the sleep-waking cycle. J Neurosci 1(8):876–886

Aston-Jones G, Ennis M, Pieribone VA, Nickell WT, Shipley MT (1986) The brain nucleus locus coeruleus: restricted afferent control of a broad efferent network. Science 234(4777):734–737. https://doi.org/10.1126/science.3775363

Aston-Jones G, Rajkowski J, Cohen J (1999) Role of locus coeruleus in attention and behavioral flexibility. Biol Psychiatry 46(9):1309–1320. https://doi.org/10.1016/s0006-3223(99)00140-7

Balemans MG, Mans D, Smith I, Van Benthem J (1983) The influence of GABA on the synthesis of N-acetylserotonin, melatonin, O-acetyl-5-hydroxytryptophol and O-acetyl-5-methoxytryptophol in the pineal gland of the male Wistar rat. Reprod Nutr Dev 23(1):151–160. https://doi.org/10.1051/rnd:19830114

Baraban JM, Aghajanian GK (1980) Suppression of firing activity of 5-HT neurons in the dorsal raphe by alpha-adrenoceptor antagonists. Neuropharmacology 19(4):355–363. https://doi.org/10.1016/0028-3908(80)90187-2

Baude A, Nusser Z, Roberts JD, Mulvihill E, McIlhinney RA, Somogyi P (1993) The metabotropic glutamate receptor (mGluR1 alpha) is concentrated at perisynaptic membrane of neuronal subpopulations as detected by immunogold reaction. Neuron 11(4):771–787. https://doi.org/10.1016/0896-6273(93)90086-7

Bean AJ, During MJ, Roth RH (1990) Effects of dopamine autoreceptor stimulation on the release of colocalized transmitters: in vivo release of dopamine and neurotensin from rat prefrontal cortex. Neurosci Lett 108(1–2):143–148. https://doi.org/10.1016/0304-3940(90)90721-k

Bean AJ, Roth RH (1991) Extracellular dopamine and neurotensin in rat prefrontal cortex in vivo: effects of median forebrain bundle stimulation frequency, stimulation pattern, and dopamine autoreceptors. J Neurosci 11(9):2694–2702

Becquet D, Faudon M, Hery F (1990) The role of serotonin release and autoreceptors in the dorsal raphe nucleus in the control of serotonin release in the cat caudate nucleus. Neuroscience 39(3):639–647. https://doi.org/10.1016/0306-4522(90)90248-3

Benarroch EE (2007) Neurosteroids: endogenous modulators of neuronal excitability and plasticity. Neurology 68(12):945–947. https://doi.org/10.1212/01.wnl.0000257836.09570.e1

Beninato M, Spencer RF (1987) A cholinergic projection to the rat substantia nigra from the pedunculopontine tegmental nucleus. Brain Res 412(1):169–174. https://doi.org/10.1016/0006-8993(87)91455-7

Bespalov A, Muller R, Relo AL, Hudzik T (2016) Drug tolerance: a known unknown in translational neuroscience. Trends Pharmacol Sci 37(5):364–378. https://doi.org/10.1016/j.tips.2016.01.008

Blandina P, Munari L, Provensi G, Passani MB (2012) Histamine neurons in the tuberomammillary nucleus: a whole center or distinct subpopulations? Front Syst Neurosci 6:33. https://doi.org/10.3389/fnsys.2012.00033

Blier P (2001) Pharmacology of rapid-onset antidepressant treatment strategies. J Clin Psychiatry 62(Suppl 15):12–17

Blue ME, Yagaloff KA, Mamounas LA, Hartig PR, Molliver ME (1988) Correspondence between 5-HT2 receptors and serotonergic axons in rat neocortex. Brain Res 453(1–2):315–328. https://doi.org/10.1016/0006-8993(88)90172-2

Bolam JP, Francis CM, Henderson Z (1991) Cholinergic input to dopaminergic neurons in the substantia nigra: a double immunocytochemical study. Neuroscience 41(2–3):483–494. https://doi.org/10.1016/0306-4522(91)90343-m

Bolam JP, Smith Y (1990) The GABA and substance P input to dopaminergic neurons in the substantia nigra of the rat. Brain Res 529(1–2):57–78. https://doi.org/10.1016/0006-8993(90)90811-o

Bonci A, Grillner P, Siniscalchi A, Mercuri NB, Bernardi G (1997) Glutamate metabotropic receptor agonists depress excitatory and inhibitory transmission on rat mesencephalic principal neurons. Eur J Neurosci 9(11):2359–2369. https://doi.org/10.1111/j.1460-9568.1997.tb01653.x

Breslau N, Kilbey MM, Andreski P (1993) Nicotine dependence and major depression. New evidence from a prospective investigation. Arch Gen Psychiatry 50(1):31–35. https://doi.org/10.1001/archpsyc.1993.01820130033006

Briley M, Moret C (1993) Neurobiological mechanisms involved in antidepressant therapies. Clin Neuropharmacol 16(5):387–400. https://doi.org/10.1097/00002826-199310000-00002

Brown RE, Stevens DR, Haas HL (2001) The physiology of brain histamine. Prog Neurobiol 63(6):637–672. https://doi.org/10.1016/s0301-0082(00)00039-3

Bruno V, Copani A, Knopfel T, Kuhn R, Casabona G, Dell'Albani P, Condorelli DF, Nicoletti F (1995) Activation of metabotropic glutamate receptors coupled to inositol phospholipid hydrolysis amplifies NMDA-induced neuronal degeneration in cultured cortical cells. Neuropharmacology 34(8):1089–1098. https://doi.org/10.1016/0028-3908(95)00077-j

Buisson A, Yu SP, Choi DW (1996) DCG-IV selectively attenuates rapidly triggered NMDA-induced neurotoxicity in cortical neurons. Eur J Neurosci 8(1):138–143. https://doi.org/10.1111/j.1460-9568.1996.tb01174.x

Bunney BS, Walters JR, Roth RH, Aghajanian GK (1973) Dopaminergic neurons: effect of antipsychotic drugs and amphetamine on single cell activity. J Pharmacol Exp Ther 185(3):560–571

Calabresi P, Lacey MG, North RA (1989) Nicotinic excitation of rat ventral tegmental neurones in vitro studied by intracellular recording. Br J Pharmacol 98(1):135–140. https://doi.org/10.1111/j.1476-5381.1989.tb16873.x

Calabresi P, Mercuri N, Stanzione P, Stefani A, Bernardi G (1987) Intracellular studies on the dopamine-induced firing inhibition of neostriatal neurons in vitro: evidence for D1 receptor involvement. Neuroscience 20(3):757–771. https://doi.org/10.1016/0306-4522(87)90239-9

Campagna JA, Miller KW, Forman SA (2003) Mechanisms of actions of inhaled anesthetics. N Engl J Med 348(21):2110–2124. https://doi.org/10.1056/NEJMra021261

Cardenas L, Tremblay LK, Naranjo CA, Herrmann N, Zack M, Busto UE (2002) Brain reward system activity in major depression and comorbid nicotine dependence. J Pharmacol Exp Ther 302(3):1265–1271. https://doi.org/10.1124/jpet.302.3.1265

Carlsson A (1977) Dopaminergic autoreceptors: background and implications. Adv Biochem Psychopharmacol 16:439–441

Carr DB, O'Donnell P, Card JP, Sesack SR (1999) Dopamine terminals in the rat prefrontal cortex synapse on pyramidal cells that project to the nucleus accumbens. J Neurosci 19(24):11049–11060

Carr DB, Sesack SR (2000) Projections from the rat prefrontal cortex to the ventral tegmental area: target specificity in the synaptic associations with mesoaccumbens and mesocortical neurons. J Neurosci 20(10):3864–3873

Cartmell J, Schoepp DD (2000) Regulation of neurotransmitter release by metabotropic glutamate receptors. J Neurochem 75(3):889–907. https://doi.org/10.1046/j.1471-4159.2000.0750889.x

Castillo PE, Malenka RC, Nicoll RA (1997) Kainate receptors mediate a slow postsynaptic current in hippocampal CA3 neurons. Nature 388(6638):182–186. https://doi.org/10.1038/40645

Cechetto DF, Saper CB (1988) Neurochemical organization of the hypothalamic projection to the spinal cord in the rat. J Comp Neurol 272(4):579–604. https://doi.org/10.1002/cne.902720410

Chapouthier G, Venault P (2001) A pharmacological link between epilepsy and anxiety? Trends Pharmacol Sci 22(10):491–493. https://doi.org/10.1016/s0165-6147(00)01807-1

Chaput Y, de Montigny C, Blier P (1986) Effects of a selective 5-HT reuptake blocker, citalopram, on the sensitivity of 5-HT autoreceptors: electrophysiological studies in the rat brain. Naunyn Schmiedeberg's Arch Pharmacol 333(4):342–348. https://doi.org/10.1007/BF00500007

Chaput Y, Lesieur P, de Montigny C (1990) Effects of short-term serotonin depletion on the efficacy of serotonin neurotransmission: electrophysiological studies in the rat central nervous system. Synapse 6(4):328–337. https://doi.org/10.1002/syn.890060404

Charpak S, Gahwiler BH, Do KQ, Knopfel T (1990) Potassium conductances in hippocampal neurons blocked by excitatory amino-acid transmitters. Nature 347(6295):765–767. https://doi.org/10.1038/347765a0

Chavis P, Fagni L, Bockaert J, Lansman JB (1995) Modulation of calcium channels by metabotropic glutamate receptors in cerebellar granule cells. Neuropharmacology 34(8):929–937. https://doi.org/10.1016/0028-3908(95)00082-h

Cheng F, McLaughlin PJ, Verderame MF, Zagon IS (2009) The OGF-OGFr axis utilizes the p16INK4a and p21WAF1/CIP1 pathways to restrict normal cell proliferation. Mol Biol Cell 20(1):319–327. https://doi.org/10.1091/mbc.E08-07-0681

Cheng F, Zagon IS, Verderame MF, McLaughlin PJ (2007) The opioid growth factor (OGF)-OGF receptor axis uses the p16 pathway to inhibit head and neck cancer. Cancer Res 67(21):10511–10518. https://doi.org/10.1158/0008-5472.CAN-07-1922

Chergui K, Charlety PJ, Akaoka H, Saunier CF, Brunet JL, Buda M, Svensson TH, Chouvet G (1993) Tonic activation of NMDA receptors causes spontaneous burst discharge of rat midbrain dopamine neurons in vivo. Eur J Neurosci 5(2):137–144. https://doi.org/10.1111/j.1460-9568.1993.tb00479.x

Chergui K, Nomikos GG, Mathe JM, Gonon F, Svensson TH (1996) Burst stimulation of the medial forebrain bundle selectively increase Fos-like immunoreactivity in the limbic forebrain of the rat. Neuroscience 72(1):141–156. https://doi.org/10.1016/0306-4522(95)00513-7

Chergui K, Svenningsson P, Nomikos GG, Gonon F, Fredholm BB, Svennson TH (1997) Increased expression of NGFI-A mRNA in the rat striatum following burst stimulation of the medial forebrain bundle. Eur J Neurosci 9(11):2370–2382. https://doi.org/10.1111/j.1460-9568.1997.tb01654.x

Chiodo LA, Bannon MJ, Grace AA, Roth RH, Bunney BS (1984) Evidence for the absence of impulse-regulating somatodendritic and synthesis-modulating nerve terminal autoreceptors on subpopulations of mesocortical dopamine neurons. Neuroscience 12(1):1–16. https://doi.org/10.1016/0306-4522(84)90133-7

Choi DW (1994) Glutamate receptors and the induction of excitotoxic neuronal death. Prog Brain Res 100:47–51. https://doi.org/10.1016/s0079-6123(08)60767-0

Choi DW, Yokoyama M, Koh J (1988) Zinc neurotoxicity in cortical cell culture. Neuroscience 24(1):67–79. https://doi.org/10.1016/0306-4522(88)90312-0

Choi S, Lovinger DM (1996) Metabotropic glutamate receptor modulation of voltage-gated Ca2+ channels involves multiple receptor subtypes in cortical neurons. J Neurosci 16(1):36–45

Christie MJ, Bridge S, James LB, Beart PM (1985) Excitotoxin lesions suggest an aspartatergic projection from rat medial prefrontal cortex to ventral tegmental area. Brain Res 333(1):169–172. https://doi.org/10.1016/0006-8993(85)90140-4

Clarke PB, Hommer DW, Pert A, Skirboll LR (1987) Innervation of substantia nigra neurons by cholinergic afferents from pedunculopontine nucleus in the rat: neuroanatomical and electrophysiological evidence. Neuroscience 23(3):1011–1019. https://doi.org/10.1016/0306-4522(87)90176-x

Cortes R, Palacios JM (1986) Muscarinic cholinergic receptor subtypes in the rat brain. I. Quantitative autoradiographic studies. Brain Res 362(2):227–238. https://doi.org/10.1016/0006-8993(86)90448-8

Coyle JT, Puttfarcken P (1993) Oxidative stress, glutamate, and neurodegenerative disorders. Science 262(5134):689–695. https://doi.org/10.1126/science.7901908

Crepel V, Aniksztejn L, Ben-Ari Y, Hammond C (1994) Glutamate metabotropic receptors increase a Ca(2+)-activated nonspecific cationic current in CA1 hippocampal neurons. J Neurophysiol 72(4):1561–1569. https://doi.org/10.1152/jn.1994.72.4.1561

Crespi F, Garratt JC, Sleight AJ, Marsden CA (1990) In vivo evidence that 5-hydroxytryptamine (5-HT) neuronal firing and release are not necessarily correlated with 5-HT metabolism. Neuroscience 35(1):139–144. https://doi.org/10.1016/0306-4522(90)90128-q

Damaskos C, Garmpis N, Karatzas T, Kostakis ID, Nikolidakis L, Kostakis A, Kouraklis G (2014) Nuclear receptors in pancreatic tumor cells. Anticancer Res 34(12):6897–6911

Deakin J (2013) The origins of "5-HT and mechanisms of defence" by Deakin and Graeff: a personal perspective. J Psychopharmacol 27(12):1084–1089. https://doi.org/10.1177/0269881113503508

Deakin JF, Graeff FG (1991) 5-HT and mechanisms of defence. J Psychopharmacol 5(4):305–315. https://doi.org/10.1177/026988119100500414

Desai MA, Conn PJ (1991) Excitatory effects of ACPD receptor activation in the hippocampus are mediated by direct effects on pyramidal cells and blockade of synaptic inhibition. J Neurophysiol 66(1):40–52. https://doi.org/10.1152/jn.1991.66.1.40

Di Chiara G (1995) The role of dopamine in drug abuse viewed from the perspective of its role in motivation. Drug Alcohol Depend 38(2):95–137. https://doi.org/10.1016/0376-8716(95)01118-i

Di Chiara G, Porceddu ML, Morelli M, Mulas ML, Gessa GL (1979) Evidence for a GABAergic projection from the substantia nigra to the ventromedial thalamus and to the superior colliculus of the rat. Brain Res 176(2):273–284. https://doi.org/10.1016/0006-8993(79)90983-1

Donahue RN, McLaughlin PJ, Zagon IS (2009) Cell proliferation of human ovarian cancer is regulated by the opioid growth factor-opioid growth factor receptor axis. Am J Physiol Regul Integr Comp Physiol 296(6):R1716–R1725. https://doi.org/10.1152/ajpregu.00075.2009

Dray A, Gonye TJ, Oakley NR, Tanner T (1976) Evidence for the existence of a raphe projection to the substantia nigra in rat. Brain Res 113(1):45–57. https://doi.org/10.1016/0006-8993(76)90005-6

Drevets WC (2000) Neuroimaging studies of mood disorders. Biol Psychiatry 48(8):813–829. https://doi.org/10.1016/s0006-3223(00)01020-9

Edwards AC, Maes HH, Pedersen NL, Kendler KS (2011) A population-based twin study of the genetic and environmental relationship of major depression, regular tobacco use and nicotine dependence. Psychol Med 41(2):395–405. https://doi.org/10.1017/S0033291710000589

Ehlert F, Roeske W, Yamamura H (1994) Molecular biology, pharmacology and brain distribution of subtypes of the muscarinic receptor. In: Bloom F, Kupfer D (eds) Psychopharmacology: the fourth generation of Progress. Raven Press, New York, pp 111–124

Engel SR, Grant KA (2001) Neurosteroids and behavior. Int Rev Neurobiol 46:321–348. https://doi.org/10.1016/s0074-7742(01)46067-3

Erhardt S, Mathe JM, Chergui K, Engberg G, Svensson TH (2002) GABA(B) receptor-mediated modulation of the firing pattern of ventral tegmental area dopamine neurons in vivo. Naunyn Schmiedeberg's Arch Pharmacol 365(3):173–180. https://doi.org/10.1007/s00210-001-0519-5

Ferre S, Cortes R, Artigas F (1994) Dopaminergic regulation of the serotonergic raphe-striatal pathway: microdialysis studies in freely moving rats. J Neurosci 14(8):4839–4846

Fink K, Schlicker E, Neise A, Gothert M (1990) Involvement of presynaptic H3 receptors in the inhibitory effect of histamine on serotonin release in the rat brain cortex. Naunyn Schmiedeberg's Arch Pharmacol 342(5):513–519. https://doi.org/10.1007/BF00169038

Fisher JM, Sossin W, Newcomb R, Scheller RH (1988) Multiple neuropeptides derived from a common precursor are differentially packaged and transported. Cell 54(6):813–822. https://doi.org/10.1016/s0092-8674(88)91131-2

Fonnum F, Gottesfeld Z, Grofova I (1978) Distribution of glutamate decarboxylase, choline acetyltransferase and aromatic amino acid decarboxylase in the basal ganglia of normal and operated rats. Evidence for striatopallidal, striatoentopeduncular and striatonigral GABAergic fibres. Brain Res 143(1):125–138. https://doi.org/10.1016/0006-8993(78)90756-4

Fontanilla D, Johannessen M, Hajipour AR, Cozzi NV, Jackson MB, Ruoho AE (2009) The hallucinogen N,N-dimethyltryptamine (DMT) is an endogenous sigma-1 receptor regulator. Science 323(5916):934–937. https://doi.org/10.1126/science.1166127

Foote S, Aston-Jones G (1995) Pharmacology and physiology of central noradrenergic systems. In: Bloom F, Kupfer D (eds) Psychopharmacology: the fourth generation of Progress. Raven Press, New York, pp 335–345

Foote SL, Bloom FE, Aston-Jones G (1983) Nucleus locus coeruleus: new evidence of anatomical and physiological specificity. Physiol Rev 63(3):844–914. https://doi.org/10.1152/physrev.1983.63.3.844

Foster AC, Kemp JA (2006) Glutamate- and GABA-based CNS therapeutics. Curr Opin Pharmacol 6(1):7–17. https://doi.org/10.1016/j.coph.2005.11.005

Foster AC, Mena EE, Monaghan DT, Cotman CW (1981) Synaptic localization of kainic acid binding sites. Nature 289(5793):73–75. https://doi.org/10.1038/289073a0

Frye CA (2009) Neurosteroids' effects and mechanisms for social, cognitive, emotional, and physical functions. Psychoneuroendocrinology 34(Suppl 1):S143–S161. https://doi.org/10.1016/j.psyneuen.2009.07.005

Fu Q, Heath AC, Bucholz KK, Lyons MJ, Tsuang MT, True WR, Eisen SA (2007) Common genetic risk of major depression and nicotine dependence: the contribution of antisocial traits in a United States veteran male twin cohort. Twin Res Hum Genet 10(3):470–478. https://doi.org/10.1375/twin.10.3.470

Gereau RW, Conn PJ (1995) Multiple presynaptic metabotropic glutamate receptors modulate excitatory and inhibitory synaptic transmission in hippocampal area CA1. J Neurosci 15(10):6879–6889

Gilman AG (1990) Regulation of adenylyl cyclase by G proteins. Adv Second Messenger Phosphoprotein Res 24:51–57

Gobert A, Rivet JM, Audinot V, Newman-Tancredi A, Cistarelli L, Millan MJ (1998) Simultaneous quantification of serotonin, dopamine and noradrenaline levels in single frontal cortex dialysates of freely-moving rats reveals a complex pattern of reciprocal auto- and heteroreceptor-mediated control of release. Neuroscience 84(2):413–429. https://doi.org/10.1016/s0306-4522(97)00565-4

Gobert A, Rivet JM, Cistarelli L, Melon C, Millan MJ (1997) Alpha2-adrenergic receptor blockade markedly potentiates duloxetine- and fluoxetine-induced increases in noradrenaline, dopamine, and serotonin levels in the frontal cortex of freely moving rats. J Neurochem 69(6):2616–2619. https://doi.org/10.1046/j.1471-4159.1997.69062616.x

Gonon FG (1988) Nonlinear relationship between impulse flow and dopamine released by rat midbrain dopaminergic neurons as studied by in vivo electrochemistry. Neuroscience 24(1):19–28. https://doi.org/10.1016/0306-4522(88)90307-7

Grace AA, Bunney BS (1983) Intracellular and extracellular electrophysiology of nigral dopaminergic neurons--1. Identification and characterization. Neuroscience 10(2):301–315. https://doi.org/10.1016/0306-4522(83)90135-5

Grace AA, Bunney BS (1984) The control of firing pattern in nigral dopamine neurons: burst firing. J Neurosci 4(11):2877–2890

Graeff FG, Silveira MC, Nogueira RL, Audi EA, Oliveira RM (1993) Role of the amygdala and periaqueductal gray in anxiety and panic. Behav Brain Res 58(1–2):123–131. https://doi.org/10.1016/0166-4328(93)90097-a

Grenhoff J, Aston-Jones G, Svensson TH (1986) Nicotinic effects on the firing pattern of midbrain dopamine neurons. Acta Physiol Scand 128(3):351–358. https://doi.org/10.1111/j.1748-1716.1986.tb07988.x

Grenhoff J, North RA, Johnson SW (1995) Alpha 1-adrenergic effects on dopamine neurons recorded intracellularly in the rat midbrain slice. Eur J Neurosci 7(8):1707–1713. https://doi.org/10.1111/j.1460-9568.1995.tb00692.x

Grenhoff J, Svensson TH (1989) Clonidine modulates dopamine cell firing in rat ventral tegmental area. Eur J Pharmacol 165(1):11–18. https://doi.org/10.1016/0014-2999(89)90765-6

Grenhoff J, Svensson TH (1993) Prazosin modulates the firing pattern of dopamine neurons in rat ventral tegmental area. Eur J Pharmacol 233(1):79–84. https://doi.org/10.1016/0014-2999(93)90351-h

Grenhoff J, Tung CS, Svensson TH (1988a) The excitatory amino acid antagonist kynurenate induces pacemaker-like firing of dopamine neurons in rat ventral tegmental area in vivo. Acta Physiol Scand 134(4):567–568. https://doi.org/10.1111/j.1748-1716.1998.tb08535.x

Grenhoff J, Ugedo L, Svensson TH (1988b) Firing patterns of midbrain dopamine neurons: differences between A9 and A10 cells. Acta Physiol Scand 134(1):127–132. https://doi.org/10.1111/j.1748-1716.1988.tb08468.x

Gresch PJ, Sved AF, Zigmond MJ, Finlay JM (1995) Local influence of endogenous norepinephrine on extracellular dopamine in rat medial prefrontal cortex. J Neurochem 65(1):111–116. https://doi.org/10.1046/j.1471-4159.1995.65010111.x

Grofova I, Deniau JM, Kitai ST (1982) Morphology of the substantia nigra pars reticulata projection neurons intracellularly labeled with HRP. J Comp Neurol 208(4):352–368. https://doi.org/10.1002/cne.902080406

Guerineau NC, Gahwiler BH, Gerber U (1994) Reduction of resting K+ current by metabotropic glutamate and muscarinic receptors in rat CA3 cells: mediation by G-proteins. J Physiol 474(1):27–33. https://doi.org/10.1113/jphysiol.1994.sp019999

Guitart X, Codony X, Monroy X (2004) Sigma receptors: biology and therapeutic potential. Psychopharmacology 174(3):301–319. https://doi.org/10.1007/s00213-004-1920-9

Gulat-Marnay C, Lafitte A, Arrang JM, Schwartz JC (1989) Modulation of histamine release and synthesis in the brain mediated by alpha 2-adrenoceptors. J Neurochem 53(2):513–518. https://doi.org/10.1111/j.1471-4159.1989.tb07364.x

Heinz A, Mann K, Weinberger DR, Goldman D (2001) Serotonergic dysfunction, negative mood states, and response to alcohol. Alcohol Clin Exp Res 25(4):487–495

Hertel P, Mathe JM, Nomikos GG, Iurlo M, Mathe AA, Svensson TH (1995) Effects of D-amphetamine and phencyclidine on behavior and extracellular concentrations of neurotensin and dopamine in the ventral striatum and the medial prefrontal cortex of the rat. Behav Brain Res 72(1–2):103–114. https://doi.org/10.1016/0166-4328(96)00138-6

Herve D, Pickel VM, Joh TH, Beaudet A (1987) Serotonin axon terminals in the ventral tegmental area of the rat: fine structure and synaptic input to dopaminergic neurons. Brain Res 435(1–2):71–83. https://doi.org/10.1016/0006-8993(87)91588-5

Hodges H, Green S, Glenn B (1987) Evidence that the amygdala is involved in benzodiazepine and serotonergic effects on punished responding but not on discrimination. Psychopharmacology 92(4):491–504. https://doi.org/10.1007/BF00176484

Hokfelt T, Skirboll L, Rehfeld JF, Goldstein M, Markey K, Dann O (1980) A subpopulation of mesencephalic dopamine neurons projecting to limbic areas contains a cholecystokinin-like peptide: evidence from immunohistochemistry combined with retrograde tracing. Neuroscience 5(12):2093–2124. https://doi.org/10.1016/0306-4522(80)90127-x

Holden JE, Jeong Y, Forrest JM (2005) The endogenous opioid system and clinical pain management. AACN Clin Issues 16(3):291–301. https://doi.org/10.1097/00044067-200507000-00003

Huang YS, Lu HL, Zhang LJ, Wu Z (2014) Sigma-2 receptor ligands and their perspectives in cancer diagnosis and therapy. Med Res Rev 34(3):532–566. https://doi.org/10.1002/med.21297

Innis RB, Aghajanian GK (1987) Pertussis toxin blocks autoreceptor-mediated inhibition of dopaminergic neurons in rat substantia nigra. Brain Res 411(1):139–143. https://doi.org/10.1016/0006-8993(87)90690-1

Jacobs BL, Azmitia EC (1992) Structure and function of the brain serotonin system. Physiol Rev 72(1):165–229. https://doi.org/10.1152/physrev.1992.72.1.165

Jacobs BL, Fornal CA (1995) Serotonin and behavior: a general hypothesis. In: Bloom FE, Kupfer DJ (eds) Psychopharmacology: the fourth generation of Progress. Raven Press, New York, pp 461–469

Jacques S (1979) Brain stimulation and reward: "pleasure centers" after twenty-five years. Neurosurgery 5(2):277–283. https://doi.org/10.1227/00006123-197908000-00017

Jodo E, Chiang C, Aston-Jones G (1998) Potent excitatory influence of prefrontal cortex activity on noradrenergic locus coeruleus neurons. Neuroscience 83(1):63–79. https://doi.org/10.1016/s0306-4522(97)00372-2

Johnson SW, North RA (1992a) Opioids excite dopamine neurons by hyperpolarization of local interneurons. J Neurosci 12(2):483–488

Johnson SW, North RA (1992b) Two types of neuron in the rat ventral tegmental area and their synaptic inputs. J Physiol 450:455–468. https://doi.org/10.1113/jphysiol.1992.sp019136

Joshi S, Rajasekaran K, Kapur J (2013) GABAergic transmission in temporal lobe epilepsy: the role of neurosteroids. Exp Neurol 244:36–42. https://doi.org/10.1016/j.expneurol.2011.10.028

Joyce MP, Rayport S (2000) Mesoaccumbens dopamine neuron synapses reconstructed in vitro are glutamatergic. Neuroscience 99(3):445–456. https://doi.org/10.1016/s0306-4522(00)00219-0

Kaur G, Kulkarni SK (2001) Subchronic studies on modulation of feeding behavior and body weight by neurosteroids in female mice. Methods Find Exp Clin Pharmacol 23(3):115–119. https://doi.org/10.1358/mf.2001.23.3.627942

Kehr W, Carlsson A, Lindqvist M, Magnusson T, Atack C (1972) Evidence for a receptor-mediated feedback control of striatal tyrosine hydroxylase activity. J Pharm Pharmacol 24(9):744–747. https://doi.org/10.1111/j.2042-7158.1972.tb09104.x

Khanna A, Walcott BP, Kahle KT (2013) Limitations of current GABA agonists in neonatal seizures: toward GABA modulation via the targeting of neuronal Cl(−) transport. Front Neurol 4:78. https://doi.org/10.3389/fneur.2013.00078

Killen JD, Fortmann SP, Schatzberg A, Hayward C, Varady A (2003) Onset of major depression during treatment for nicotine dependence. Addict Behav 28(3):461–470. https://doi.org/10.1016/s0306-4603(01)00266-0

Kilpatrick GJ, Hagan RM, Gale JD (1996) 5-HT3 and 5-HT4 receptors in terminal regions of the mesolimbic system. Behav Brain Res 73(1–2):11–13. https://doi.org/10.1016/0166-4328(96)00063-0

King SR (2008) Emerging roles for neurosteroids in sexual behavior and function. J Androl 29(5):524–533. https://doi.org/10.2164/jandrol.108.005660
Koch HP, Chamberlin AR, Bridges RJ (1999) Nontransportable inhibitors attenuate reversal of glutamate uptake in synaptosomes following a metabolic insult. Mol Pharmacol 55(6):1044–1048. https://doi.org/10.1124/mol.55.6.1044
Koob GF, Sanna PP, Bloom FE (1998) Neuroscience of addiction. Neuron 21(3):467–476. https://doi.org/10.1016/s0896-6273(00)80557-7
Kuriyama K, Sze PY (1971) Blood-brain barrier to H3-gamma-aminobutyric acid in normal and amino oxyacetic acid-treated animals. Neuropharmacology 10(1):103–108. https://doi.org/10.1016/0028-3908(71)90013-x
Lacey MG, Calabresi P, North RA (1990) Muscarine depolarizes rat substantia nigra zona compacta and ventral tegmental neurons in vitro through M1-like receptors. J Pharmacol Exp Ther 253(1):395–400
Lacey MG, Mercuri NB, North RA (1987) Dopamine acts on D2 receptors to increase potassium conductance in neurones of the rat substantia nigra zona compacta. J Physiol 392:397–416. https://doi.org/10.1113/jphysiol.1987.sp016787
Lambert PD, Gross R, Nemeroff CB, Kilts CD (1995) Anatomy and mechanisms of neurotensin-dopamine interactions in the central nervous system. Ann N Y Acad Sci 757:377–389. https://doi.org/10.1111/j.1749-6632.1995.tb17496.x
Lanctot KL, Herrmann N, Mazzotta P (2001) Role of serotonin in the behavioral and psychological symptoms of dementia. J Neuropsychiatry Clin Neurosci 13(1):5–21. https://doi.org/10.1176/jnp.13.1.5
Lavin A, Grace AA (2001) Stimulation of D1-type dopamine receptors enhances excitability in prefrontal cortical pyramidal neurons in a state-dependent manner. Neuroscience 104(2):335–346. https://doi.org/10.1016/s0306-4522(01)00096-3
Le Moal M (1994) Mesocorticolimbic dopaminergic neurons: functional and regulatory roles. In: Bloom F, Kupfer D (eds) Psychopharmacology: the fourth generation of Progress. Raven Press, New York, pp 283–293
Le Moal M, Simon H (1991) Mesocorticolimbic dopaminergic network: functional and regulatory roles. Physiol Rev 71(1):155–234. https://doi.org/10.1152/physrev.1991.71.1.155
LeDoux JE, Cicchetti P, Xagoraris A, Romanski LM (1990) The lateral amygdaloid nucleus: sensory interface of the amygdala in fear conditioning. J Neurosci 10(4):1062–1069
Leonard BE (2004) Sigma receptors and sigma ligands: bjackground to a pharmacological enigma. Pharmacopsychiatry 37(S3):S166–S170. https://doi.org/10.1055/s-2004-832674
Lievens JC, Dutertre M, Forni C, Salin P, Kerkerian-Le Goff L (1997) Continuous administration of the glutamate uptake inhibitor L-trans-pyrrolidine-2,4-dicarboxylate produces striatal lesion. Brain Res Mol Brain Res 50(1–2):181–189. https://doi.org/10.1016/s0169-328x(97)00182-4
Linden DJ, Smeyne M, Connor JA (1994) Trans-ACPD, a metabotropic receptor agonist, produces calcium mobilization and an inward current in cultured cerebellar Purkinje neurons. J Neurophysiol 71(5):1992–1998. https://doi.org/10.1152/jn.1994.71.5.1992
Longo N, Shuster RC, Griffin LD, Langley SD, Elsas LJ (1992) Activation of insulin receptor signaling by a single amino acid substitution in the transmembrane domain. J Biol Chem 267(18):12416–12419
Longuemare MC, Swanson RA (1995) Excitatory amino acid release from astrocytes during energy failure by reversal of sodium-dependent uptake. J Neurosci Res 40(3):379–386. https://doi.org/10.1002/jnr.490400312
Lujan R, Roberts JD, Shigemoto R, Ohishi H, Somogyi P (1997) Differential plasma membrane distribution of metabotropic glutamate receptors mGluR1 alpha, mGluR2 and mGluR5, relative to neurotransmitter release sites. J Chem Neuroanat 13(4):219–241. https://doi.org/10.1016/s0891-0618(97)00051-3
Luthi A, Gahwiler BH, Gerber U (1996) A slowly inactivating potassium current in CA3 pyramidal cells of rat hippocampus in vitro. J Neurosci 16(2):586–594
Lyons M, Hitsman B, Xian H, Panizzon MS, Jerskey BA, Santangelo S, Grant MD, Rende R, Eisen S, Eaves L, Tsuang MT (2008) A twin study of smoking, nicotine dependence, and major depression in men. Nicotine Tob Res 10(1):97–108. https://doi.org/10.1080/14622200701705332

Malendowicz LK, Rebuffat P, Tortorella C, Nussdorfer GG, Ziolkowska A, Hochol A (2005) Effects of met-enkephalin on cell proliferation in different models of adrenocortical-cell growth. Int J Mol Med 15(5):841–845. https://doi.org/10.3892/ijmm.15.5.841

Mamounas LA, Mullen CA, O'Hearn E, Molliver ME (1991) Dual serotoninergic projections to forebrain in the rat: morphologically distinct 5-HT axon terminals exhibit differential vulnerability to neurotoxic amphetamine derivatives. J Comp Neurol 314(3):558–586. https://doi.org/10.1002/cne.903140312

Martire M, Fuxe K, Pistritto G, Preziosi P, Agnati LF (1989) Neuropeptide Y increases the inhibitory effects of clonidine on potassium evoked 3H-noradrenaline but not 3H-5-hydroxytryptamine release from synaptosomes of the hypothalamus and the frontoparietal cortex of the male Sprague-Dawley rat. J Neural Transm Gen Sect 78(1):61–72. https://doi.org/10.1007/BF01247114

McBean GJ, Roberts PJ (1985) Neurotoxicity of L-glutamate and DL-threo-3-hydroxyaspartate in the rat striatum. J Neurochem 44(1):247–254. https://doi.org/10.1111/j.1471-4159.1985.tb07137.x

McEwen BS (1999) Stress and hippocampal plasticity. Annu Rev Neurosci 22(1):105–122. https://doi.org/10.1146/annurev.neuro.22.1.105

McLaughlin PJ, Levin RJ, Zagon IS (1999) Regulation of human head and neck squamous cell carcinoma growth in tissue culture by opioid growth factor. Int J Oncol 14(5):991–998. https://doi.org/10.3892/ijo.14.5.991

Meldrum BS (2000) Glutamate as a neurotransmitter in the brain: review of physiology and pathology. J Nutr 130(4S Suppl):1007S–1015S. https://doi.org/10.1093/jn/130.4.1007S

Mercuri NB, Grillner P, Bernardi G (1996) N-methyl-D-aspartate receptors mediate a slow excitatory postsynaptic potential in the rat midbrain dopaminergic neurons. Neuroscience 74(3):785–792. https://doi.org/10.1016/0306-4522(96)00189-3

Miczek KA, Fish EW, De Bold JF (2003) Neurosteroids, GABAA receptors, and escalated aggressive behavior. Horm Behav 44(3):242–257. https://doi.org/10.1016/j.yhbeh.2003.04.002

Miller JJ, Richardson TL, Fibiger HC, McLennan H (1975) Anatomical and electrophysiological identification of a projection from the mesencephalic raphe to the caudate-putamen in the rat. Brain Res 97(1):133–136. https://doi.org/10.1016/0006-8993(75)90920-8

Miller NS, Dackis CA, Gold MS (1987) The relationship of addiction, tolerance, and dependence to alcohol and drugs: a neurochemical approach. J Subst Abus Treat 4(3–4):197–207. https://doi.org/10.1016/s0740-5472(87)80014-4

Mineff E, Valtschanoff J (1999) Metabotropic glutamate receptors 2 and 3 expressed by astrocytes in rat ventrobasal thalamus. Neurosci Lett 270(2):95–98. https://doi.org/10.1016/s0304-3940(99)00484-x

Moga MM, Saper CB, Gray TS (1990) Neuropeptide organization of the hypothalamic projection to the parabrachial nucleus in the rat. J Comp Neurol 295(4):662–682. https://doi.org/10.1002/cne.902950409

Mongeau R, Blier P, de Montigny C (1993) In vivo electrophysiological evidence for tonic activation by endogenous noradrenaline of alpha 2-adrenoceptors on 5-hydroxytryptamine terminals in the rat hippocampus. Naunyn Schmiedeberg's Arch Pharmacol 347(3):266–272. https://doi.org/10.1007/BF00167444

Monshipouri M, Jiang H, Lazarovici P (2000) NGF stimulation of erk phosphorylation is impaired by a point mutation in the transmembrane domain of trkA receptor. J Mol Neurosci 14(1–2):69–76. https://doi.org/10.1385/JMN:14:1-2:069

Montgomery SA, Fineberg N (1989) Is there a relationship between serotonin receptor subtypes and selectivity of response in specific psychiatric illnesses? Br J Psychiatry Suppl 8:63–69

Moore RY, Bloom FE (1979) Central catecholamine neuron systems: anatomy and physiology of the norepinephrine and epinephrine systems. Annu Rev Neurosci 2:113–168. https://doi.org/10.1146/annurev.ne.02.030179.000553

Muller EE, Locatelli V, Cocchi D (1999) Neuroendocrine control of growth hormone secretion. Physiol Rev 79(2):511–607. https://doi.org/10.1152/physrev.1999.79.2.511

Murphy BL, Arnsten AF, Goldman-Rakic PS, Roth RH (1996) Increased dopamine turnover in the prefrontal cortex impairs spatial working memory performance in rats and monkeys. Proc Natl Acad Sci U S A 93(3):1325–1329. https://doi.org/10.1073/pnas.93.3.1325

Nishino H, Ono T, Muramoto K, Fukuda M, Sasaki K (1987) Neuronal activity in the ventral tegmental area (VTA) during motivated bar press feeding in the monkey. Brain Res 413(2):302–313. https://doi.org/10.1016/0006-8993(87)91021-3

O'Donnell P, Grace AA (1996) Dopaminergic reduction of excitability in nucleus accumbens neurons recorded in vitro. Neuropsychopharmacology 15(1):87–97. https://doi.org/10.1016/0893-133X(95)00177-F

Olney JW (1989) Excitatory amino acids and neuropsychiatrie disorders. Biol Psychiatry 26(5):505–525. https://doi.org/10.1016/0006-3223(89)90072-3

Onn SP, West AR, Grace AA (2000) Dopamine-mediated regulation of striatal neuronal and network interactions. Trends Neurosci 23(10 Suppl):S48–S56. https://doi.org/10.1016/s1471-1931(00)00020-3

Panula P, Chazot PL, Cowart M, Gutzmer R, Leurs R, Liu WL, Stark H, Thurmond RL, Haas HL (2015) International union of basic and clinical pharmacology. XCVIII. Histamine receptors. Pharmacol Rev 67(3):601–655. https://doi.org/10.1124/pr.114.010249

Petralia RS, Wang YX, Niedzielski AS, Wenthold RJ (1996a) The metabotropic glutamate receptors, mGluR2 and mGluR3, show unique postsynaptic, presynaptic and glial localizations. Neuroscience 71(4):949–976. https://doi.org/10.1016/0306-4522(95)00533-1

Petralia RS, Wang YX, Zhao HM, Wenthold RJ (1996b) Ionotropic and metabotropic glutamate receptors show unique postsynaptic, presynaptic, and glial localizations in the dorsal cochlear nucleus. J Comp Neurol 372(3):356–383. https://doi.org/10.1002/(SICI)1096-9861(19960826)372:3<356::AID-CNE3>3.0.CO;2-1

Phillipson OT (1979) Afferent projections to the ventral tegmental area of Tsai and interfascicular nucleus: a horseradish peroxidase study in the rat. J Comp Neurol 187(1):117–143. https://doi.org/10.1002/cne.901870108

Picciotto MR, Zoli M, Rimondini R, Lena C, Marubio LM, Pich EM, Fuxe K, Changeux JP (1998) Acetylcholine receptors containing the beta2 subunit are involved in the reinforcing properties of nicotine. Nature 391(6663):173–177. https://doi.org/10.1038/34413

Pidoplichko VI, DeBiasi M, Williams JT, Dani JA (1997) Nicotine activates and desensitizes midbrain dopamine neurons. Nature 390(6658):401–404. https://doi.org/10.1038/37120

Pineyro G, Blier P (1999) Autoregulation of serotonin neurons: role in antidepressant drug action. Pharmacol Rev 51(3):533–591

Portas CM, McCarley RW (1994) Behavioral state-related changes of extracellular serotonin concentration in the dorsal raphe nucleus: a microdialysis study in the freely moving cat. Brain Res 648(2):306–312. https://doi.org/10.1016/0006-8993(94)91132-0

Powers ME, Yarrow JF, McCoy SC, Borst SE (2008) Growth hormone isoform responses to GABA ingestion at rest and after exercise. Med Sci Sports Exerc 40(1):104–110. https://doi.org/10.1249/mss.0b013e318158b518

Pozzo Miller LD, Petrozzino JJ, Connor JA (1995) G protein-coupled receptors mediate a fast excitatory postsynaptic current in CA3 pyramidal neurons in hippocampal slices. J Neurosci 15(12):8320–8330

Qin K, Dong C, Wu G, Lambert NA (2011) Inactive-state preassembly of G(q)-coupled receptors and G(q) heterotrimers. Nat Chem Biol 7(10):740–747. https://doi.org/10.1038/nchembio.642

Quirion R, Chiueh CC, Everist HD, Pert A (1985) Comparative localization of neurotensin receptors on nigrostriatal and mesolimbic dopaminergic terminals. Brain Res 327(1–2):385–389. https://doi.org/10.1016/0006-8993(85)91542-2

Raiteri M, Maura G, Folghera S, Cavazzani P, Andrioli GC, Schlicker E, Schalnus R, Gothert M (1990) Modulation of 5-hydroxytryptamine release by presynaptic inhibitory alpha 2-adrenoceptors in the human cerebral cortex. Naunyn Schmiedeberg's Arch Pharmacol 342(5):508–512. https://doi.org/10.1007/BF00169037

Rajkowska G (2000) Histopathology of the prefrontal cortex in major depression: what does it tell us about dysfunctional monoaminergic circuits? Prog Brain Res 126:397–412. https://doi.org/10.1016/S0079-6123(00)26026-3

Ranjbar-Slamloo Y, Fazlali Z (2019) Dopamine and noradrenaline in the brain; overlapping or dissociate functions? Front Mol Neurosci 12(334):334. https://doi.org/10.3389/fnmol.2019.00334

Raymond LA (2017) Striatal synaptic dysfunction and altered calcium regulation in Huntington disease. Biochem Biophys Res Commun 483(4):1051–1062. https://doi.org/10.1016/j.bbrc.2016.07.058

Redgrave P, Prescott TJ, Gurney K (1999) Is the short-latency dopamine response too short to signal reward error? Trends Neurosci 22(4):146–151. https://doi.org/10.1016/s0166-2236(98)01373-3

Reiner P, Fibiger C (1994) Functional heterogeneity of central cholinergic systems. In: Bloom F, Kupfer D (eds) Psychopharmacology: the fourth generation of Progress. Raven Press, New York, pp 147–153

Reisine TD, Soubrie P, Artaud F, Glowinski J (1982) Involvement of lateral habenula-dorsal raphe neurons in the differential regulation of striatal and nigral serotonergic transmission cats. J Neurosci 2(8):1062–1071

Robinson MB, Djali S, Buchhalter JR (1993) Inhibition of glutamate uptake with L-trans-pyrrolidine-2,4-dicarboxylate potentiates glutamate toxicity in primary hippocampal cultures. J Neurochem 61(6):2099–2103. https://doi.org/10.1111/j.1471-4159.1993.tb07447.x

Rodriguez-Moreno A, Herreras O, Lerma J (1997) Kainate receptors presynaptically downregulate GABAergic inhibition in the rat hippocampus. Neuron 19(4):893–901. https://doi.org/10.1016/s0896-6273(00)80970-8

Ross CA, Pearlson GD (1996) Schizophrenia, the heteromodal association neocortex and development: potential for a neurogenetic approach. Trends Neurosci 19(5):171–176. https://doi.org/10.1016/s0166-2236(96)10022-9

Roth R, Elsworth J (1994) Biochemical pharmacology of midbrain dopamine neurons. In: Bloom F, Kupfer D (eds) Psychopharmacology: the fourth generation of Progress. Raven Press, New York, pp 227–243

Rothe T, Bigl V, Grantyn R (1994) Potentiating and depressant effects of metabotropic glutamate receptor agonists on high-voltage-activated calcium currents in cultured retinal ganglion neurons from postnatal mice. Pflugers Arch 426(1–2):161–170. https://doi.org/10.1007/BF00374684

Rothman SM, Olney JW (1987) Excitotoxicity and the NMDA receptor. Trends Neurosci 10(7):299–302. https://doi.org/10.1016/0166-2236(87)90177-9

Rothstein JD (1995) Excitotoxicity and neurodegeneration in amyotrophic lateral sclerosis. Clin Neurosci 3(6):348–359

Sahara Y, Westbrook GL (1993) Modulation of calcium currents by a metabotropic glutamate receptor involves fast and slow kinetic components in cultured hippocampal neurons. J Neurosci 13(7):3041–3050

Salt TE, Eaton SA (1994) The function of metabotropic excitatory amino acid receptors in synaptic transmission in the thalamus: studies with novel phenylglycine antagonists. Neurochem Int 24(5):451–458. https://doi.org/10.1016/0197-0186(94)90093-0

Sassani JW, Zagon IS, McLaughlin PJ (2003) Opioid growth factor modulation of corneal epithelium: uppers and downers. Curr Eye Res 26(5):249–262. https://doi.org/10.1076/ceyr.26.4.249.15427

Sato S, Yin C, Teramoto A, Sakuma Y, Kato M (2008) Sexually dimorphic modulation of GABA(A) receptor currents by melatonin in rat gonadotropin-releasing hormone neurons. J Physiol Sci 58(5):317–322. https://doi.org/10.2170/physiolsci.RP006208

Sawaguchi T, Goldman-Rakic PS (1994) The role of D1-dopamine receptor in working memory: local injections of dopamine antagonists into the prefrontal cortex of rhesus monkeys performing an oculomotor delayed-response task. J Neurophysiol 71(2):515–528. https://doi.org/10.1152/jn.1994.71.2.515

Sawchenko PE, Swanson LW (1982) Immunohistochemical identification of neurons in the paraventricular nucleus of the hypothalamus that project to the medulla or to the spinal cord in the rat. J Comp Neurol 205(3):260–272. https://doi.org/10.1002/cne.902050306
Schalling M, Friberg K, Seroogy K, Riederer P, Bird E, Schiffmann SN, Mailleux P, Vanderhaeghen JJ, Kuga S, Goldstein M et al (1990) Analysis of expression of cholecystokinin in dopamine cells in the ventral mesencephalon of several species and in humans with schizophrenia. Proc Natl Acad Sci U S A 87(21):8427–8431. https://doi.org/10.1073/pnas.87.21.8427
Schultz W (1986) Responses of midbrain dopamine neurons to behavioral trigger stimuli in the monkey. J Neurophysiol 56(5):1439–1461. https://doi.org/10.1152/jn.1986.56.5.1439
Schultz W (1997) Dopamine neurons and their role in reward mechanisms. Curr Opin Neurobiol 7(2):191–197. https://doi.org/10.1016/s0959-4388(97)80007-4
Schultz W (1998) Predictive reward signal of dopamine neurons. J Neurophysiol 80(1):1–27. https://doi.org/10.1152/jn.1998.80.1.1
Schultz W, Apicella P, Ljungberg T (1993) Responses of monkey dopamine neurons to reward and conditioned stimuli during successive steps of learning a delayed response task. J Neurosci 13(3):900–913
Schwartz JC, Diaz J, Pilon C, Sokoloff P (2000) Possible implications of the dopamine D(3) receptor in schizophrenia and in antipsychotic drug actions. Brain Res Brain Res Rev 31(2–3):277–287. https://doi.org/10.1016/s0165-0173(99)00043-0
Schwartz RD (1986) Autoradiographic distribution of high affinity muscarinic and nicotinic cholinergic receptors labeled with [3H]acetylcholine in rat brain. Life Sci 38(23):2111–2119. https://doi.org/10.1016/0024-3205(86)90210-9
Servan-Schreiber D, Printz H, Cohen JD (1990) A network model of catecholamine effects: gain, signal-to-noise ratio, and behavior. Science 249(4971):892–895. https://doi.org/10.1126/science.2392679
Sesack SR, Deutch AY, Roth RH, Bunney BS (1989) Topographical organization of the efferent projections of the medial prefrontal cortex in the rat: an anterograde tract-tracing study with Phaseolus vulgaris leucoagglutinin. J Comp Neurol 290(2):213–242. https://doi.org/10.1002/cne.902900205
Sesack SR, Hawrylak VA, Matus C, Guido MA, Levey AI (1998) Dopamine axon varicosities in the prelimbic division of the rat prefrontal cortex exhibit sparse immunoreactivity for the dopamine transporter. J Neurosci 18(7):2697–2708
Shiah IS, Yatham LN (2000) Serotonin in mania and in the mechanism of action of mood stabilizers: a review of clinical studies. Bipolar Disord 2(2):77–92. https://doi.org/10.1034/j.1399-5618.2000.020201.x
Shigemoto R, Kinoshita A, Wada E, Nomura S, Ohishi H, Takada M, Flor PJ, Neki A, Abe T, Nakanishi S, Mizuno N (1997) Differential presynaptic localization of metabotropic glutamate receptor subtypes in the rat hippocampus. J Neurosci 17(19):7503–7522
Skerry TM, Genever PG (2001) Glutamate signalling in non-neuronal tissues. Trends Pharmacol Sci 22(4):174–181. https://doi.org/10.1016/s0165-6147(00)01642-4
Skuza G, Rogoz Z (2002) A potential antidepressant activity of SA4503, a selective sigma 1 receptor agonist. Behav Pharmacol 13(7):537–543. https://doi.org/10.1097/00008877-200211000-00003
Skuza G, Rogoz Z (2006) The synergistic effect of selective sigma receptor agonists and uncompetitive NMDA receptor antagonists in the forced swim test in rats. J Physiol Pharmacol 57(2):217–229
Smiley JF, Levey AI, Ciliax BJ, Goldman-Rakic PS (1994) D1 dopamine receptor immunoreactivity in human and monkey cerebral cortex: predominant and extrasynaptic localization in dendritic spines. Proc Natl Acad Sci U S A 91(12):5720–5724. https://doi.org/10.1073/pnas.91.12.5720
Smith Y, Bolam JP (1990) The output neurones and the dopaminergic neurones of the substantia nigra receive a GABA-containing input from the globus pallidus in the rat. J Comp Neurol 296(1):47–64. https://doi.org/10.1002/cne.902960105
Sprouse JS, Aghajanian GK (1988) Responses of hippocampal pyramidal cells to putative serotonin 5-HT1A and 5-HT1B agonists: a comparative study with dorsal raphe neurons. Neuropharmacology 27(7):707–715. https://doi.org/10.1016/0028-3908(88)90079-2

Stahl SM (2008) Stahl's essential psychopharmacology: neuroscientific basis and practical applications, 3rd edn. Cambridge University Press, Cambridge

Stanford IM, Lacey MG (1996) Differential actions of serotonin, mediated by 5-HT1B and 5-HT2C receptors, on GABA-mediated synaptic input to rat substantia nigra pars reticulata neurons in vitro. J Neurosci 16(23):7566–7573

Stephenson RP (1956) A modification of receptor theory. Br J Pharmacol Chemother 11(4):379–393. https://doi.org/10.1111/j.1476-5381.1956.tb00006.x

Suaud-Chagny MF, Chergui K, Chouvet G, Gonon F (1992) Relationship between dopamine release in the rat nucleus accumbens and the discharge activity of dopaminergic neurons during local in vivo application of amino acids in the ventral tegmental area. Neuroscience 49(1):63–72. https://doi.org/10.1016/0306-4522(92)90076-e

Sulzer D, Joyce MP, Lin L, Geldwert D, Haber SN, Hattori T, Rayport S (1998) Dopamine neurons make glutamatergic synapses in vitro. J Neurosci 18(12):4588–4602

Summers KL, Giacobini E (1995) Effects of local and repeated systemic administration of (–) nicotine on extracellular levels of acetylcholine, norepinephrine, dopamine, and serotonin in rat cortex. Neurochem Res 20(6):753–759. https://doi.org/10.1007/BF01705545

Svensson TH (1987) Peripheral, autonomic regulation of locus coeruleus noradrenergic neurons in brain: putative implications for psychiatry and psychopharmacology. Psychopharmacology 92(1):1–7. https://doi.org/10.1007/BF00215471

Svensson TH (2000) Brain noradrenaline and the mechanisms of action of antidepressant drugs. Acta Psychiatr Scand Suppl 402:18–27. https://doi.org/10.1034/j.1600-0447.2000.02604.x

Svensson TH, Bunney BS, Aghajanian GK (1975) Inhibition of both noradrenergic and serotonergic neurons in brain by the alpha-adrenergic agonist clonidine. Brain Res 92(2):291–306. https://doi.org/10.1016/0006-8993(75)90276-0

Swanson J, Gupta S, Guinta D, Flynn D, Agler D, Lerner M, Williams L, Shoulson I, Wigal S (1999) Acute tolerance to methylphenidate in the treatment of attention deficit hyperactivity disorder in children. Clin Pharmacol Ther 66(3):295–305. https://doi.org/10.1016/S0009-9236(99)70038-X

Takahashi M, Billups B, Rossi D, Sarantis M, Hamann M, Attwell D (1997) The role of glutamate transporters in glutamate homeostasis in the brain. J Exp Biol 200(Pt 2):401–409

Tazi A, Dantzer R, Le Moal M (1988) Schedule-induced polydipsia experience decreases locomotor response to amphetamine. Brain Res 445(2):211–215. https://doi.org/10.1016/0006-8993(88)91180-8

Tazi A, Dantzer R, Mormede P, Le Moal M (1986) Pituitary-adrenal correlates of schedule-induced polydipsia and wheel running in rats. Behav Brain Res 19(3):249–256. https://doi.org/10.1016/0166-4328(86)90025-2

Tong Y (2014) Seizures caused by pyridoxine (vitamin B6) deficiency in adults: a case report and literature review. Intractable Rare Dis Res 3(2):52–56. https://doi.org/10.5582/irdr.2014.01005

Trendelenburg AU, Starke K, Limberger N (1994) Presynaptic alpha 2A-adrenoceptors inhibit the release of endogenous dopamine in rabbit caudate nucleus slices. Naunyn Schmiedeberg's Arch Pharmacol 350(5):473–481. https://doi.org/10.1007/BF00173016

Trombley PQ, Westbrook GL (1992) L-AP4 inhibits calcium currents and synaptic transmission via a G-protein-coupled glutamate receptor. J Neurosci 12(6):2043–2050

Ugedo L, Grenhoff J, Svensson TH (1989) Ritanserin, a 5-HT2 receptor antagonist, activates midbrain dopamine neurons by blocking serotonergic inhibition. Psychopharmacology 98(1):45–50. https://doi.org/10.1007/BF00442004

Valentino R, Aston-Jones G (1995) Physiological and anatomical determinants of locus coeruleus discharge. In: Bloom F, Kupfer D (eds) Psychopharmacology: the fourth generation of Progress. Raven Press, New York, pp 373–385

Vallee M, Mayo W, Koob GF, Le Moal M (2001) Neurosteroids in learning and memory processes. Int Rev Neurobiol 46:273–320. https://doi.org/10.1016/s0074-7742(01)46066-1

Van Bockstaele EJ, Cestari DM, Pickel VM (1994) Synaptic structure and connectivity of serotonin terminals in the ventral tegmental area: potential sites for modulation of mesolimbic dopamine neurons. Brain Res 647(2):307–322. https://doi.org/10.1016/0006-8993(94)91330-7

van Waarde A, Rybczynska AA, Ramakrishnan NK, Ishiwata K, Elsinga PH, Dierckx RA (2015) Potential applications for sigma receptor ligands in cancer diagnosis and therapy. Biochim Biophys Acta 1848(10 Pt B):2703–2714. https://doi.org/10.1016/j.bbamem.2014.08.022
AM VD (2009) Biology of the NMDA receptor. In: Frontiers in neuroscience. CRC Press, Boca Raton
Vignes M, Bleakman D, Lodge D, Collingridge GL (1997) The synaptic activation of the GluR5 subtype of kainate receptor in area CA3 of the rat hippocampus. Neuropharmacology 36(11–12):1477–1481. https://doi.org/10.1016/s0028-3908(97)00158-5
Wada E, Wada K, Boulter J, Deneris E, Heinemann S, Patrick J, Swanson LW (1989) Distribution of alpha 2, alpha 3, alpha 4, and beta 2 neuronal nicotinic receptor subunit mRNAs in the central nervous system: a hybridization histochemical study in the rat. J Comp Neurol 284(2):314–335. https://doi.org/10.1002/cne.902840212
Wang J, Mack AL, Coop A, Matsumoto RR (2007) Novel sigma (sigma) receptor agonists produce antidepressant-like effects in mice. Eur Neuropsychopharmacol 17(11):708–716. https://doi.org/10.1016/j.euroneuro.2007.02.007
Wang Q, Ren L, Wan Y, Prud'homme GJ (2019) GABAergic regulation of pancreatic islet cells: physiology and antidiabetic effects. J Cell Physiol. https://doi.org/10.1002/jcp.28214
Weiner WJ, Koller WC, Perlik S, Nausieda PA, Klawans HL (1980) Drug holiday and management of Parkinson disease. Neurology 30(12):1257–1261. https://doi.org/10.1212/wnl.30.12.1257
Weissman AD, Su TP, Hedreen JC, London ED (1988) Sigma receptors in post-mortem human brains. J Pharmacol Exp Ther 247(1):29–33
Westerfield JM, Barrera FN (2020) Membrane receptor activation mechanisms and transmembrane peptide tools to elucidate them. J Biol Chem 295(7):1792–1814. https://doi.org/10.1074/jbc.REV119.009457
Westerink BH, de Vries JB (1989) On the mechanism of neuroleptic induced increase in striatal dopamine release: brain dialysis provides direct evidence for mediation by autoreceptors localized on nerve terminals. Neurosci Lett 99(1–2):197–202. https://doi.org/10.1016/0304-3940(89)90289-9
Whitaker-Azmitia PM, Azmitia EC (1989) Stimulation of astroglial serotonin receptors produces culture media which regulates growth of serotonergic neurons. Brain Res 497(1):80–85. https://doi.org/10.1016/0006-8993(89)90972-4
Whitaker-Azmitia PM, Azmitia EC (1991) Serotonin trophic factors in development, plasticity and aging. In: Serotonin: molecular biology, receptors and functional effects. Birkhäuser, Basel. https://doi.org/10.1007/978-3-0348-7259-1_5
White FJ, Wang RY (1986) Electrophysiological evidence for the existence of both D-1 and D-2 dopamine receptors in the rat nucleus accumbens. J Neurosci 6(1):274–280
Wise RA (2008) Dopamine and reward: the anhedonia hypothesis 30 years on. Neurotox Res 14(2–3):169–183. https://doi.org/10.1007/BF03033808
Wouters MM, Vicario M, Santos J (2016) The role of mast cells in functional GI disorders. Gut 65(1):155–168. https://doi.org/10.1136/gutjnl-2015-309151
Wu Y, McLaughlin PJ, Zagon IS (1998) Ontogeny of the opioid growth factor, [Met5]-enkephalin, preproenkephalin gene expression, and the ζ opioid receptor in the developing and adult aorta of rat. Dev Dyn 211(4):327–337. https://doi.org/10.1002/(sici)1097-0177(199804)211:4<327::Aid-aja4>3.0.Co;2-j
Yanai K, Tashiro M (2007) The physiological and pathophysiological roles of neuronal histamine: an insight from human positron emission tomography studies. Pharmacol Ther 113(1):1–15. https://doi.org/10.1016/j.pharmthera.2006.06.008
Yang S, Bhardwaj A, Cheng J, Alkayed NJ, Hurn PD, Kirsch JR (2007) Sigma receptor agonists provide neuroprotection in vitro by preserving bcl-2. Anesth Analg 104(5):1179–1184, tables of contents. https://doi.org/10.1213/01.ane.0000260267.71185.73
Yeomans J, Baptista M (1997) Both nicotinic and muscarinic receptors in ventral tegmental area contribute to brain-stimulation reward. Pharmacol Biochem Behav 57(4):915–921. https://doi.org/10.1016/s0091-3057(96)00467-4

Yeomans JS, Kofman O, McFarlane V (1985) Cholinergic involvement in lateral hypothalamic rewarding brain stimulation. Brain Res 329(1–2):19–26. https://doi.org/10.1016/0006-8993(85)90508-6

Yeomans JS, Mathur A, Tampakeras M (1993) Rewarding brain stimulation: role of tegmental cholinergic neurons that activate dopamine neurons. Behav Neurosci 107(6):1077–1087. https://doi.org/10.1037//0735-7044.107.6.1077

Young MR, Fleetwood-Walker SM, Mitchell R, Dickinson T (1995) The involvement of metabotropic glutamate receptors and their intracellular signalling pathways in sustained nociceptive transmission in rat dorsal horn neurons. Neuropharmacology 34(8):1033–1041. https://doi.org/10.1016/0028-3908(95)00071-d

Zagon IS, Smith JP, McLaughlin PJ (1999a) Human pancreatic cancer cell proliferation in tissue culture is tonically inhibited by opioid growth factor. Int J Oncol 14(3):577–584. https://doi.org/10.3892/ijo.14.3.577

Zagon IS, Verderame MF, McLaughlin PJ (2002) The biology of the opioid growth factor receptor (OGFr). Brain Res Brain Res Rev 38(3):351–376. https://doi.org/10.1016/s0165-0173(01)00160-6

Zagon IS, Wu Y, McLaughlin PJ (1999b) Opioid growth factor and organ development in rat and human embryos. Brain Res 839(2):313–322. https://doi.org/10.1016/s0006-8993(99)01753-9

Zetterstrom T, Ungerstedt U (1984) Effects of apomorphine on the in vivo release of dopamine and its metabolites, studied by brain dialysis. Eur J Pharmacol 97(1–2):29–36. https://doi.org/10.1016/0014-2999(84)90509-0

Zhang H, Cuevas J (2005) Sigma receptor activation blocks potassium channels and depresses neuroexcitability in rat intracardiac neurons. J Pharmacol Exp Ther 313(3):1387–1396. https://doi.org/10.1124/jpet.105.084152

Zheng F, Gallagher JP (1992) Metabotropic glutamate receptors are required for the induction of long-term potentiation. Neuron 9(1):163–172. https://doi.org/10.1016/0896-6273(92)90231-2

Zhou FM, Hablitz JJ (1999) Activation of serotonin receptors modulates synaptic transmission in rat cerebral cortex. J Neurophysiol 82(6):2989–2999. https://doi.org/10.1152/jn.1999.82.6.2989

Zuo Y, Lu H, Vaupel DB, Zhang Y, Chefer SI, Rea WR, Moore AV, Yang Y, Stein EA (2011) Acute nicotine-induced tachyphylaxis is differentially manifest in the limbic system. Neuropsychopharmacology 36(12):2498–2512. https://doi.org/10.1038/npp.2011.139

Second Messenger Systems

8

Contents

As described in previous chapters, the binding of a neurotransmitter on a postsynaptic receptor triggers a cascade of membrane-bound and intracellular processes mediated by second messenger systems. These systems trigger changes in the physiology of postsynaptic neurons.

8.1 From the Synapsis to the Second Messenger

The signal is transmitted intracellularly mainly with the following mechanisms:

- Transmission from the receptor to the second messenger via a *G-protein*.
- Transmission from the receptor to the second messenger via various mechanisms and molecules, for example, *arachidonic acid*.
- Activation of the receptor causes directly an *influx of Ca^{2+} ions*, which play the role of the second messenger.
- *Diffusion of gaseous second messengers* directly through the cellular membrane without the need for any other receptor or mechanism.

K. N. Fountoulakis, *The Human Connectome*,
https://doi.org/10.1007/978-3-031-10351-3_8

From the Synapsis to the Second Messenger

- G-protein.
- Through various mechanisms and molecules, for example, arachidonic acid.
- Direct influx of Ca^{2+} ions.
- Diffusion of gaseous second messengers.
- Adjusting proteins.
- Scaffold proteins.
- Cross-communication.

Proteins without any enzymatic function participate also in the transmission pathways. These proteins usually function as adjusting proteins and lead the organization of signaling protein complexes. In these mechanisms, proteins that have significant role in interacting with many other proteins simultaneously are called *scaffold proteins*. Cross-communication is a basic characteristic of these pathways and leads to extremely high degrees of complexity. The molecules that mediate the above mechanisms and functions are the following:

8.1.1 G-Protein

This mechanism starts with the interaction between a receptor and a G-protein, which binds a guanine nucleotide and subsequently is linked with the various intracellular enzyme systems of second messengers. The G-protein and the receptors form the *G-protein-linked receptor* (GPLR) (Qin et al. 2011). These proteins belong to a larger group called GTPases, they act as internal switches inside cells, and their activity is regulated by factors which determine their ability to bind and hydrolyze the nucleotide guanosine triphosphate (GTP) to guanosine diphosphate (GDP). When linked to GTP, they are in active state (on), while when they are linked to GDP, they are inactive (off). There are three types of G-proteins, the small monomeric, the heterotrimeric G-protein complexes which are composed of alpha (α)-, beta (β)-, and gamma (γ)-subunits, and the dimeric complexes which are made of beta- and gamma-subunits (Hurowitz et al. 2000; Clapham and Neer 1997; Neves et al. 2002).

G-Protein

- Protein that binds a guanine nucleotide.
- Linked with the various intracellular enzyme systems.
- G-protein-linked receptor (GPLR) complex.
- Excitatory regulating G-protein (Gs).
- Inhibitory regulating G-protein G (Gi).

- Active state (on).
- Inactive state (off).
- Small monomeric G-proteins.
- Heterotrimeric G-protein complexes (subunits α, β, and γ).
- Dimeric complexes (subunits β and γ).
- With the neurotransmitter binding on the receptor.
  - G-protein exchanges GDP with GTP.
  - Detachment of alpha-subunit.
  - The free alpha-subunit moves freely on the inner surface of the cell membrane until it meets the main effector.
  - The main effector produces the second messenger signal.

With the neurotransmitter binding on the receptor (Fig. 8.1), G-protein exchanges GDP which was linked to its alpha-subunit with a GTP molecule, and this results in the detachment of this alpha-subunit from the protein complex. The main function of beta- and gamma-subunits is the binding and deactivation of the alpha-subunit (Clapham and Neer 1997). After its release, the free alpha-subunit can move freely on the inner surface of the cell membrane, until it meets another protein which constitutes the *main effector* (e.g., adenylate cyclase or phospholipase C) and produces the second messenger signal which is then widely diffused inside the cell, often with the activation of a *secondary effector* also (Qin et al. 2011). All subunits remain attached on the inner surface of the cell membrane. Subunits beta and gamma exert a neurotransmitting function also, either independently or in collaboration with the alpha-subunit, and their effect on ion channels has been proven, especially on those

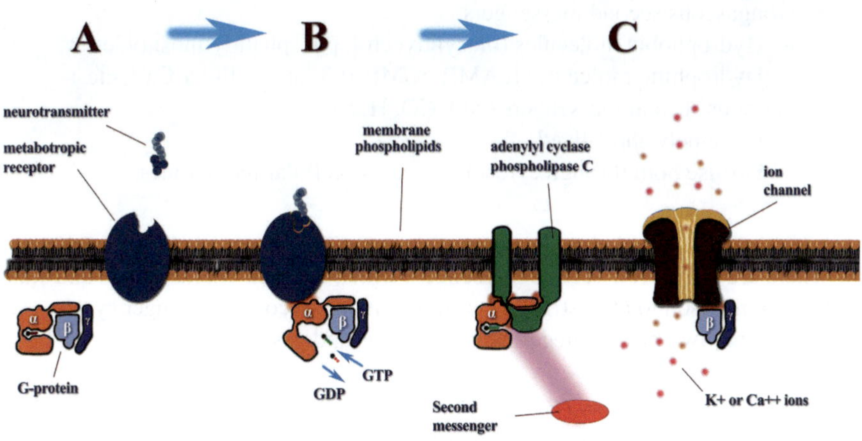

Fig. 8.1 (**a**) The neurotransmitter is not bound on the receptor and the G-protein is in its inactive (off) form with all three subunits bound together. (**b**) With the binding of the neurotransmitter on the receptor, the G-protein exchanges GDP with GTP and the alpha-subunit is released. (**c**) The alpha-subunit is linked with the effector and a second messenger is activated, while the beta- and gamma-subunits exert their action on an ion channel (copyright KN Fountoulakis 2021)

of potassium (Logothetis et al. 1987) and calcium (Ikeda 1996; Gulati et al. 2018), but also on the regulation of cAMP activity (Tang and Gilman 1991). There are two G-protein types, the excitatory (Gs) and the inhibitory (Gi), which bind to excitatory or inhibitory receptors, respectively, and they activate or inhibit their target enzyme.

8.1.2 Effectors

The most important effectors are two, adenylate cyclase and phospholipase C:

- *Adenylate cyclase (adenylyl cyclase)* is a 12-segment transmembrane glycoprotein which catalyzes the production of cAMP from ATP with the assistance of Mg^{2+} or Mn^{2+}. The resulting cAMP constitutes the second messenger, and it is an allosteric activator of protein kinase A.
- The activation of *phospholipase C* leads to the production of 1,4,5-inositol trisphosphate (IP3) and diacylglycerol (DAG) from phosphatidylinositol 4,5-bisphosphate (PIP2), which constitute two distinct second messengers which in turn increase the intracellular concentration of calcium ions and activate protein kinase C (Kadamur and Ross 2013; Ellis et al. 1998; Balla 2013).

8.2 Second Messengers

8.2.1 Classes of Second Messenger Molecules

Second Messengers

- Nongaseous second messengers.
  - Hydrophobic molecules (diacylglycerol, phosphatidylinositol, etc.)
  - Hydrophilic molecules (cAMP, cGMP, IP3, intracellular Ca^{2+}, etc.)
- Gaseous second messengers (NO, CO, H2S).
  - Extremely short-lived.
  - Diffuse both through cytosol and across cellular membranes.

There are only a few second messengers which are well characterized and studied, in comparison to almost 100 neurotransmitters. Second messenger systems are classified into two broad categories: nongaseous and gaseous:

- The *nongaseous second messengers* include *hydrophobic* (water-insoluble) molecules like diacylglycerol and phosphatidylinositol and *hydrophilic* (water-soluble) molecules like cAMP, cGMP, IP3, and intracellular Ca^{2+}. Hydrophobic messengers are membrane-associated and diffuse from the plasma membrane into the intermembrane space, while hydrophilic messengers are located within the cytosol (Majerus 1992; Arachidonate related lipid mediators 1990; Needleman et al. 1986). The most studied nongaseous second messenger systems are those of cAMP, phosphatidylinositol, arachidonic acid, cGMP, and tyrosine kinase systems. One interesting property of arachidonic acid and its metabolites is that they are *highly lipid soluble*, and thus they diffuse through membranes. In this way, they can act also on neighboring cells, acting as a type of transcellular synaptic messengers (Tallima and El Ridi 2018).

- The *gaseous second messengers* (Mustafa et al. 2009) are highly diffusible and include nitric oxide (NO), carbon monoxide (CO), and hydrogen sulfide (H2S). They are extremely short-lived. All of them can diffuse both through cytosol and across cellular membranes. The criteria of "gasotransmitter" are the following (Wang 2002; Wang 2004):
  - Small molecule of a gas.
  - Freely permeable to membranes and its effects do not rely on membrane receptors.
  - It is endogenously and enzymatically generated, and its production is regulated.
  - It has well-defined and specific functions at physiologically relevant concentrations.
  - Functions of this endogenous gas can be mimicked by its exogenously applied counterpart.
  - Its cellular effects may or may not be mediated by second messengers but should have specific cellular and molecular targets.

8.2.2 Second Messenger Pathways and Systems

Second Messenger Pathways and Systems

- cAMP.
  - Protein kinase A (PKA).
  - Exchange protein activated by cAMP (EPAC).
  - GTPase.
  - Regulatory effect directly on cyclic nucleotide-gated channels (CNGCs).
- cGMP.
  - Protein kinase G (PKG).
- Phosphatidylinositol.
  - Phospholipase C.
  - Production of 1,4,5-inositol trisphosphate (IP3) and diacylglycerol (DAG) from phosphatidylinositol 4,5-bisphosphate (PIP2).
  - Increase of intracellular Ca^{2+}.
  - Activation of protein kinase C (PKC).
- Ca^{2+} ions.
  - Collaboration with DAG for the activation of PKC and the calcium-modulated protein (calmodulin or Cam) kinase pathway.

As already said, there are several systems—second messenger pathways, but the most important and the better studied are those of cAMP, cGMP, phosphatidylinositol, της diacylglycerol, arachidonic acid, and calcium (Fig. 8.2):

- The *cAMP pathway* (Gilman 1987) is linked either to an excitatory (Gs) or an inhibitory regulatory G-protein (Gi). Most of its actions are due to the activation of protein kinase A (Sassone-Corsi 2012). Catalysis of cAMP by a cAMP phosphodiesterase leads to the deactivation of the kinase. Another enzyme which is activated by cAMP is the exchange protein activated by cAMP (EPAC), which in turn activates a GTPase. cAMP exerts a regulatory action directly on a group of cation channels with a tetramer structure and binding sites for cAMP and cGMP, called cyclic nucleotide-gated channels (CNGCs). The binding of cAMP opens these channels and leads to the depolarization of the cell membrane.
- *cGMP* is responsible for the regulation of a number of functions of the tissues anywhere in the body as well as in neurons. In contrast to other second messen-

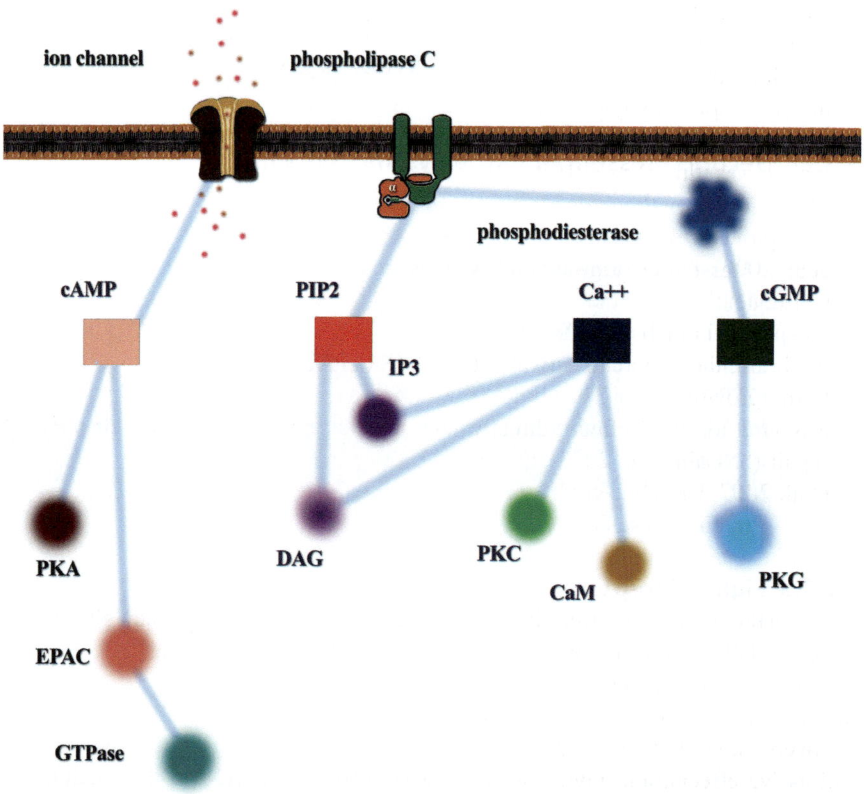

Fig. 8.2 Chart of the main second messenger systems and pathways (copyright KN Fountoulakis 2021)

gers, cGMP is not directly regulated by G-proteins but indirectly through phosphodiesterases. It activates protein kinase G (PKG) which is also a serine/threonine kinase (Francis et al. 2010).

- The *phosphatidylinositol pathway* includes the activation of phospholipase C by the G protein-receptor complex, and this results in the production of *1,4,5-inositol trisphosphate (IP3)* and *diacylglycerol (DAG)* from phosphatidylinositol 4,5-bisphosphate (PIP2), which constitute two distinct second messengers. They in turn increase the intracellular concentration of calcium ions and activate protein kinase C (Kadamur and Ross 2013; Ellis et al. 1998; Balla 2013; Gilman 1987):
  - *IP3* is bound to specific receptors on the membrane of the smooth endoplasmic reticulum and of the mitochondria and opens Ca^{2+} ion channels, thus releasing these ions from their intracellular reservoirs. This abrupt increase in intracellular calcium triggers the activation of several enzymes and exerts a regulatory effect on cellular functions, including glucagon breakdown and controlled exocytosis (Hughes and Putney 1990; Wedegaertner et al. 1995; Yoshida and Imai 1997).
  - *DAG*, in contrast to IP3, is a lipophilic molecule and remains on the cellular membrane. Its main action concerns the activation of protein kinase C (PKC), which is a serine/threonine kinase (Mochly-Rosen et al. 2012) with the ability to modify the catalytic properties of various proteins, by phosphorylating them. At resting state, PKC can be found in the cytoplasm in an inactive form; however, with the hydrolysis of PIP2 and the combined production of DAG and the increase in cytoplasmic Ca^{2+} because of IP3 activity, PKC is activated through a complex mechanism that also includes autophosphorylation (self-phosphorylation).
- The second messenger pathway which concerns the intracellular action of *Ca^{2+} ions* is involved in a variety of actions that include the collaboration with DAG for the activation of PKC and the calcium-modulated protein (calmodulin or Cam) kinase pathway. These two pathways are connected by the complex Ca^{2+}-CaM, which constitutes a regulatory subunit of adenyl cyclase and phosphodiesterase in the cAMP pathway. Calcium ions are normally stored in intracellular reservoirs in the endoplasmic reticulum and elsewhere and are released after specific stimulation of the neuron. Their intracellular density increases also with the opening of their channels and the subsequent influx of ions from the extracellular space.

Eventually, all second messenger systems produce changes in *specific target proteins* within the cell. The phosphorylation mediated by protein kinases and the action of phospholipases are central in the whole process (Tanaka and Nishizuka 1994). Protein kinases can also induce the synthesis of new proteins by altering gene expression. This leads to long-term changes which are important for neuronal development and for the long-term memory. To make things more complex, the effects of second messenger systems are broadened through interaction between them with parallel, convergent, and antagonistic actions.

References

Arachidonate related lipid mediators (1990) Methods Enzymol 187:1–628

Balla T (2013) Phosphoinositides: tiny lipids with giant impact on cell regulation. Physiol Rev 93(3):1019–1137. https://doi.org/10.1152/physrev.00028.2012

Clapham DE, Neer EJ (1997) G protein beta gamma subunits. Annu Rev Pharmacol Toxicol 37:167–203. https://doi.org/10.1146/annurev.pharmtox.37.1.167

Ellis MV, James SR, Perisic O, Downes CP, Williams RL, Katan M (1998) Catalytic domain of phosphoinositide-specific phospholipase C (PLC). Mutational analysis of residues within the active site and hydrophobic ridge of plcdelta1. J Biol Chem 273(19):11650–11659. https://doi.org/10.1074/jbc.273.19.11650

Francis SH, Busch JL, Corbin JD, Sibley D (2010) cGMP-dependent protein kinases and cGMP phosphodiesterases in nitric oxide and cGMP action. Pharmacol Rev 62(3):525–563. https://doi.org/10.1124/pr.110.002907

Gilman AG (1987) G proteins: transducers of receptor-generated signals. Annu Rev Biochem 56:615–649. https://doi.org/10.1146/annurev.bi.56.070187.003151

Gulati S, Jin H, Masuho I, Orban T, Cai Y, Pardon E, Martemyanov KA, Kiser PD, Stewart PL, Ford CP, Steyaert J, Palczewski K (2018) Targeting G protein-coupled receptor signaling at the G protein level with a selective nanobody inhibitor. Nat Commun 9(1):1996. https://doi.org/10.1038/s41467-018-04432-0

Hughes AR, Putney JW Jr (1990) Inositol phosphate formation and its relationship to calcium signaling. Environ Health Perspect 84:141–147. https://doi.org/10.1289/ehp.9084141

Hurowitz EH, Melnyk JM, Chen YJ, Kouros-Mehr H, Simon MI, Shizuya H (2000) Genomic characterization of the human heterotrimeric G protein alpha, beta, and gamma subunit genes. DNA Res 7(2):111–120. https://doi.org/10.1093/dnares/7.2.111

Ikeda SR (1996) Voltage-dependent modulation of N-type calcium channels by G-protein beta gamma subunits. Nature 380(6571):255–258. https://doi.org/10.1038/380255a0

Kadamur G, Ross EM (2013) Mammalian phospholipase C. Annu Rev Physiol 75:127–154. https://doi.org/10.1146/annurev-physiol-030212-183750

Logothetis DE, Kurachi Y, Galper J, Neer EJ, Clapham DE (1987) The beta gamma subunits of GTP-binding proteins activate the muscarinic K^+ channel in heart. Nature 325(6102):321–326. https://doi.org/10.1038/325321a0

Majerus PW (1992) Inositol phosphate biochemistry. Annu Rev Biochem 61:225–250. https://doi.org/10.1146/annurev.bi.61.070192.001301

Mochly-Rosen D, Das K, Grimes KV (2012) Protein kinase C, an elusive therapeutic target? Nat Rev Drug Discov 11(12):937–957. https://doi.org/10.1038/nrd3871

Mustafa AK, Gadalla MM, Snyder SH (2009) Signaling by gasotransmitters. Sci Signal 2(68):re2. https://doi.org/10.1126/scisignal.268re2

Needleman P, Turk J, Jakschik BA, Morrison AR, Lefkowith JB (1986) Arachidonic acid metabolism. Annu Rev Biochem 55:69–102. https://doi.org/10.1146/annurev.bi.55.070186.000441

Neves SR, Ram PT, Iyengar R (2002) G protein pathways. Science 296(5573):1636–1639. https://doi.org/10.1126/science.1071550

Qin K, Dong C, Wu G, Lambert NA (2011) Inactive-state preassembly of G(q)-coupled receptors and G(q) heterotrimers. Nat Chem Biol 7(10):740–747. https://doi.org/10.1038/nchembio.642

Sassone-Corsi P (2012) The cyclic AMP pathway. Cold Spring Harb Perspect Biol 4(12):a011148. https://doi.org/10.1101/cshperspect.a011148

Tallima H, El Ridi R (2018) Arachidonic acid: physiological roles and potential health benefits—a review. J Adv Res 11:33–41. https://doi.org/10.1016/j.jare.2017.11.004

Tanaka C, Nishizuka Y (1994) The protein kinase C family for neuronal signaling. Annu Rev Neurosci 17:551–567. https://doi.org/10.1146/annurev.ne.17.030194.003003

Tang WJ, Gilman AG (1991) Type-specific regulation of adenylyl cyclase by G protein beta gamma subunits. Science 254(5037):1500–1503. https://doi.org/10.1126/science.1962211

Wang R (2002) Two's company, three's a crowd: can H2S be the third endogenous gaseous transmitter? FASEB J 16(13):1792–1798. https://doi.org/10.1096/fj.02-0211hyp
Wang R (2004) Signal transduction and the Gasotransmitters: NO, CO and H2S in biology and medicine. Humana Press, New Jersey, USA
Wedegaertner PB, Wilson PT, Bourne HR (1995) Lipid modifications of trimeric G proteins. J Biol Chem 270(2):503–506. https://doi.org/10.1074/jbc.270.2.503
Yoshida Y, Imai S (1997) Structure and function of inositol 1,4,5-trisphosphate receptor. Jpn J Pharmacol 74(2):125–137. https://doi.org/10.1254/jjp.74.125

Neurotrophic Factors

9

Contents

9.1 Basics for Neurotrophic Factors

Basics for Neurotrophic Factors

- Development.
- Survival.
- General functioning of neurons.
- Cell differentiation.
- Synapse formation.
- Nervous system plasticity.
- Neurogenesis.

Neurotrophic factors or *neurotrophins* is a family of proteins that play an important role in the development, survival, and general function of neurons in the brain as well as in the peripheral nervous system (Skaper 2018; Thoenen 1995; Hempstead 2006; Reichardt 2006; Allen and Dawbarn 2006). Apart from preventing neuronal death, they play an important role in the differentiation of cells, the formation of synapses, and the general plasticity of the nervous system (Lewin and Barde 1996;

K. N. Fountoulakis, *The Human Connectome*,
https://doi.org/10.1007/978-3-031-10351-3_9

Bibel and Barde 2000; Dechant and Barde 2002; Huang et al. 2001; Poo 2001; Lu et al. 2005). Their main central action is the balancing between the volume of the target tissue with the number of afferent neuronal terminals.

It is interesting that in spite of the fact that the overwhelming majority of neurons in the brain of mammals are shaped already before birth, brain regions (e.g., the hippocampus) keep their ability to differentiate new neurons from undifferentiated stem cells. This process is called *neurogenesis*, and neurotrophic factors play an important role in it (Eriksson et al. 1998).

9.2 Neurotrophic Factor Molecules

The following are four main neurotrophic factors identified so far (Malenka et al. 2009; Sanes et al. 2019):

- The *nerve growth factor (NGF)* is a protein with important role in the survival and maintenance of neurons, mainly of the sympathetic peripheral nervous system and the sensory neurons of the brain (Levi-Montalcini 1987; Dreyfus 1989; Hamburger and Levi-Montalcini 1949). During development, the target cell secretes this factor which binds on the TrkA receptors. Subsequently, endocytosis follows, and afterward the NGF is transferred across the axon to the neuronal body of the neuron whose synapse with the target cell aims to rescue (Ginty and Segal 2002; Hamburger et al. 1981; Harrington and Ginty 2013).
- The *brain-derived neurotrophic factor (BDNF)* was originally identified in the brain, but later it was found that it is also expressed in the peripheral body, with action in the brain as well as in the peripheral nervous system. In the central nervous system, it can be found in the cortex, the hippocampus, and the cerebellum and plays a role in memory, learning, and higher cognitive function. It helps in the survival of already existing neurons but also in the development of new ones by facilitating the differentiation of stem cells and the formation of new axis, dendrites, and synapses (Barde et al. 1982; Leibrock et al. 1989).
- The *neurotrophin-3 (NT3)* manifests its action in the brain but also in the peripheral nervous system and plays a role in the survival of already existing neurons but also in the development of new ones by facilitating the differentiation of stem cells and the formation of new synapses. It is an extremely dynamic molecule since it can activate two tyrosine kinase receptors (TrkC and TrkB) (Bothwell 1995; Ibanez 1995).
- The *neurotrophin-4 (NT4)* is mainly bound on TrkB of tyrosine kinase receptors, and it has also been labeled with numbers 5 and 4/5 (Bothwell 1995; Ibanez 1995; Ip et al. 1992; Berkemeier et al. 1991; Malenka et al. 2009).

Neurotrophic Factor Molecules

Main neurotrophic factors

- Nerve growth factor (NGF).
- Brain-derived neurotrophic factor (BDNF).
- Neurotrophin-3 (NT3).
- Neurotrophin-4 (NT4).

Other substances with similar action

- Ciliary neurotrophic factor (CNTF).
- Agents belonging to the family produced by the glial cells.
- Insulin-like growth factor.
- Basic fibroblast growth factor.
- Transforming growth factor-β (TGFβ).
- Sonic hedgehog.

Additionally, neurotrophins 6 and 7 have been identified but so far only in the zebra fish (Gotz et al. 1994; Lai et al. 1998; Sanes et al. 2019). There are a number of molecules including the ciliary neurotrophic factor (CNTF) or agents belonging to the family produced by the glial cells (including the glial cell line-derived neurotrophic factor, GDNF), the insulin-like growth factor, the basic fibroblast growth factor, the transforming growth factor-β (TGFβ), and the sonic hedgehog, which play a role in survival and the development of neurons through their binding on tyrosine kinase receptors (Lindsay and Yancopoulos 1996; Barde 1994; Kullander et al. 1997; Butte 2001; Ip and Yancopoulos 1996; Dore et al. 1997; Walicke et al. 1986; Farkas et al. 2003; Miao et al. 1997).

9.3 Neurotrophic Factor Receptors

All four basic neurotrophic factors exert their action through the activation of the neurotrophin receptor p75 (p75NTR), which belongs to the tumor necrosis factor (TNF) family of receptors, but more specifically through binding with one or more of the tropomyosin-related kinase receptors (TrkA, TrkB, και TrkC) (Arevalo and Wu 2006; Bothwell 2016; Radeke et al. 1987; Rodriguez-Tebar et al. 1990; Rodriguez-Tebar et al. 1992; He and Garcia 2004).

Neurotrophic Factor Receptors

- Neurotrophin receptor p75 (p75NTR).
  - NF-κB pathway.
  - Jun-kinase pathway.
  - Rho (rhodopsin) family GTPase pathway.
- Family of tropomyosin-related kinase (Trk) receptors.
  - TrkA binds NGF.
  - TrkB binds BDNF and NT-4.
  - TrkC binds NT-3.
  - Ras pathway.
  - Phosphoinositide 3-kinase (PI3) pathway through Ras or GAB1.
  - Phospholipase C-γ1 (PLC-γ1) pathway.
- The Trk receptor function is regulated by the p75LNTR receptor at multiple levels.

- The *p75 neurotrophin receptor (p75NTR)* was initially identified as a nerve growth factor receptor, but later it was proven that it binds all neurotrophic factors (Rodriguez-Tebar et al. 1990; Frade and Barde 1998). This receptor is related to three signaling pathways. The NF-κB as well as that of the Jun kinase leads to the transcription of a number of genes, many of whom are related to the survival of neurons (NF-κB) (Heckscher et al. 2007; Kaltschmidt et al. 2006; Boersma et al. 2011) and apoptosis (Jun kinase) (Ip and Davis 1998; Waetzig and Herdegen 2005; Van Meter et al. 2016). The Rho (rhodopsin) family GTPase pathway controls the mobility of the axis growth cone (Dickson 2001; Gujar et al. 2019). As a response to an insult, the expression the p75LNTR receptor is increased (Ibanez and Simi 2012).
- The family of *tropomyosin-related kinase (Trk) receptors* includes the *TrkA* that binds the NGF (Ibanez and Simi 2012; Klein et al. 1991a), the *TrkB* that binds the BDNF and NT-4 (Klein et al. 1991b; Squinto et al. 1991), and the *TrkC* that binds the NT-3 (Lamballe et al. 1991) which however can also bind less strongly on TrkA and TrkB (Squinto et al. 1991). Each of these receptors regulates three major signaling pathways. The activation of the Ras pathway leads to the activation of signaling through mitogen-activated protein kinase (MAP kinase), resulting in the accentuation of neuronal differentiation and the development of the axis (Avruch et al. 2001; Downward 2003; Orton et al. 2005). The activation of the phosphoinositide 3-kinase (PI3) pathway through Ras or GAB1 (GRB2-associated binding protein 1) assists in the survival and the increase of neurons and other cells (Knight et al. 2006; Jean and Kiger 2014; Vanhaesebroeck et al. 2012; Man et al. 2003). Activation of the phospholipase C-γ1 (PLC-γ1) results in the activation of pathways which are controlled by $Ca2+$ ions and by protein kinase C and increase synaptic plasticity (Burgess et al. 1990; Kang et al. 2020; Phillips-Mason et al. 2011). All the above pathways also increase the expression of genes.

The Trk receptor function is regulated by the p75LNTR receptor at multiple levels, from the accentuation of the binding with a ligand to the development of the axis, thus resulting in the better access to neurotrophins. It also increases endocytosis and the retrograde transportation through membranes (Bradshaw et al. 2015; Benedetti et al. 1993; Bibel et al. 1999; Makkerh et al. 2005). Additionally, the p75NTR receptor inhibits the activation of Trk by unwanted neurotrophins (Benedetti et al. 1993; Bibel et al. 1999; Brennan et al. 1999) and promotes the strong binding of NGF (Mahadeo et al. 1994; Davies et al. 1993; Esposito et al. 2001).

9.4 Neurotrophic Factor Actions

Neurotrophic Factor Actions

- Neuronal differentiation.
  - Structural specialization.
  - Functional specialization.
  - Selection of neurotransmitter(s).
  - Spatial and temporal expression of neurotransmitters.
  - Balance between neurotransmitters and other molecules.
- Survival and repair of neurons.
- Regulation of the neuronal response to.
  - Environmental stimuli.
  - Stress.
  - Damage.
- Apoptotic mechanisms.
- Dysfunction or insufficiency of neurotrophins leads to neuronal death.
- BDNF also plays an important role in.
  - Memory.
  - Learning.
  - Response and adjustment to pain.

During development, the brain is formed, and neurons differentiate as a result of genetic programming in interaction with environmental factors. Neuronal differentiation includes their structural as well as their functional specialization with the selection of which neurotransmitters they will use, their topographical location, and their relationship to neighboring neurons (Patterson and Nawa 1993). This developmental and differentiation process concerns a variety of tissues (Lessmann et al. 2003; Reichardt 2006) and brain areas (Aloe et al. 1994; Ojeda et al. 1991; Lahtinen et al. 1989), as well as an interaction with the rest of the body (Otten et al. 1979; Scaccianoce et al. 1993), and strongly relies on various neurotrophic factors (Sofroniew et al. 2001; Hempstead 2014; Mitre et al. 2017; Carnahan and Nawa 1995; Hayashi et al. 1999; Maisonpierre et al. 1990; Thoenen 1991; Kerekes et al.

2000; Loudes et al. 2000), which determine the spatial and temporal expression of neurotransmitters (Wagner and Kostyk 1990), the balance between neurotransmitters and other molecules (Kornhuber and Weller 1994), the maintenance and plasticity of synapses (Lo 1995; Lindvall et al. 1994; Lindsay et al. 1994), and the survival and repair of neurons (Dekkers et al. 2013; Gash et al. 1996; Hagg and Varon 1993; Lindvall et al. 1994). An important part of their function is the regulation of the organism's response and specifically that of neurons to environmental stimuli (Thoenen 1995), stress (Duman et al. 1997; Bayer et al. 2000), and particularly damage. Their role in apoptotic processes is also important. *Apoptosis* is the programmed neuronal death which is essential for the maintenance of functional equilibrium (Sanes et al. 2019; Squire et al. 2013; Bamji et al. 1998). Dysfunction or insufficiency of neurotrophins leads to neuronal death (Squire et al. 2013; Lu et al. 2005; Huang and Reichardt 2001), and particularly the neurons in the substantia nigra and the thalamus are sensitive to such conditions (Squire et al. 2013).

BDNF and NT-5 increase the synthesis of NPY and somatostatin in GABAergic neurons in the cortex as well as of CKK during development. These substances, in turn, exert a neurotrophic action and assist in the consolidation of neuronal circuits (Nawa et al. 1993; Hayashi et al. 1990; Yamashita et al. 1990; Yamashita 1992; Schwartz 1992). Especially the BDNF plays an important role in the formation of the dopaminergic system (Hyman et al. 1991).

The BDNF plays an additional important role in the development of functions pertaining to memory and learning, by exerting action in the development of long-term potentiation (LTP) in the hippocampus (Korte et al. 1998; Korte et al. 1996; Malenka and Nicoll 1999; Silva et al. 1992) as well as pertaining to the reaction and adjustment to pain (Carroll et al. 1998; Skaper 2017).

9.5 Other Neurotrophic Mechanisms and Substances

Apart from neurotrophins, there are additional molecules with an action of neurotrophic type in the brain (Lu et al. 2005; Lessmann et al. 2003). These include dehydroepiandrosterone (DHEA) and its sulfate ester (DHEA-S), which are antagonistically bound on the neurotrophin receptors with high affinity, and therefore, they are called "microneurotrophins" (Prough et al. 2016; Lazaridis et al. 2011; Pediaditakis et al. 2015; Gravanis et al. 2012).

Treatment with antidepressants seems to exert a neurotrophic effect, especially on the hippocampus. This kind of treatment causes neurogenesis in areas of the hippocampus and more specifically proliferation and improved survival of neurons (Malberg et al. 2000). The mechanism behind this effect is essentially unknown, but it most probably involves the BDNF (Nibuya et al. 1995), which is closely related to the serotonergic system (Mamounas et al. 1995), the hippocampus (Kang and Schuman 1995), and the response to stress (Smith et al. 1995), while there are reports that BDNF exerts an antidepressant action by itself (Siuciak et al. 1997).

References

Allen SJ, Dawbarn D (2006) Clinical relevance of the neurotrophins and their receptors. Clin Sci (Lond) 110(2):175–191. https://doi.org/10.1042/CS20050161

Aloe L, Skaper SD, Leon A, Levi-Montalcini R (1994) Nerve growth factor and autoimmune diseases. Autoimmunity 19(2):141–150. https://doi.org/10.3109/08916939409009542

Arevalo JC, Wu SH (2006) Neurotrophin signaling: many exciting surprises! Cell Mol Life Sci 63(13):1523–1537. https://doi.org/10.1007/s00018-006-6010-1

Avruch J, Khokhlatchev A, Kyriakis JM, Luo Z, Tzivion G, Vavvas D, Zhang XF (2001) Ras activation of the Raf kinase: tyrosine kinase recruitment of the MAP kinase cascade. Recent Prog Horm Res 56:127–155. https://doi.org/10.1210/rp.56.1.127

Bamji SX, Majdan M, Pozniak CD, Belliveau DJ, Aloyz R, Kohn J, Causing CG, Miller FD (1998) The p75 neurotrophin receptor mediates neuronal apoptosis and is essential for naturally occurring sympathetic neuron death. J Cell Biol 140(4):911–923. https://doi.org/10.1083/jcb.140.4.911

Barde YA (1994) Neurotrophic factors: an evolutionary perspective. J Neurobiol 25(11):1329–1333. https://doi.org/10.1002/neu.480251102

Barde YA, Edgar D, Thoenen H (1982) Purification of a new neurotrophic factor from mammalian brain. EMBO J 1(5):549–553

Bayer TA, Schramm M, Feldmann N, Knable MB, Falkai P (2000) Antidepressant drug exposure is associated with mRNA levels of tyrosine receptor kinase B in major depressive disorder. Prog Neuro-Psychopharmacol Biol Psychiatry 24(6):881–888. https://doi.org/10.1016/s0278-5846(00)00115-9

Benedetti M, Levi A, Chao MV (1993) Differential expression of nerve growth factor receptors leads to altered binding affinity and neurotrophin responsiveness. Proc Natl Acad Sci U S A 90(16):7859–7863. https://doi.org/10.1073/pnas.90.16.7859

Berkemeier LR, Winslow JW, Kaplan DR, Nikolics K, Goeddel DV, Rosenthal A (1991) Neurotrophin-5: a novel neurotrophic factor that activates trk and trkB. Neuron 7(5):857–866. https://doi.org/10.1016/0896-6273(91)90287-a

Bibel M, Barde YA (2000) Neurotrophins: key regulators of cell fate and cell shape in the vertebrate nervous system. Genes Dev 14(23):2919–2937. https://doi.org/10.1101/gad.841400

Bibel M, Hoppe E, Barde YA (1999) Biochemical and functional interactions between the neurotrophin receptors trk and p75NTR. EMBO J 18(3):616–622. https://doi.org/10.1093/emboj/18.3.616

Boersma MC, Dresselhaus EC, De Biase LM, Mihalas AB, Bergles DE, Meffert MK (2011) A requirement for nuclear factor-kappaB in developmental and plasticity-associated synaptogenesis. J Neurosci 31(14):5414–5425. https://doi.org/10.1523/JNEUROSCI.2456-10.2011

Bothwell M (1995) Functional interactions of neurotrophins and neurotrophin receptors. Annu Rev Neurosci 18:223–253. https://doi.org/10.1146/annurev.ne.18.030195.001255

Bothwell M (2016) Recent advances in understanding neurotrophin signaling. F1000Res 5. https://doi.org/10.12688/f1000research.8434.1

Bradshaw RA, Pundavela J, Biarc J, Chalkley RJ, Burlingame AL, Hondermarck H (2015) NGF and ProNGF: regulation of neuronal and neoplastic responses through receptor signaling. Adv Biol Regul 58:16–27. https://doi.org/10.1016/j.jbior.2014.11.003

Brennan C, Rivas-Plata K, Landis SC (1999) The p75 neurotrophin receptor influences NT-3 responsiveness of sympathetic neurons in vivo. Nat Neurosci 2(8):699–705. https://doi.org/10.1038/11158

Burgess WH, Dionne CA, Kaplow J, Mudd R, Friesel R, Zilberstein A, Schlessinger J, Jaye M (1990) Characterization and cDNA cloning of phospholipase C-gamma, a major substrate for heparin-binding growth factor 1 (acidic fibroblast growth factor)-activated tyrosine kinase. Mol Cell Biol 10(9):4770–4777. https://doi.org/10.1128/mcb.10.9.4770-4777.1990

Butte MJ (2001) Neurotrophic factor structures reveal clues to evolution, binding, specificity, and receptor activation. Cell Mol Life Sci 58(8):1003–1013. https://doi.org/10.1007/PL00000915

Carnahan J, Nawa H (1995) Regulation of neuropeptide expression in the brain by neurotrophins. Potential role in vivo. Mol Neurobiol 10(2–3):135–149. https://doi.org/10.1007/BF02740672

Carroll P, Lewin GR, Koltzenburg M, Toyka KV, Thoenen H (1998) A role for BDNF in mechanosensation. Nat Neurosci 1(1):42–46. https://doi.org/10.1038/242

Davies AM, Lee KF, Jaenisch R (1993) p75-deficient trigeminal sensory neurons have an altered response to NGF but not to other neurotrophins. Neuron 11(4):565–574. https://doi.org/10.1016/0896-6273(93)90069-4

Dechant G, Barde YA (2002) The neurotrophin receptor p75(NTR): novel functions and implications for diseases of the nervous system. Nat Neurosci 5(11):1131–1136. https://doi.org/10.1038/nn1102-1131

Dekkers MP, Nikoletopoulou V, Barde YA (2013) Cell biology in neuroscience: death of developing neurons: new insights and implications for connectivity. J Cell Biol 203(3):385–393. https://doi.org/10.1083/jcb.201306136

Dickson BJ (2001) Rho GTPases in growth cone guidance. Curr Opin Neurobiol 11(1):103–110. https://doi.org/10.1016/s0959-4388(00)00180-x

Dore S, Kar S, Quirion R (1997) Rediscovering an old friend, IGF-I: potential use in the treatment of neurodegenerative diseases. Trends Neurosci 20(8):326–331. https://doi.org/10.1016/s0166-2236(96)01036-3

Downward J (2003) Targeting RAS signalling pathways in cancer therapy. Nat Rev Cancer 3(1):11–22. https://doi.org/10.1038/nrc969

Dreyfus CF (1989) Effects of nerve growth factor on cholinergic brain neurons. Trends Pharmacol Sci 10(4):145–149. https://doi.org/10.1016/0165-6147(89)90166-1

Duman RS, Heninger GR, Nestler EJ (1997) A molecular and cellular theory of depression. Arch Gen Psychiatry 54(7):597–606. https://doi.org/10.1001/archpsyc.1997.01830190015002

Eriksson PS, Perfilieva E, Bjork-Eriksson T, Alborn AM, Nordborg C, Peterson DA, Gage FH (1998) Neurogenesis in the adult human hippocampus. Nat Med 4(11):1313–1317. https://doi.org/10.1038/3305

Esposito D, Patel P, Stephens RM, Perez P, Chao MV, Kaplan DR, Hempstead BL (2001) The cytoplasmic and transmembrane domains of the p75 and Trk A receptors regulate high affinity binding to nerve growth factor. J Biol Chem 276(35):32687–32695. https://doi.org/10.1074/jbc.M011674200

Farkas LM, Dunker N, Roussa E, Unsicker K, Krieglstein K (2003) Transforming growth factor-beta(s) are essential for the development of midbrain dopaminergic neurons in vitro and in vivo. J Neurosci 23(12):5178–5186. https://doi.org/10.1523/JNEUROSCI.23-12-05178.2003

Frade JM, Barde YA (1998) Nerve growth factor: two receptors, multiple functions. BioEssays 20(2):137–145. https://doi.org/10.1002/(SICI)1521-1878(199802)20:2<137::AID-BIES6>3.0.CO;2-Q

Gash DM, Zhang Z, Ovadia A, Cass WA, Yi A, Simmerman L, Russell D, Martin D, Lapchak PA, Collins F, Hoffer BJ, Gerhardt GA (1996) Functional recovery in parkinsonian monkeys treated with GDNF. Nature 380(6571):252–255. https://doi.org/10.1038/380252a0

Ginty DD, Segal RA (2002) Retrograde neurotrophin signaling: Trk-ing along the axon. Curr Opin Neurobiol 12(3):268–274. https://doi.org/10.1016/s0959-4388(02)00326-4

Gotz R, Koster R, Winkler C, Raulf F, Lottspeich F, Schartl M, Thoenen H (1994) Neurotrophin-6 is a new member of the nerve growth factor family. Nature 372(6503):266–269. https://doi.org/10.1038/372266a0

Gravanis A, Calogeropoulou T, Panoutsakopoulou V, Thermos K, Neophytou C, Charalampopoulos I (2012) Neurosteroids and microneurotrophins signal through NGF receptors to induce prosurvival signaling in neuronal cells. Sci Signal 5(246):pt8. https://doi.org/10.1126/scisignal.2003387

Gujar MR, Stricker AM, Lundquist EA (2019) RHO-1 and the Rho GEF RHGF-1 interact with UNC-6/Netrin signaling to regulate growth cone protrusion and microtubule organization in Caenorhabditis elegans. PLoS Genet 15(6):e1007960. https://doi.org/10.1371/journal.pgen.1007960

Hagg T, Varon S (1993) Ciliary neurotrophic factor prevents degeneration of adult rat substantia nigra dopaminergic neurons in vivo. Proc Natl Acad Sci U S A 90(13):6315–6319. https://doi.org/10.1073/pnas.90.13.6315

Hamburger V, Brunso-Bechtold JK, Yip JW (1981) Neuronal death in the spinal ganglia of the chick embryo and its reduction by nerve growth factor. J Neurosci 1(1):60–71. https://doi.org/10.1523/jneurosci.01-01-00060.1981

Hamburger V, Levi-Montalcini R (1949) Proliferation, differentiation and degeneration in the spinal ganglia of the chick embryo under normal and experimental conditions. J Exp Zool 111(3):457–501. https://doi.org/10.1002/jez.1401110308

Harrington AW, Ginty DD (2013) Long-distance retrograde neurotrophic factor signalling in neurons. Nat Rev Neurosci 14(3):177–187. https://doi.org/10.1038/nrn3253

Hayashi M, Mitsunaga F, Ohira K, Shimizu K, Yamashita A (1999) Development of full-length Trk B-immunoreactive structures in the hippocampal formation of the macaque monkey. Anat Embryol 199(6):529–537. https://doi.org/10.1007/s004290050250

Hayashi M, Yamashita A, Shimizu K, Sogawa K, Fujii Y (1990) Somatostatin gene expression in the developing monkey frontal and cerebellar cortices. Brain Res Dev Brain Res 57(1):37–41. https://doi.org/10.1016/0165-3806(90)90182-x

He XL, Garcia KC (2004) Structure of nerve growth factor complexed with the shared neurotrophin receptor p75. Science 304(5672):870–875. https://doi.org/10.1126/science.1095190

Heckscher ES, Fetter RD, Marek KW, Albin SD, Davis GW (2007) NF-kappaB, IkappaB, and IRAK control glutamate receptor density at the Drosophila NMJ. Neuron 55(6):859–873. https://doi.org/10.1016/j.neuron.2007.08.005

Hempstead BL (2006) Dissecting the diverse actions of pro- and mature neurotrophins. Curr Alzheimer Res 3(1):19–24. https://doi.org/10.2174/156720506775697061

Hempstead BL (2014) Deciphering proneurotrophin actions. Handb Exp Pharmacol 220:17–32. https://doi.org/10.1007/978-3-642-45106-5_2

Huang EJ, Liu W, Fritzsch B, Bianchi LM, Reichardt LF, Xiang M (2001) Brn3a is a transcriptional regulator of soma size, target field innervation and axon pathfinding of inner ear sensory neurons. Development 128(13):2421–2432

Huang EJ, Reichardt LF (2001) Neurotrophins: roles in neuronal development and function. Annu Rev Neurosci 24:677–736. https://doi.org/10.1146/annurev.neuro.24.1.677

Hyman C, Hofer M, Barde YA, Juhasz M, Yancopoulos GD, Squinto SP, Lindsay RM (1991) BDNF is a neurotrophic factor for dopaminergic neurons of the substantia nigra. Nature 350(6315):230–232. https://doi.org/10.1038/350230a0

Ibanez CF (1995) Neurotrophic factors: from structure-function studies to designing effective therapeutics. Trends Biotechnol 13(6):217–227. https://doi.org/10.1016/S0167-7799(00)88949-0

Ibanez CF, Simi A (2012) p75 neurotrophin receptor signaling in nervous system injury and degeneration: paradox and opportunity. Trends Neurosci 35(7):431–440. https://doi.org/10.1016/j.tins.2012.03.007

Ip NY, Ibanez CF, Nye SH, McClain J, Jones PF, Gies DR, Belluscio L, Le Beau MM, Espinosa R 3rd, Squinto SP et al (1992) Mammalian neurotrophin-4: structure, chromosomal localization, tissue distribution, and receptor specificity. Proc Natl Acad Sci U S A 89(7):3060–3064. https://doi.org/10.1073/pnas.89.7.3060

Ip NY, Yancopoulos GD (1996) The neurotrophins and CNTF: two families of collaborative neurotrophic factors. Annu Rev Neurosci 19:491–515. https://doi.org/10.1146/annurev.ne.19.030196.002423

Ip YT, Davis RJ (1998) Signal transduction by the c-Jun N-terminal kinase (JNK)--from inflammation to development. Curr Opin Cell Biol 10(2):205–219. https://doi.org/10.1016/s0955-0674(98)80143-9

Jean S, Kiger AA (2014) Classes of phosphoinositide 3-kinases at a glance. J Cell Sci 127(Pt 5):923–928. https://doi.org/10.1242/jcs.093773

Kaltschmidt B, Ndiaye D, Korte M, Pothion S, Arbibe L, Prullage M, Pfeiffer J, Lindecke A, Staiger V, Israel A, Kaltschmidt C, Memet S (2006) NF-kappaB regulates spatial memory formation and synaptic plasticity through protein kinase a/CREB signaling. Mol Cell Biol 26(8):2936–2946. https://doi.org/10.1128/MCB.26.8.2936-2946.2006

Kang DS, Kim IS, Baik JH, Kim D, Cocco L, Suh PG (2020) The function of PLCgamma1 in developing mouse mDA system. Adv Biol Regul 75:100654. https://doi.org/10.1016/j.jbior.2019.100654

Kang H, Schuman EM (1995) Long-lasting neurotrophin-induced enhancement of synaptic transmission in the adult hippocampus. Science 267(5204):1658–1662. https://doi.org/10.1126/science.7886457

Kerekes N, Landry M, Lundmark K, Hokfelt T (2000) Effect of NGF, BDNF, bFGF, aFGF and cell density on NPY expression in cultured rat dorsal root ganglion neurones. J Auton Nerv Syst 81(1–3):128–138. https://doi.org/10.1016/s0165-1838(00)00115-6

Klein R, Jing SQ, Nanduri V, O'Rourke E, Barbacid M (1991a) The trk proto-oncogene encodes a receptor for nerve growth factor. Cell 65(1):189–197. https://doi.org/10.1016/0092-8674(91)90419-y

Klein R, Nanduri V, Jing SA, Lamballe F, Tapley P, Bryant S, Cordon-Cardo C, Jones KR, Reichardt LF, Barbacid M (1991b) The trkB tyrosine protein kinase is a receptor for brain-derived neurotrophic factor and neurotrophin-3. Cell 66(2):395–403. https://doi.org/10.1016/0092-8674(91)90628-c

Knight ZA, Gonzalez B, Feldman ME, Zunder ER, Goldenberg DD, Williams O, Loewith R, Stokoe D, Balla A, Toth B, Balla T, Weiss WA, Williams RL, Shokat KM (2006) A pharmacological map of the PI3-K family defines a role for p110alpha in insulin signaling. Cell 125(4):733–747. https://doi.org/10.1016/j.cell.2006.03.035

Kornhuber J, Weller M (1994) Neuroleptic malignant syndrome. Curr Opin Neurol 7(4):353–357. https://doi.org/10.1097/00019052-199408000-00013

Korte M, Kang H, Bonhoeffer T, Schuman E (1998) A role for BDNF in the late-phase of hippocampal long-term potentiation. Neuropharmacology 37(4–5):553–559. https://doi.org/10.1016/s0028-3908(98)00035-5

Korte M, Staiger V, Griesbeck O, Thoenen H, Bonhoeffer T (1996) The involvement of brain-derived neurotrophic factor in hippocampal long-term potentiation revealed by gene targeting experiments. J Physiol Paris 90(3–4):157–164. https://doi.org/10.1016/s0928-4257(97)81415-5

Kullander K, Carlson B, Hallbook F (1997) Molecular phylogeny and evolution of the neurotrophins from monotremes and marsupials. J Mol Evol 45(3):311–321. https://doi.org/10.1007/pl00006235

Lahtinen T, Soinila S, Lakshmanan J (1989) Biological demonstration of nerve growth factor in the rat pituitary gland. Neuroscience 30(1):165–170. https://doi.org/10.1016/0306-4522(89)90362-x

Lai KO, Fu WY, Ip FC, Ip NY (1998) Cloning and expression of a novel neurotrophin, NT-7, from carp. Mol Cell Neurosci 11(1–2):64–76. https://doi.org/10.1006/mcne.1998.0666

Lamballe F, Klein R, Barbacid M (1991) trkC, a new member of the trk family of tyrosine protein kinases, is a receptor for neurotrophin-3. Cell 66(5):967–979. https://doi.org/10.1016/0092-8674(91)90442-2

Lazaridis I, Charalampopoulos I, Alexaki VI, Avlonitis N, Pediaditakis I, Efstathopoulos P, Calogeropoulou T, Castanas E, Gravanis A (2011) Neurosteroid dehydroepiandrosterone interacts with nerve growth factor (NGF) receptors, preventing neuronal apoptosis. PLoS Biol 9(4):e1001051. https://doi.org/10.1371/journal.pbio.1001051

Leibrock J, Lottspeich F, Hohn A, Hofer M, Hengerer B, Masiakowski P, Thoenen H, Barde YA (1989) Molecular cloning and expression of brain-derived neurotrophic factor. Nature 341(6238):149–152. https://doi.org/10.1038/341149a0

Lessmann V, Gottmann K, Malcangio M (2003) Neurotrophin secretion: current facts and future prospects. Prog Neurobiol 69(5):341–374. https://doi.org/10.1016/s0301-0082(03)00019-4

Levi-Montalcini R (1987) The nerve growth factor 35 years later. Science 237(4819):1154–1162. https://doi.org/10.1126/science.3306916

Lewin GR, Barde YA (1996) Physiology of the neurotrophins. Annu Rev Neurosci 19:289–317. https://doi.org/10.1146/annurev.ne.19.030196.001445

Lindsay RM, Wiegand SJ, Altar CA, DiStefano PS (1994) Neurotrophic factors: from molecule to man. Trends Neurosci 17(5):182–190. https://doi.org/10.1016/0166-2236(94)90099-x

Lindsay RM, Yancopoulos GD (1996) GDNF in a bind with known orphan: accessory implicated in new twist. Neuron 17(4):571–574. https://doi.org/10.1016/s0896-6273(00)80189-0

Lindvall O, Kokaia Z, Bengzon J, Elmer E, Kokaia M (1994) Neurotrophins and brain insults. Trends Neurosci 17(11):490–496. https://doi.org/10.1016/0166-2236(94)90139-2

Lo DC (1995) Neurotrophic factors and synaptic plasticity. Neuron 15(5):979–981. https://doi.org/10.1016/0896-6273(95)90085-3

Loudes C, Petit F, Kordon C, Faivre-Bauman A (2000) Brain-derived neurotrophic factor but not neurotrophin-3 enhances differentiation of somatostatin neurons in hypothalamic cultures. Neuroendocrinology 72(3):144–153. https://doi.org/10.1159/000054581

Lu B, Pang PT, Woo NH (2005) The yin and yang of neurotrophin action. Nat Rev Neurosci 6(8):603–614. https://doi.org/10.1038/nrn1726

Mahadeo D, Kaplan L, Chao MV, Hempstead BL (1994) High affinity nerve growth factor binding displays a faster rate of association than p140trk binding. Implications for multi-subunit polypeptide receptors. J Biol Chem 269(9):6884–6891

Maisonpierre PC, Belluscio L, Friedman B, Alderson RF, Wiegand SJ, Furth ME, Lindsay RM, Yancopoulos GD (1990) NT-3, BDNF, and NGF in the developing rat nervous system: parallel as well as reciprocal patterns of expression. Neuron 5(4):501–509. https://doi.org/10.1016/0896-6273(90)90089-x

Makkerh JP, Ceni C, Auld DS, Vaillancourt F, Dorval G, Barker PA (2005) p75 neurotrophin receptor reduces ligand-induced Trk receptor ubiquitination and delays Trk receptor internalization and degradation. EMBO Rep 6(10):936–941. https://doi.org/10.1038/sj.embor.7400503

Malberg JE, Eisch AJ, Nestler EJ, Duman RS (2000) Chronic antidepressant treatment increases neurogenesis in adult rat hippocampus. J Neurosci 20(24):9104–9110

Malenka R, Nestler E, Hyman S (2009) Chapter 8: atypical neurotransmitters. In: Sydor A, Brown R (eds) Molecular neuropharmacology: a Foundation for Clinical Neuroscience, 2nd edn. McGraw-Hill Medical, New York, pp 199–215

Malenka RC, Nicoll RA (1999) Long-term potentiation--a decade of progress? Science 285(5435):1870–1874. https://doi.org/10.1126/science.285.5435.1870

Mamounas LA, Blue ME, Siuciak JA, Altar CA (1995) Brain-derived neurotrophic factor promotes the survival and sprouting of serotonergic axons in rat brain. J Neurosci 15(12):7929–7939

Man HY, Wang Q, Lu WY, Ju W, Ahmadian G, Liu L, D'Souza S, Wong TP, Taghibiglou C, Lu J, Becker LE, Pei L, Liu F, Wymann MP, MacDonald JF, Wang YT (2003) Activation of PI3-kinase is required for AMPA receptor insertion during LTP of mEPSCs in cultured hippocampal neurons. Neuron 38(4):611–624. https://doi.org/10.1016/s0896-6273(03)00228-9

Miao N, Wang M, Ott JA, D'Alessandro JS, Woolf TM, Bumcrot DA, Mahanthappa NK, Pang K (1997) Sonic hedgehog promotes the survival of specific CNS neuron populations and protects these cells from toxic insult in vitro. J Neurosci 17(15):5891–5899. https://doi.org/10.1523/jneurosci.17-15-05891.1997

Mitre M, Mariga A, Chao MV (2017) Neurotrophin signalling: novel insights into mechanisms and pathophysiology. Clin Sci (Lond) 131(1):13–23. https://doi.org/10.1042/CS20160044

Nawa H, Bessho Y, Carnahan J, Nakanishi S, Mizuno K (1993) Regulation of neuropeptide expression in cultured cerebral cortical neurons by brain-derived neurotrophic factor. J Neurochem 60(2):772–775. https://doi.org/10.1111/j.1471-4159.1993.tb03216.x

Nibuya M, Morinobu S, Duman RS (1995) Regulation of BDNF and trkB mRNA in rat brain by chronic electroconvulsive seizure and antidepressant drug treatments. J Neurosci 15(11):7539–7547

Ojeda SR, Hill DF, Katz KH (1991) The genes encoding nerve growth factor and its receptor are expressed in the developing female rat hypothalamus. Brain Res Mol Brain Res 9(1–2):47–55. https://doi.org/10.1016/0169-328x(91)90129-l

Orton RJ, Sturm OE, Vyshemirsky V, Calder M, Gilbert DR, Kolch W (2005) Computational modelling of the receptor-tyrosine-kinase-activated MAPK pathway. Biochem J 392(Pt 2):249–261. https://doi.org/10.1042/BJ20050908

Otten U, Baumann JB, Girard J (1979) Stimulation of the pituitary-adrenocortical axis by nerve growth factor. Nature 282(5737):413–414. https://doi.org/10.1038/282413a0

Patterson PH, Nawa H (1993) Neuronal differentiation factors/cytokines and synaptic plasticity. Cell 72(Suppl):123–137. https://doi.org/10.1016/s0092-8674(05)80032-7

Pediaditakis I, Iliopoulos I, Theologidis I, Delivanoglou N, Margioris AN, Charalampopoulos I, Gravanis A (2015) Dehydroepiandrosterone: an ancestral ligand of neurotrophin receptors. Endocrinology 156(1):16–23. https://doi.org/10.1210/en.2014-1596

Phillips-Mason PJ, Kaur H, Burden-Gulley SM, Craig SE, Brady-Kalnay SM (2011) Identification of phospholipase C gamma1 as a protein tyrosine phosphatase mu substrate that regulates cell migration. J Cell Biochem 112(1):39–48. https://doi.org/10.1002/jcb.22710

Poo MM (2001) Neurotrophins as synaptic modulators. Nat Rev Neurosci 2(1):24–32. https://doi.org/10.1038/35049004

Prough RA, Clark BJ, Klinge CM (2016) Novel mechanisms for DHEA action. J Mol Endocrinol 56(3):R139–R155. https://doi.org/10.1530/JME-16-0013

Radeke MJ, Misko TP, Hsu C, Herzenberg LA, Shooter EM (1987) Gene transfer and molecular cloning of the rat nerve growth factor receptor. Nature 325(6105):593–597. https://doi.org/10.1038/325593a0

Reichardt LF (2006) Neurotrophin-regulated signalling pathways. Philos Trans R Soc Lond Ser B Biol Sci 361(1473):1545–1564. https://doi.org/10.1098/rstb.2006.1894

Rodriguez-Tebar A, Dechant G, Barde YA (1990) Binding of brain-derived neurotrophic factor to the nerve growth factor receptor. Neuron 4(4):487–492. https://doi.org/10.1016/0896-6273(90)90107-q

Rodriguez-Tebar A, Dechant G, Gotz R, Barde YA (1992) Binding of neurotrophin-3 to its neuronal receptors and interactions with nerve growth factor and brain-derived neurotrophic factor. EMBO J 11(3):917–922

Sanes D, Reh T, Harris W, Landgraf M (2019) Development of the nervous system, 4th edn. Academic Press, Cambridge

Scaccianoce S, Cigliana G, Nicolai R, Muscolo LA, Porcu A, Navarra D, Perez-Polo JR, Angelucci L (1993) Hypothalamic involvement in the activation of the pituitary-adrenocortical axis by nerve growth factor. Neuroendocrinology 58(2):202–209. https://doi.org/10.1159/000126534

Schwartz JP (1992) Neurotransmitters as neurotrophic factors: a new set of functions. Int Rev Neurobiol 34:1–23. https://doi.org/10.1016/s0074-7742(08)60096-3

Silva AJ, Stevens CF, Tonegawa S, Wang Y (1992) Deficient hippocampal long-term potentiation in alpha-calcium-calmodulin kinase II mutant mice. Science 257(5067):201–206. https://doi.org/10.1126/science.1378648

Siuciak JA, Lewis DR, Wiegand SJ, Lindsay RM (1997) Antidepressant-like effect of brain-derived neurotrophic factor (BDNF). Pharmacol Biochem Behav 56(1):131–137. https://doi.org/10.1016/S0091-3057(96)00169-4

Skaper SD (2017) Nerve growth factor: a neuroimmune crosstalk mediator for all seasons. Immunology 151(1):1–15. https://doi.org/10.1111/imm.12717

Skaper SD (2018) Neurotrophic factors: an overview. Methods Mol Biol 1727:1–17. https://doi.org/10.1007/978-1-4939-7571-6_1

Smith MA, Makino S, Kvetnansky R, Post RM (1995) Effects of stress on neurotrophic factor expression in the rat brain. Ann N Y Acad Sci 771(1 stress):234–239. https://doi.org/10.1111/j.1749-6632.1995.tb44684.x

Sofroniew MV, Howe CL, Mobley WC (2001) Nerve growth factor signaling, neuroprotection, and neural repair. Annu Rev Neurosci 24:1217–1281. https://doi.org/10.1146/annurev.neuro.24.1.1217

Squinto SP, Stitt TN, Aldrich TH, Davis S, Bianco SM, Radziejewski C, Glass DJ, Masiakowski P, Furth ME, Valenzuela DM et al (1991) trkB encodes a functional receptor for brain-derived neurotrophic factor and neurotrophin-3 but not nerve growth factor. Cell 65(5):885–893. https://doi.org/10.1016/0092-8674(91)90395-f

Squire L, Berg D, Bloom F, du Lac S, Ghosh A, Spitzer N (2013) Fundamental Neuroscience. Academic Press, Cambridge

Thoenen H (1991) The changing scene of neurotrophic factors. Trends Neurosci 14(5):165–170. https://doi.org/10.1016/0166-2236(91)90097-e

Thoenen H (1995) Neurotrophins and neuronal plasticity. Science 270(5236):593–598. https://doi.org/10.1126/science.270.5236.593

Van Meter M, Simon M, Tombline G, May A, Morello TD, Hubbard BP, Bredbenner K, Park R, Sinclair DA, Bohr VA, Gorbunova V, Seluanov A (2016) JNK phosphorylates SIRT6 to stimulate DNA double-Strand break repair in response to oxidative stress by recruiting PARP1 to DNA breaks. Cell Rep 16(10):2641–2650. https://doi.org/10.1016/j.celrep.2016.08.006

Vanhaesebroeck B, Stephens L, Hawkins P (2012) PI3K signalling: the path to discovery and understanding. Nat Rev Mol Cell Biol 13(3):195–203. https://doi.org/10.1038/nrm3290

Waetzig V, Herdegen T (2005) Context-specific inhibition of JNKs: overcoming the dilemma of protection and damage. Trends Pharmacol Sci 26(9):455–461. https://doi.org/10.1016/j.tips.2005.07.006

Wagner JA, Kostyk SK (1990) Regulation of neural cell survival and differentiation by peptide growth factors. Curr Opin Cell Biol 2(6):1050–1057. https://doi.org/10.1016/0955-0674(90)90155-8

Walicke P, Cowan WM, Ueno N, Baird A, Guillemin R (1986) Fibroblast growth factor promotes survival of dissociated hippocampal neurons and enhances neurite extension. Proc Natl Acad Sci U S A 83(9):3012–3016. https://doi.org/10.1073/pnas.83.9.3012

Yamashita A (1992) Ontogeny of cholecystokinin-immunoreactive structures in the primate cerebral neocortex. Int J Neurosci 64(1–4):139–151. https://doi.org/10.3109/00207459209000540

Yamashita A, Shimizu K, Hayashi M (1990) Ontogeny of substance P-immunoreactive structures in the primate cerebral neocortex. Brain Res Dev Brain Res 57(2):197–207. https://doi.org/10.1016/0165-3806(90)90046-2

Functional Organization of the Brain 10

Contents

The human brain has a complex but highly efficient organization (Gazzaniga 1989, 1995). Throughout the brain, neuronal cells are more or less the same both morphologically and also in terms of function. It is their interconnections and the networks that arise, which produce the differences in function between brain areas, and they give birth to behavior itself (Mishkin 1993; Changeux and Dehaene 1993; Lewis and Oeth 1995). Thus, this *fundamental simplicity of structure* supports a *great complexity* through astronomical numbers and exponential power. The end result is currently beyond our comprehension, even at its middle levels (e.g., coding and storing of external stimuli) not to mention the higher levels (e.g., consciousness). Some aspects of brain function are possibly beyond not only our current ability, which is due to a lack of knowledge, but even beyond our essential ability to comprehend independently of available knowledge.

The *following seven principles* seem to govern brain functional anatomy and brain topography:

- *Independent sensory and motor processing.*
- *Relative localization and distribution of functions.* The various functions manifest localization, that is, there are brain areas that have to do more with the specific function (e.g., the visual cortex). However, this specialization is not absolute neither for the function nor for the area.

K. N. Fountoulakis, *The Human Connectome*,
https://doi.org/10.1007/978-3-031-10351-3_10

- *Lateralization and symmetry* (simultaneous symmetry-asymmetry). Similar to the previous principle, there is a localization of every function, especially in the one of the two hemispheres (lateralization); however, also the other hemisphere contributes significantly as well (symmetry).
- Organization in *hierarchical and parallel processing* of the same input of information.
- *Overlapping of the same function by different brain regions*. This overlapping is probably a residue of evolution, and as higher and more evolved structures obtained control of a specific function, the more ancient regions did not lose control over the function completely.
- Processing of the same incoming information at *different levels and with a different way and aim.*
- *Economy of functioning*. This principle secures that all brain parts are topographically arranged in the best possible way in terms of ergonomics and the highest possible output is achieved with the lowest possible energy consumption. This principle is the result of evolutionary pressure which act "just so."

10.1 Basic Functional Anatomy of the Brain

The *basic operationalized conceptualization of the human brain* is that of a *helmet with spheres at its basis* (Feneis and Dauber 2000; Kandel 2000; Putz and Pabst 2006; Standing 2016; Ungerleider 1995) (Fig. 10.1). Both the helmet and the spheres are made of gray matter, while the space between them, that is, the internal space of the helmet, is full of white matter. The helmet corresponds to the cortex and the

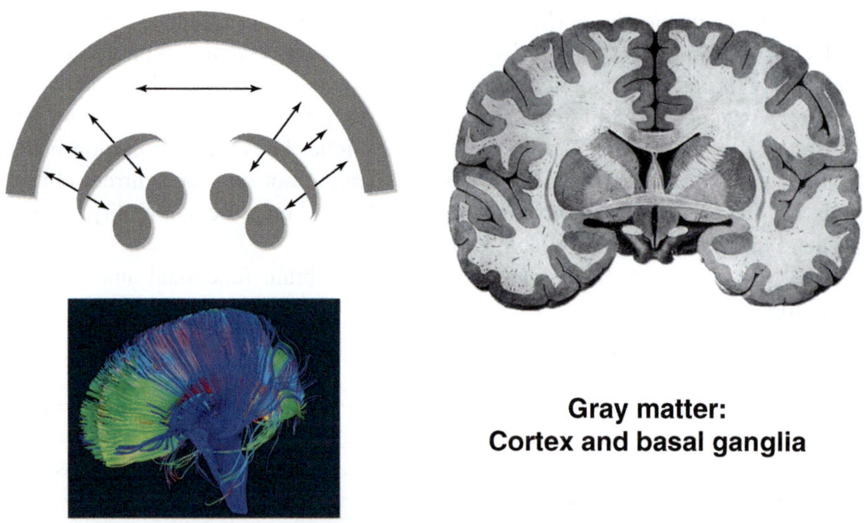

Fig. 10.1 Basic simplified functional depiction of the structure of human brain in the form of helmet with spheres at its base (copyright KN Fountoulakis 2021)

spheres to the thalamus, the hypothalamus, the amygdala, the hippocampus, and the basal ganglia (nucleus caudatus, putamen, and globus pallidus). Gray matter consists of the cell bodies of neurons, which are the processors where the information is being analyzed and processed, while the white matter corresponds to the axons, which are the cables connecting the various processors. The spheres/basal ganglia deal with more primitive and primary processes (basic emotions, instincts, vital functions, and motor coordination), while the helmet/cortex deals with higher cognitive function especially in the frontal areas (detailed process and analysis, abstract thinking, and planning). This whole design is extremely ergonomic; it permits the circulation of information in the fastest, most efficient, and economical way possible.

Basic Functional Anatomy of the Brain

- Helmet with spheres at its basis.
- Gray matter: cell bodies of neurons (processors).
- White matter: axons (cables).
- Cortex: higher mental functions.
- Deep brain structures: more primitive and primary processes.
- Information chiasm.
- Dominant hemisphere.
- Left hemisphere, digital; right hemisphere, fuzzy.
- Brain ventricles: cooling and drainage.

An important feature of the brain organization is that most input *cross over* to the opposite (contralateral) side of the brain or spinal cord. This is called *chiasmus* from the Greek letter *chi* (X). Subsequently, the sensory and motor activities are mediated by the opposite side brain hemisphere. The reason for this is not yet understood; however, it could serve the keeping of longer distance between the danger, the organism faces (and is confronted by the motor-dominant half of the body which is the right), and the dominant hemisphere of the brain (left). What is however also known is that this asymmetry is accentuated in humans in comparison to primates because of the presence of complex and symbolic thought and behavior.

The *ventricles* serve as the cooling system, and they also assist in the removal of toxic and metabolic waste (Fig. 10.2).

While this is true, it is also a fact that most activities engage both hemispheres but with a different contribution. Brain functions seem to be supported mainly by either hemisphere, and this phenomenon is called *lateralization*. Classically, the *dominant hemisphere* is more concerned with detailed thought, abstract thinking, and speech and subsequently with intentional behavior, while the nondominant hemisphere is more concerned with emotional aspects of behavior and spatial analysis. In some way, it could be said that the dominant hemisphere is the "digital" while the nondominant is the "analogue" or better the "fuzzy." In the normal brain, there is communication between the two hemispheres, and there is now evidence that the capacity of one hemisphere to perform a particular task may deteriorate after commissurotomy.

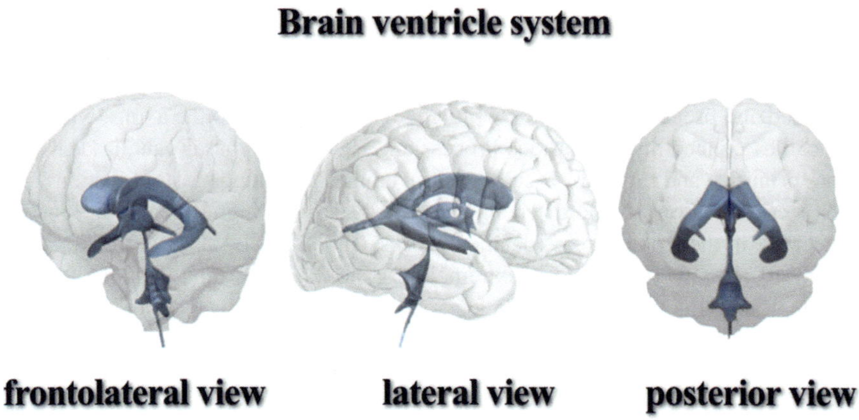

Fig. 10.2 The brain ventricle system (copyright KN Fountoulakis 2021)

10.2 Perception

The first component of behavior is the sense of the environment, external but also internal. The human brain utilizes five main *senses*: sight, hearing, smell, taste, and touch (Fig. 10.3). It also utilizes pain and the sensation of body movements. A number of other senses also provide information concerning the internal environment, which is of course the human body. Overall, the senses give information both for the external and the internal environments.

Concerning the external environment, specialized apparatus and receptor cells on the body surface receive stimuli and encode information. Essentially, they *transform external natural phenomena (*e.g., *sound, light, pressure,* etc.*) into neuronal phenomena*, and in this way, they register their intensity and their temporal and spatial characteristics.

Sensory Perception and Cortical Representation

- Five main senses.
  - Sight.
  - Hearing.
  - Smell.
  - Taste.
  - Touch.
- Pain, body movement sense, and sense of the internal of the body.
- Sensory organs.
- Receptor cells that transform external natural phenomena into neuronal phenomena.
- Analysis of stimuli to their characteristic components and charting of them on the brain cortex.

Fig. 10.3 The "Allegory of the Five Senses." Painting by Pietro Paolini (1603–1681)

An important characteristic of these "sensors" is that they tend to *focus on and prioritize* specific characteristics of the stimuli. For example, sight has high resolution only in a small fraction of the visual field (spatial focus, priority to vision directly in front), while the visual color spectrum is limited (other species have different visual spectrum probably as a result of selection pressure). Also, the senses from the face or the fingers are much more detailed in terms of information in comparison to the trunk or legs. As a result, the receptor cells in the sensory organs are arranged in a *topographical way*, reflecting this focus and prioritization, and this topographical arrangement is kept through the successive stages of processing; in this way, an orderly neural map of information from the receptive surface is retained at each successive level in the brain (Fig. 10.4). This map analyzes the visual information to its fundamental elements which include the spatial perception on the basis of the visual field, movement, color, and other similar characteristics (Penfield and Rasmussen 1950). Similarly, the other senses utilize mapping on the cortex which corresponds to their unique characteristics, for example, the cochlea analyzes the sound frequency or the body senses are mapped vs. a human body map (Fig. 10.5).

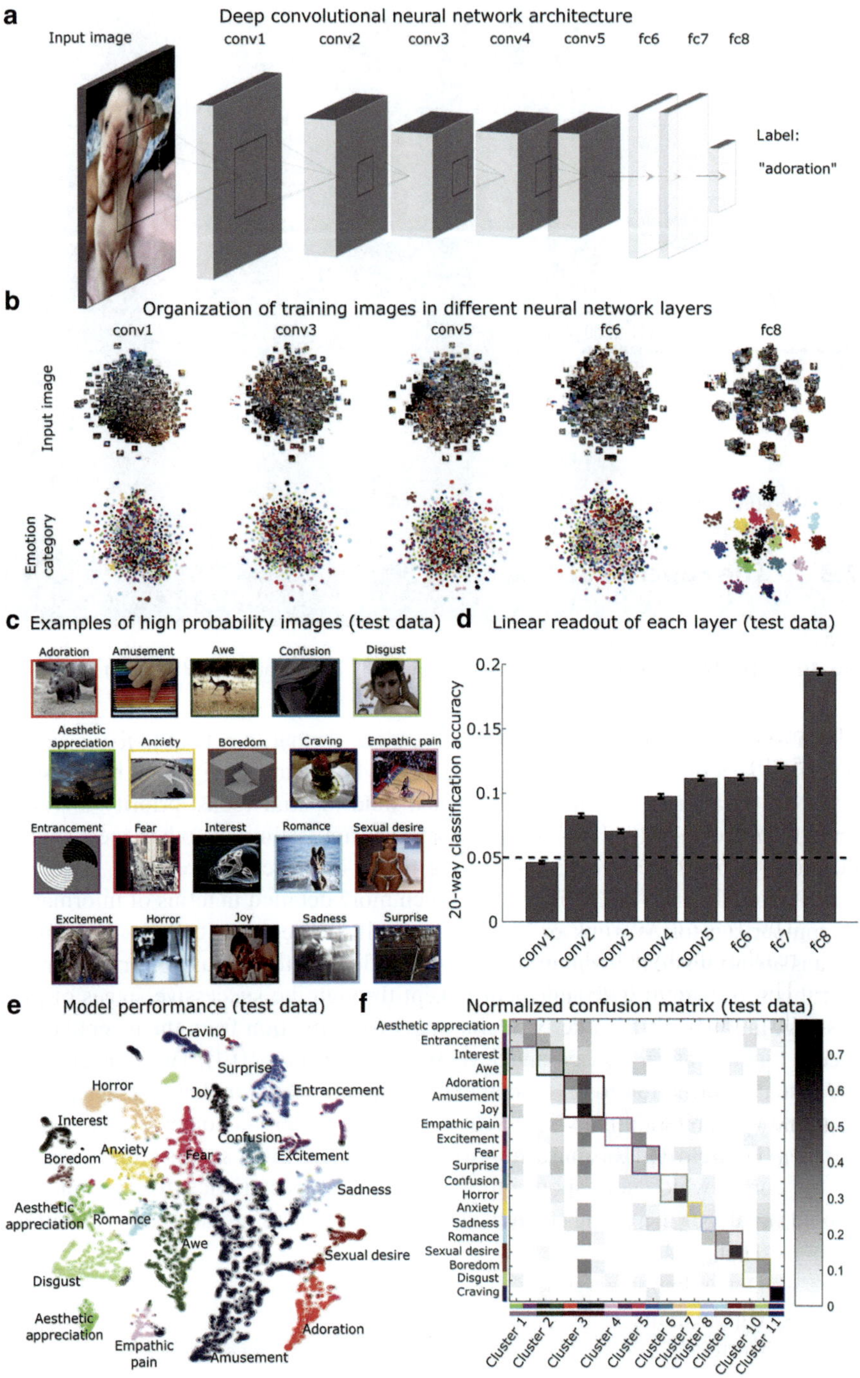

Fig. 10.4 Neuronal maps of visual perception at various levels of the human visual system (Kragel et al., Science Advances 2019;5(7):eaaw4358, κατόπιν αδείας)

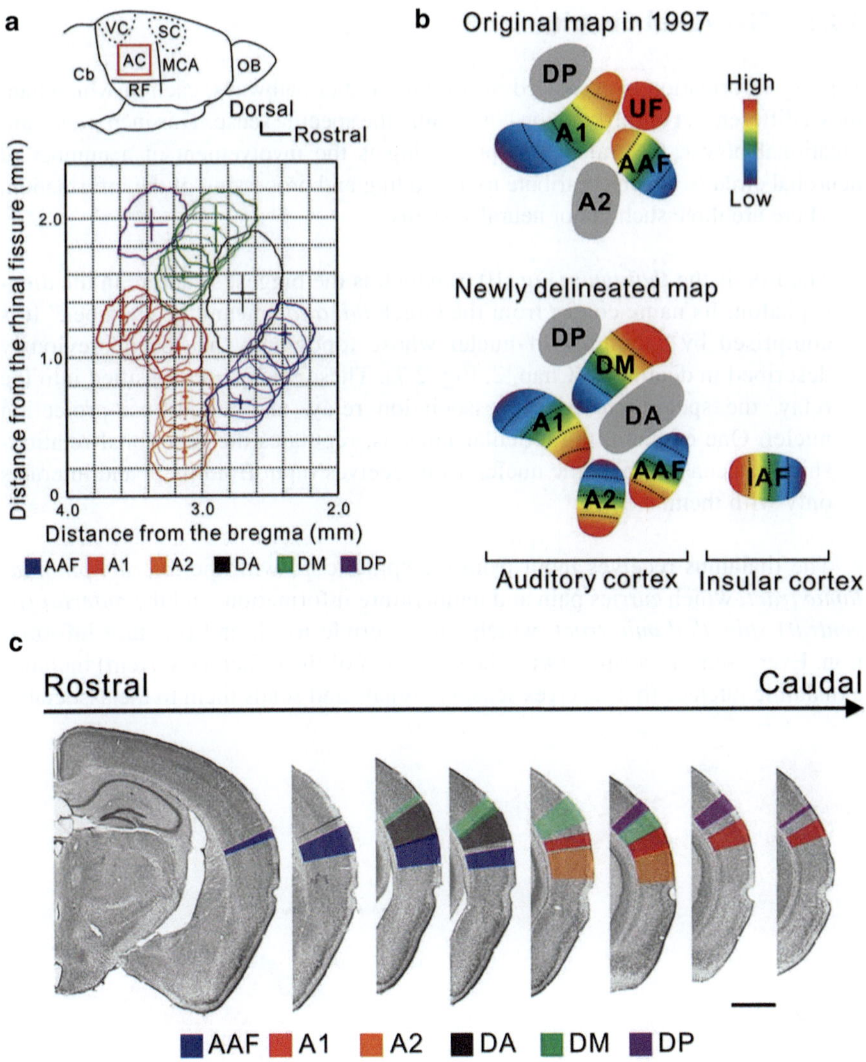

Fig. 10.5 Neuronal maps at various levels of the auditory system in rodents (Tsukano et al., Front. Neural Circuits, 2017;11:14, κατόπιν αδείας)

10.3 Neuronal Circuits

Sensory information is processed in several parallel pathways, each of which handles a different set of information concerning the specific sense. This increases computational power. Central to this processing is the involvement of a number of neuronal circuits which contribute to the coding and processing of the information.

There are three such major neural systems:

- The first is the *thalamus* (Fig. 10.6) which is the biggest structure in the diencephalon. Its name comes from the Greek *θάλαμος* meaning "chamber." It is comprised by a number of nuclei whose topography has been previously described in detail (see Chap. 2, Fig. 2.7). These nuclei are classified into the relay, the specific relay, the association relay, and the diffuse-projection nuclei. One of them, the reticular nucleus, regulates the functional relationship between the thalamic nuclei, as it receives input from them and interacts only with them.

The thalamus receives input from the spinal cord with the *lateral spinothalamic tract*, which carries pain and temperature information, and the *anterior (or ventral) spinothalamic tract*, which carries crude touch and pressure information. Every sensory system (with the exception of the olfactory system) includes a thalamic nucleus that receives sensory signals and sends them to the associated

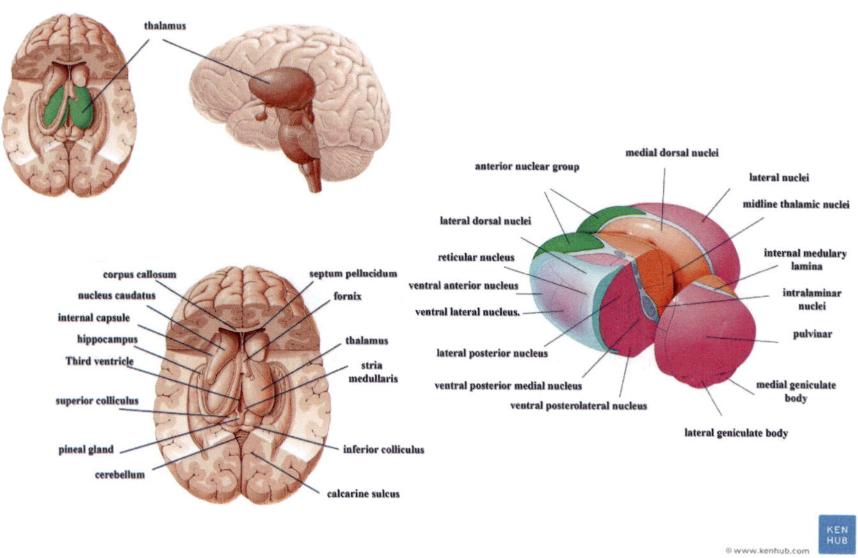

Fig. 10.6 The thalamus and its nuclei (after permission by kenhub.com)

primary cortical area. The thalamus is manifoldly connected to the hippocampus via the *mammillothalamic tract* (Carlesimo et al. 2011; Stein et al. 2000), and this connection serves the link between memory mechanisms and new stimuli. The basal ganglia provide the thalamus with input information concerning movement.

The projection to the cerebral cortex is carried via the *thalamocortical radiations* (Briggs and Usrey 2008). These pathways provide information input to the cortex from the senses, but also the thalamus relays input from the basal ganglia and cerebellum to the cortical motor areas (Asanuma et al. 1983; Evarts and Thach 1969). To summarize, the thalamus filters and processes both the sensory input to the cortex during the first steps of the process of information coming from the environment, but also it filters and processes the motor responses from the basal ganglia before they enter into the motor cortex during the early stages of response, after the process of the sensory information is completed. With its link to the hippocampus, it allows the brain to keep record of these processes in the form of a type of memory.

Thalamus

- Relay.
- Specific relay.
- Association relay.
- Diffuse-projection nuclei (e.g., reticular nucleus).
- Input from.
  - Spinal cord (lateral spinothalamic tract and anterior (or ventral) spinothalamic tract).
  - Sensory systems (olfactory excluded).
  - Hippocampus (via the mammillothalamic tract).
  - Basic ganglia.
- Projection to the cerebral cortex via the thalamocortical radiations.

- The *basal ganglia* (Fig. 10.7) (Lanciego et al. 2012; Ward et al. 2013; O'Connor 1998) are the nuclei of the telencephalon, and they constitute groups of neurons (gray matter) at the base of each brain hemisphere, and their role is to control body posture and voluntary movement. They include the following:
- The striatum (nucleus caudatus and lentiform or lenticular nucleus which is composed by the putamen and the globus pallidus).
- The claustrum.
- The amygdala (amygdaloid body).

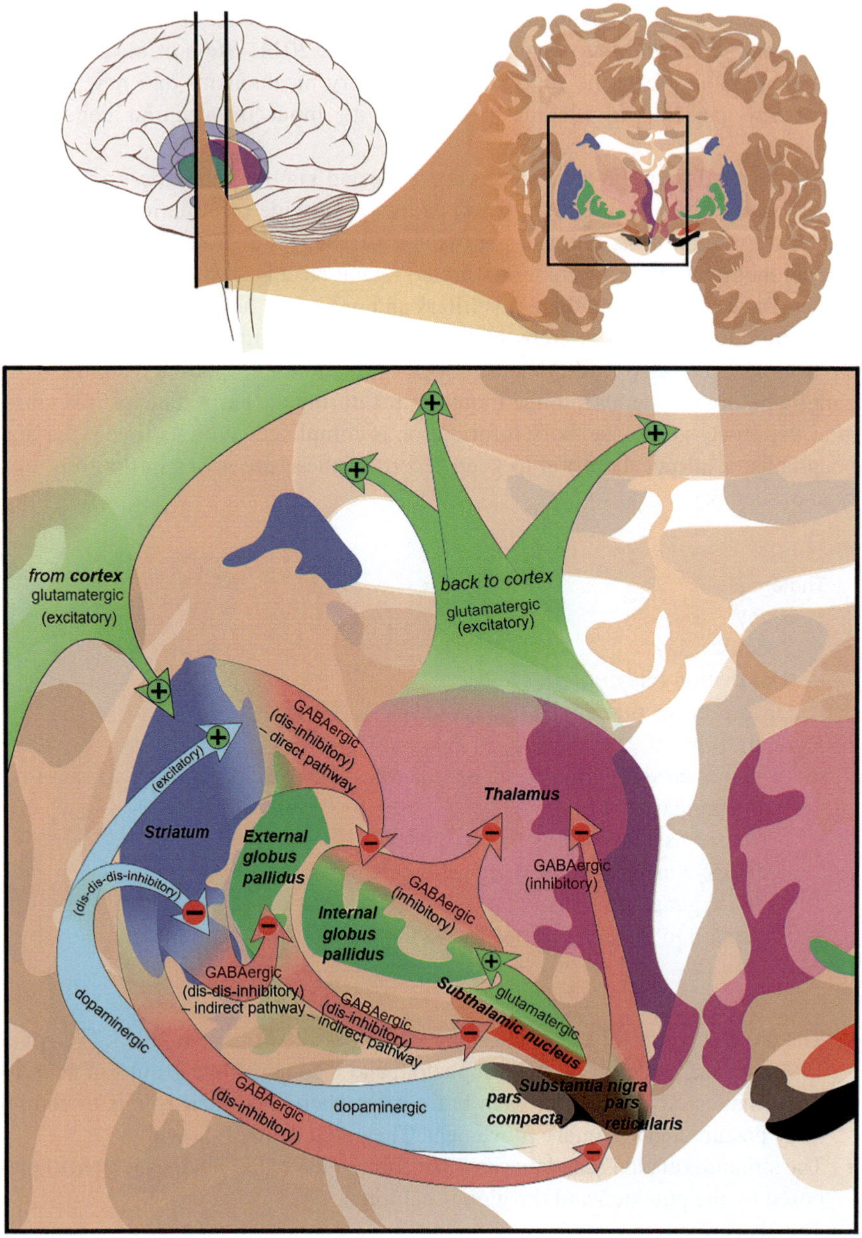

Fig. 10.7 The basal ganglia (image by Mikael Häggström, based on images by Patrick J. Lynch and Andrew Gillies, 2010). The green arrows correspond to excitatory glutamate pathways, the red to inhibitory GABA pathways, and the light blue to dopaminergic pathways which are sometimes inhibitory and other times excitatory

Different subdivisions concern the neostriatum (head of the nucleus caudatus and putamen), the paleostriatum (globus pallidus), and the archistriatum (amygdala or amygdaloid body).

Basal Ganglia

- Striatum.
  - Nucleus caudatus.
  - Lentiform nucleus.
    - Putamen.
    - Globus pallidus.
- Claustrum.
- Amygdala (amygdaloid body).
- Neostriatum.
  - Head of the nucleus caudatus.
  - Putamen.
- Paleostriatum.
  - Globus pallidus.
- Archistriatum.
  - Amygdala or amygdaloid body.

- Depending on the author, the *limbic system* (Fig. 10.8) includes the cingulate gyrus, the olfactory bulbs, the hippocampus, the hypothalamus, the amygdala, the fornix, the mammillary body, the septum pellucidum, the habenular commissure, the parahippocampal gyrus, and the limbic midbrain areas (Morgane et al. 2005). Essentially it includes the structures where the production of mood (amygdala), emotion (cingulate gyrus), and memory (hippocampus) is seated.

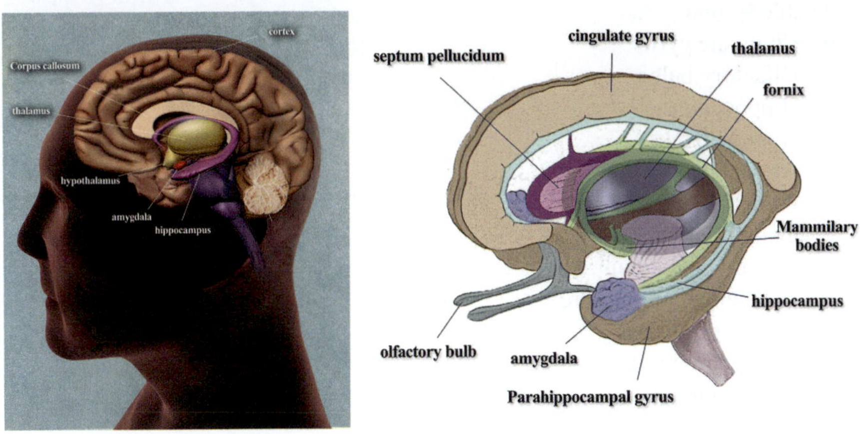

Fig. 10.8 The limbic system (left image from the National Institute on Aging, National Institutes of Health)

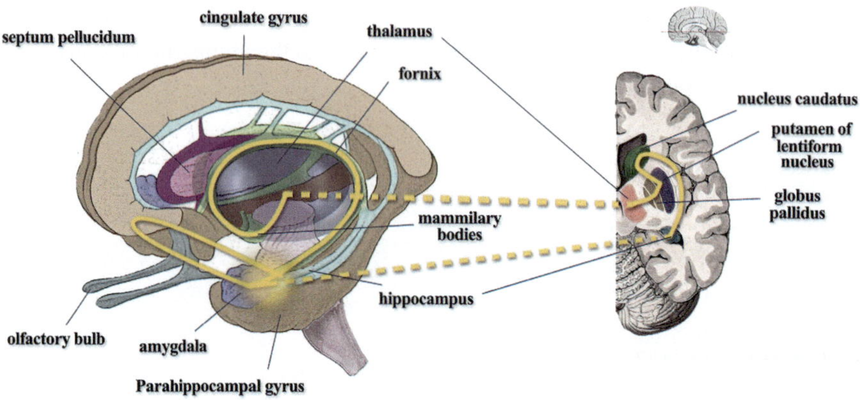

Fig. 10.9 The connections among parts of the limbic system (copyright KN Fountoulakis 2021)

The hippocampus complex includes the dentate gyrus, the hippocampus, and the subiculum. The very important entorhinal cortex is part of the parahippocampal gyrus.

There are a large number of connections among these structures and the thalamus and the frontal association cortex, and some of them are reciprocal, thus creating two main oscillatory circuits which at the century they have the hippocampus-amygdala complex and one of them involves the basal ganglia while the other serves the communication with the cortex and specifically with the anterior cingulate gyrus (Fig. 10.9).

Limbic System

- Cingulate gyrus.
- Olfactory bulbs.
- Hippocampus.
- Hypothalamus.
- Amygdala.
- Fornix.
- mammillary bodies,
- Septum pellucidum.
- Habenular commissure.
- Parahippocampal gyrus.
- Limbic midbrain.
- Mood (amygdala).
- Affect (cingulate gyrus).
- Memory (hippocampus).

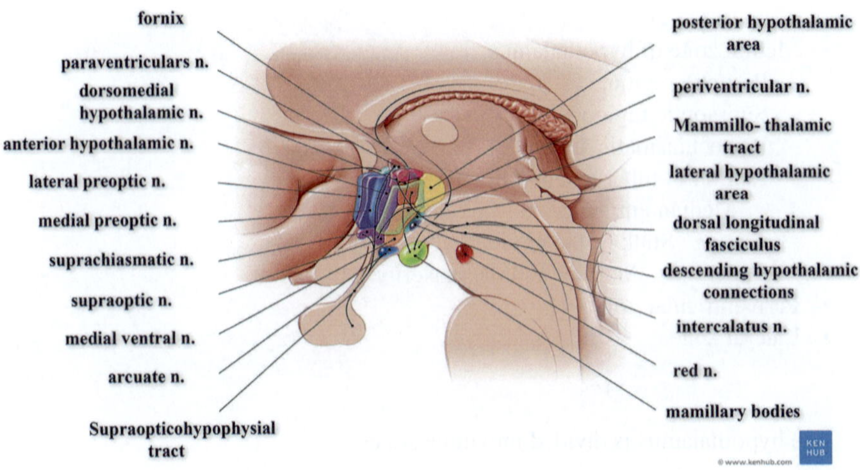

Fig. 10.10 Hypothalamus and its nuclei (after permission by kenhub.com)

- The *hypothalamus* (Fig. 10.10) is located under the thalamus and includes a large number of nuclei with a variety of functions. Its main role is the liaison of the brain with the endocrine system and is responsible for the regulation of processes pertaining to metabolism, including food and liquid intake, sleep, fatigue, and circadian rhythms. Hypothalamus plays also a role in thermoregulation as well as in behaviors related to parenthood and attachment (Saper et al. 2002, 2005; Adan et al. 2008; Chaudhri et al. 2006; Benarroch 2010; Williams et al. 2001; Gao and Horvath 2007; Kampe et al. 2009). In its nuclei, the hypothalamic hormones are synthesized. These hormones stimulate or inhibit the secretion of hormones from the pituitary, which in turn regulate the functioning of the endocrine glands of the body.

Hypothalamus
- Liaison of the brain with the endocrine system.
- Food and liquid intake.
- Sleep.
- Fatigue.
- Circadian rhythms.
- Thermoregulation.
- Behaviors related to parenthood and attachment.
- Secretion of hypothalamic hormones.

- Medial zone of hypothalamus.
  - Preoptic region.
  - Supraoptic region.
  - Retrochiasmatic area.
  - Region of tuber cinereum.
    Median eminence.
      - Stalk of the pituitary.
        - Anterior and the posterior pituitary.
- Periventricular zone.
- Lateral zone.

The hypothalamus is divided into three zones:

- The *medial zone* which includes the paraventricular, the supraoptic, the ventromedial, the dorsomedial, the supramammillary, the lateral mammillary, and the medial preoptic nuclei, the anterior nucleus of the hypothalamus, the retrochiasmatic area, the medial part of the medial mammillary nucleus, and the lateral part of the medial mammillary nucleus. The area of the tuber cinereum continues downward and creates the median eminence which is gradually turned into the stalk of the pituitary gland, which includes the anterior and the posterior pituitary.
- The *periventricular zone* which includes the median, anteroventral, intermediate, and periventricular preoptic nuclei, the suprachiasmatic, the arcuate, and the posterior periventricular nuclei, the organum vasculosum, the neurohypophysis, and the posterior hypothalamic area, while some authors include also the subfornical organ.
- The *lateral zone* includes the lateral preoptic area and the lateral hypothalamic area.
- The *default mode network (DMN)* concerns the automatic function of the brain, and anatomically, it is also mentioned as medial frontoparietal network (M-FPN). It is comprised by the median prefrontal cortex, the posterior cingulate gyrus, and the uncinate gyrus. It is active when the person does not act but instead it is absorbed in him/herself, meditates, and is daydreaming or planning for the future (Buckner et al. 2008; Sormaz et al. 2018; Spreng 2012; Murphy et al. 2018; Uddin et al. 2019; Broyd et al. 2009), and it is deactivated when the person acts. Other names for this network are the "language network," the "semantic system," and the "limbic network." Another and rather wrong term is "task-negative network" (Spreng 2012).
- At the opposite pole of this network, there is the *dorsal attention network (DAN)* which is also termed dorsal frontoparietal network (D-FPN). It includes the intraparietal sulcus and the frontal eye fields (FEF). Its major role is the voluntary orientation of visuospatial attention in order to control action. It is also termed "task-positive network" (Fox et al. 2005, 2006; Farrant and Uddin 2015; Kincade et al. 2005; Corbetta and Shulman 2002).

Topographically, the location of the *internal capsule* is extremely important, since it concerns the passage of white matter fibers between the thalamus, the nucleus caudatus, and the lentiform nucleus. This passage contains fibers that concern all of the body and at high density. As a result, lesions in this area might cause severe neurological signs and symptoms.

10.4 Internal Representation of the External Environment

After the initial transformation of the external stimuli to neuronal signals by the sensory organs and the entrance of information into the brain, this information is processed and transformed at every step. At each stage, every postsynaptic neuron typically receives inputs from thousands of presynaptic neurons. Eventually, the sensory information are projected onto the cortex at areas specific for each sense. These areas in the cortex are called *primary association areas*, and their main characteristic is that they receive input from a single sense, and because of this, they are called *unimodal*. The areas where each of the five senses projects are well known, and it is the pole of the occipital lobe (calcarine sulcus) for sight and the Heschl's gyri in the temporal lobe for hearing and the posterior central gyrus in the parietal lobe for somatosensory senses. Next to or surrounding the primary association cortex, an adjacent, higher-order area, the *secondary association cortex* which is also unimodal is located. The concept of "association areas" was introduced by John Hughlings Jackson (1835–1911).

Association Areas

- Primary association areas (unimodal).
- Secondary association areas (unimodal).
  - Projection into the cortex is done in a sort of a topographical map.
  - This map is characterized by differential resolution.
- Tertiary major multimodal association areas (multiphasic).
  - Posterior association area.
  - Limbic association area.
  - Anterior association area (prefrontal cortex).
  - Integration of two or more sensory modalities.
  - Internal neuronal representation of the external environment is developed.

An important characteristic is that the projection into the cortex is done in an orderly manner, thus creating a sort of a *topographical map*. This map corresponds to the characteristics of the specific sense (e.g., light touch vs. pain on the same area of the skin), and it is spatial for sight and somatic senses and frequency-oriented for sounds, whose spatial origin however is able to locate (Clark et al. 1988; Kaas et al. 1979; Jenkins et al. 1990; Meldrum 2000). It also corresponds to the differential

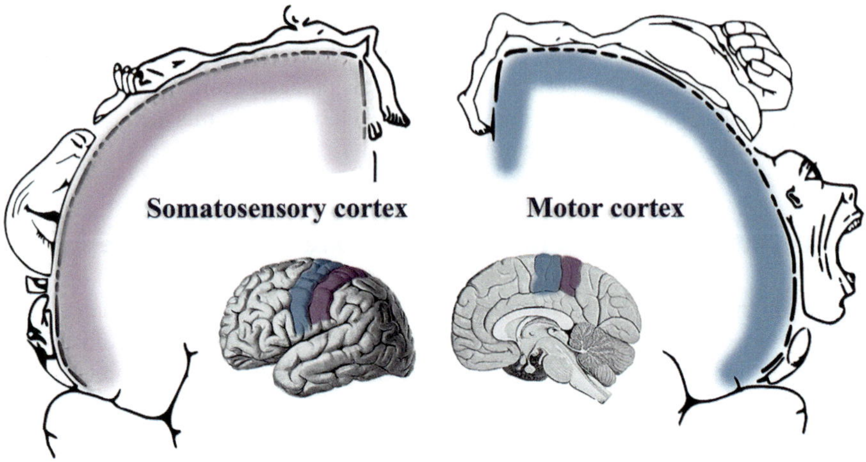

Fig. 10.11 Topographical maps on the somatosensory and motor cortex (copyright KN Fountoulakis 2021)

resolution (and subsequently importance) given to each part of the sensory stimulus by the external sensors. Thus, the topographical map is not an accurate representation of the physical stimulus but rather a distorted one, according to the weight given by the senses. It reflects both the *physical topography but also the density of the perceived information and their resolution*. For example, in the somatosensory cortex, the face, tongue, and fingers occupy a much larger surface (and neurons are subsequently the focus in processing) in comparison to the trunk or legs (Fig. 10.11). In the visual field, the center of vision (the fovea) occupies much larger area in comparison to the peripheral parts of the visual field (Fox et al. 1987; Penfield and Rasmussen 1950). These maps change with experience (Mogilner et al. 1993; Jenkins et al. 1990; Ramachandran 1993).

All secondary association areas project to one of the three *tertiary major multimodal association areas*. In these areas, the integration of two or more sensory modalities takes place, and a more comprehensive internal neuronal representation of the external environment is developed (Van Hoesen 1993).

Three multimodal association areas are particularly important:

- The *posterior association area*, at the margin of the parietal, temporal, and occipital lobes. It receives information from several sensory modalities (Fig. 10.12).
- The *limbic association area*, which includes the anterior pole of the temporal lobe and the medial edge of the cerebral hemisphere. It is concerned with emotion and memory storage.
- The *anterior association area (prefrontal cortex)* which is located rostral to the postcentral gyrus (Fig. 10.12). It is concerned with the executive functions, judgment, planning for the future, and holding and organizing events from memory for prospective action (working memory) as well as with the planning of movement (Goldman-Rakic 1992, 1996).

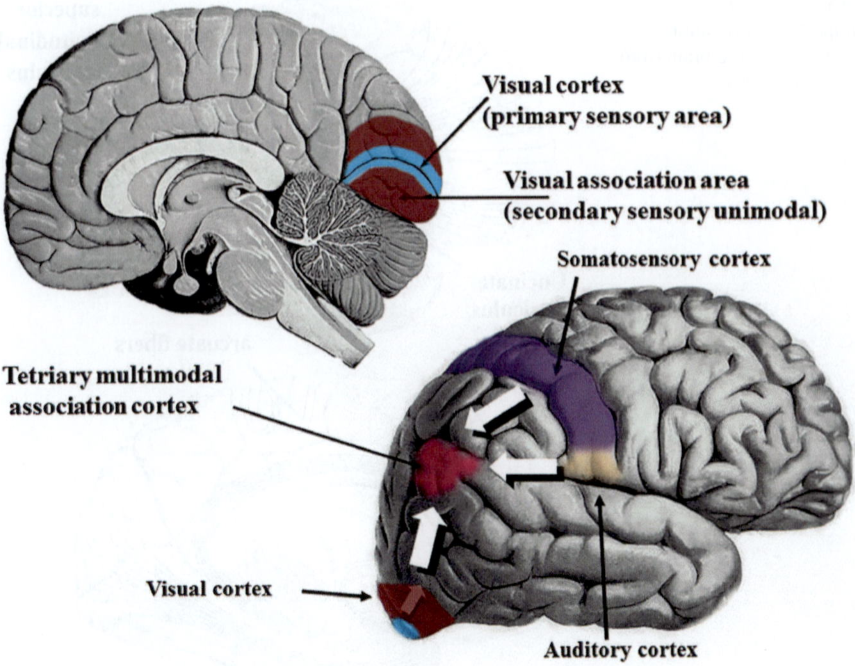

Fig. 10.12 Association areas and information flow (copyright KN Fountoulakis 2021)

Association processing and the consolidation of the mechanisms are served by a network of axons in tracts which interconnect remote brain areas but also neighboring ones (Fig. 10.13). Sensory information are projected to the *amygdala* which adds an *emotional component* which is essentially a primitive kind of assessment and decision (e.g., fear in the sight of a snake) and to the *hippocampus*, which stores aspects of perception in long-term memory and also provides with feedback from previous *experiences and memories* (Fig. 10.14) (Fountoulakis et al. 2008). It seems that the *anterior cingulate* hosts the processes of *affect* while the posterior serves cognitive and psychomotor aspects related to mood and affect (Mayberg 1997; Mayberg et al. 1999). In this way, a comprehensive internal representation of the external environment is formulated. The hippocampus is responsible for the formation of long-term memories, but it is not the permanent storage site of memories. If damaged, the subject loses the ability to form new memories but keeps old ones. Overall, the roles of these connections are to input affect on the information process and the decision-making and to develop and store memories and to selectively retrieve them on the basis of current sensory input and emotional state (Van Hoesen 1993). Because of this complex emotional-mnemonic function, it has been proposed as the seat of the *social brain* (Brotehrs 1990).

In primates, in various areas of the brain, a functional battery of neurons has been discovered and labeled *mirror neurons*. These neurons seem to trigger when the

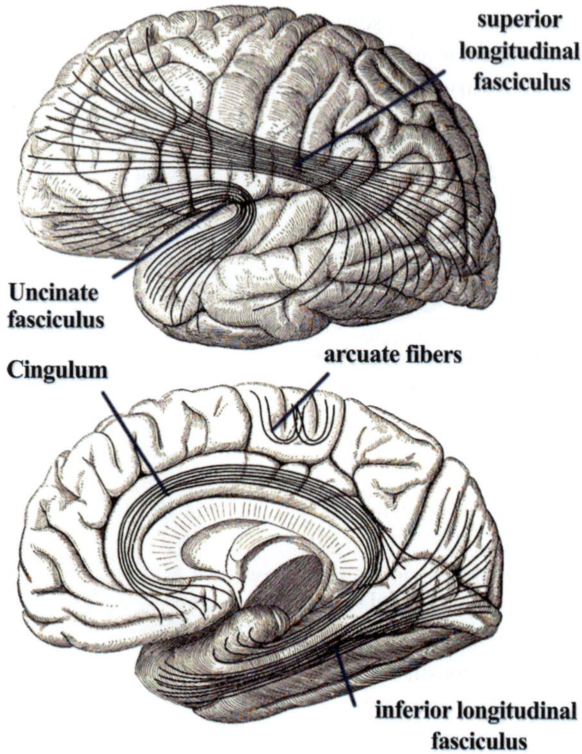

Fig. 10.13 The most important association pathways of the brain (από Sobotta, 1909)

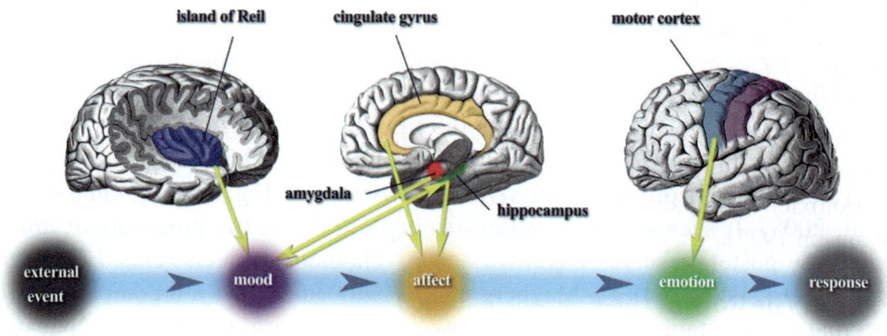

Fig. 10.14 From the entering of the stimulus to the final decision through various stages of processing with the participation of mnemonic registrations and affects (copyright KN Fountoulakis 2021)

animal manifests a behavior but also when it observes the same behavior (Prather et al. 2008; Keysers 2009; Rizzolatti and Craighero 2004). In humans, such neurons have been identified in the premotor cortex, the primary somatosensory area, and the inferior parietal lobe (Molenberghs et al. 2009), but their true role in humans is under dispute (Dinstein et al. 2008; Hickok 2009).

With this complex way and the connection between processed information from the perception of external stimuli with previous mnemonic registrations and emotional components, a comprehensive internal representation of the external environment is formed, and the progress to the decision-making is prepared.

10.5 Decision-Making and Action

As mentioned above, most information go through the thalamus which functions as a *gatekeeper of the information flow to the brain cortex*. The thalamus together with the limbic system precludes or enhances the passage of specific information, the retrieval of relevant memories, and the storage of new ones, depending on the behavioral status of the organism.

Decision

- Information entering into the cortex via the thalamus and the limbic system.
- Frontal lobe.
  - Outer surface (dorsal prefrontal association area): working memory and tertiary association area.
  - Medial surface (medial prefrontal cortex): affect.
  - Orbital surface (ventral orbitofrontal cortex): decision.

A simplistic functional description of the frontal lobe could consider it as a *trilateral pyramid* with the base toward the parietal lobe and with its top corresponding to the frontal pole. One of its lateral surfaces is horizontal and corresponds to the orbital cortex, while the other two correspond to the outer and the medial surface of the lobe. The two lateral surfaces corresponding to the orbital and the medial surface are joined in a 90° angle (Fig. 10.15):

- The outer surface *(dorsal prefrontal association area)* corresponds to what is closer to RAM in modern computers and hosts working memory as well as the tertiary association area.
- The medial surface (*medial prefrontal cortex*) concerns the production of affect.
- The orbital surface (*ventral orbitofrontal cortex*) hosts the decision circuits.

The decision is taken exactly on the basis of the interaction among these three brain areas which because of their triangular spatial relationship they manifest an extremely high ability of interaction.

After the decision is made, projections to the *premotor cortex* are triggered (Fig. 10.16). Every decision implies the initiation of a behavior which almost always has a motor part, either to act or to stand still. A topographical map exists also in the motor cortex (Fig. 10.11). This map, like the sensory maps, does not represent every part of the body equally, but it puts specific emphasis on individual body parts

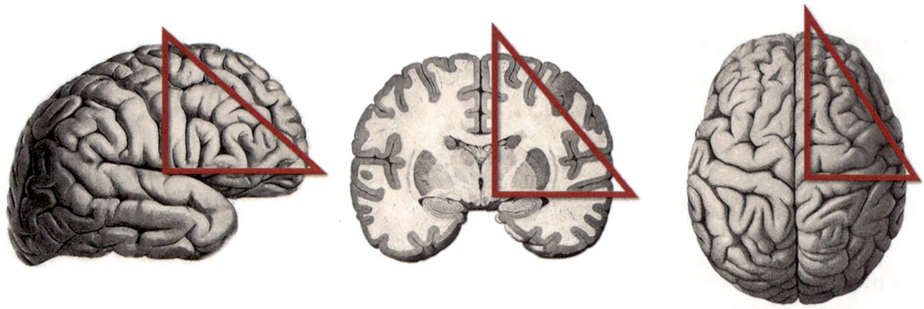

Fig. 10.15 The simplified pyramid shape of the frontal lobe (copyright KN Fountoulakis 2021)

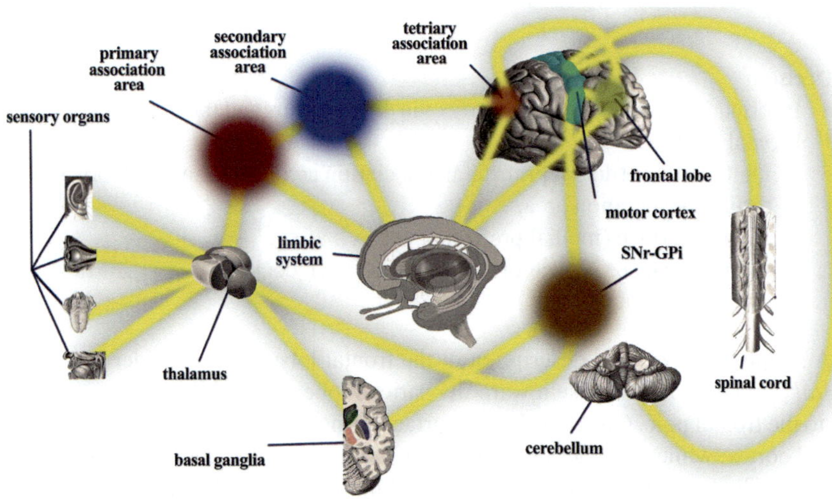

Fig. 10.16 Abstract representation of the flow of information and its processing in different brain areas (copyright KN Fountoulakis 2021)

reflecting the fineness of control required for the motor function of these particular body parts.

The *primary motor cortex* probably contains the "software" of movement and is located rostral to the central sulcus and is associated with the motor systems of the spinal cord. The human corticospinal tract consists of about one million axons, and about 40% of them originate in the motor cortex. The entire projection is also called the *pyramidal tract*.

In this process, there is input to the motor cortex from the basal ganglia which constitute key structures for motor planning. There are two major pathways with projections from and to the basal ganglia (Fig. 10.16):

- The *direct pathway*, which carries projections from cortical cells to the striatum via the SNr-GPi complex, which in turn projects to the thalamus through the inhibitory ansa lenticularis pathway, and eventually the thalamus projects to the motor cortex which in turn projects to the brain stem and ultimately mobilizes muscle fibers via the lateral corticospinal tract.

- The *indirect pathway* also originates from the striatum, and after stimulation by the cortex, it projects to the globus pallidus externa, which in turn projects to the subthalamic nucleus which also in turn projects to the SNr-GPi complex which inhibits the thalamus. This indirect pathway regulates the inhibition of the motor cortex by the thalamus. Both pathways are based on a serial sequence of inhibitory and excitatory projections which produce a fine-tuned end-output. Overall, the direct and the indirect pathways are antagonistic in their functions, and this antagonism is modulated by the substantia nigra pars compacta. The antagonism serves balance and fine-tuning of motor movement.

It is important for the long-term adaptation and survival that those decisions, choices, and responses with a positive outcome for the individual be *ranked* higher in the consideration of future resolutions and given more weight and speed in situations of ambiguity. In other words, behaviors, which are successful, are also rewarding, and this is registered not only in terms of essential mnemonic registrations but also at a lower unconscious level. In general, dopaminergic neurons in the midbrain mediate these rewarding aspects of behavior.

Learning of Behaviors

- Successful behaviors are also rewarding.
- The globus pallidus and the substantia nigra inhibit the thalamus.
- Selection circuit.
  - Neostriatum, nucleus accumbens, and VTA.
  - Memories and templates in the prefrontal cortex.
  - Affect.
  - The above circuit tests a variety of incoming options against templates.
- Result
  - Removal of inhibition by the globus pallidus and the substantia nigra of selective areas of the thalamus.

The basal ganglia circuit probably serves the *selection and enforcement of behaviors that lead to reward.* This could be achieved by using past experience to predict which patterns of input from the neocortex to the thalamus will lead to reward. In order to execute this function, the cells in the globus pallidus and the substantia nigra fire tonically producing a constant inhibition of neurons in the thalamus and the superior colliculus. In case this inhibition pauses, then the neurons in these areas are released and respond to excitatory inputs that would otherwise be subthreshold. The central location of the globus pallidus in the basal ganglia formation implies also a central role in the functioning. It constitutes the inner component of the lentiform nucleus, and with the putamen, it forms a cone-like structure, with its tip directed medially. The majority of projections from the neostriatum (which receives input from the neocortex, the thalamus, and the substantia nigra) are to the globus pallidus which serves as an intermediate processing center (Wilson 2004). A role for the nucleus accumbens and the VTA is particularly well established. Numerous things that people find rewarding, including addictive drugs, good-tasting food, and sex, have been shown to elicit activation of the ventral tegmental area dopamine system.

This above network probably tests a variety of inputs concerning both the interpretation and the response options on the basis of templates and maybe error memories provided by the prefrontal cortex to choose those that systematically and reliably elicit reward reactions in the brain. The prefrontal cortex provides the templates to test against but on the other hand also inhibits or modifies the "resolutions" of the basal ganglia. This is why although mood, affect, and initial tendencies to act are not under volitional control and cannot be influenced, the final output (externalized behavior) is under volitional control through effortful regulation.

References

Adan RA, Vanderschuren LJ, la Fleur SE (2008) Anti-obesity drugs and neural circuits of feeding. Trends Pharmacol Sci 29(4):208–217. https://doi.org/10.1016/j.tips.2008.01.008

Asanuma C, Thach WT, Jones EG (1983) Cytoarchitectonic delineation of the ventral lateral thalamic region in the monkey. Brain Res 286(3):219–235. https://doi.org/10.1016/0165-0173(83)90014-0

Benarroch EE (2010) Neural control of feeding behavior: overview and clinical correlations. Neurology 74(20):1643–1650. https://doi.org/10.1212/WNL.0b013e3181df0a3f

Briggs F, Usrey WM (2008) Emerging views of corticothalamic function. Curr Opin Neurobiol 18(4):403–407. https://doi.org/10.1016/j.conb.2008.09.002

Brotehrs L (1990) The social brain: a project for integrating primate behavior and neurophysiology in a new domain. Concepts Neurosci 1:27–51

Broyd SJ, Demanuele C, Debener S, Helps SK, James CJ, Sonuga-Barke EJ (2009) Default-mode brain dysfunction in mental disorders: a systematic review. Neurosci Biobehav Rev 33(3):279–296. https://doi.org/10.1016/j.neubiorev.2008.09.002

Buckner RL, Andrews-Hanna JR, Schacter DL (2008) The brain's default network: anatomy, function, and relevance to disease. Ann N Y Acad Sci 1124:1–38. https://doi.org/10.1196/annals.1440.011

Carlesimo GA, Lombardi MG, Caltagirone C (2011) Vascular thalamic amnesia: a reappraisal. Neuropsychologia 49(5):777–789. https://doi.org/10.1016/j.neuropsychologia.2011.01.026

Changeux J, Dehaene S (1993) Formal neuronal models for cognitive functions associated with the prefrontal cortex. In: Poggio T, Glaser D (eds) Exploring brain functions- models in neuroscience. Wiley, Chichester, pp 249–267

Chaudhri O, Small C, Bloom S (2006) Gastrointestinal hormones regulating appetite. Philos Trans R Soc Lond Ser B Biol Sci 361(1471):1187–1209. https://doi.org/10.1098/rstb.2006.1856

Clark SA, Allard T, Jenkins WM, Merzenich MM (1988) Receptive fields in the body-surface map in adult cortex defined by temporally correlated inputs. Nature 332(6163):444–445. https://doi.org/10.1038/332444a0

Corbetta M, Shulman GL (2002) Control of goal-directed and stimulus-driven attention in the brain. Nat Rev Neurosci 3(3):201–215. https://doi.org/10.1038/nrn755

Dinstein I, Thomas C, Behrmann M, Heeger DJ (2008) A mirror up to nature. Curr Biol 18(1):R13–R18. https://doi.org/10.1016/j.cub.2007.11.004

Evarts EV, Thach WT (1969) Motor mechanisms of the CNS: cerebrocerebellar interrelations. Annu Rev Physiol 31:451–498. https://doi.org/10.1146/annurev.ph.31.030169.002315

Farrant K, Uddin LQ (2015) Asymmetric development of dorsal and ventral attention networks in the human brain. Dev Cogn Neurosci 12:165–174. https://doi.org/10.1016/j.dcn.2015.02.001

Feneis H, Dauber W (2000) Pocket atlas of human anatomy, 4th edn. Thieme, Stuttgart, New York

Fountoulakis KN, Giannakopoulos P, Kovari E, Bouras C (2008) Assessing the role of cingulate cortex in bipolar disorder: neuropathological, structural and functional imaging data. Brain Res Rev 59(1):9–21. https://doi.org/10.1016/j.brainresrev.2008.04.005

Fox PT, Miezin FM, Allman JM, Van Essen DC, Raichle ME (1987) Retinotopic organization of human visual cortex mapped with positron-emission tomography. J Neurosci Off J Soc Neurosci 7(3):913–922

Fox MD, Snyder AZ, Vincent JL, Corbetta M, Van Essen DC, Raichle ME (2005) The human brain is intrinsically organized into dynamic, anticorrelated functional networks. Proc Natl Acad Sci U S A 102(27):9673–9678. https://doi.org/10.1073/pnas.0504136102

Fox MD, Corbetta M, Snyder AZ, Vincent JL, Raichle ME (2006) Spontaneous neuronal activity distinguishes human dorsal and ventral attention systems. Proc Natl Acad Sci U S A 103(26):10046–10051. https://doi.org/10.1073/pnas.0604187103

Gao Q, Horvath TL (2007) Neurobiology of feeding and energy expenditure. Annu Rev Neurosci 30:367–398. https://doi.org/10.1146/annurev.neuro.30.051606.094324

Gazzaniga MS (1989) Organization of the human brain. Science 245(4921):947–952. https://doi.org/10.1126/science.2672334

Gazzaniga MS (1995) Principles of human brain organization derived from split-brain studies. Neuron 14(2):217–228. https://doi.org/10.1016/0896-6273(95)90280-5

Goldman-Rakic PS (1992) Working memory and the mind. Sci Am 267(3):110–117. https://doi.org/10.1038/scientificamerican0992-110

Goldman-Rakic PS (1996) Regional and cellular fractionation of working memory. Proc Natl Acad Sci U S A 93(24):13473–13480. https://doi.org/10.1073/pnas.93.24.13473

Hickok G (2009) Eight problems for the mirror neuron theory of action understanding in monkeys and humans. J Cogn Neurosci 21(7):1229–1243. https://doi.org/10.1162/jocn.2009.21189

Jenkins WM, Merzenich MM, Ochs MT, Allard T, Guic-Robles E (1990) Functional reorganization of primary somatosensory cortex in adult owl monkeys after behaviorally controlled tactile stimulation. J Neurophysiol 63(1):82–104. https://doi.org/10.1152/jn.1990.63.1.82

Kaas JH, Nelson RJ, Sur M, Lin CS, Merzenich MM (1979) Multiple representations of the body within the primary somatosensory cortex of primates. Science 204(4392):521–523. https://doi.org/10.1126/science.107591

Kampe J, Tschop MH, Hollis JH, Oldfield BJ (2009) An anatomic basis for the communication of hypothalamic, cortical and mesolimbic circuitry in the regulation of energy balance. Eur J Neurosci 30(3):415–430. https://doi.org/10.1111/j.1460-9568.2009.06818.x

Kandel E (2000) The brain and behavior. In: Kandel E, Schwarts J, Jessell T (eds) Principles of neural science. McGraw-Hill, pp 5–18

Keysers C (2009) Mirror neurons. Curr Biol 19(21):R971–R973. https://doi.org/10.1016/j.cub.2009.08.026

Kincade JM, Abrams RA, Astafiev SV, Shulman GL, Corbetta M (2005) An event-related functional magnetic resonance imaging study of voluntary and stimulus-driven orienting of attention. J Neurosci Off J Soc Neurosci 25(18):4593–4604. https://doi.org/10.1523/JNEUROSCI.0236-05.2005

Lanciego JL, Luquin N, Obeso JA (2012) Functional neuroanatomy of the basal ganglia. Cold Spring Harb Perspect Med 2(12):a009621. https://doi.org/10.1101/cshperspect.a009621

Lewis D, Oeth K (1995) Functional neuroanatomy. In: Kaplan H, Sadock B (eds) Comprehensive textbook of psychiatry, 6th edn. Williams and Wilkins, Baltimore, pp 4–24

Mayberg HS (1997) Limbic-cortical dysregulation: a proposed model of depression. J Neuropsychiatry Clin Neurosci 9(3):471–481. https://doi.org/10.1176/jnp.9.3.471

Mayberg HS, Liotti M, Brannan SK, McGinnis S, Mahurin RK, Jerabek PA, Silva JA, Tekell JL, Martin CC, Lancaster JL, Fox PT (1999) Reciprocal limbic-cortical function and negative mood: converging PET findings in depression and normal sadness. Am J Psychiatry 156(5):675–682. https://doi.org/10.1176/ajp.156.5.675

Meldrum BS (2000) Glutamate as a neurotransmitter in the brain: review of physiology and pathology. J Nutr 130(4S Suppl):1007S–1015S. https://doi.org/10.1093/jn/130.4.1007S

Mishkin M (1993) Cerebral memory circuits. In: Poggio T, Glaser D (eds) Exploring brain functions- models in neuroscience. Wiley, Chichester, pp 113–127

Mogilner A, Grossman JA, Ribary U, Joliot M, Volkmann J, Rapaport D, Beasley RW, Llinas RR (1993) Somatosensory cortical plasticity in adult humans revealed by magnetoencephalography. Proc Natl Acad Sci U S A 90(8):3593–3597. https://doi.org/10.1073/pnas.90.8.3593

Molenberghs P, Cunnington R, Mattingley JB (2009) Is the mirror neuron system involved in imitation? A short review and meta-analysis. Neurosci Biobehav Rev 33(7):975–980. https://doi.org/10.1016/j.neubiorev.2009.03.010

Morgane PJ, Galler JR, Mokler DJ (2005) A review of systems and networks of the limbic forebrain/limbic midbrain. Prog Neurobiol 75(2):143–160. https://doi.org/10.1016/j.pneurobio.2005.01.001

Murphy C, Jefferies E, Rueschemeyer SA, Sormaz M, Wang HT, Margulies DS, Smallwood J (2018) Distant from input: evidence of regions within the default mode network supporting perceptually-decoupled and conceptually-guided cognition. NeuroImage 171:393–401. https://doi.org/10.1016/j.neuroimage.2018.01.017

O'Connor WT (1998) Functional neuroanatomy of the basal ganglia as studied by dual-probe microdialysis. Nucl Med Biol 25(8):743–746. https://doi.org/10.1016/s0969-8051(98)00066-3

Penfield W, Rasmussen T (1950) The cerebral cortex of man: a clinical study of localization of function. Macmillan, New York

Prather JF, Peters S, Nowicki S, Mooney R (2008) Precise auditory-vocal mirroring in neurons for learned vocal communication. Nature 451(7176):305–310. https://doi.org/10.1038/nature06492

Putz R, Pabst R (2006) Sobotta atlas of human anatomy, 22nd edn. Elsevier GmbH, Munich

Ramachandran VS (1993) Behavioral and magnetoencephalographic correlates of plasticity in the adult human brain. Proc Natl Acad Sci U S A 90(22):10413–10420. https://doi.org/10.1073/pnas.90.22.10413

Rizzolatti G, Craighero L (2004) The mirror-neuron system. Annu Rev Neurosci 27:169–192. https://doi.org/10.1146/annurev.neuro.27.070203.144230

Saper CB, Chou TC, Elmquist JK (2002) The need to feed: homeostatic and hedonic control of eating. Neuron 36(2):199–211. https://doi.org/10.1016/s0896-6273(02)00969-8

Saper CB, Scammell TE, Lu J (2005) Hypothalamic regulation of sleep and circadian rhythms. Nature 437(7063):1257–1263. https://doi.org/10.1038/nature04284

Sormaz M, Murphy C, Wang HT, Hymers M, Karapanagiotidis T, Poerio G, Margulies DS, Jefferies E, Smallwood J (2018) Default mode network can support the level of detail in experience during active task states. Proc Natl Acad Sci U S A 115(37):9318–9323. https://doi.org/10.1073/pnas.1721259115

Spreng RN (2012) The fallacy of a "task-negative" network. Front Psychol 3:145. https://doi.org/10.3389/fpsyg.2012.00145

Standing S (2016) Gray's anatomy, 41st edn. Elsevier

Stein T, Moritz C, Quigley M, Cordes D, Haughton V, Meyerand E (2000) Functional connectivity in the thalamus and hippocampus studied with functional MR imaging. AJNR Am J Neuroradiol 21(8):1397–1401

Uddin LQ, Yeo BTT, Spreng RN (2019) Towards a universal taxonomy of macro-scale functional human brain networks. Brain Topogr 32(6):926–942. https://doi.org/10.1007/s10548-019-00744-6

Ungerleider LG (1995) Functional brain imaging studies of cortical mechanisms for memory. Science 270(5237):769–775. https://doi.org/10.1126/science.270.5237.769

Van Hoesen GW (1993) The modern concept of association cortex. Curr Opin Neurobiol 3(2):150–154. https://doi.org/10.1016/0959-4388(93)90202-a

Ward P, Seri And S, Cavanna AE (2013) Functional neuroanatomy and behavioural correlates of the basal ganglia: evidence from lesion studies. Behav Neurol 26(4):219–223. https://doi.org/10.3233/BEN-2012-120264

Williams G, Bing C, Cai XJ, Harrold JA, King PJ, Liu XH (2001) The hypothalamus and the control of energy homeostasis: different circuits, different purposes. Physiol Behav 74(4–5):683–701. https://doi.org/10.1016/s0031-9384(01)00612-6

Wilson C (2004) Basal ganglia. In: Shepherd G (ed) The synaptic organization of the brain. Oxford University Press, Oxford, pp 361–413

Higher Mental Functions

11

Contents

11.1 Attention and Concentration

11.1.1 Neuropsychological Models of Attention

Attention is defined as the brain function and its accompanying behavior that concern the selective focus on a distinct part of a stimulus or an information and could be either external/objective or internal/subjective. Attention also leads to the selection of these parts which the brain will neglect (James 1890; Homskaya 2001; Posner and Petersen 1990). It is through attention that the limited mental resources are invested in real time, as the brain has limited ability to process the incoming information (Anderson 2004). Sustaining attention for a prolonged period is called *concentration*.

K. N. Fountoulakis, *The Human Connectome*,
https://doi.org/10.1007/978-3-031-10351-3_11

Dividing attention on many information simultaneously, or the simultaneous execution of many works, leads to more errors and lower efficacy (Spelke et al. 1976). Probably this happens because the mechanism of attention is single and indivisible (Kahneman 1973); however, more recent research suggests that there are functional subunits specialized in different kinds of objects and information (Wahn and Konig 2017).

Attention

- Selective focus of processing.
- Dividing attention increases errors.
- Concentration is the sustaining of attention.
- Attention, concentration, and immediate memory constitute a single clinical parameter.
- Spotlight model.
- Zoom-lens model.
- Two-stage model.
  - Pre-attentive stage.
  - Focused attention stage.
- Hierarchical model.
  - Focusing of attention.
  - Sustained attention (vigilance and concentration).
  - Selective attention.
  - Alternation of focusing attention.
  - Divided attention.
- Four processes model.
  - Working memory.
  - Competitive selection.
  - Top-down sensitivity control.
  - Bottom-up filtering.
- Three functional parts model.
  - Alertness.
  - Orientation.
  - Executive attention.

From a practical-clinical view, *attention cannot be separated from concentration and immediate memory*, simply because in order to assess attention, it is necessary the subject to concentrate for a minimum duration of time and to comprehend the directions and keep them in memory again for at least a minimum time duration, before executing them. In case, any of the above mechanisms dysfunction, assessing attention is impossible.

Fig. 11.1 Attention could be focused a much as a spotlight. In 1999, Simons and Chabris showed a video to a group of subjects. In that video, a group of players were passing the ball. The observers were asked to count how many times three players in white shirts passed the ball. Thirty seconds later and while the players were passing the ball, a woman in a gorilla suit entered the stage, turned and faced the camera, hit her chest, and left. Half of the observers did not notice her at all, a finding that supports the concept of "attentional blindness." (The image is shown after license by Daniel Simons, while the complete video is available at www.dansimons.com and www.theinvisiblegorilla.com From Simons, D. J., & Chabris, C. F. (1999). Perception, 28, 1059–1074)

There are several models which describe the proposed mechanisms on whom attention is based:

- The *spotlight model* which is based on the theories of William James who considered that attention has a focus point with a margin which separates it from the low-resolution fringe. The "invisible gorilla" experiment was based on this theory (Fig. 11.1) (Eriksen and Hoffman 1972; Simons and Chabris 1999).
- The *zoom-lens model*, which adds to the spotlight model, the ability of changing the size of the focus point. This possible leads to a compromise between size and efficacy, as the capacity of attention is considered to be fixed and limited (Eriksen and St James 1986; Castiello and Umilta 1990).
- According to the *two-stage model* of Treisman, attention includes:
  - The pre-attentive stage which concerns the detection of the object and the identification of its characteristics at the unconscious level (Treisman et al. 1992).
  - The focused attention stage, during which attention focuses on any of the characteristics of the object so that subsequently this particular characteristic will be included in short-term memory (Raftopoulos 2007).
- The *hierarchical model* of Sohlberg and Mateer (2001) which was based on the way of recovery of patients with brain damage, and describes the following functional characteristics of attention, in hierarchical order from the more basic to the more complex and laborious:
  - Focusing of attention.
  - Sustained attention (vigilance and concentration).
  - Selective attention.
  - Alternation of focusing attention.
  - Divided attention.

- The *four processes model* of Knudsen (Knudsen 2007; Pattyn et al. 2008) which includes:
  - Working memory, which stores information temporarily for immediate processing.
  - Competitive selection, which concerns which information will be held by working memory.
  - Top-down sensitivity control, which regulates the strength of information as they compete to each other to enter into working memory.
  - Bottom-up filtering for salient stimuli, which automatically amplifies the response to unusual stimuli or to stimuli of known value.
- The *three functional parts model* of Posner, which proposes that attention includes (Rueda et al. 2005; Posner and Petersen 1990; Posner and Boies 1971; Trautwein et al. 2016; Fan et al. 2009; Callejas et al. 2004):
  - Alertness.
  - Orientation.
  - Executive attention, which largely corresponds to working memory.

11.1.2 Neurobiology of Attention

A meta-analysis reported that the *dorsal part of the anterior cingulate* was activated by Stroop attention tasks as well as by similar tasks testing sustained, divided, and complex focused attention on distinct stimuli. It is interesting that this particular area was deactivated by emotional tasks (Bush et al. 2000).

In contrast, tasks that relate to emotional content activated *the rostral ventral portion of the anterior cingulate*, whereas this region was deactivated by cognitively challenging tasks. These two parts of the cingulate gyrus manifest a synchronized activation-inhibition depending on the nature of the task (Bench et al. 1992, 1993; Mayberg 1997; George et al. 1994a, b, c). It also seems that the left cingulate gyrus was preferentially activated (George et al. 1997).

11.2 Memory and Learning

Memory is the process according to which information is initially encoded, then retained, and finally stored in the brain, so that there is the option to retrieve it in order to participate in analyses at a later time, in interaction with novel information which have just entered into the system. At the human level, memory gives humans the ability to remember the past and to utilize it in the form of experience in order to assess the present and plan for the future (Arnold 2013).

Learning is closely connected with memory in conceptual terms, but also because they share neural circuits and mechanisms. One could say that learning is the result of memory but the reverse could also be stated (Engel 1999). The concept of learning reflects the aspect of memory which is *closely linked to behavior* and to the effect memory has on it, but also the reverse. According to this view, it is important that memory and learning are closely linked to higher mental functions and play a decisive role in decision-making. They do not only determine the ability of the person to survive, but especially in humans, they give the ability to widen his/her mental horizons, to self-define in space and time, and to develop personal but also collective history and culture (Benfenati 2007; Mazur 2015; Rescorla 2014). At least some forms of learning demand effort and depend much on the degree of fatigue of the mental apparatus (Fig. 11.2).

Memory and Learning

- Sensory memory.
- Working (short-term) memory.
- Long-term memory.
  - Declarative (or explicit) memory.
    - Semantic memory.
    - Episodic or autobiographic memory.
  - Non-declarative (implicit) memory.
    - Procedural memory.
    - Priming.
    - Habits.
      - Classic conditioning.
      - Operant conditioning.
      - Observational learning.
  - Nonassociative learning.
    - Habituation.
    - Sensitization.
- Long-term memory.
  - Permanent changes in neuronal connections.
  - Synthesis of new proteins.
  - Long-term potentiation, LTP.
  - DNA methyliosis.
  - Prion genes.
  - Hippocampus: spatial learning and declarative memory.
  - Amygdala: emotional mnemonic registrations.
  - Limbic system.
  - Dominant hemisphere: verbal memory.
  - Nonverbal hemisphere: spatial memory.

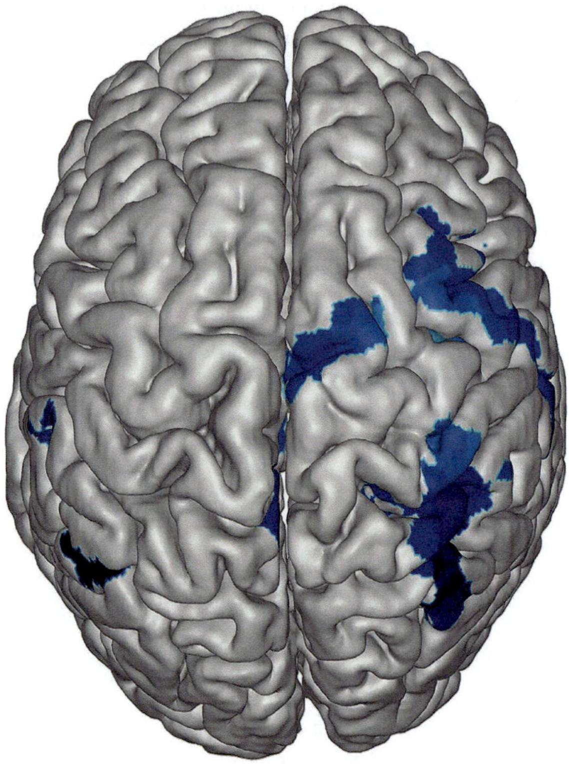

Fig. 11.2 In a healthy volunteers study, it has shown that the development of new skills is improved with short periods of rest (image from the Cohen Lab/National Institute of Neurological Disorders and Stroke, NIH)

11.2.1 Categories of Memory and Learning

In terms of function, memory seems to be composed of three distinct systems (Baddeley 2007; Squire 2004):

- *Sensory processor and sensory memory*: It codes external stimuli by transforming them to neuronal function simultaneously with an initial process which analyzes their fundamental characteristics as it has been described in the previous chapter. This subsystem also includes a form of memory which is modality-specific and has very short duration (less than a second). There are three types of sensory memory, iconic memory, echoic memory, and haptic memory (Anderson 2000; Sperling 1960, 1963; Neisser 1967; Mewhort et al. 1981; Coltheart 1980; Di Lollo 1977; Dick 1974; Guttman and Julesz 1963). This memory is located in the association areas as described in Chap. 6.
- *Working (short-term) memory*: Although for some authors the short-term and the working memory reflect different functions, with short-term being "passive" and the working memory to actively participate in the process, in fact from the point of view of psychological assessment, there can be no distinction of them. This type of memory temporary holds information at the conscious level (something similar to the RAM of computers). These information could

concern past registrations or present stimuli, so as to be available for processing at real time (Baddeley and Hitch 1974; James 1950; Baddeley 2003). It is extremely laborious and tends to lose holding of the information at a fast rate. Its capacity seems to be limited, and in humans it is around four to five independent arithmetic digits, although in the past, it was believed it was up to 7 ± 2, and the duration is no more than 30 s. The concept of the "digit" is related to the characteristics of the information, and mainly with the relationship its partial components have with each other, as well as with the way they are grouped (Miller 1956; Gobet et al. 2001; Cowan 2001; Conrad 1964). This type of memory seems to play an important role to what is generally conceptualized as "intelligence," is located in the dorsolateral prefrontal cortex, and participates in the decision-making (Nitschke et al. 2000; Pelosi et al. 2000) as has been described in Chap. 10.

- *Long-term memory*: It stores information for prolonged periods of time, aiming their use in the remote future. In contrast to the sensory and working memory, its capacity is enormous, and the storage duration could be lifelong. There are several theories which distinguish various types of long-term memory depending on the approach, but research data do not seem to definitely support any of them, which are the following (McDougall 1924; Anderson 1976):
  - *Declarative (or explicit) memory*: concerns the laborious and conscious recall of memory recordings, which may take the form of verbal or other kind of precise expression. It seems that the role of the hippocampus is very important in the processes concerning this type of memory (Tulving and Markowitsch 1998). This type of memory is further subdivided into (Tulving 1972, 2002; Squire and Zola 1998):

    Semantic memory which refers to the information which is stored and retrieved in an abstract way and out of its specific frame. It concerns general knowledge and abstract concepts.

    Episodic or autobiographic memory refers to the process involved in recalling specific experiences throughout a person's life, and it is based on the relationship of events with persons, time, and space, and the event is placed on the line of perception and memory of the person concerning his/her life and also contributes to the formation of the sense of self.
  - *Non-declarative (implicit) memory* concerns mnemonic registrations which can be recalled unconsciously without effort, and they cannot take the form of a verbal or precise expression (Squire and Zola-Morgan 1988; Squire and Zola 1996). It can be divided into:

    Procedural memory, which involves behaviors and skills which are executed at the preconscious level, and they can be obtained through training (e.g., driving).

    Priming is a term referring to the improved probability of recalling a memory registration as a response to a stimulus when it has been preceded by another specific stimuli which orients the system toward a specific direction of response.

    Habits, concern behaviors which have been established by conditioning:
    - Classic conditioning (Pavlov et al. 1929).

- Operant conditioning (Mackintosh 1983; Thorndike 1898, 1927; Skinner 1953, 1990).
- Observational learning (Bandura 1969; Bandura and Walters 1977; Ramachandran 2000).

Various other behaviors which have been established by nonassociative learning and are based on the mere repeating of the stimuli (Mazur 2015; Rescorla 2014). They include:

- Habituation (Pinsker et al. 1970; Kandel et al. 1976; Carew et al. 1972; Castellucci et al. 1970; Ezzeddine and Glanzman 2003; Esdin et al. 2010).
- Sensitization (Landsteiner and Jacobs 1935; Pinsker et al. 1970, 1973; Castellucci and Kandel 1976; Kandel et al. 1976; Bristol et al. 2004; Squire and Kandel 2000).

11.2.2 Neurobiology of Memory and Learning

The long-term memory is based on the consolidation of permanent changes in neuronal connections, probably in a diffuse way throughout the whole brain. It also seems to relate with the synthesis of new proteins which play a role in the modification of the communication among neurons though synapses and their parts. This is achieved through the *long-term potentiation (LTP)*, which is a complex process leading to long-term changes (Bliss and Lomo 1973; Lynch 2004; Costa-Mattioli and Sonenberg 2008; Kullmann and Lamsa 2008; Cooke and Bliss 2006; Lomo 2003; Ouardouz and Sastry 2000; Pastalkova et al. 2006; Whitlock et al. 2006; Bekinschtein et al. 2008; Meyer et al. 2014; Yin and Tully 1996; Lamprecht and LeDoux 2004). The neurobiological substrate of long-term memory might also include DNA methyliosis and prion genes (Buchmann et al. 2008; Papassotiropoulos et al. 2005; Miller and Sweatt 2007).

The *hippocampus* seems to participate in spatial learning and declarative memory and is connected with the association areas, thus playing a pivotal role probably in the transferring of information from short- to long-term memory.

The *amygdala* plays a role in emotional mnemonic registrations (LaBar and Cabeza 2006; Adolphs et al. 1997; Cahill et al. 1995). In general, the limbic system seems to play an important role too (Papez 1937; Markowitsch 2005; Milner et al. 1968, 1998; Smith and Milner 1981), while the dominant hemisphere contributes to verbal memory, and the other hemisphere to spatial memory (De Renzi et al. 1977; Ojemann and Dodrill 1985; Habib et al. 2003).

The synchronized discharge of neurons might lead to the facilitation of learning via a process called *Hebb's learning*. This simultaneous activation of neurons increases the synaptic power between them, so that eventually the activation of a single one leads to the activation of the other (Lomo 2003; Hebb 1949, 2005; Mitchison and Swindale 1999; Keysers and Perrett 2004; Bauer et al. 2001). From a psychological viewpoint, learning can be facilitated or diminished on the basis of the existence or nonexistence of a motive (Mazur 2015).

It is interesting that persons with developmental disorders or variations of development, which usually manifest a decrease of their overall mental ability, could manifest unusually strong memory (what laypersons call "photographic memory"), like the famous Kim Peek and Stephen Wiltshire, with the first one being able to execute complex arithmetic calculations.

11.3 Visuospatial Function

Visuospatial function includes all these cognitive processes which identify external objects; they chart them in space, take consideration of their movement, as well as assess the spatial relationships among them. The *three-dimensional perception* of space is supported by the existence of double visual and auditory sensory organs, and as a result, the perception of visual depth and the localization of the source of sounds are possible, which were both extremely important information for the environment the humankind was living in, several thousands of years ago.

The spatial representation of the environment and its charting around the human body seems to be the result of the synthesis of visual, auditory, and somatosensory information in the tertiary association cortex at the margins of the parietal, temporal, and occipital lobes.

The human visuospatial ability is probably limited to the perception and processing of two-dimensional space, although very close to the human body a three-dimensional perception could be possible, with reference to the body and its parts. The general limitation to two-dimensional perception is probably a consequence of the fact that humans were evolved for moving on a surface, while, in contrast, e.g., birds move in a three-dimensional space. An example of the limitations because of the adaptation to a 2D processing is the fact that even a small child could shoot a straight shot against a target, while ballistic or parabolic shooting is difficult, cannot be achieved "automatically," and demands training. An example of straightforward shooting is the shooting of an arrow with a bow, while an example of a parabolic shooting is the classic shooting in basketball, where the shot is even more difficult to calculate if it includes hitting the backboard (Fig. 11.3).

Fig. 11.3 Left: Arrow shooting with a bow in a straight shot. Right: Ballistic shot in basketball, which needs technical instructions

Visuospatial Function

- Synthesis of visual, auditory, and somatosensory information in the tertiary association cortex at the margins of the parietal, temporal, and occipital lobes.
- Spatial representation of the environment.
- Charting of the environment around the human body.
- Three-dimensional perception of space—origin.
- Pairs of sensory organs for hearing and vision.
- It serves:
  - The need to move in space.
  - Recognition of opportunities and threats.
  - Assessment of cost and ability to perform each movement.

Visuospatial function serves the need for moving in space but also the recognition of chances as well as of threats and the assessment of the cost and the chances of completion of each move.

In general, it is believed that there is a better visuospatial function in males, and this is probably because it supported their role as hunters during the Pleistocene, the period during which the mental and cognitive characteristics of humankind developed and matured.

It is important to note that the internal representation of the environment includes many elements that come from the mental association processes and not from the stimuli per se. This serves the faster processing and the addition of meaning on the elements of the environment; however, it is not uncommon that the modern environment leads to false perceptions and wrong conclusions (Fig. 11.4).

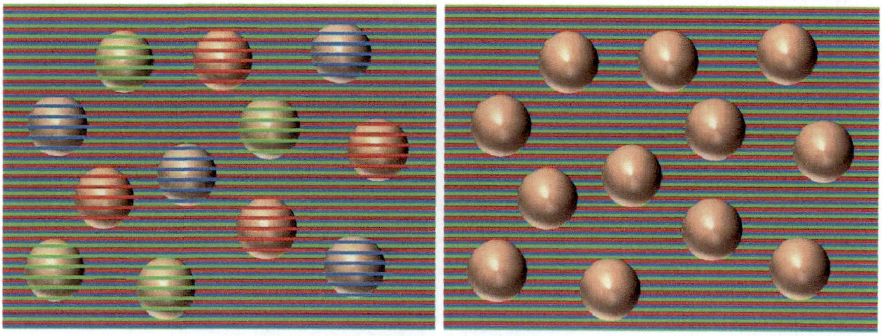

Fig. 11.4 The internal representation of the environment includes many elements that come from mental associations rather than the external stimuli. This could lead to erroneous perception as in the example of this picture. While all the spheres have the same color, masking with stripes changes visual perception. Even the knowledge that the spheres have the same color is not sufficient to correct the illusion. (The image was created by David Novick and is used after his permission)

11.4 Praxis

The term *praxis* refers to the ability of the brain to initiate motor response voluntarily, to monitor, guide, coordinate, and regulate them in a way that the selected task is accomplished with the best possible accuracy, efficacy, and resource sparing. It includes the ability of the brain to initiate and complete the motor part of bodily movements, facial expressions or gestures, object handling and use of tools, and control over and securing of the proper execution of partial movements and their sequential execution. It also controls and regulates the orientation and the positioning of movements in space and time and in relationship with the whole body as well as with objects or persons in the environment.

11.5 Language

An interesting ability of modern humans is that they can *talk*, *read*, and *write*. Children learn to speak early, without any special training or intervention, and they are taught to read and write after special training, but this is done early and is achieved rather easily. However, reading/writing and maybe speech were not inherent abilities of all our ancestors, although the use of sounds and gestures/behaviors is a common way for animals to communicate and exchange crucial information.

There are several theories describing the evolutionary pathways and mechanisms for the evolution of language, but all remain under dispute (Workman and Reader 2008; Fitch 2010; Deacon 1997), while even the mythologies of various people around the world have dealt with the issue (Haugen 1973) (Fig. 11.5). What is well documented is that language is a universal characteristic of *Homo sapiens* and that children, although they are not born with this capability, learn to talk and understand language somewhere between the first and fourth year of life, in a seemingly easy-going way and without systematic training of any kind. These characteristics of language definitely suggest that there is a human adaptation, and it was evolved in parallel with the body organs that support it (Pinker and Bloom 1990) and maybe on the basis of special needs and circumstances the specific human group or society was dealing with (Evans and Levinson 2009).

The fact that there has never been found on earth a group of humans without an oral language, no matter the level of civilization advancement, supports the conclusion that oral language is an inherent brain function. There is a large debate concerning its precise nature and the pathway through which it came to evolve as a characteristic; however, in modern *Homo sapiens*, it should be *taken as a granted mental ability* (Pinker 2010; Arbib and Bota 2003). This does not mean that the whole of mental activity related to language is inherent; on the contrary, it is developed and formatted depending on the needs of the human society, which in turn it affects via feedback. Such an example is the incorporation of words reflecting colors in language. The Greek language is the oldest documented from the modern spoken languages, and therefore, it constitutes an excellent example for study. The language of the Homeric epics includes a limited number of colors (black, white,

Fig. 11.5 The Tower of Babel, painting by Pieter Bruegel the Elder (1563). The biblical myth refers to the explanation of the origin of the many and different languages of antiquity. Different such myths exist in different civilizations

green-yellow—which is used even for blood—and purple which reflects a dark hue of a vivid color) which reflect properties rather than true color. It is of particular interest that they are often related to metals and their properties, for example, the sky is "bronze" to denote that it is shining, while green-yellow is the color of vividness. In the Eskimo language, the different hue of the snow is reflected in many words, while in all languages, it seems that the first color which is incorporated is the red, while blue and brown are the last. The modern color vocabulary and the concepts of colors were completed rather recently and most probably during the Late Middle Ages, in spite of the fact that colors were the focus of study by Aristotle but also by Hippocrates (Loreto et al. 2012; Kay and Maffi 1999).

In contrast to the inherent character of oral language, things are different with writing and more specifically with the particularly abstract and genuine alphabetical writing, which was discovered only once in human history, in the region of Middle East, and was further developed through ancient Greece and Rome (which adopted the Greek Chalcidian alphabet). In no other moment in history and in no other civilization no matter how evolved it was, alphabetical writing evolved from scratch, while even today, advanced countries of the Far East still utilize difficult-to-use iconographic or syllabic writing systems. This fact definitely suggests that alphabetical writing is clearly a cultural product of civilization, and to be precise, it constitutes a unique and difficult technological discovery (Fig. 11.6).

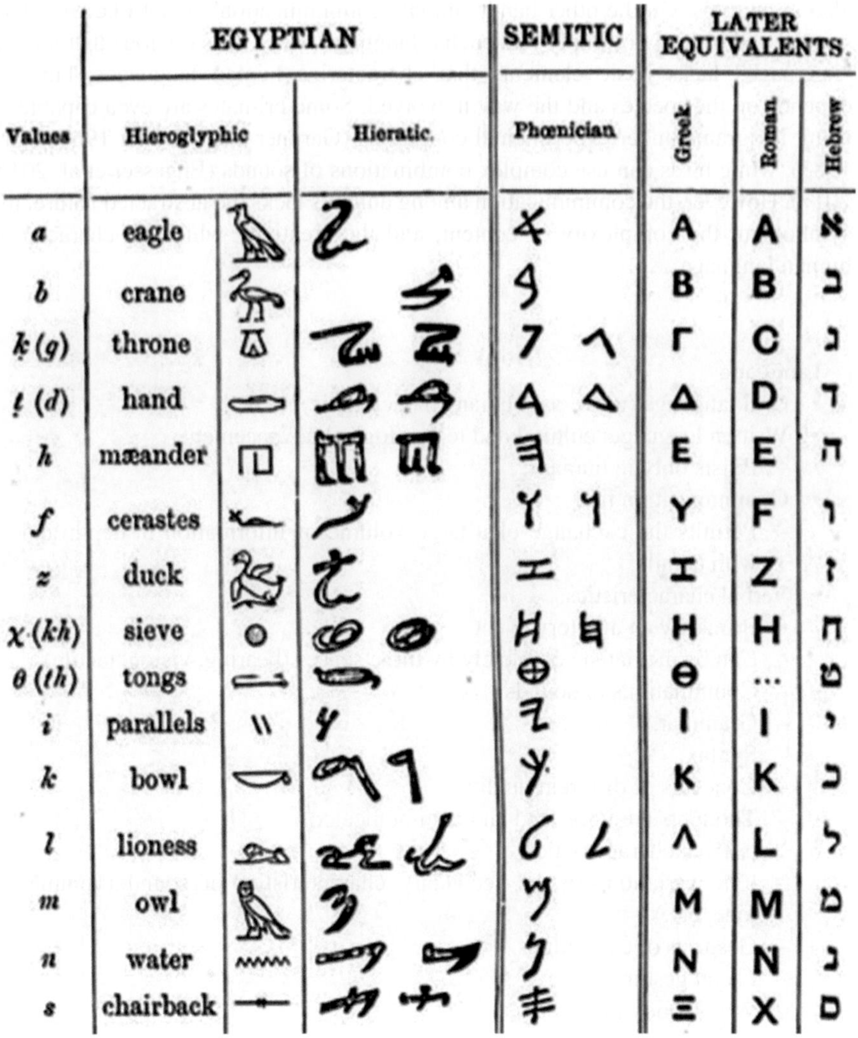

Fig. 11.6 The evolution of writing from the hieroglyphics of ancient Egypt and through Phoenicia to the modern alphabets of ancient Greece (Attic-Ionian) and Rome (Chalcidian)

In spite of the existence of different theoretical approaches concerning the components and the foundations on which language developed, the above are in accord with the theories by Steven Pinker and Noam Chomsky which propose the existence of some type of *inherent ability concerning language* (Pinker 1994; Workman and Reader 2008; Chomsky and McGilvray 2012) which is definitely genetically determined (Chomsky 2000).

Opposing views argue that although human languages share many common elements and structure, the universality of the nature of the human language has not been proven beyond doubt, as kinds of "linguistic communication" can be found also in animals. On the other hand, animal "communication" is not like "real language" and it does not imply that such a language exists. It is obvious that animal "language" lacks basic elements that characterize human languages, but this depends on the species and the way it evolved. Some primates are even capable of using lexigrams under experimental conditions (Gardner and Gardner 1969; Ward 1983), while birds can use complex combinations of sounds (Engesser et al. 2015, 2017). However, the communication among animals lacks the abstracted nature, the symbolism, the complexity of content, and the creativity which all characterize human language.

Language

- Oral language (universal human characteristic).
- Written language: cultural and technological advancement.
  - Exists only in humans.
- Communication tool.
  - Permits the exchange of a large volume of information in depth and with details.
- Verbal characteristics.
  - Primarily an auditory event.
  - Can be mediated completely by three senses (hearing, vision, tactile).
  - Combinations of sounds.
  - Grammar.
  - Syntax.
  - Concepts of different quality.
  - Thoughts are imprinted and communicated.
- Nonverbal characteristics.
  - Fine variations in the secondary characteristics of sound (volume, tone, etc.)
  - Gestures of extremities.
  - Facial gestures.
  - Whole body movements.
- Wernicke area.
- Broca area.

At any case, for humans, language constitutes a tool of communication which permits the exchange of a large volume of information in depth and with details, as it can include not only a simple detailed description of objects or events but additionally it can include abstracted concepts, and through nonverbal elements (e.g., sound volume, tone), it can provide information concerning the inner mental state of the person.

The human language is *primarily an auditory event*, and so far, no other version of linguistic communication (e.g., written) was able to include completely all the information the sounds of language can carry. For this reason, the full development of those organs which are able to produce complex sounds (tongue, larynx, etc.) and of the sensory organs which can receive and analyze those sounds (acoustic apparatus) as well as of those brain structures which can translate thinking into a series of sounds and vice versa the series of sounds into mental concepts is absolutely necessary. The exact genetic substrate is not well known; however, a role for gene FOXP2 is highly possible (Newbury and Monaco 2010; Reuter et al. 2017).

On the basis of the above, the human language includes a number of *verbal characteristics* including *grammar* and *syntax* with combination of sounds (words) which denote concepts of different quality (objects, actions, situations, events) which in turn could reflect the experiencing of objects (e.g., external real objects or persons) or abstract ideas (e.g., aesthetics) or even the internal mental state of the speaking person (e.g., intentions) and can be put in a frame which differs much from the here-and-now reality (conditional, future, relative, etc.). Grammar and syntax connect groups of words into conceptual mental representations in a way that thinking is imprinted as information to be communicated. All human languages include a "subject," a "verb," and an "object." It is of particular importance that the "words" that seem to constitute the fundamental structural components of verbal communication, while they originate and have an essence based on sound, can also be communicated via a nonauditory way, without the quantity or the quality of information to deteriorate. In other words, the verbal component of the language can be mediated completely by three senses (hearing, vision, tactile).

However, the human language includes also *nonverbal characteristics*, which accompany verbal characteristics, either in the form of fine variations in the secondary characteristics of sound (volume, tone, etc.) or in the form of gestures of the face or of the extremities, but also with the contribution of the totality of the body movements. Nonverbal characteristics load additional information on the verbally communicated information, and it seems that females have higher on average ability to use these characteristics. This nonverbal information can be included in written language only partially, and only recently the audiovisual recording made possible their relatively precise reproduction and communication, but still, this is not perfect.

The brain processes related to language are not fully understood. It is not even known what the exact relationship between thought and language is, and to which extent logical thinking prerequisites language, or which is the precise way persons with developmental disorders or autism really think. Our basic knowledge concerns two main areas (Fig. 11.7); however, much of our knowledge concerning their role comes from the study of patients with various types of aphasia (Marcus and Rabagliati 2006):

- The *Wernicke area* is located at the posterior part of the superior temporal gyrus on the dominant hemisphere. It is considered to be the area with a significant role in the mental preparation of verbal expression and the comprehension of language. This area is embedded inside the wider area of the tertiary association area (Fig. 10.12).

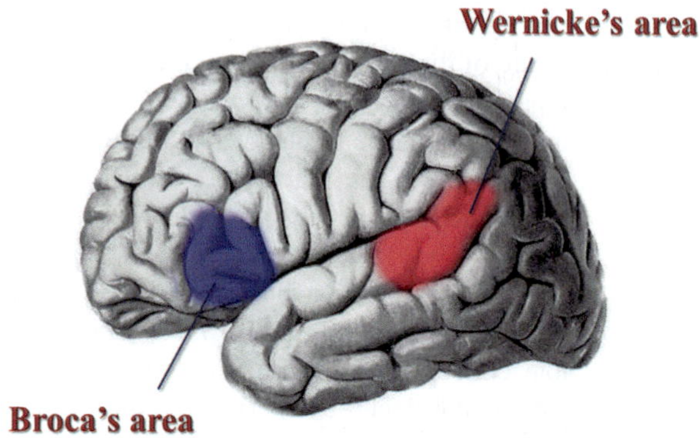

Fig. 11.7 The Wernicke (mental preparation of linguistic expression and comprehension) and the Broca (motor programming of linguistic expression) areas

- The *Broca area* is located at the anterior inferior frontal gyrus of the dominant hemisphere. It is related to the motor planning of linguistic expression and is located within the wider motor cortex of the anterior brain.

It is interesting that lesions in these areas affect both oral and sign language in a similar way, thus leading to the conclusion that these areas are related to language at a high level, and with language as a wider ability, and not with the physiology of the production of this specific form of oral language (Sandler and Lillo-Martin 2001).

11.6 Arithmetic Calculations

Arithmetic calculations in the form the modern life demands them do not seem to be an inherent function but rather a *product of training* in the use of a technological development (arithmetic). This is the conclusion that comes out both from the low ability for arithmetic calculations by low education persons and from the need to learn the "times table" by heart as a prerequisite for the execution of complex arithmetic operations. The ability to execute arithmetic calculations in a spontaneous way and completely mentally is extremely limited and seems to be based on an optical representation of objects (most often of hand fingers).

That arithmetic calculations are the product of training is also proven by imaging studies which show an activation of wide areas of the brain especially in the dominant hemisphere (Arsalidou et al. 2018; Amalric and Dehaene 2016; Price et al. 2013).

11.7 Abstract Thinking

Abstract thinking is the *process of identifying general common or on the contrary different properties or procedures* through the observation of different objects or events (Fig. 11.8). Its process and its product seem to depend on, and guided to, a significant degree by some aim, because of which the analysis itself had started

Fig. 11.8 Identification of similarities and differences concerning the form or the origin. Two standard questions/tests
In which way an apple is similar to a ball (left)
In which way a lake differs from a swimming pool (right)

(e.g., recognition of possible utility of the object). The basis of the process concerns the identification of similarities and differences and the mental construct of a *model template mental object* which correspond to them only selectively (Chaitin 2006).

Normal human abstract thinking has the ability to move mental processes to *both directions*, that is, both to the development of abstract models from a plethora of heterogenous observations and to the application of a preexisting template on a new observation. However, this latter process includes the risk of bias and of systematic error.

Abstract thinking differs completely from the *simplistic-naïve approach*. The first demands high cognitive processing and eventually it leads to a mental construct which derives from the analysis of more complex objects. On the contrary, the latter is an unavoidably limited processing of objects, based on their absolutely necessary characteristics as a result of the inability to utilize higher and more intense levels of cognitive processing (Fig. 11.9).

Abstract processing seems to correlate with the activation of parts of the frontal orbital cortex as well as of the neighboring regions of the temporal lobe, especially at the dominant hemisphere. Naïve processing seems to be limited to the activation of the cingulate and the parahippocampal gyri (Lewis 2006; Wang et al. 2010). After taking into consideration the differential role these regions play in the process of the creation of an internal representation of the environment and in decision-making, the differences in the final product can be easily comprehended.

The relationship between abstract thinking and language has been the focus of studies and debates. For many authors, the two functions are closely associated and neither of them can exist without the other one.

Fig. 11.9 Up: Abstract thinking as shown in the paintings by Pablo Picasso (1881–1973) in self-portraits at the age of 15 (left), 25 (center), and 90 (right). The form has been completely conquered since a very young age and evolves by keeping complexity in a transformed way, by selectively putting the emphasis and bringing out details and characteristics. Down: The evolving simplification in the technique of Carolus Horn (1921–1992) because of Alzheimer's disease. The form becomes crude in a horizontal and not selective way, complexity decays, and details disappear

11.8 Executive Functions and Neurocognitive Control

11.8.1 Neuropsychological Definition of Executive Function

The term *executive functions* refers to a group of neurocognitive functions which exert mental control over behavior, that is, the selection of the appropriate behaviors for manifestation, and the supervision of their proper execution, so that the aim is achieved. These procedures include concentration, continuous update and prioritization of targets (Morris and Jones 1990), as well as the selection of the focus of attention (Alivisatos and Milner 1989), its change (Monsell 1996), or the distribution of attention (Cicerone et al. 1983; Tipper 2001), working memory (Anderson and Green 2001), self-monitoring (Luria 1966), inhibition, and control (Miyake et al. 2000; Aron and Poldrack 2006). They concern also the ability to change the procedure at any level, including higher strategies, which means they reflect a kind of mental flexibility (Monsell 1996; Diamond 2013; Chan et al. 2008; Barkley 1997). These functions even include verbal fluency as well as "metacognition." Executive functions involve three important target functions: *shifting, updating, and inhibition* (Miyake et al. 2000).

Executive Functions

- Mental control of behavior.
  - Selection of appropriate behaviors.
  - Supervision of their proper execution.
  - Prioritization of aims.
  - Selection of the target of attention and shifting or dividing of attention.
  - Working memory.
  - Inhibition, control, and capacity to change strategy.
- Three surfaces of frontal lobes.

11.8.2 Neurobiology of Executive Functions

A large part of the processes concerning executive functions take place in the frontal lobes, but the whole brain and especially the association areas participate at least partially, depending on the region and the function (Banich 1997; Engels et al. 2007; Heller et al. 1997; Diamond 2013; Malenka et al. 2009; Alvarez and Emory 2006; Stuss and Alexander 2000; Burgess and Stuss 2017). The *three surfaces of the frontal lobe* (Fig. 10.5), that is, the association area of the dorsolateral prefrontal cortex, the anterior region of the cingulate gyrus, and the orbital cortex, seem to play a particularly vital role (Lezak et al. 2004; Clark et al. 2008; Rolls and Grabenhorst 2008; Allman et al. 2001). It is interesting that even the cerebellum seems to play an important role (Koziol et al. 2012; Noroozian 2014).

There is a bulk of literature on the study of activation or deactivation of brain areas during executive function tests in depressed patients, but the findings are difficult to interpret and to utilize them for the understanding of executive functions themselves, both because of the underlying mental disorder and because of the related pharmacotherapy (Dolan et al. 1994; Grasby et al. 1993; Drevets and Raichle 1995, 1998; Bowen et al. 1989; Coffey et al. 1993; Compton et al. 2003; Heller et al. 1997; Miller 1996; Nitschke et al. 2000).

Recent functional imaging studies have shown the importance of interaction and connectivity among many brain areas with different role and contribution in the consolidation of executive functions, and especially in the synthesis of analytical-neurocognitive and affective processes (Shokri-Kojori et al. 2012; Gazzaley et al. 2004; Fuster et al. 1985; Barcelo et al. 2000; Miller and D'Esposito 2005; Kastner et al. 1999; Liu et al. 2003). The biggest part of these functions has already been described in Chap. 10.

11.9 Social Cognition and Theory of Mind

Theory of mind is a term which refers to the ability of people to experience insight and to understand the psychological world of the others, that is, to attribute beliefs, thoughts, intentions, affects to others, as well as anything that concerns internally experienced mental states and processes. It is because of this ability that humans can perceive similarities and differences among themselves and the others, and on the basis of them to develop *social behavior* (Fig. 11.10).

It is clear that brain functions are not directly observable and this is why the attribution of specific psychological states to others is achieved through observation of expressed behavior and mainly through spoken language, in combination with a significant degree of guessing, based on experience and the general way the humankind behaves (Call and Tomasello 2008; Premack and Woodruff 2010; Gordon 1996). In this sense, the theory of mind leads to the *development of internal hypothetical representations of other humans*, concerning their cognitive-psychological state, their motives, as well as the prediction of their possible behavior (Courtin 2000; Courtin and Melot 2005). The correct interpretation of these processes is that they model mental representations which emerge inside the subject's brain, as a result of the perception of other humans as external stimuli, aiming to the support of the development of social behavior.

Fig. 11.10 "The goalkeeper's fear of the penalty"

Social Cognition and Theory of Mind

- Insight: affective, automatic, unconscious process.
- Theory of mind: neurocognitive, partially voluntary, and conscious.
- Attribute to others:
  - Beliefs.
  - Thoughts.
  - Intentions.
  - Affects.
  - Internally experienced mental states and processes.
- Development of internal hypothetical representations of other humans.
  - Through observation of expressed behavior.
  - In combination with hypothetical approaches.
- On the basis of them, social behavior is developed.

The difference between insight and theory of mind is that the former concerns mainly emotional/affective and more automatic preconscious or even unconscious processes, while the latter concerns more neurocognitively based and often at least partially conscious and effortful processes (Hynes et al. 2006; Kalbe et al. 2010). This is in accord with the observation that theory of mind seems to correlate strongly and specifically with the linguistic development in humans (Moore et al. 1990; Woolfe et al. 2002; Ruffman et al. 2002; Miller 2006; Milligan et al. 2007).

As these functions are extremely complex, they essentially involve most of the brain with regions containing *mirror neurons* playing a particular role, while every region and circuit contributes, depending on its specialized function as this has already been described in Chap. 10 (Sanfey et al. 2015; Haroush and Williams 2015; Samson et al. 2004; Pelphrey et al. 2004; Scholz et al. 2009; Decety and Lamm 2007; Saxe and Kanwisher 2003).

References

Adolphs R, Cahill L, Schul R, Babinsky R (1997) Impaired declarative memory for emotional material following bilateral amygdala damage in humans. Learn Mem 4(3):291–300. https://doi.org/10.1101/lm.4.3.291

Alivisatos B, Milner B (1989) Effects of frontal or temporal lobectomy on the use of advance information in a choice reaction time task. Neuropsychologia 27(4):495–503. https://doi.org/10.1016/0028-3932(89)90054-7

Allman JM, Hakeem A, Erwin JM, Nimchinsky E, Hof P (2001) The anterior cingulate cortex. The evolution of an interface between emotion and cognition. Ann N Y Acad Sci 935:107–117

Alvarez JA, Emory E (2006) Executive function and the frontal lobes: a meta-analytic review. Neuropsychol Rev 16(1):17–42. https://doi.org/10.1007/s11065-006-9002-x

Amalric M, Dehaene S (2016) Origins of the brain networks for advanced mathematics in expert mathematicians. Proc Natl Acad Sci U S A 113(18):4909–4917. https://doi.org/10.1073/pnas.1603205113

Anderson JR (1976) Language, memory, and thought. L. Erlbaum Associates; Distributed by the Halsted Press Division of Wiley, Hillsdale, NJ
Anderson JR (2000) Learning and memory. Wiley, Hoboken, NJ
Anderson J (2004) Cognitive psychology and its implications, 6th edn. Worth Publishers
Anderson MC, Green C (2001) Suppressing unwanted memories by executive control. Nature 410(6826):366–369. https://doi.org/10.1038/35066572
Arbib M, Bota M (2003) Language evolution: neural homologies and neuroinformatics. Neural Netw 16(9):1237–1260. https://doi.org/10.1016/j.neunet.2003.08.002
Arnold MB (2013) Memory and the brain. Psychology Press, New York
Aron AR, Poldrack RA (2006) Cortical and subcortical contributions to stop signal response inhibition: role of the subthalamic nucleus. J Neurosci Off J Soc Neurosci 26(9):2424–2433. https://doi.org/10.1523/JNEUROSCI.4682-05.2006
Arsalidou M, Pawliw-Levac M, Sadeghi M, Pascual-Leone J (2018) Brain areas associated with numbers and calculations in children: meta-analyses of fMRI studies. Dev Cogn Neurosci 30:239–250. https://doi.org/10.1016/j.dcn.2017.08.002
Baddeley A (2003) Working memory: looking back and looking forward. Nat Rev Neurosci 4(10):829–839. https://doi.org/10.1038/nrn1201
Baddeley A (2007) Working memory, thought, and action. Oxford University Press, Oxford
Baddeley AD, Hitch G (1974) Working memory. Psychol Learn Motiv 8:47–89
Bandura A (1969) Principles of behavior modification. Holt, Rinehart and Winston, New York
Bandura A, Walters RH (1977) Social learning theory
Banich MT (1997) Neuropsychology: the neural bases of mental function. Houghton Mifflin, Boston, MA
Barcelo F, Suwazono S, Knight RT (2000) Prefrontal modulation of visual processing in humans. Nat Neurosci 3(4):399–403. https://doi.org/10.1038/73975
Barkley RA (1997) Behavioral inhibition, sustained attention, and executive functions: constructing a unifying theory of ADHD. Psychol Bull 121(1):65–94. https://doi.org/10.1037/0033-2909.121.1.65
Bauer EP, LeDoux JE, Nader K (2001) Fear conditioning and LTP in the lateral amygdala are sensitive to the same stimulus contingencies. Nat Neurosci 4(7):687–688. https://doi.org/10.1038/89465
Bekinschtein P, Cammarota M, Katche C, Slipczuk L, Rossato JI, Goldin A, Izquierdo I, Medina JH (2008) BDNF is essential to promote persistence of long-term memory storage. Proc Natl Acad Sci U S A 105(7):2711–2716. https://doi.org/10.1073/pnas.0711863105
Bench CJ, Friston KJ, Brown RG, Scott LC, Frackowiak RS, Dolan RJ (1992) The anatomy of melancholia--focal abnormalities of cerebral blood flow in major depression. Psychol Med 22(3):607–615. https://doi.org/10.1017/s003329170003806x
Bench CJ, Frith CD, Grasby PM, Friston KJ, Paulesu E, Frackowiak RSJ, Dolan RJ (1993) Investigations of the functional anatomy of attention using the stroop test. Neuropsychologia 31(9):907–922. https://doi.org/10.1016/0028-3932(93)90147-r
Benfenati F (2007) Synaptic plasticity and the neurobiology of learning and memory. Acta Biomed 78 Suppl 1:58–66
Bliss TV, Lomo T (1973) Long-lasting potentiation of synaptic transmission in the dentate area of the anaesthetized rabbit following stimulation of the perforant path. J Physiol 232(2):331–356. https://doi.org/10.1113/jphysiol.1973.sp010273
Bowen DM, Najlerahim A, Procter AW, Francis PT, Murphy E (1989) Circumscribed changes of the cerebral cortex in neuropsychiatric disorders of later life. Proc Natl Acad Sci U S A 86(23):9504–9508. https://doi.org/10.1073/pnas.86.23.9504
Bristol AS, Sutton MA, Carew TJ (2004) Neural circuit of tail-elicited siphon withdrawal in Aplysia. I. Differential lateralization of sensitization and dishabituation. J Neurophysiol 91(2):666–677
Buchmann A, Mondadori CR, Hanggi J, Aerni A, Vrticka P, Luechinger R, Boesiger P, Hock C, Nitsch RM, de Quervain DJ, Papassotiropoulos A, Henke K (2008) Prion protein M129V poly-

morphism affects retrieval-related brain activity. Neuropsychologia 46(9):2389–2402. https://doi.org/10.1016/j.neuropsychologia.2008.03.002
Burgess PW, Stuss DT (2017) Fifty years of prefrontal cortex research: impact on assessment. J Int Neuropsychol Soc 23(9–10):755–767. https://doi.org/10.1017/S1355617717000704
Bush G, Luu P, Posner MI (2000) Cognitive and emotional influences in anterior cingulate cortex. Trends Cogn Sci 4(6):215–222. https://doi.org/10.1016/s1364-6613(00)01483-2
Cahill L, Babinsky R, Markowitsch HJ, McGaugh JL (1995) The amygdala and emotional memory. Nature 377(6547):295–296. https://doi.org/10.1038/377295a0
Call J, Tomasello M (2008) Does the chimpanzee have a theory of mind? 30 years later. Trends Cogn Sci 12(5):187–192. https://doi.org/10.1016/j.tics.2008.02.010
Callejas A, Lupianez J, Tudela P (2004) The three attentional networks: on their independence and interactions. Brain Cogn 54(3):225–227. https://doi.org/10.1016/j.bandc.2004.02.012
Carew TJ, Pinsker HM, Kandel ER (1972) Long-term habituation of a defensive withdrawal reflex in aplysia. Science 175(4020):451–454. https://doi.org/10.1126/science.175.4020.451
Castellucci V, Kandel E (1976) Presynaptic facilitation as a mechanism for behavioral sensitization in Aplysia. Science 194(4270):1176–1178
Castellucci V, Pinsker H, Kupfermann I, Kandel ER (1970) Neuronal mechanisms of habituation and dishabituation of the gill-withdrawal reflex in Aplysia. Science 167(3926):1745–1748. https://doi.org/10.1126/science.167.3926.1745
Castiello U, Umilta C (1990) Size of the attentional focus and efficiency of processing. Acta Psychol 73(3):195–209. https://doi.org/10.1016/0001-6918(90)90022-8
Chaitin G (2006) The limits of reason. Sci Am 294(3):74–81. https://doi.org/10.1038/scientificamerican0306-74
Chan RC, Shum D, Toulopoulou T, Chen EY (2008) Assessment of executive functions: review of instruments and identification of critical issues. Arch Clin Neuropsychol 23(2):201–216. https://doi.org/10.1016/j.acn.2007.08.010
Chomsky N (2000) The architecture of language. Oxford University Press, Oxford
Chomsky N, McGilvray J (2012) The science of language. Cambridge University Press, Cambridge
Cicerone KD, Lazar RM, Shapiro WR (1983) Effects of frontal lobe lesions on hypothesis sampling during concept formation. Neuropsychologia 21(5):513–524. https://doi.org/10.1016/0028-3932(83)90007-6
Clark L, Bechara A, Damasio H, Aitken MR, Sahakian BJ, Robbins TW (2008) Differential effects of insular and ventromedial prefrontal cortex lesions on risky decision-making. Brain 131(Pt 5):1311–1322. https://doi.org/10.1093/brain/awn066
Coffey CE, Wilkinson WE, Weiner RD, Parashos IA, Djang WT, Webb MC, Figiel GS, Spritzer CE (1993) Quantitative cerebral anatomy in depression. A controlled magnetic resonance imaging study. Arch Gen Psychiatry 50(1):7–16. https://doi.org/10.1001/archpsyc.1993.01820130009002
Coltheart M (1980) Iconic memory and visible persistence. Percept Psychophys 27(3):183–228. https://doi.org/10.3758/bf03204258
Compton RJ, Banich MT, Mohanty A, Milham MP, Herrington J, Miller GA, Scalf PE, Webb A, Heller W (2003) Paying attention to emotion: an fMRI investigation of cognitive and emotional stroop tasks. Cogn Affect Behav Neurosci 3(2):81–96. https://doi.org/10.3758/cabn.3.2.81
Conrad R (1964) Acoustic confusions in immediate memory. Brit J Psychol 55(1):75–84. https://doi.org/10.1111/j.2044-8295.1964.tb00899.x
Cooke SF, Bliss TV (2006) Plasticity in the human central nervous system. Brain 129(Pt 7):1659–1673. https://doi.org/10.1093/brain/awl082
Costa-Mattioli M, Sonenberg N (2008) Translational control of gene expression: a molecular switch for memory storage. Prog Brain Res 169:81–95. https://doi.org/10.1016/S0079-6123(07)00005-2
Courtin C (2000) The impact of sign language on the cognitive development of deaf children: the case of theories of mind. J Deaf Stud Deaf Educ 5(3):266–276. https://doi.org/10.1093/deafed/5.3.266

Courtin C, Melot AM (2005) Metacognitive development of deaf children: lessons from the appearance-reality and false belief tasks. Dev Sci 8(1):16–25. https://doi.org/10.1111/j.1467-7687.2005.00389.x

Cowan N (2001) The magical number 4 in short-term memory: a reconsideration of mental storage capacity. Behav Brain Sci 24(1):87–114; discussion 114-185. https://doi.org/10.1017/s0140525x01003922

De Renzi E, Faglioni P, Previdi P (1977) Spatial memory and hemispheric locus of lesion. Cortex 13(4):424–433. https://doi.org/10.1016/s0010-9452(77)80022-1

Deacon T (1997) The symbolic species: the co-evolution of language and the brain. W.W. Norton & Co, London

Decety J, Lamm C (2007) The role of the right temporoparietal junction in social interaction: how low-level computational processes contribute to meta-cognition. The Neuroscientist: a review journal bringing neurobiology, neurology and psychiatry 13(6):580–593. https://doi.org/10.1177/1073858407304654

Di Lollo V (1977) Temporal characteristics of iconic memory. Nature 267(5608):241–243. https://doi.org/10.1038/267241a0

Diamond A (2013) Executive functions. Annu Rev Psychol 64:135–168. https://doi.org/10.1146/annurev-psych-113011-143750

Dick A (1974) Iconic memory and its relation to perceptual processing and other memory mechanisms. Percept Psychophys 16(3):575–596

Dolan RJ, Bench CJ, Brown RG, Scott LC, Frackowiak RS (1994) Neuropsychological dysfunction in depression: the relationship to regional cerebral blood flow. Psychol Med 24(4):849–857. https://doi.org/10.1017/s0033291700028944

Drevets WC, Raichle ME (1995) Positron emission tomographic imaging studies of human emotional disorders. In: Gazzaniga MS (ed) The cognitive neurosciences. MIT Press, Cambridge, MA, pp 1153–1164

Drevets WC, Raichle ME (1998) Suppression of regional cerebral blood during emotional versus higher Cognitive implications for interactions between emotion and cognition. Cogn Emotion 12(3):353–385. https://doi.org/10.1080/026999398379646

Engel S (1999) Context is everything: the nature of memory. WH Freeman/Times Books/Henry Holt & Co, New York

Engels AS, Heller W, Mohanty A, Herrington JD, Banich MT, Webb AG, Miller GA (2007) Specificity of regional brain activity in anxiety types during emotion processing. Psychophysiology 44(3):352–363. https://doi.org/10.1111/j.1469-8986.2007.00518.x

Engesser S, Crane JM, Savage JL, Russell AF, Townsend SW (2015) Experimental evidence for phonemic contrasts in a nonhuman vocal system. PLoS Biol 13(6):e1002171. https://doi.org/10.1371/journal.pbio.1002171

Engesser S, Ridley AR, Townsend SW (2017) Element repetition rates encode functionally distinct information in pied babbler 'clucks' and 'purrs'. Anim Cogn 20(5):953–960. https://doi.org/10.1007/s10071-017-1114-6

Eriksen CW, Hoffman JE (1972) Temporal and spatial characteristics of selective encoding from visual displays. Percept Psychophys 12(2):201–204. https://doi.org/10.3758/bf03212870

Eriksen CW, St James JD (1986) Visual attention within and around the field of focal attention: a zoom lens model. Percept Psychophys 40(4):225–240. https://doi.org/10.3758/bf03211502

Esdin J, Pearce K, Glanzman DL (2010) Long-term habituation of the gill-withdrawal reflex in aplysia requires gene transcription, calcineurin and L-type voltage-gated calcium channels. Front Behav Neurosci 4:181. https://doi.org/10.3389/fnbeh.2010.00181

Evans N, Levinson SC (2009) The myth of language universals: language diversity and its importance for cognitive science. Behav Brain Sci 32(5):429–448; discussion 448-494. https://doi.org/10.1017/S0140525X0999094X

Ezzeddine Y, Glanzman DL (2003) Prolonged habituation of the gill-withdrawal reflex in Aplysia depends on protein synthesis, protein phosphatase activity, and postsynaptic glutamate receptors. J Neurosci Off J Soc Neurosci 23(29):9585–9594

Fan J, Gu X, Guise KG, Liu X, Fossella J, Wang H, Posner MI (2009) Testing the behavioral interaction and integration of attentional networks. Brain Cogn 70(2):209–220. https://doi.org/10.1016/j.bandc.2009.02.002

Fitch T (2010) The evolution of language. Cambridge University Press, Cambridge

Fuster JM, Bauer RH, Jervey JP (1985) Functional interactions between inferotemporal and prefrontal cortex in a cognitive task. Brain Res 330(2):299–307. https://doi.org/10.1016/0006-8993(85)90689-4

Gardner RA, Gardner BT (1969) Teaching sign language to a chimpanzee. Science 165(3894):664–672. https://doi.org/10.1126/science.165.3894.664

Gazzaley A, Rissman J, D'Esposito M (2004) Functional connectivity during working memory maintenance. Cogn Affect Behav Neurosci 4(4):580–599. https://doi.org/10.3758/cabn.4.4.580

George MS, Ketter TA, Parekh PI, Rosinsky N, Ring H, Casey BJ, Trimble MR, Horwitz B, Herscovitch P, Post RM (1994a) Regional brain activity when selecting a response despite interference: an H2 (15) O PET study of the stroop and an emotional stroop. Hum Brain Mapp 1(3):194–209. https://doi.org/10.1002/hbm.460010305

George MS, Ketter TA, Post RM (1994b) An overview of brain activation studies in mood disorders. Biol Psychiatry 35(9):658. https://doi.org/10.1016/0006-3223(94)90814-1

George MS, Ketter TA, Post RM (1994c) Prefrontal cortex dysfunction in clinical depression. Depression 2(2):59–72. https://doi.org/10.1002/depr.3050020202

George MS, Ketter TA, Parekh PI, Rosinsky N, Ring HA, Pazzaglia PJ, Marangell LB, Callahan AM, Post RM (1997) Blunted left cingulate activation in mood disorder subjects during a response interference task (the Stroop). J Neuropsychiatry Clin Neurosci 9(1):55–63. https://doi.org/10.1176/jnp.9.1.55

Gobet F, Lane PC, Croker S, Cheng PC, Jones G, Oliver I, Pine JM (2001) Chunking mechanisms in human learning. Trends Cogn Sci 5(6):236–243. https://doi.org/10.1016/s1364-6613(00)01662-4

Gordon RM (1996) 'Radical' simulationism. In: Carruthers P, Smith P (eds) Theories of theories of mind. Cambridge University Press, Cambridge

Grasby PM, Frith CD, Friston KJ, Bench C, Frackowiak RS, Dolan RJ (1993) Functional mapping of brain areas implicated in auditory--verbal memory function. Brain 116(Pt 1):1–20. https://doi.org/10.1093/brain/116.1.1

Guttman N, Julesz B (1963) Lower limits of auditory periodicity analysis. J Acoust Soc Am 35(4):610–610

Habib R, Nyberg L, Tulving E (2003) Hemispheric asymmetries of memory: the HERA model revisited. Trends Cogn Sci 7(6):241–245. https://doi.org/10.1016/s1364-6613(03)00110-4

Haroush K, Williams ZM (2015) Neuronal prediction of opponent's behavior during cooperative social interchange in primates. Cell 160(6):1233–1245. https://doi.org/10.1016/j.cell.2015.01.045

Haugen E (1973) The curse of babel. Daedalus 102(3):47–57

Hebb D (1949) The organization of behavior. Wiley, New York

Hebb DO (2005) The organization of behavior: a neuropsychological theory. Psychology Press

Heller W, Nitschke JB, Etienne MA, Miller GA (1997) Patterns of regional brain activity differentiate types of anxiety. J Abnorm Psychol 106(3):376–385. https://doi.org/10.1037//0021-843x.106.3.376

Homskaya E (2001) Alexander Romanovich Luria: a scientific biography. Springer, New York

Hynes CA, Baird AA, Grafton ST (2006) Differential role of the orbital frontal lobe in emotional versus cognitive perspective-taking. Neuropsychologia 44(3):374–383. https://doi.org/10.1016/j.neuropsychologia.2005.06.011

James W (1890) The principles of psychology. Henry Holt, New York

James W (1950) The principles of psychology: in 2 volumes. Dover Publications, Newburyport

Kahneman D (1973) Attention and effort. Prentice-Hall, Englewood Cliffs, NJ

Kalbe E, Schlegel M, Sack AT, Nowak DA, Dafotakis M, Bangard C, Brand M, Shamay-Tsoory S, Onur OA, Kessler J (2010) Dissociating cognitive from affective theory of mind: a TMS study. Cortex 46(6):769–780. https://doi.org/10.1016/j.cortex.2009.07.010

Kandel E, Brunelli M, Byrne J, Castellucci VA (1976) Common presynaptic locus for the synaptic changes underlying short-term habituation and sensitization of the gill-withdrawal reflex in Aplysia. Cold Spring Harb Symp Quant Biol 40:465–482

Kastner S, Pinsk MA, De Weerd P, Desimone R, Ungerleider LG (1999) Increased activity in human visual cortex during directed attention in the absence of visual stimulation. Neuron 22(4):751–761. https://doi.org/10.1016/s0896-6273(00)80734-5

Kay P, Maffi L (1999) Color appearance and the emergence and evolution of basic color lexicons. Am Anthropol 101(4):743–760. https://doi.org/10.1525/aa.1999.101.4.743

Keysers C, Perrett DI (2004) Demystifying social cognition: a Hebbian perspective. Trends Cogn Sci 8(11):501–507. https://doi.org/10.1016/j.tics.2004.09.005

Knudsen EI (2007) Fundamental components of attention. Annu Rev Neurosci 30:57–78. https://doi.org/10.1146/annurev.neuro.30.051606.094256

Koziol LF, Budding DE, Chidekel D (2012) From movement to thought: executive function, embodied cognition, and the cerebellum. Cerebellum 11(2):505–525. https://doi.org/10.1007/s12311-011-0321-y

Kullmann DM, Lamsa K (2008) Roles of distinct glutamate receptors in induction of anti-Hebbian long-term potentiation. J Physiol 586(6):1481–1486. https://doi.org/10.1113/jphysiol.2007.148064

LaBar KS, Cabeza R (2006) Cognitive neuroscience of emotional memory. Nat Rev Neurosci 7(1):54–64. https://doi.org/10.1038/nrn1825

Lamprecht R, LeDoux J (2004) Structural plasticity and memory. Nat Rev Neurosci 5(1):45–54. https://doi.org/10.1038/nrn1301

Landsteiner K, Jacobs J (1935) Studies on the sensitization of animals with simple chemical compounds. J Exp Med 61(5):643–656. https://doi.org/10.1084/jem.61.5.643

Lewis JW (2006) Cortical networks related to human use of tools. Neuroscientist 12(3):211–231. https://doi.org/10.1177/1073858406288327

Lezak M, Howieson D, Loring D (2004) Neuropsychological assessment, 4th edn. Oxford University Press, New York

Liu T, Slotnick SD, Serences JT, Yantis S (2003) Cortical mechanisms of feature-based attentional control. Cereb Cortex 13(12):1334–1343. https://doi.org/10.1093/cercor/bhg080

Lomo T (2003) The discovery of long-term potentiation. Philos Trans R Soc Lond Ser B Biol Sci 358(1432):617–620. https://doi.org/10.1098/rstb.2002.1226

Loreto V, Mukherjee A, Tria F (2012) On the origin of the hierarchy of color names. Proc Natl Acad Sci U S A 109(18):6819–6824. https://doi.org/10.1073/pnas.1113347109

Luria AR (1966) Higher cortical functions in man. Basic Books, New York

Lynch MA (2004) Long-term potentiation and memory. Physiol Rev 84(1):87–136. https://doi.org/10.1152/physrev.00014.2003

Mackintosh NJ (1983) Conditioning and associative learning. Clarendon Press, Oxford

Malenka R, Nestler E, Hyman S (2009) Chapter 13: higher Cognitive function and behavioral control. In: Sydor A, Brown R (eds) Molecular neuropharmacology: a Foundation for Clinical Neuroscience, 2nd edn. McGraw-Hill Medical, New York, pp 313–321

Marcus G, Rabagliati H (2006) What developmental disorders can tell us about the nature and origins of language. Nat Neurosci 9(10):1226–1229. https://doi.org/10.1038/nn1766

Markowitsch HJ (2005) The neuroanatomy of memory. In: Halligan PW, Wade DT (eds) The effectiveness of rehabilitation for cognitive deficits. Oxford University Press, Oxford, p 105

Mayberg HS (1997) Limbic-cortical dysregulation: a proposed model of depression. J Neuropsychiatry Clin Neurosci 9(3):471–481. https://doi.org/10.1176/jnp.9.3.471

Mazur JE (2015) Learning and behavior. Psychology Press, New York

McDougall W (1924) Outline of psychology. Methuen, London

Mewhort D, Campbell A, Marchetti FM, Campbell JI (1981) Identification, localization, and "iconic memory": an evaluation of the bar-probe task. Mem Cogn 9(1):50–67

Meyer D, Bonhoeffer T, Scheuss V (2014) Balance and stability of synaptic structures during synaptic plasticity. Neuron 82(2):430–443. https://doi.org/10.1016/j.neuron.2014.02.031

Miller GA (1956) The magical number seven plus or minus two: some limits on our capacity for processing information. Psychol Rev 63(2):81–97

Miller GA (1996) How we think about cognition, emotion, and biology in psychopathology. Psychophysiology 33(6):615–628. https://doi.org/10.1111/j.1469-8986.1996.tb02356.x

Miller CA (2006) Developmental relationships between language and theory of mind. Am J Speech Lang Pathol 15(2):142–154. https://doi.org/10.1044/1058-0360(2006/014)

Miller BT, D'Esposito M (2005) Searching for "the top" in top-down control. Neuron 48(4):535–538. https://doi.org/10.1016/j.neuron.2005.11.002

Miller CA, Sweatt JD (2007) Covalent modification of DNA regulates memory formation. Neuron 53(6):857–869. https://doi.org/10.1016/j.neuron.2007.02.022

Milligan K, Astington JW, Dack LA (2007) Language and theory of mind: meta-analysis of the relation between language ability and false-belief understanding. Child Dev 78(2):622–646. https://doi.org/10.1111/j.1467-8624.2007.01018.x

Milner B, Corkin S, Teuber H-L (1968) Further analysis of the hippocampal amnesic syndrome: 14-year follow-up study of HM. Neuropsychologia 6(3):215–234

Milner B, Squire LR, Kandel ER (1998) Cognitive neuroscience and the study of memory. Neuron 20(3):445–468. https://doi.org/10.1016/s0896-6273(00)80987-3

Mitchison GJ, Swindale NV (1999) Can Hebbian volume learning explain discontinuities in cortical maps? Neural Comput 11(7):1519–1526. https://doi.org/10.1162/089976699300016115

Miyake A, Friedman NP, Emerson MJ, Witzki AH, Howerter A, Wager TD (2000) The unity and diversity of executive functions and their contributions to complex "frontal lobe" tasks: a latent variable analysis. Cogn Psychol 41(1):49–100. https://doi.org/10.1006/cogp.1999.0734

Monsell S (1996) Control of mental processes. In: Bruce V (ed) Unsolved mysteries of the mind: tutorial essays in cognition. Erlbaum Taylor and Francis, Hove, pp 93–148

Moore C, Pure K, Furrow D (1990) Children's understanding of the modal expression of speaker certainty and uncertainty and its relation to the development of a representational theory of mind. Child Dev 61(3):722–730

Morris N, Jones DM (1990) Memory updating in working memory: the role of the central executive. Br J Psychol 81(2):111–121. https://doi.org/10.1111/j.2044-8295.1990.tb02349.x

Neisser U (1967) Cognitive psychology. Appleton-Century-Crofts, New York

Newbury DF, Monaco AP (2010) Genetic advances in the study of speech and language disorders. Neuron 68(2):309–320. https://doi.org/10.1016/j.neuron.2010.10.001

Nitschke JB, Heller W, Miller GA (2000) Anxiety, stress, and cortical brain function. In: Borod JC (ed) The neuropsychology of emotion. Oxford University Press, New York, pp 298–319

Noroozian M (2014) The role of the cerebellum in cognition: beyond coordination in the central nervous system. Neurol Clin 32(4):1081–1104. https://doi.org/10.1016/j.ncl.2014.07.005

Ojemann GA, Dodrill CB (1985) Verbal memory deficits after left temporal lobectomy for epilepsy. Mechanism and intraoperative prediction. J Neurosurg 62(1):101–107. https://doi.org/10.3171/jns.1985.62.1.0101

Ouardouz M, Sastry BR (2000) Mechanisms underlying LTP of inhibitory synaptic transmission in the deep cerebellar nuclei. J Neurophysiol 84(3):1414–1421. https://doi.org/10.1152/jn.2000.84.3.1414

Papassotiropoulos A, Wollmer MA, Aguzzi A, Hock C, Nitsch RM, de Quervain DJ (2005) The prion gene is associated with human long-term memory. Hum Mol Genet 14(15):2241–2246. https://doi.org/10.1093/hmg/ddi228

Papez JW (1937) A proposed mechanism of emotion. Arch Neurol Psychiatr 38(4):725–743

Pastalkova E, Serrano P, Pinkhasova D, Wallace E, Fenton AA, Sacktor TC (2006) Storage of spatial information by the maintenance mechanism of LTP. Science 313(5790):1141–1144. https://doi.org/10.1126/science.1128657

Pattyn N, Neyt X, Henderickx D, Soetens E (2008) Psychophysiological investigation of vigilance decrement: boredom or cognitive fatigue? Physiol Behav 93(1–2):369–378. https://doi.org/10.1016/j.physbeh.2007.09.016

Pavlov IP, Gantt WH, Volborth G, Cannon W (1929) Lectures on conditioned reflexes. International Publishers, New York

Pelosi L, Slade T, Blumhardt LD, Sharma VK (2000) Working memory dysfunction in major depression: an event-related potential study. Clin Neurophysiol 111(9):1531–1543. https://doi.org/10.1016/s1388-2457(00)00354-0

Pelphrey KA, Morris JP, McCarthy G (2004) Grasping the intentions of others: the perceived intentionality of an action influences activity in the superior temporal sulcus during social perception. J Cogn Neurosci 16(10):1706–1716. https://doi.org/10.1162/0898929042947900

Pinker S (1994) The language instinct. William Morrow and Company, New York

Pinker S (2010) The language instinct: how the mind creates language. HarperCollins e-books, New York; Illustrated edition

Pinker S, Bloom P (1990) Natural language and natural selection. Behav Brain Sci 13(4):707–727

Pinsker H, Kupfermann I, Castellucci V, Kandel E (1970) Habituation and dishabituation of the GM-withdrawal reflex in Aplysia. Science 167(3926):1740–1742

Pinsker HM, Hening WA, Carew TJ, Kandel ER (1973) Long-term sensitization of a defensive withdrawal reflex in Aplysia. Science 182(4116):1039–1042. https://doi.org/10.1126/science.182.4116.1039

Posner MI, Boies SJ (1971) Components of attention. Psychol Rev 78(5):391–408. https://doi.org/10.1037/h0031333

Posner MI, Petersen SE (1990) The attention system of the human brain. Annu Rev Neurosci 13:25–42. https://doi.org/10.1146/annurev.ne.13.030190.000325

Premack D, Woodruff G (2010) Does the chimpanzee have a theory of mind? Behav Brain Sci 1(4):515–526. https://doi.org/10.1017/s0140525x00076512

Price GR, Mazzocco MM, Ansari D (2013) Why mental arithmetic counts: brain activation during single digit arithmetic predicts high school math scores. J Neurosci Off J Soc Neurosci 33(1):156–163. https://doi.org/10.1523/JNEUROSCI.2936-12.2013

Raftopoulos A (2007) Visual processing and attention: cognition and perception. Oxford University Press, Oxford

Ramachandran VS (2000) Mirror neurons and imitation learning as the driving force behind "the great leap forward" in human evolution. Edge, New York, NY

Rescorla RA (2014) Pavlovian second-order conditioning (psychology revivals): studies in associative learning. Psychology Press, New York

Reuter MS, Riess A, Moog U, Briggs TA, Chandler KE, Rauch A, Stampfer M, Steindl K, Glaser D, Joset P, DDD Study, Krumbiegel M, Rabe H, Schulte-Mattler U, Bauer P, Beck-Wodl S, Kohlhase J, Reis A, Zweier C (2017) FOXP2 variants in 14 individuals with developmental speech and language disorders broaden the mutational and clinical spectrum. J Med Genet 54(1):64–72. https://doi.org/10.1136/jmedgenet-2016-104094

Rolls ET, Grabenhorst F (2008) The orbitofrontal cortex and beyond: from affect to decision-making. Prog Neurobiol 86(3):216–244. https://doi.org/10.1016/j.pneurobio.2008.09.001

Rueda MR, Rothbart MK, McCandliss BD, Saccomanno L, Posner MI (2005) Training, maturation, and genetic influences on the development of executive attention. Proc Natl Acad Sci U S A 102(41):14931–14936. https://doi.org/10.1073/pnas.0506897102

Ruffman T, Slade L, Crowe E (2002) The relation between children's and mothers' mental state language and theory-of-mind understanding. Child Dev 73(3):734–751. https://doi.org/10.1111/1467-8624.00435

Samson D, Apperly IA, Chiavarino C, Humphreys GW (2004) Left temporoparietal junction is necessary for representing someone else's belief. Nat Neurosci 7(5):499–500. https://doi.org/10.1038/nn1223

Sandler W, Lillo-Martin D (2001) Natural sign languages. In: Aronoff M, Rees-Miller J (eds) The handbook of linguistics. Blackwell, Hoboken, pp 533–563

Sanfey AG, Civai C, Vavra P (2015) Predicting the other in cooperative interactions. Trends Cogn Sci 19(7):364–365. https://doi.org/10.1016/j.tics.2015.05.009

Saxe R, Kanwisher N (2003) People thinking about thinking people. The role of the temporoparietal junction in "theory of mind". NeuroImage 19(4):1835–1842. https://doi.org/10.1016/s1053-8119(03)00230-1

Scholz J, Triantafyllou C, Whitfield-Gabrieli S, Brown EN, Saxe R (2009) Distinct regions of right temporo-parietal junction are selective for theory of mind and exogenous attention. PLoS One 4(3):e4869. https://doi.org/10.1371/journal.pone.0004869

Shokri-Kojori E, Motes MA, Rypma B, Krawczyk DC (2012) The network architecture of cortical processing in visuo-spatial reasoning. Sci Rep 2:411. https://doi.org/10.1038/srep00411

Simons DJ, Chabris CF (1999) Gorillas in our midst: sustained inattentional blindness for dynamic events. Perception 28(9):1059–1074. https://doi.org/10.1068/p281059

Skinner BF (1953) Science and human behavior. Simon and Schuster, New York

Skinner BF (1990) The behavior of organisms: an experimental analysis. BF Skinner Foundation

Smith ML, Milner B (1981) The role of the right hippocampus in the recall of spatial location. Neuropsychologia 19(6):781–793. https://doi.org/10.1016/0028-3932(81)90090-7

Sohlberg MM, Mateer CA (2001) Cognitive rehabilitation: an integrative neuropsychological approach. Guilford Press, New York, NY

Spelke E, Hirst W, Neisser UJC (1976) Skills of divided attention. Cognition 4:215–230

Sperling G (1960) The information available in brief visual presentations. Psychol Monogr Gen Appl 74(11):1

Sperling G (1963) A model for visual memory tasks. Hum Factors 5:19–31. https://doi.org/10.1177/001872086300500103

Squire LR (2004) Memory systems of the brain: a brief history and current perspective. Neurobiol Learn Mem 82(3):171–177. https://doi.org/10.1016/j.nlm.2004.06.005

Squire LR, Kandel ER (2000) Memory: from mind to molecules, vol 69. Macmillan

Squire LR, Zola SM (1996) Structure and function of declarative and nondeclarative memory systems. Proc Natl Acad Sci U S A 93(24):13515–13522. https://doi.org/10.1073/pnas.93.24.13515

Squire LR, Zola SM (1998) Episodic memory, semantic memory, and amnesia. Hippocampus 8(3):205–211. https://doi.org/10.1002/(SICI)1098-1063(1998)8:3<205::AID-HIPO3>3.0.CO;2-I

Squire LR, Zola-Morgan S (1988) Memory: brain systems and behavior. Trends Neurosci 11(4):170–175. https://doi.org/10.1016/0166-2236(88)90144-0

Stuss DT, Alexander MP (2000) Executive functions and the frontal lobes: a conceptual view. Psychol Res 63(3–4):289–298. https://doi.org/10.1007/s004269900007

Thorndike EL (1898) Animal intelligence: an experimental study of the associative processes in animals. Psychol Rev 2(4):i

Thorndike EL (1927) The law of effect. Am J Psychol 39(1/4):212–222

Tipper SP (2001) Does negative priming reflect inhibitory mechanisms? A review and integration of conflicting views. Q J Exp Psychol A 54(2):321–343. https://doi.org/10.1080/713755969

Trautwein FM, Singer T, Kanske P (2016) Stimulus-driven reorienting impairs executive control of attention: evidence for a common bottleneck in anterior insula. Cereb Cortex 26(11):4136–4147. https://doi.org/10.1093/cercor/bhw225

Treisman A, Vieira A, Hayes A (1992) Automaticity and preattentive processing. Am J Psychol 105(2):341–362

Tulving E (1972) Episodic and semantic memory. In: E Tulving, W DonaldsonOrganization of memory London: Academic 381 (4):382–404

Tulving E (2002) Episodic memory: from mind to brain. Annu Rev Psychol 53(1):1–25. https://doi.org/10.1146/annurev.psych.53.100901.135114

Tulving E, Markowitsch HJ (1998) Episodic and declarative memory: role of the hippocampus. Hippocampus 8(3):198–204. https://doi.org/10.1002/(SICI)1098-1063(1998)8:3<198::AID-HIPO2>3.0.CO;2-G

Wahn B, Konig P (2017) Is attentional resource allocation across sensory modalities task-dependent? Adv Cogn Psychol 13(1):83–96. https://doi.org/10.5709/acp-0209-2

Wang J, Conder JA, Blitzer DN, Shinkareva SV (2010) Neural representation of abstract and concrete concepts: a meta-analysis of neuroimaging studies. Hum Brain Mapp 31(10):1459–1468. https://doi.org/10.1002/hbm.20950

Ward EF (1983) Teaching sign language to a chimpanzee: some historical references. J Exp Anal Behav 40(3):341–342. https://doi.org/10.1901/jeab.1983.40-341
Whitlock JR, Heynen AJ, Shuler MG, Bear MF (2006) Learning induces long-term potentiation in the hippocampus. Science 313(5790):1093–1097. https://doi.org/10.1126/science.1128134
Woolfe T, Want SC, Siegal M (2002) Signposts to development: theory of mind in deaf children. Child Dev 73(3):768–778. https://doi.org/10.1111/1467-8624.00437
Workman L, Reader W (2008) Evolutionary psychology: an introduction. Cambridge University Press, Cambridge
Yin JC, Tully T (1996) CREB and the formation of long-term memory. Curr Opin Neurobiol 6(2):264–268. https://doi.org/10.1016/s0959-4388(96)80082-1

Thought

12

Contents

12.1 Basic Characteristics of Human Thinking

After the description of cognitive functions, a discussion on "thoughts" and "thinking" could seem redundant. However, *thinking* is the end product of brain function and constitutes what the term "mind" includes as its essence. The philosophical and analytical approaches of this issue are not an aim for the current book. However, it is important to discuss what are the macroscopic features of the process of thinking and the concept of *intelligence*, which constitute an expansion of what has been mentioned and described so far in previous chapters.

Thinking
- Complex process of intelligence in extremely short time.
- Not capable of performing less-demanding processing.
- Due to evolutionary reasons and because of pressure by the needs, the basic architecture and the basic principles on which brain function is based are completely different from those of electronic computers and more specifically from "Turing machines".

K. N. Fountoulakis, *The Human Connectome*,
https://doi.org/10.1007/978-3-031-10351-3_12

- Cognitive biases.
  - Were embedded in brain function when humans were living in a completely different environment.
  - Systematic model of deviation of judgment from usual or deviation from rational reasoning.
  - Personal version of the world and especially of social reality.
  - Essentially adaptation role.
  - Fast decision-making.
  - Heuristics.
  - Often severe and systematic errors in final decision.
- Comparative way of perception with central corrections.
- Preexisting templates and algorithms which "shortcut".

One astonishing characteristic of the human brain is that, while it is capable of performing *complex processing of information* in extremely short periods of time, it *is incapable of performing much less-demanding processing of information.* It is as if it is calibrated to *perform only on high demands.* For example, the human brain, apart from its ability to comprehend and analyze abstract concepts and meanings, in everyday life, is able, to perform the complex visuospatial and motor calculations necessary to throw a spear to hit a moving target at a distance or to perform complex body acrobatics, while the processing demands of social interaction is beyond our ability to calculate. All these are capabilities, which modern engineering and computing find it difficult if not impossible to fully simulate. In contrast, the human brain is unable to perform a multiplication between two five-digit numbers, which is one of the easiest tasks for electronic-based calculators even at credit-card size.

This is because the human brain, only during the last few centuries, was found in the need to perform this kind of calculations, while other kinds of processing used to be essential for survival. As a result, and because of evolutionary reasons and the pressure by the needs, the essential architecture and the fundamental principles on which the brain function is based are completely different from that of electronic computers and specifically from what is called "Turing machines." Again, this topic is beyond the scope of the current chapter; however it is useful to note that only recently we have started understanding the principles of the so-called non-Turing machines and of quantum computers which utilize a probabilistic system instead of a binary code.

One probable characteristic of some of these computational systems is their complex coding at input and that they are *extremely relying on templates and pre-acquired information.* This is also a characteristic of the brain, and at least some of these pre-acquired templates emerge in the form of archetypical thoughts and impulses, as well as with the form of cognitive biases.

The concept of cognitive biases was introduced by Amos Tversky and Daniel Kahneman in 1972 (Kahneman and Shane 2002). Kahneman received the Nobel Prize in Economic Sciences in 2002, and his overall work challenges the assumption of human rationality which was prevailing in modern economic theory and decision-making. Cognitive biases refer to a systematic pattern of deviation from norm or from rationality in judgment. This leads to inferences about the environment and to events in an illogical fashion (Haselton et al. 2005). This way of thinking leads individuals to create *their own subjective version of the world* and especially of the social reality, and very often, it is the preexisting biases rather than the objective information input that determine the response and overall behavior (Morewedge and Kahneman 2010; Kahneman and Klein 2009; Kahneman and Frederick 2007; Kahneman 2003; Kahneman and Tversky 1982, 1996; Tversky and Kahneman 1981; Gigerenzer and Goldstein 1996). Approximately 200 such biases have been described so far, and the most important of them are shown in Table 12.1.

Although some of these biases are the unwanted result of limitations in the process capacity of the brain, in principle, the role of biases is profoundly adaptive, since they enable faster decisions *when timeliness are more valuable than accuracy.* This is illustrated in *heuristics* (from the Greek verb "ευρίσκω," meaning "I find"), which concern processing of information with the use of processing shortcuts which are simple for the brain to compute but sometimes introduce severe and systematic errors in the final output.

The problem is that, for most of biases, their basic principles were embedded in the human brain when humans were living in a much different environment than today, with different everyday needs. In modern human society, they might not be always able

Table 12.1 Most important cognitive biases

• Agent detection (Gray and Wegner 2010)
• Ambiguity effect (Baron 1994)
• Anthropocentric thinking (Coley and Tanner 2012)
• Anthropomorphism (Dacey 2017) vs. dehumanization (Harris and Fiske 2011)
• Availability heuristic (Wänke et al. 1995)
• Clustering illusion (Iverson et al. 2008)
• Confirmation bias (Risen and Giovich 2007)
• Conservatism bias (DuCharme 1970)
• Declinism or rosy retrospection (Edgerton 2011)
• Dunning-Kruger phenomenon (Kruger and Dunning 1999)
• "Halo" effect (Baron 1994)
• Hindsight bias (Pohl 2004)
• Illusion of control (Thompson 1999)
• Mood-congruent memory bias (Jaffe et al. 2019)
• Naturalistic fallacy (Moore 1903)
• Normalcy bias (Omer and Alon 1994)
• Optimism or pessimism bias (Baron 1994)
• Overconfidence effect (Hoffrage 2004)
• Proportionality bias (Leman and Cinnirella 2007)
• Self-serving bias (Campbell and Sedikides 1999)
• Stereotyping (Whiteley and Kite 2010)

to play an adaptive role since they can activate stereotypes and inaccurate judgments of others. It is important to note that cognitive biases can be controlled and are influenced by training and education, but often it seems that intelligent and educated persons use their knowledge and skills only to further strengthen and support their biases, instead of correcting them and adjusting them to the real contemporary needs.

Apart from biases, the way the brain perceives the environment plays an important role for the final result. This way of perceiving is to a significant extent *based on comparisons with central corrections*. The brain has not been evolved and developed aiming at the absolutely objective and accurate perception and registration of the environment. Its processing capability is so enormous that in case such an accurate and detailed imaging was important and would have had an effect on survival, it is certain that the brain would have had achieved it. However, the brain, like the form of all life on earth, is based on *economy and sparing of resources and energy, and this is the product of evolutionary pressure*. There is nothing unnecessary in the way of functioning and no waste of resources happens. Evolutionary pressure pushed the organization toward the direction with maximum efficacy concerning the survival of the species and simultaneously maximum sparing of resources.

The obvious compromise which developed was the achievement of high speed in decision-making but at the cost of basing this decision on preexisting templates and processing algorithms, with "shortcuts" which put the emphasis on important features of the objects and neglect the rest and "guess" in order to correct errors and fill gaps of the sensory information. This often leads to an erroneous perception of the environment (Fig. 10.12).

12.2 General Intelligence

While, on one hand, the everyday human thinking is characterized by the seeking for the general picture and the prediction of future events and situations (and particularly of future threats), under conditions of imperfect information and in a chaotic environment, on the other hand, there is a concept which refers to the "objective" ability to solve specific problems, many of which did not exist as such in the environment where Homo sapiens emerged from, during the Pleistocene.

This ability is called "intelligence" and has been defined in many different ways, including the ability to reason and to utilize abstract thinking, comprehension, self-knowledge, learning, emotional insight and awareness, planning, creativity, and problem-solving. Eventually, various approaches have been adopted concerning human intelligence, and its definition remains a matter of debate (Legg and Hutter 2007; Gottfredson 1997; Neisser et al. 1996). As a scientific topic, it emerged as the result of research on anthropological differences among individuals, during the late eighteenth and the early nineteenth centuries (Goldstein et al. 2015). Further elaboration on the issue is beyond the scope of the present chapter.

The word "intelligence" comes from the Latin intelligentia or intellēctus, which came from the merging of "inter" (in-between) and "legere" (select, collect, select words-speak).

Intelligence

- General factor *g*.
- Theory of multiple intelligences by Howard Gardner.
  - Naturalistic.
  - Musical-rhythmic.
  - Visual-spatial.
  - Linguistic-verbal.
  - Bodily-kinesthetic.
  - Logical-mathematical.
  - Intrapersonal.
  - Interpersonal.
- Theory of David Wechsler, on distinct domains of intelligence which are not completely independent to each other.
- Inconclusive empirical data.

The dominant fundamental approach concerns the existence of a *general intelligence factor or index (mental capability) also known as "g"* which constitutes the main substrate on which all the partial neurocognitive abilities of the person rely, and this is the reason why they manifest a significant correlation to each other (Reader et al. 2011).

Howard Gardner argued for a completely different approach and suggested the presence of a number of distinct and independent from each other "structural intelligence units" (Gardner 2011) but has received criticism on the basis that his theory is based more on personal opinion rather than research findings (Waterhouse 2006). In spite of that, it reflects to a significant degree the observed variability in the patterns of human neurocognitive ability.

Eight specific distinct intelligences have been proposed by Gardner, which could be arranged in four dipoles (Fig. 12.1):

- *Naturalistic*, which refers to the special ability to perceive natural objects and their characteristics and properties, recognize their similarities and differences, and handle them more easily.
- *Musical-rhythmic and harmonic*, which refers to the mental processing of sound and in modern society is expressed through musical talent.
- *Visual-spatial* which manifests mainly thought achievements in dancing and sports.
- *Linguistic-verbal* which concerns the use of language at all levels and ways from the ability to narrate and write to memorization of texts and speeches. It is one of the most recognized factors and significantly contributes to the concept of the *g* factor.
- *Bodily-kinesthetic* refers to the perception and movement of the body as well as to the handling of objects and, together with visual-spatial, leads to high performance in tasks demanding motion in space and body-environment interaction.

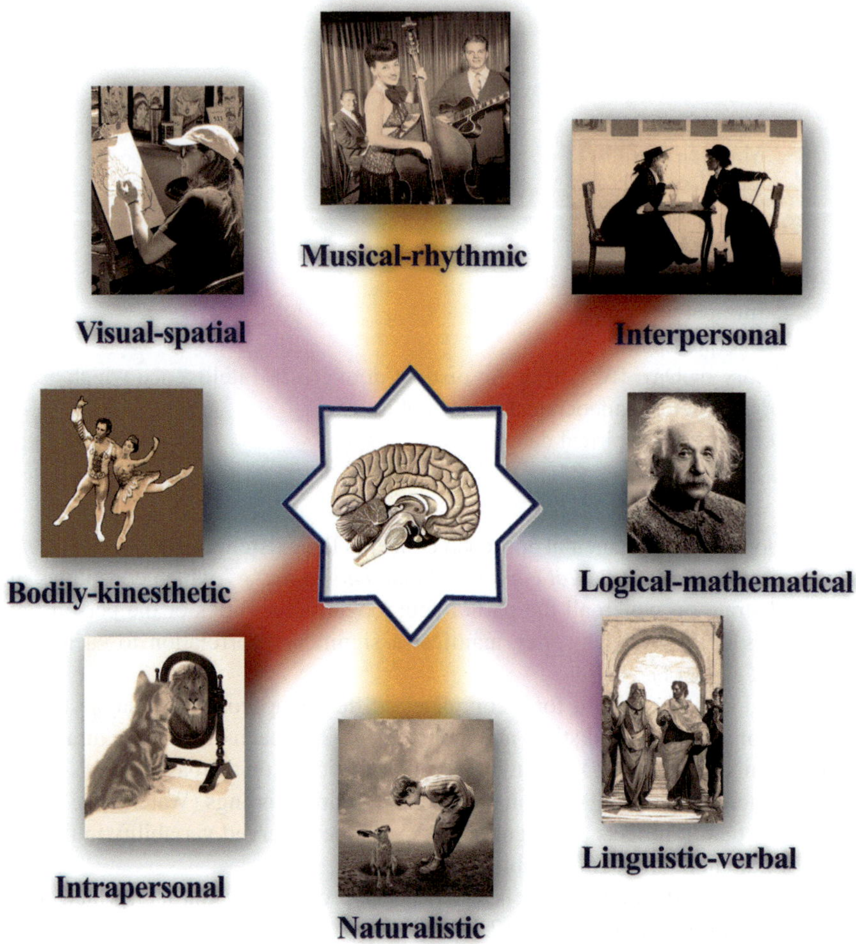

Fig. 12.1 Types of intelligence according to the theory of Howard Gardner (copyright by KN Fountoulakis, 2021)

- *Logical-mathematical* is related to abstract thinking and logical analysis, and also it is one of the most recognized factors. It also contributes significantly to the concept of the *g* factor.
- *Intrapersonal*, which concerns the ability of the person to be aware of him/herself, to introspect and analyze by understanding his/her weaknesses and talents, and recruit meta-cognitive processes.
- *Interpersonal* which refers to the ability of the person to perceive and respond in an appropriate way to interpersonal and social stimuli from others and to be able to collaborate in a more efficacious way with them.

The problem of the theory of multiple intelligences is that, on one hand, the psychometric data show a high correlation among the supposed types of intelligence, and this is in accord with the existence of the general index *g*, who essentially drives intelligence, and, on the other hand, educational programs based on multiple intelligences have not given results beyond reasonable doubt, and the related research is of poor quality. On top of all these, it seems there is a kind of "war" concerning the use of concepts and terms (e.g., what the difference between intelligence and talent is, what the relationship between personality and intelligence is), and this war, from the viewpoint of the strictly scientific researcher who follows the data, has zero value, since it is based almost exclusively on theoretical arguments and in an attempt of each side to label as "pseudoscience" the opposite side, by publishing in journals with low scientific prestige.

The internationally most accepted approach today is based on the use of the *intelligence scales of David Wechsler*, who developed them on the basis of the theory that there are distinct domains of intelligence, which however are not entirely independent from each other, while at the same time intelligence is not a unidimensional concept (Wechsler 1939). This approach is dominant since it is strongly in accord with the everyday experience of psychiatrists and psychologists.

References

Baron J (1994) Thinking and deciding, 2nd edn. Cambridge University Press

Campbell W, Sedikides C (1999) Self-threat magnifies the self-serving bias: a meta-analytic integration. Rev Gen Psychol 3(1):23–43

Coley JD, Tanner KD (2012) Common origins of diverse misconceptions: cognitive principles and the development of biology thinking. CBE Life Sci Educ 11(3):209–215. https://doi.org/10.1187/cbe.12-06-0074

Dacey M (2017) Anthropomorphism as cognitive bias. Philos Sci 84(5):1152–1164. https://doi.org/10.1086/694039

DuCharme W (1970) Response bias explanation of conservative human inference. J Exp Psychol Gen 85(1):66–74

Edgerton D (2011) The decline of Declinism. Business History Rev 71(2):201–206. https://doi.org/10.2307/3116157

Gardner H (2011) Frames of mind: the theory of multiple intelligences, 3rd edn. Basic Books, New York

Gigerenzer G, Goldstein DG (1996) Reasoning the fast and frugal way: models of bounded rationality. Psychol Rev 103(4):650–669. https://doi.org/10.1037/0033-295x.103.4.650

Goldstein S, Princiotta D, Naglieri J (eds) (2015) Handbook of intelligence: evolutionary theory, historical perspective, and current concepts. Springer, New York

Gottfredson LS (1997) Mainstream science on intelligence: an editorial with 52 signatories, history, and bibliography. Intelligence 24(1):13–23. https://doi.org/10.1016/s0160-2896(97)90011-8

Gray K, Wegner DM (2010) Blaming god for our pain: human suffering and the divine mind. Personal Soc Psychol Rev 14(1):7–16. https://doi.org/10.1177/1088868309350299

Harris LT, Fiske ST (2011) Dehumanized perception: a psychological means to facilitate atrocities, torture, and genocide? Z Psychol 219(3):175–181. https://doi.org/10.1027/2151-2604/a000065

Haselton MG, Nettle D, Andrews PW (2005) The evolution of cognitive bias. In: Buss D (ed) The handbook of evolutionary psychology. Wiley, Hoboken, NJ, pp 724–746

Hoffrage U (2004) Overconfidence. In: Pohl R (ed) Cognitive illusions: a handbook on fallacies and biases in thinking, judgement and memory. Psychology Press, Hove

Iverson G, Brooks B, Holdnack J (2008) Misdiagnosis of cognitive impairment in forensic neuropsychology. In: Heilbronner R (ed) Neuropsychology in the courtroom: expert analysis of reports and testimony. Guilford Press, New York

Jaffe AE, Blayney JA, Bedard-Gilligan M, Kaysen D (2019) Are trauma memories state-dependent? Intrusive memories following alcohol-involved sexual assault. Eur J Psychotraumatol 10(1):1634939. https://doi.org/10.1080/20008198.2019.1634939
Kahneman D (2003) A perspective on judgment and choice: mapping bounded rationality. Am Psychol 58(9):697–720. https://doi.org/10.1037/0003-066X.58.9.697
Kahneman D, Frederick S (2007) Frames and brains: elicitation and control of response tendencies. Trends Cogn Sci 11(2):45–46. https://doi.org/10.1016/j.tics.2006.11.007
Kahneman D, Klein G (2009) Conditions for intuitive expertise: a failure to disagree. Am Psychol 64(6):515–526. https://doi.org/10.1037/a0016755
Kahneman D, Shane F (2002) Representativeness revisited: attribute substitution in intuitive judgment. In: Gilovich T, Griffin D, Kahneman D (eds) Heuristics and biases: the psychology of intuitive judgment. Cambridge University Press, Cambridge, pp 51–52
Kahneman D, Tversky A (1982) Variants of uncertainty. Cognition 11(2):143–157. https://doi.org/10.1016/0010-0277(82)90023-3
Kahneman D, Tversky A (1996) On the reality of cognitive illusions. Psychol Rev 103(3):582–591; discusion 592-586. https://doi.org/10.1037/0033-295x.103.3.582
Kruger J, Dunning D (1999) Unskilled and unaware of it: how difficulties in recognizing one's own incompetence lead to inflated self-assessments. J Pers Soc Psychol 77(6):1121–1134. https://doi.org/10.1037//0022-3514.77.6.1121
Legg S, Hutter M (2007) A collection of definitions of intelligence. Paper presented at the proceedings of the 2007 conference on advances in artificial general intelligence: concepts, architectures and algorithms: proceedings of the AGI Workshop 2006
Leman P, Cinnirella M (2007) A major event has a major cause: evidence for the role of heuristics in reasoning about conspiracy theories. Soc Psychol Rev 9(2):18–28
Moore G (1903) Principia Ethica. Project Gutenberg
Morewedge CK, Kahneman D (2010) Associative processes in intuitive judgment. Trends Cogn Sci 14(10):435–440. https://doi.org/10.1016/j.tics.2010.07.004
Neisser U, Boodoo G, Bouchard TJ, Boykin AW, Brody N, Ceci SJ, Halpern DF, Loehlin JC, Perloff R, Sternberg RJ, Urbina S (1996) Intelligence: knowns and unknowns. Am Psychol 51(2):77–101. https://doi.org/10.1037/0003-066x.51.2.77
Omer H, Alon N (1994) The continuity principle: a unified approach to disaster and trauma. Am J Community Psychol 22(2):273–287. https://doi.org/10.1007/BF02506866
Pohl R (2004) Hindsight bias. In: Pohl R (ed) Cognitive illusions: a handbook on fallacies and biases in thinking, judgement and memory. Psychology Press, Hove, pp 363–378
Reader SM, Hager Y, Laland KN (2011) The evolution of primate general and cultural intelligence. Philos Trans R Soc Lond Ser B Biol Sci 366(1567):1017–1027. https://doi.org/10.1098/rstb.2010.0342
Risen J, Giovich T (2007) Informal logical fallacies. In: Sternberg R, Roediger H III, Halpern D (eds) Critical thinking in psychology. Cambridge University Press, pp 110–130
Thompson SC (1999) Illusions of control: how we overestimate our personal influence. Curr Dir Psychol Sci 8(6):187–190
Tversky A, Kahneman D (1981) The framing of decisions and the psychology of choice. Science 211(4481):453–458. https://doi.org/10.1126/science.7455683
Wänke M, Schwarz N, Bless H (1995) The availability heuristic revisited: experienced ease of retrieval in mundane frequency estimates. Acta Psychol 89(1):83–90. https://doi.org/10.1016/0001-6918(93)e0072-a
Waterhouse L (2006) Multiple intelligences, the Mozart effect, and emotional intelligence: a critical review. Educ Psychol 41(4):207–225. https://doi.org/10.1207/s15326985ep4104_1
Wechsler D (1939) The measurement of adult intelligence. Williams & Witkins, Baltimore, MD
Whiteley B, Kite M (2010) The psychology of prejudice and discrimination, 2nd edn. Wadsworth Cengage Learning, Belmont, CA

Mood, Affect, and Emotions 13

Contents

13.1 Mood, Affect, and Emotions

The contemporary approach distinguishes among the following:

- *Mood* is the long-term internal affective tone, and to a significant degree it constitutes a trait.
- *Affect* is a term which refers to the general affective state of the last few recent days or weeks, and it is related to recent events.
- *Emotion* corresponds to a temporary affective state which is manifested through movement behaviors (face expressions, body movements, complex behaviors) and concerns reaction to internal or external stimuli or events. It is determined by mood and is mediated by affect.

Mood has an enduring nature; it tends not to be focused and it is diffused. It includes expectations for the future, and it is manifested in subtle ways, while, on the contrary, emotions tend to be short-lived and have a specific focus.

In the current chapter, the term "emotion" is also used in a generic way, to denote affective feelings and sentiments when a general and not a specific reference to this domain of internal experience and function was deemed necessary.

K. N. Fountoulakis, *The Human Connectome*,
https://doi.org/10.1007/978-3-031-10351-3_13

13.2 The Role of Mood, Affect, and Emotions

Mood, Affect, and Emotions Serve Two Main Aims:

- The first concerns the internal functioning and provides the person with the ability of *fast decision-making*, which serves the survival of the individual as well as the species. Some of these decisions are easily comprehensible (e.g., fear of animals), but others are rather difficult to understand (e.g., aesthetics and attraction by the opposite sex). In the same frame, emotions provide with feedback concerning the behavior of the individual, and in this way, they amplify the expression of specific behaviors, or they preclude their future manifestation. For example, sorrow corresponds to an emotional response to loss, frustration, disappointment, and other adversities. Its adaptive function is to permit withdrawal with the aim to preserve energy, asking for support by significant others while the possible coexistence of an autonomic nervous excitation facilitates the seeking for the lost object or of some adequate substitute.
- The second aim is to *communicate the internal affective state of the individual* to others. This can be achieved through facial expressions, gestures, body posture and movements, as well as through the verbal and non-verbal elements of speech. These ways of communication differ to a certain extent from culture to culture, but the most part of the repertoire is common for all humans. The emotions of the individual constitute the main source of interaction with the others, as they exert an influence on the feelings, the thoughts, as well as the behaviors of others, they trigger either positive or negative feedback, and they give birth to circles of future interactions and mutual influences. The significance of feelings as a method of interpersonal communication is evident also from the adoption of special emotional icons (emoticons) by the messenger services (Fig. 13.1).

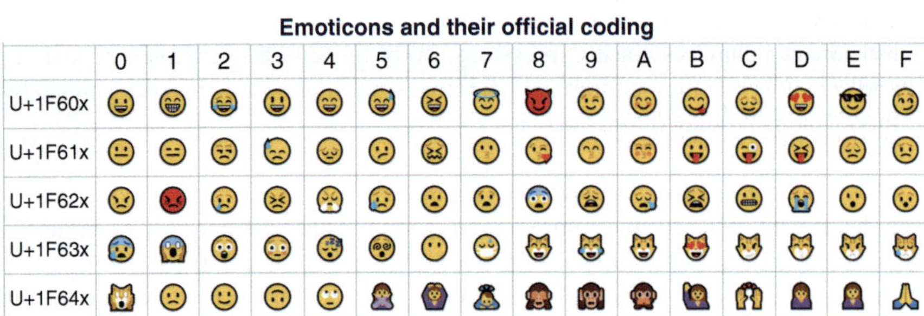

Fig. 13.1 Emoticons and their official coding for use in messaging

13.3 Neurobiology of Mood, Affect, and Emotions

The first modern neurobiological theory of emotions was independently developed by William James (1842–1910) and Carl Lange (1834–1900) in the 1880s. This theory suggests that emotions are the end result of changes in body physiology and they are caused by various stimuli, but they are not their direct result.

A contemporary approach of this issue goes through a fundamental consideration of brain function which suggests that there are *two distinct cognitive processes*: the slow logical-analytic thinking and the fast decision-making with emotions playing the determining role. While animals have emotions also, logical reasoning is primarily a human characteristic. However, elements of logical reasoning also exist in the behavior of primates, while reversely, in spite of considering humans as beings characterized by reason, *in fact logical and analytical reasoning plays only a minor role in the total of decisions made by humans*.

Traditionally, it is considered that the left hemisphere is the seat of logical reasoning while the right serves as the production of emotions. A consequence of the above localizing hypothesis is that dysfunctions that lead to schizophrenia are considered to reside in the left hemisphere while those leading to depression in the right. Although this hypothesis is supported also by some neuropsychological findings and especially from the study of patients who had undergone brain surgery or had suffered brain damage, it is almost certain that there is no complete and oppositional lateralization between reasoning and emotions.

Mood-Affect-Emotion

- Fast decision-making.
- Communication with other individuals.
- Slow logical-analytical thinking and laborious, precise, and accurate assessment of the situation.
- Fast decision-making—Emotions play a determining role, automatic and with dominant effect on behavior, biases that aim to the survival of the species.
- The left hemisphere is the seat of reason.
- The right hemisphere subserves emotions.
- Mood—amygdala and the isle of Reil.
- Affects—anterior cingulate gyrus and prefrontal cortex.
- Emotions—anterior cingulate gyrus (areas 24 and mainly 25).
- Universally common characteristics of emotions in humans.
- Common characteristics in the emotions of humans and animals.

A very simplified neurobiological model (Fig. 10.14) could propose that mood emerges from processes that take place mainly in the *amygdala and the isle of Reil* and emotions mainly from the *anterior cingulate* and more specifically from area 25. However, area 24 and the dorsolateral prefrontal cortex are probably involved in their laborious regulation. Between mood and emotions,

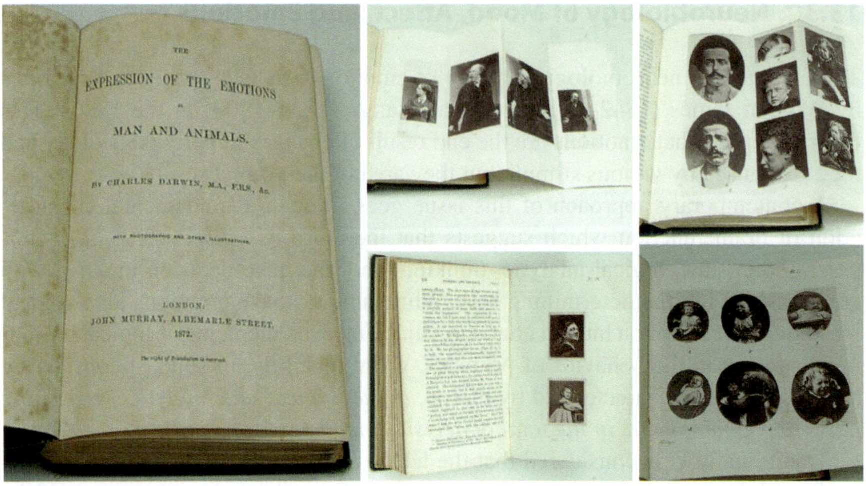

Fig. 13.2 The book *The Expression of the Emotions in Man and Animals* by Charles Darwin (1872) (after permission by Botterweg Auctions Amsterdam)

affects exist which come partially from the anterior cingulate and partially from other areas of the brain, including the prefrontal cortex (Phillips et al. 2003; Fountoulakis et al. 2008).

Interest on emotions from an evolutionary perspective came from the publication of the book *The Expression of the Emotions in Man and Animals* by Charles Darwin (1809–1872) in 1872 (Fig. 13.2). In that book, Darwin stresses the *universal nature of emotions* and the connection between mental states and the neurological organization of movement. It is clear that there is an enormous difference not only concerning the size but also concerning the functional capacity of animal brain in comparison to the human brain (Defelipe 2011) (Figs. 13.3 and 13.4).

Carnivora:
Bear (*Ursus arctos*, 289 g)
Lion (*Panthera leo*, 165 g)
Cheetah (*Acinonyx jubatus*, 119 g)
Dog (*Canis familiaris*, 95 g)
Cat (*Felis catus*, 32 g)
Artiodactyla:
Giraffe (*Giraffa camelopardalis*, 700 g)
Kudu (*Tragelaphus strepsiceros*, 166 g)
Mouflon (*Ovis musimon*, 118 g)
Goat (*Capra pyrenaica*, 115 g)
Wallaby (*Notamacropus*, 41 g)
Marsupials
Peccary (*Tayassu pecari*, 28 g)
Lagomorpha
Rabbit (*Oryctolagus cuniculus*, 5.2 g)
Rodents
Rat (*Rattus rattus*, 2.6 g)
Mouse (*Mus musculus*, 0.5 g)

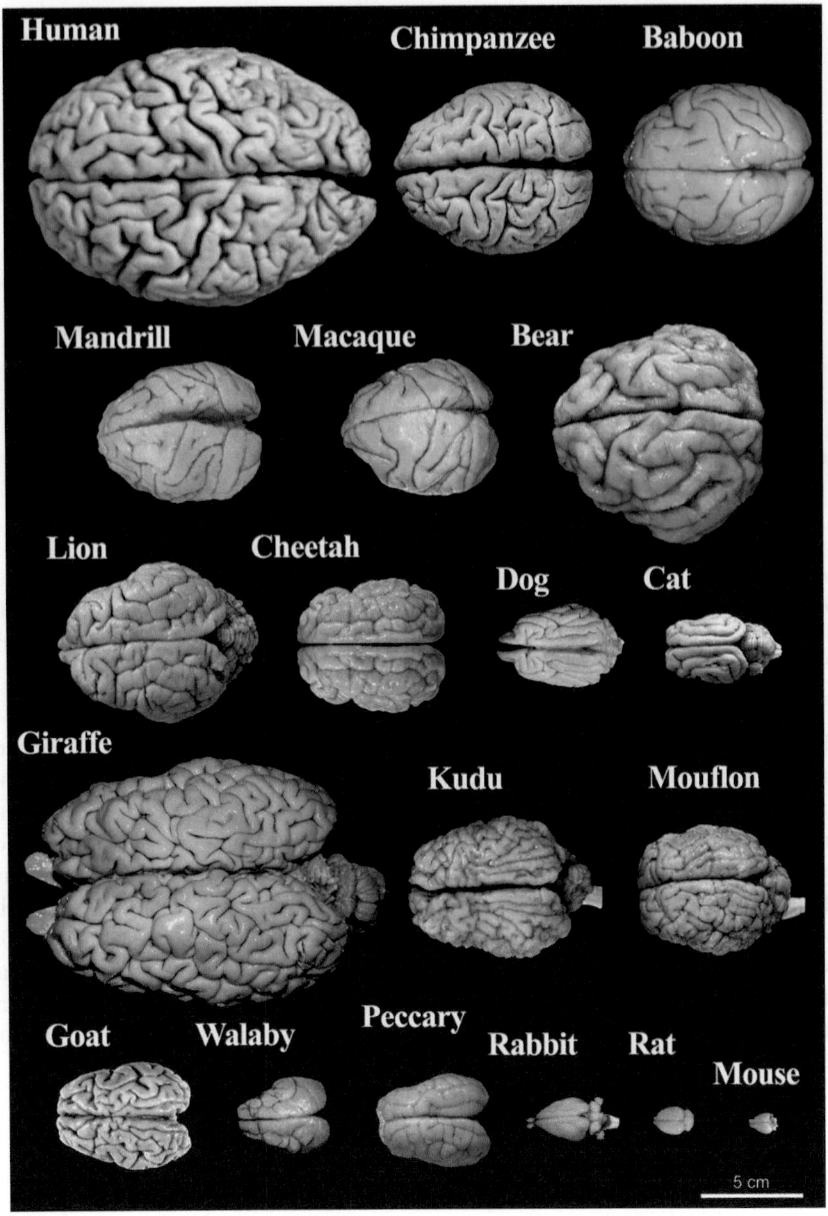

Fig. 13.3 The human brain in comparison to the brain of other mammals (image from Defelipe 2011, after permission)
Primates:
Human (*Homo sapiens*, 1.176 kg)
Chimpanzee (*Pan troglodytes*, 273 g)
Baboon (*Papio cynocephalus*, 151 g)
Mandrill (*Mandrillus sphinx*, 123 g)
Macaque (*Macaca tonkeana*, 110 g)

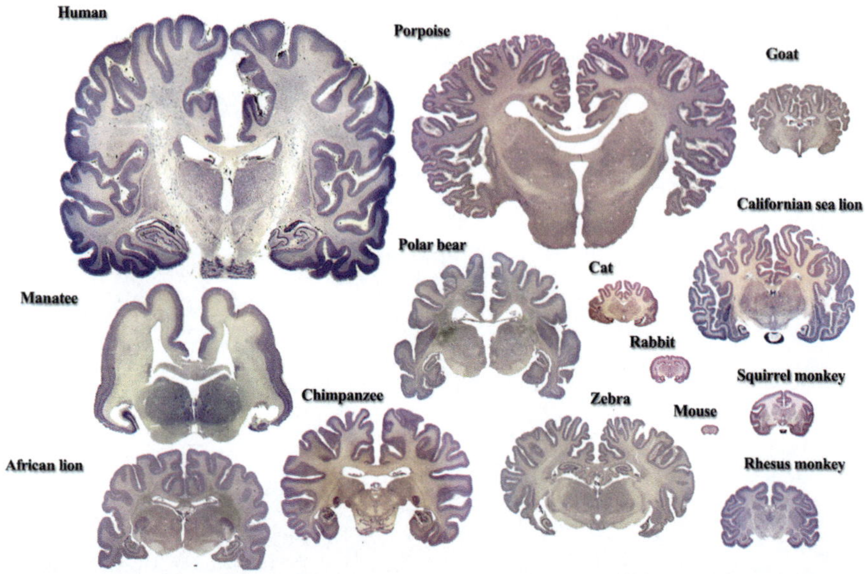

Fig. 13.4 Variability in brain size and patterns of coevolution of different mammals (thionine stain) (image from Defelipe 2011, after permission from http://www.brains.rad.msu.edu, supported by the US National Science Foundation)

However, central to his understanding was that there is a common human and animal origin, a fact that came in bitter opposition with the claims of his time, that there were divinely created human muscles for the expression of uniquely human emotions. The original suggestion by Darwin was that emotions developed through natural selection, and therefore they have universally transcultural equivalents, a proposal which was confirmed almost a century later, mainly through the work of Paul Ekman and others (Ekman 1965, 1980, 1992a, b, 1993, 1994, 2003, 2009, 2016; Ekman and Friesen 1967, 1971; Ekman et al. 1969, 1987; Guo et al. 2019), but still the debate on universality of emotions is far from over (Jack et al. 2012). Additionally, animals seem to have emotions similar to those of humans (Figs. 13.5 and 13.6).

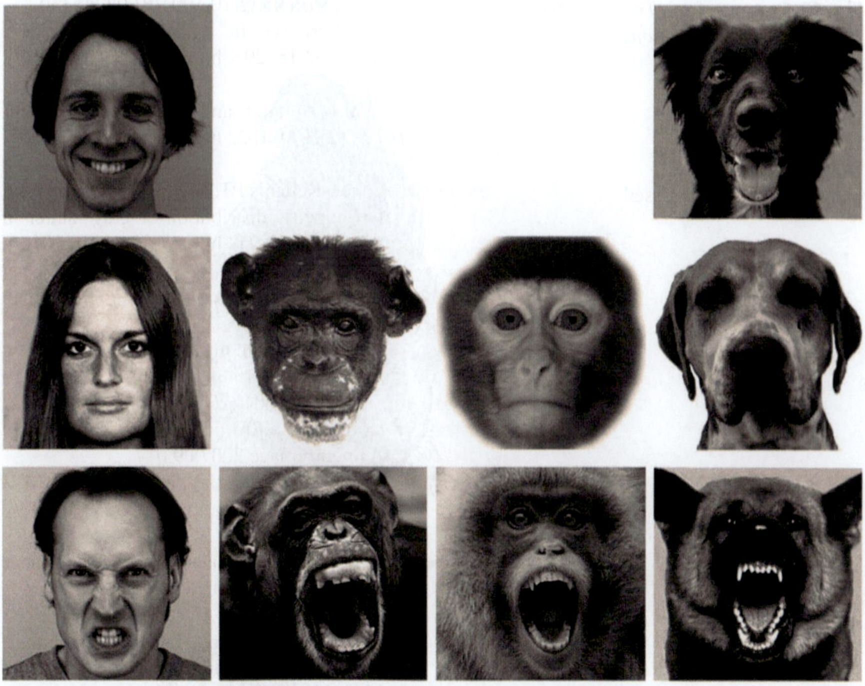

Fig. 13.5 From left to right: faces of humans, chimpanzees, monkeys and dogs, manifesting positive, relaxed/neutral, and negative expressions (after permission from Guo et al. Experimental Brain Research 2019;237(8): 2045–2059)

Fig. 13.6 Image from *The Expression of the Emotions in Man and Animals* by Charles Darwin (1872) Up: *Cynopithecus* with calm expression Down: *Cynopithecus*, pleased after been caressed

CHAP. V. MONKEYS.

Fig. 16. *Cynopithecus niger*, in a placid condition. Drawn from life by Mr. Wolf.

Fig. 17. The same, when pleased by being caressed.

13.4 Comparison of Rational and Emotional Processing

- The mechanism of emotions is evolutionary more ancient, and it is characterized by *speed and dominance*. It leads to fast decision-making on the basis of preexisting assumptions and by taking into consideration the general features of the situation. For example, fear is triggered almost immediately and almost before the conscious realization of the stimulus (automatically via non-conscious processes) and leads to the fast manifestation of adjustment behaviors (fight or flight). A snake will always cause fear, no matter whether it is poisonous or not.

- Decisions based on emotions are stronger than those based on reason, and when the two collide, the person faces a difficult dilemma, because it is very difficult for reason to overcome emotional pressure.
- On the contrary, reason is slow, demands the conscious process of stimuli, demands concentration and effort (it is laborious), and is not as dominant as emotions are in terms of the effect on behavior.
- Emotions manifest bias with a tendency to activate those behaviors which serve the survival of the individual and the species, while logical reasoning aims at a more objective assessment of the situation. In artificial intelligence terms, the best description would be that of a fuzzy vs. a digital system.
- The library of the assumptions subserving emotion is of unknown origin. It is probably partially inherited, is probably characteristic of the species, and is partially obtained through experience. Logical reasoning is mainly based on education.
- The two processes are essentially independent; however, they interact and have an effect on each other. The emotional state causes biases to intrude into logical reasoning, and rational analysis triggers affects, depending on the positive or negative result. This interaction is most probable to take place at multiple levels (e.g., selective memory recall, amplification through novel analysis, biased selection of possible solutions, etc.).

References

Defelipe J (2011) The evolution of the brain, the human nature of cortical circuits, and intellectual creativity. Front Neuroanat 5(29):29. https://doi.org/10.3389/fnana.2011.00029

Ekman P (1965) Differential communication of affect by head and body cues. J Pers Soc Psychol 2(5):726–735. https://doi.org/10.1037/h0022736

Ekman P (1980) Asymmetry in facial expression. Science 209(4458):833–834. https://doi.org/10.1126/science.7403851

Ekman P (1992a) Are there basic emotions? Psychol Rev 99(3):550–553. https://doi.org/10.1037/0033-295x.99.3.550

Ekman P (1992b) Facial expressions of emotion: an old controversy and new findings. Philos Trans R Soc Lond Ser B Biol Sci 335(1273):63–69. https://doi.org/10.1098/rstb.1992.0008

Ekman P (1993) Facial expression and emotion. Am Psychol 48(4):384–392. https://doi.org/10.1037//0003-066x.48.4.384

Ekman P (1994) Strong evidence for universals in facial expressions: a reply to Russell's mistaken critique. Psychol Bull 115(2):268–287. https://doi.org/10.1037/0033-2909.115.2.268

Ekman P (2003) Emotions inside out. 130 years after Darwin's "the expression of the emotions in man and animal". Ann N Y Acad Sci 1000:1–6. https://doi.org/10.1196/annals.1280.002

Ekman P (2009) Darwin's contributions to our understanding of emotional expressions. Philos Trans R Soc Lond Ser B Biol Sci 364(1535):3449–3451. https://doi.org/10.1098/rstb.2009.0189

Ekman P (2016) What scientists who study emotion agree about. Perspect Psychol Sci 11(1):31–34. https://doi.org/10.1177/1745691615596992

Ekman P, Friesen WV (1967) Head and body cues in the judgment of emotion: a reformulation. Percept Mot Skills 24(3):711–724. https://doi.org/10.2466/pms.1967.24.3.711

Ekman P, Friesen WV (1971) Constants across cultures in the face and emotion. J Pers Soc Psychol 17(2):124–129. https://doi.org/10.1037/h0030377

Ekman P, Sorenson ER, Friesen WV (1969) Pan-cultural elements in facial displays of emotion. Science 164(3875):86–88. https://doi.org/10.1126/science.164.3875.86

Ekman P, Friesen WV, O'Sullivan M, Chan A, Diacoyanni-Tarlatzis I, Heider K, Krause R, LeCompte WA, Pitcairn T, Ricci-Bitti PE et al (1987) Universals and cultural differences in the judgments of facial expressions of emotion. J Pers Soc Psychol 53(4):712–717. https://doi.org/10.1037//0022-3514.53.4.712

Fountoulakis KN, Giannakopoulos P, Kovari E, Bouras C (2008) Assessing the role of cingulate cortex in bipolar disorder: neuropathological, structural and functional imaging data. Brain Res Rev 59(1):9–21. https://doi.org/10.1016/j.brainresrev.2008.04.005

Guo K, Li Z, Yan Y, Li W (2019) Viewing heterospecific facial expressions: an eye-tracking study of human and monkey viewers. Exp Brain Res 237(8):2045–2059. https://doi.org/10.1007/s00221-019-05574-3

Jack RE, Garrod OGB, Yu H, Caldara R, Schyns PG (2012) Facial expressions of emotion are not culturally universal. Proc Natl Acad Sci U S A 109(19):7241–7244. https://doi.org/10.1073/pnas.1200155109

Phillips ML, Drevets WC, Rauch SL, Lane R (2003) Neurobiology of emotion perception II: implications for major psychiatric disorders. Biol Psychiatry 54(5):515–528. https://doi.org/10.1016/s0006-3223(03)00171-9

Sleep

14

Contents

14.1 General Comments on Sleep

An outdated and lay person's conceptualization and understanding of sleep suggests that wakefulness is actively maintained by sensory stimulation and when fatigue comes, together with a relative sensory deprivation, sleep occurs as a phase of recovery from the labor of the day. In religion and popular culture as well as in the arts, dreams have a prominent position as this has been expressed in the 1655 painting by Antonio de Pereda. The masterly way of placing objects with a deep allegorical meaning expresses in an excellent way the internal conflict in the conscious as well as at the unconscious level, which often appears through dreaming (Fig. 14.1).

However, in contrast to folk beliefs, sleep is neither a passive state nor it is the brother of death. On the contrary, it is an *active state of the brain*, clearly distinct from arousal but absolutely necessary for the continuation of proper brain functioning (Stenberg 2007; Weber 2017; Cirelli and Tononi 2008).

14.2 The Structure of Sleep

There is a *circadian periodicity little longer than 24 h* (close to 25 h) for sleep and arousal. The major role in the sleep-arousal circle is played by the *reticular formation*, whose anterior part contributes to arousal but its ventral-posterior to sleep

K. N. Fountoulakis, *The Human Connectome*,
https://doi.org/10.1007/978-3-031-10351-3_14

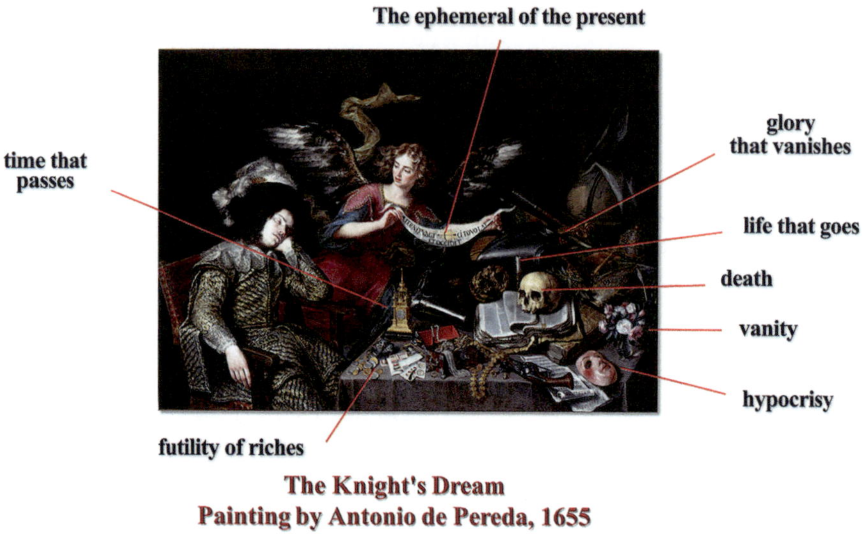

Fig. 14.1 Allegorical painting of sleep by Antonio de Pereda (1611–1678) created in 1655

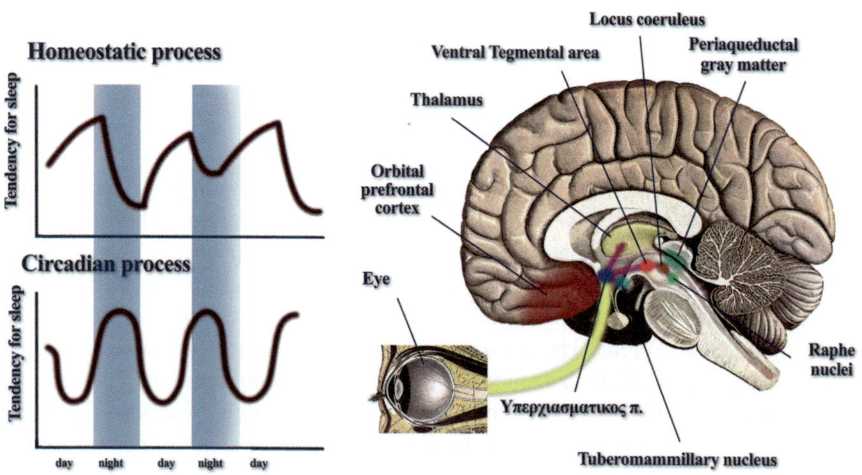

Fig. 14.2 The homeostatic and the circadian systems and the structures that play a role in the function of sleep (copyright KN Fountoulakis 2021)

(Fig. 14.2). Environmental temporal cues called *zeitgebers* (temporary pacemakers) serve the adjustment of the circle to the environment as reference points. The major environmental stimulus is sunlight and acts through the retino-hypothalamic tract on the major internal clock in the superchiasmatic nucleus of the anterior hypothalamus.

The average human sleeps approximately one third of his/her life, while the needs in duration as well as the architecture of sleep change with age. The average adult needs 6–8 h of quality sleep per 24 h, while at a later age, sleep is fragmented, and the need for sleep is decreased. On the contrary, in infants and small children, the need for sleep is greater (Reading 2013; Roffwarg et al. 1966; Parmelee Jr. 1961; Ehlers and Kupfer 1989; Iglowstein et al. 2003).

In descriptive terms, in humans, sleep has two distinct phases: The first is characterized by rapid eye movements (REM sleep) while the second is not (non-REM sleep). During sleep, the two phases alternate in a cyclic and extremely structured pattern, approximately 4–6 cycles, with 90 min duration for each period (Aserinsky and Kleitman 1953, 2003; Higgins and George 2007).

Non-REM sleep is the stage with which sleep starts and includes four distinct phases: phases 1 and 2 (light sleep) and phases 3 and 4 (deep sleep or deep-wave sleep in electroencephalographic terms). Phase 1 has the lowest awakening threshold, while phase 4 has the highest. Non-REM sleep EEG is characterized by sleep spindles, K-complexes, and slow high-voltage waves, with the deep-wave sleep being dominant during the first third of sleep in adults (Reading 2010; Carskadon and Dement 2017; Chokroverty 2013).

The *REM sleep* (also known as "paradoxical sleep") follows after one or more periods of non-REM sleep (Weber 2017), dominates the last third of sleep, and covers approximately 20% of total sleep duration, and it is the stage at which vivid dreams appear. It is characterized by loss of muscle tone, episodic bursts of rapid eye movements (REM), intense activation of the cortex and the limbic system, as well as EEG activation which is similar to that during wakefulness (Carskadon and Dement 2017; Hobson and Pace-Schott 2002; Reading 2010, 2013; Chokroverty 2013).

Sleep

- Not a passive state.
- Circadian periodicity little longer than 24 h (close to 25 h).
- Zeitgebers.
- Major external stimulus: sunlight.
- Non-REM sleep (80%).
  - Phases 1 and 2 (light sleep).
  - Phases 3 and 4 (deep sleep or deep-wave sleep in electroencephalographic terms).
- REM sleep ("paradoxical sleep," 20%).
  - Loss of muscle tone.
  - Episodic bursts of rapid eye movements (REM).
  - Intense activation of the cortex and the limbic system.
  - EEG activation similar to that during wakefulness.
- Circadian system (24-hour regulation of sleep-wakefulness).
  - Parts of the hypothalamus, mainly the suprachiasmatic nucleus.
  - Retinohypothalamic tract.
  - Basal part of frontal cortex.
  - Superior part of the brain stem (reticular formation).

- Ultradian system (alternation of REM and non-REM sleep).
  - Medial pontine reticular formation.
  - Thalamic nuclei.
  - Anterior brain.
- Homeostatic system (intensity, the quantity, and the duration of sleep).
  - Preoptic region of the hypothalamus.
- Two-process model of sleep (Process C and Process S).
- The universal presence of sleep in animals denoting a fundamental significance.
- The functional damage after sleep loss leading to death.
- Metabolism.
- Thermoregulation.
- Mental function.
- Dreams.
  - Organized in topics and are experienced as short stories.
  - Activation of regions in the brain stem and its connections with the hemisphere cortex and the limbic system.

14.3 Neurobiology of Sleep

Sleep regulation is based on a complex neuronal system which includes the retinohypothalamic tract, parts of the hypothalamus and the mesencephalon, the pons, as well as the brain stem (Fig. 14.2) (McGinty and Szymusiak 2017; Borbely and Achermann 1999; Economo 1930; Pace-Schott and Hobson 2002). This system has three functional subsystems:

- The *circadian system*, which includes parts of the hypothalamus, mainly the suprachiasmatic nucleus, the retinohypothalamic tract, and the basal part of frontal cortex and the superior part of the brain stem, and it is responsible for the regulation and the alternation of periods of alertness with periods of sleep. This system includes multiple subsystems which control alertness, and each one of them is activated relatively independently. More specifically, the reticular formation projects both to the thalamus and the hypothalamus but also to the basal ganglia, and in this way it exerts an indirect effect on the cortex (Jones 2003, 2005; McGinty and Szymusiak 2003; Zee and Manthena 2007; Alam et al. 2002; Deboer et al. 2003; Easton et al. 2004; Garcia-Rill 2002; Gerashchenko and Shiromani 2004; Koyama et al. 2002; Mendelson et al. 2003; Parmentier et al. 2002; Sakai and Crochet 2003; Takahashi et al. 2006; Vertes 2006; Datta 1995).
- The *ultradian system* is located in the medial pontine reticular formation, in the thalamic nuclei, and in the anterior brain, and it regulates the alternation of REM and non-REM sleep, mainly via cholinergic activity (Steriade and McCarley 1990; Tononi and Cirelli 2011; Hobson 1990).

- The *homeostatic system* is located in the preoptic region of the hypothalamus, and it is responsible for the regulation of the intensity, the quantity, and the duration of sleep.

Neurobiology of Sleep
- Circadian system.
- Ultradian system.
- Homeostatic system.
- Two-process model of sleep (Process C and Process S).

Essentially, the above systems reflect the mechanisms which have been proposed with the *two-process (C and S) model of sleep*. According to that model, the S process refers to a homeostatic factor which accumulates with wakefulness and is reduced with sleep, and it is the factor which presses toward sleep. It interacts with the C process which is the circadian and ultradian parameter (Borbely et al. 2016; Daan et al. 1984; Nicholson 2011; Dibner et al. 2010; Hattar et al. 2002; Welsh et al. 2010).

14.4 The Role of Sleep in Brain and Body Functions

While the precise role of sleep and the reason why all animals, humans included, need sleep (although exclusions and important differences do exist) is not known, *its universal presence in animals denotes a fundamental significance*. This significance is also proven by the fact that there is a rebound of sleep following its deprivation and also a rebound of slow-wave or REM sleep following the selective deprivation of these stages. Even more impressive is that the functional damage after sleep loss could lead to death in extreme cases.

The Role of Sleep
- Fundamental significance.
- Sleep debt.
- Neurocognitive integration.
- Dreams.

It is interesting that the total duration of sleep seems to remain *stable* from day to day, even under quite different circumstances. While the lay belief is that fatigue (both mental and physical) causes the need for prolonged sleep, this is not unequivocally supported by the data, although there are some evidence suggesting that mental or somatic fatigue indeed modify some parameters of sleep. The sense the lay person has is definitely that fatigue, both mental and physical, increases the need for sleep, making it deeper and more refreshing. However, the sole factor which beyond any doubt increases sleep is prior deprivation (*sleep debt*). There are accompanying parameters which make conclusions difficult, as the subjective experience of the person reflects a general body condition after effort and its restoration and not specifically the effect of sleep.

Although, during sleep deprivation, food intake and the creation of fatty adipose tissue increase, there are no data suggesting that energy saving is one of the functions of sleep. During sleep, metabolism is only 15% lower in comparison to wakefulness. However, the phenomenon of hibernation during winter in some mammals suggests the opposite. On the other hand, the increased food intake and the development of adipose tissue could be explained in the frame of the increased coexistent stress which is prominent during periods of sleep deprivation. Some data suggest that sleep has an effect on metabolism, thermoregulation, and cooling mechanisms, especially in mammals (Morrissey et al. 2004; Opp 2009; Van Cauter et al. 2008). But the above cannot explain the need of such functions (including physical rest) to take place in a way that leaves the animal vulnerable with its defenses down.

Additionally, it is important to note that rest without sleep leaves the individual sleepy. After several days of sleep deprivation, there are no significant physiological changes in the human body. However, it is interesting that there are mental deficits (Lansink et al. 2008; Lena et al. 2005; Maquet et al. 2003; Walker and Stickgold 2006). Even more interesting is that these deficits are reversible by medication but also by motives. This might mean that these deficits could be the result of pressure for sleep, rather than a physiological decline in mental function due to prolonged sleep deprivation; however, again the data are inconclusive. On the other hand, there is a bulk of literature arguing that during sleep there are plenty of neurocognitive activities which serve the further analysis, classification, and consolidation of memory registrations and their link to reward systems via affective connections, mainly through the activation of the hippocampus and the ventral part of the striatum and the emergence of dreaming (Gaillard and Moneme 1977; Lansink et al. 2009; Lena et al. 2005; Maquet et al. 2003; Merritt et al. 1994; Nielsen-Bohlman et al. 1991; Payne and Kensinger 2011; Perogamvros and Schwartz 2012; Schenck and Mahowald 2002; Schwartz et al. 2002; Solms 2001; Sterpenich et al. 2007).

14.5 Dreaming

Dreaming is another mental function during sleep with an unknown usefulness for the brain. Dreams are *organized in topics and are experienced as short stories*. A dream is more likely to be recalled when the person wakens during REM sleep (3/4 of awakenings) than during non-REM sleep (less than 10% of awakenings). During non-REM sleep, dreams could be shorter, but apart from that, there does not seem to be any difference among dreams in relationship to the phases of sleep (Rechtschaffen 1978, 1998; Frank 2006; Dement and Wolpert 1958; Hobson and Pace-Schott 2002), although during REM sleep they could be more bizarre and complex and much simpler and more realistic during non-REM sleep.

Their neurobiological substrate probably includes the activation of regions in the brain stem and its connections with the hemisphere cortex (mainly the prefrontal cortex) and the limbic system (Hobson and Pace-Schott 2002; Lee-Chiong 2020; Nir and Tononi 2010) (Fig. 14.2).

References

Alam MN, Gong H, Alam T, Jaganath R, McGinty D, Szymusiak R (2002) Sleep-waking discharge patterns of neurons recorded in the rat perifornical lateral hypothalamic area. J Physiol 538(Pt 2):619–631. https://doi.org/10.1113/jphysiol.2001.012888

Aserinsky E, Kleitman N (1953) Regularly occurring periods of eye motility, and concomitant phenomena, during sleep. Science 118(3062):273–274. https://doi.org/10.1126/science.118.3062.273

Aserinsky E, Kleitman N (2003) Regularly occurring periods of eye motility, and concomitant phenomena, during sleep. 1953. J Neuropsychiatry Clin Neurosci 15(4):454–455. https://doi.org/10.1176/jnp.15.4.454

Borbely AA, Achermann P (1999) Sleep homeostasis and models of sleep regulation. J Biol Rhythm 14(6):557–568. https://doi.org/10.1177/074873099129000894

Borbely AA, Daan S, Wirz-Justice A, Deboer T (2016) The two-process model of sleep regulation: a reappraisal. J Sleep Res 25(2):131–143. https://doi.org/10.1111/jsr.12371

Carskadon MA, Dement WC (2017) Normal human sleep. Principles and practice of sleep medicine. Elsevier, Philadelphia, PA. https://doi.org/10.1016/b978-0-323-24288-2.00002-7

Chokroverty S (2013) An overview of normal sleep. Sleep and movement disorders. Oxford University Press. https://doi.org/10.1093/med/9780199795161.003.0003

Cirelli C, Tononi G (2008) Is sleep essential? PLoS Biol 6(8):e216. https://doi.org/10.1371/journal.pbio.0060216

Daan S, Beersma DG, Borbely AA (1984) Timing of human sleep: recovery process gated by a circadian pacemaker. Am J Phys 246(2 Pt 2):R161–R183. https://doi.org/10.1152/ajpregu.1984.246.2.R161

Datta S (1995) Neuronal activity in the peribrachial area: relationship to behavioral state control. Neurosci Biobehav Rev 19(1):67–84. https://doi.org/10.1016/0149-7634(94)00043-z

Deboer T, Vansteensel MJ, Detari L, Meijer JH (2003) Sleep states alter activity of suprachiasmatic nucleus neurons. Nat Neurosci 6(10):1086–1090. https://doi.org/10.1038/nn1122

Dement W, Wolpert EA (1958) The relation of eye movements, body motility, and external stimuli to dream content. J Exp Psychol 55(6):543–553. https://doi.org/10.1037/h0040031

Dibner C, Schibler U, Albrecht U (2010) The mammalian circadian timing system: organization and coordination of central and peripheral clocks. Annu Rev Physiol 72(1):517–549. https://doi.org/10.1146/annurev-physiol-021909-135821

Easton A, Meerlo P, Bergmann B, Turek FW (2004) The suprachiasmatic nucleus regulates sleep timing and amount in mice. Sleep 27(7):1307–1318. https://doi.org/10.1093/sleep/27.7.1307

Economo CV (1930) Sleep as a problem of localization. J Nerv Ment Dis 71(3):249–259. https://doi.org/10.1097/00005053-193003000-00001

Ehlers CL, Kupfer DJ (1989) Effects of age on delta and REM sleep parameters. Electroencephalogr Clin Neurophysiol 72(2):118–125. https://doi.org/10.1016/0013-4694(89)90172-7

Frank MG (2006) The mystery of sleep function: current perspectives and future directions. Rev Neurosci 17(4):375–392. https://doi.org/10.1515/revneuro.2006.17.4.375

Gaillard JM, Moneme A (1977) Modification of dream content after preferential blockade of mesolimbic and mesocortical dopaminergic systems. J Psychiatr Res 13(4):247–256. https://doi.org/10.1016/s0022-3956(77)90020-6

Garcia-Rill E (2002) The sleep state-dependent midlatency auditory evoked P50 potential in various disorders. Thalamus Relat Syst 2(1):9–19. https://doi.org/10.1016/s1472-9288(02)00032-8

Gerashchenko D, Shiromani PJ (2004) Different neuronal phenotypes in the lateral hypothalamus and their role in sleep and wakefulness. Mol Neurobiol 29(1):41–60. https://doi.org/10.1385/mn:29:1:41

Hattar S, Liao HW, Takao M, Berson DM, Yau KW (2002) Melanopsin-containing retinal ganglion cells: architecture, projections, and intrinsic photosensitivity. Science 295(5557):1065–1070. https://doi.org/10.1126/science.1069609

Higgins ES, George MS (2007) Neuroscience of clinical psychiatry, the pathophysiology of behavior and mental illness, 1st edn. Lippincott Williams & Wilkins, Philadelphia, PA

Hobson JA (1990) Sleep and dreaming. J Neurosci 10(2):371–382. https://doi.org/10.1523/JNEUROSCI.10-02-00371.1990

Hobson JA, Pace-Schott EF (2002) The cognitive neuroscience of sleep: neuronal systems, consciousness and learning. Nat Rev Neurosci 3(9):679–693. https://doi.org/10.1038/nrn915

Iglowstein I, Jenni OG, Molinari L, Largo RH (2003) Sleep duration from infancy to adolescence: reference values and generational trends. Pediatrics 111(2):302–307. https://doi.org/10.1542/peds.111.2.302

Jones BE (2003) Arousal systems. Front Biosci 8(6):s438–s451. https://doi.org/10.2741/1074

Jones BE (2005) From waking to sleeping: neuronal and chemical substrates. Trends Pharmacol Sci 26(11):578–586. https://doi.org/10.1016/j.tips.2005.09.009

Koyama Y, Kodama T, Takahashi K, Okai K, Kayama Y (2002) Firing properties of neurones in the laterodorsal hypothalamic area during sleep and wakefulness. Psychiatry Clin Neurosci 56(3):339–340. https://doi.org/10.1046/j.1440-1819.2002.00975.x

Lansink CS, Goltstein PM, Lankelma JV, Joosten RN, McNaughton BL, Pennartz CM (2008) Preferential reactivation of motivationally relevant information in the ventral striatum. J Neurosci 28(25):6372–6382. https://doi.org/10.1523/JNEUROSCI.1054-08.2008

Lansink CS, Goltstein PM, Lankelma JV, McNaughton BL, Pennartz CM (2009) Hippocampus leads ventral striatum in replay of place-reward information. PLoS Biol 7(8):e1000173. https://doi.org/10.1371/journal.pbio.1000173

Lee-Chiong T (2020) Essentials of sleep medicine for the primary care provider. Sleep Med Clin 15(2):i. https://doi.org/10.1016/s1556-407x(20)30032-1

Lena I, Parrot S, Deschaux O, Muffat-Joly S, Sauvinet V, Renaud B, Suaud-Chagny MF, Gottesmann C (2005) Variations in extracellular levels of dopamine, noradrenaline, glutamate, and aspartate across the sleep--wake cycle in the medial prefrontal cortex and nucleus accumbens of freely moving rats. J Neurosci Res 81(6):891–899. https://doi.org/10.1002/jnr.20602

Maquet P, Schwartz S, Passingham R, Frith C (2003) Sleep-related consolidation of a visuomotor skill: brain mechanisms as assessed by functional magnetic resonance imaging. J Neurosci 23(4):1432–1440. https://doi.org/10.1523/JNEUROSCI.23-04-01432.2003

McGinty D, Szymusiak R (2003) Hypothalamic regulation of sleep and arousal. Front Biosci 8(6):s1074–s1083. https://doi.org/10.2741/1159

McGinty D, Szymusiak R (2017) Neural control of sleep in mammals. Elsevier, Principles and Practice of Sleep Medicine. https://doi.org/10.1016/b978-0-323-24288-2.00007-6

Mendelson WB, Bergmann BM, Tung A (2003) Baseline and post-deprivation recovery sleep in SCN-lesioned rats. Brain Res 980(2):185–190. https://doi.org/10.1016/s0006-8993(03)02896-8

Merritt JM, Stickgold R, Pace-Schott E, Williams J, Hobson JA (1994) Emotion profiles in the dreams of men and women. Conscious Cogn 3(1):46–60. https://doi.org/10.1006/ccog.1994.1004

Morrissey MJ, Duntley SP, Anch AM, Nonneman R (2004) Active sleep and its role in the prevention of apoptosis in the developing brain. Med Hypotheses 62(6):876–879. https://doi.org/10.1016/j.mehy.2004.01.014

Nicholson AN (2011) Prescription sleep medicine for aircrew. Aviat Space Environ Med 82(5):564–566. https://doi.org/10.3357/asem.2938.2011

Nielsen-Bohlman L, Knight RT, Woods DL, Woodward K (1991) Differential auditory processing continues during sleep. Electroencephalogr Clin Neurophysiol 79(4):281–290. https://doi.org/10.1016/0013-4694(91)90124-m

Nir Y, Tononi G (2010) Dreaming and the brain: from phenomenology to neurophysiology. Trends Cogn Sci 14(2):88–100. https://doi.org/10.1016/j.tics.2009.12.001

Opp MR (2009) Sleeping to fuel the immune system: mammalian sleep and resistance to parasites. BMC Evol Biol 9:8. https://doi.org/10.1186/1471-2148-9-8

Pace-Schott EF, Hobson JA (2002) The neurobiology of sleep: genetics, cellular physiology and subcortical networks. Nat Rev Neurosci 3(8):591–605. https://doi.org/10.1038/nrn895

Parmelee AH Jr (1961) Sleep patterns in infancy. A study of one infant from birth to eight months of age. Acta Paediatr 50(2):160–170. https://doi.org/10.1111/j.1651-2227.1961.tb08035.x

Parmentier R, Ohtsu H, Djebbara-Hannas Z, Valatx JL, Watanabe T, Lin JS (2002) Anatomical, physiological, and pharmacological characteristics of histidine decarboxylase knock-out mice: evidence for the role of brain histamine in behavioral and sleep-wake control. J Neurosci 22(17):7695–7711. https://doi.org/10.1523/JNEUROSCI.22-17-07695.2002

Payne JD, Kensinger EA (2011) Sleep leads to changes in the emotional memory trace: evidence from FMRI. J Cogn Neurosci 23(6):1285–1297. https://doi.org/10.1162/jocn.2010.21526

Perogamvros L, Schwartz S (2012) The roles of the reward system in sleep and dreaming. Neurosci Biobehav Rev 36(8):1934–1951. https://doi.org/10.1016/j.neubiorev.2012.05.010

Reading PJ (2010) Sleep disorders in neurology. Pract Neurol 10(5):300–309. https://doi.org/10.1136/jnnp.2010.224097

Reading P (2013) ABC of sleep medicine. Wiley-Blackwell, Chichester

Rechtschaffen A (1978) The single-mindedness and isolation of dreams. Sleep 1(1):97–109. https://doi.org/10.1093/sleep/1.1.97

Rechtschaffen A (1998) Current perspectives on the function of sleep. Perspect Biol Med 41(3):359–390. https://doi.org/10.1353/pbm.1998.0051

Roffwarg HP, Muzio JN, Dement WC (1966) Ontogenetic development of the human sleep-dream cycle. Science 152(3722):604–619. https://doi.org/10.1126/science.152.3722.604

Sakai K, Crochet S (2003) A neural mechanism of sleep and wakefulness. Sleep Biol Rhythms 1(1):29–42. https://doi.org/10.1046/j.1446-9235.2003.00004.x

Schenck CH, Mahowald MW (2002) REM sleep behavior disorder: clinical, developmental, and neuroscience perspectives 16 years after its formal identification in SLEEP. Sleep 25(2):120–138. https://doi.org/10.1093/sleep/25.2.120

Schwartz S, Maquet P, Frith C (2002) Neural correlates of perceptual learning: a functional MRI study of visual texture discrimination. Proc Natl Acad Sci U S A 99(26):17137–17142. https://doi.org/10.1073/pnas.242414599

Solms M (2001) Dreaming and REM sleep are controlled by different brain mechanisms. Sleep and dreaming. Cambridge University Press, Cambridge. https://doi.org/10.1017/cbo9780511615511.004

Stenberg D (2007) Neuroanatomy and neurochemistry of sleep. Cell Mol Life Sci 64(10):1187–1204. https://doi.org/10.1007/s00018-007-6530-3

Steriade M, McCarley RW (1990) Brainstem control of wakefulness and sleep. Springer, New York

Sterpenich V, Albouy G, Boly M, Vandewalle G, Darsaud A, Balteau E, Dang-Vu TT, Desseilles M, D'Argembeau A, Gais S, Rauchs G, Schabus M, Degueldre C, Luxen A, Collette F, Maquet P (2007) Sleep-related hippocampo-cortical interplay during emotional memory recollection. PLoS Biol 5(11):e282. https://doi.org/10.1371/journal.pbio.0050282

Takahashi K, Lin JS, Sakai K (2006) Neuronal activity of histaminergic tuberomammillary neurons during wake-sleep states in the mouse. J Neurosci 26(40):10292–10298. https://doi.org/10.1523/JNEUROSCI.2341-06.2006

Tononi G, Cirelli C (2011) The neurobiology of sleep. Neurobiology of mental illness. Oxford University Press, Oxford. https://doi.org/10.1093/med/9780199798261.003.0085

Van Cauter E, Spiegel K, Tasali E, Leproult R (2008) Metabolic consequences of sleep and sleep loss. Sleep Med 9 Suppl 1(0 1):S23–S28. https://doi.org/10.1016/S1389-9457(08)70013-3

Vertes RP (2006) Interactions among the medial prefrontal cortex, hippocampus and midline thalamus in emotional and cognitive processing in the rat. Neuroscience 142(1):1–20. https://doi.org/10.1016/j.neuroscience.2006.06.027

Walker MP, Stickgold R (2006) Sleep, memory, and plasticity. Annu Rev Psychol 57(1):139–166. https://doi.org/10.1146/annurev.psych.56.091103.070307

Weber F (2017) Modeling the mammalian sleep cycle. Curr Opin Neurobiol 46:68–75. https://doi.org/10.1016/j.conb.2017.07.009

Welsh DK, Takahashi JS, Kay SA (2010) Suprachiasmatic nucleus: cell autonomy and network properties. Annu Rev Physiol 72:551–577. https://doi.org/10.1146/annurev-physiol-021909-135919

Zee PC, Manthena P (2007) The brain's master circadian clock: implications and opportunities for therapy of sleep disorders. Sleep Med Rev 11(1):59–70. https://doi.org/10.1016/j.smrv.2006.06.001

Consciousness and Free Will 15

Contents

15.1 Theoretical Approach of Consciousness

The word *consciousness* comes from the Latin conscious, meaning "knowing, being aware" which in turn comes from conscire (con meaning "together" + scire meaning "know"). Probably this is a loan translation of the Greek word "*συνείδηση*" (sin-idisi) which comes from the verb "*σύνοιδα*" (meaning "I know well"). From the Latin scire also come the words "sense" and "science."

K. N. Fountoulakis, *The Human Connectome*,
https://doi.org/10.1007/978-3-031-10351-3_15

The Concept of Consciousness

- Most philosophers adopted a pseudo-emergentism and essentially a dualistic approach.
- Conceptualization as a state of awareness:
  - Subjectivity.
  - Unity.
  - Intentionality.
- It is not an all-or-none phenomenon.
- It puts the individual in a perspective which concerns the self in relationship with the place, time, and others.
- It probably exists also in less-developed life forms and even in pre-mammal species.

Consciousness has been the focus of philosophy concerning the mind, the self, and morality, and most philosophers essentially adopted a pseudo-emergentism (the whole cannot be reduced to its parts, mental functions cannot be reduced to neuronal interactions, and new functions emerge as a result of high complexity) and which deeply it is a dualistic approach (brain vs. soul-mind) which had been originally proposed for the first time by Rene Descartes (1596–1650, Fig. 1.18) in his books *Treatise of Man* (1647) and *The passions of the soul* (1649) (Descartes et al. 1984). However, modern developments in the field of neurosciences lead to the conclusion that *consciousness is fundamentally a brain function,* an approach which leads to the complete rejection of any metaphysical (or supernatural) belief or opinion concerning the understanding of mental life. But so far, there was no success in the efforts to identify neural mechanisms that give rise to consciousness (De Sousa 2013).

A basic approach to consciousness is its conceptualization as a *state of awareness*. Modern philosophers of mind, such as John Searle (born 1932) and Thomas Nagel (born 1937), ascribe three main features to awareness: subjectivity, unity, and intentionality.

- *Subjectivity* poses the greatest scientific challenge since it could be in direct conflict with the principles of conducting research and conceptualizing in a scientific way. It refers to the absolutely individualized experience of being conscious, an experience which is perceived as absolutely unique and different from the experience of any other individual, and it is impossible to communicate it completely and to be understood by another individual.
- *Unity* refers to the fact that experiences surface into consciousness as a unified whole, and all the various senses are merged into a single conscious experience of reality. There is unity in the experience, as the memories of the past and the wishes or the predictions for the future, together with the perception of the internal condition and of the external environment, have as common reference point the experience of "self," with which they are connected, and from which the internal mental events come from.

- *Intentionality* refers to the attribution of meaning beyond the moment of experience, that is, beyond "here and now" and beyond a simple reaction of the reflex type. Mental phenomena manifest a relationship to each other, and they occupy a place in the chronological history of the self, as they are placed in time series, which in turn create a plexus that travels in time. The relationship of the self with the external events has a central position in mental phenomena and the creating of the web of this plexus, thus giving orientation and a target to externalizing behaviors.

Consciousness is a very interesting neurobiological phenomenon. On one hand, it fulfills the criteria for *universality*, that is, it characterizes all individuals of the species but at the same time also of the complexity within the species as this was proposed by George Williams (Nichols and Grantham 2000), and, on the other hand, its presence definitely improves the collective adaptation and survival (Herron and Freeman 2013). It is possible, it constitutes the end result of a series of extremely adapting evolutional events in brain function (Eccles 1992), and probably it is *not an all-or-nothing phenomenon*. Its main aim is to put the individual in a perspective which concerns the self in relationship with the place, time, and others. Subsequently it is possible that it exists also in less-developed life forms and in a more primitive and incomplete form even in pre-mammal species (Baars 1993), with variable complexity and functionality (Gaulin and McBurney 2003).

15.2 Neurobiology of Consciousness

Self-esteem is not part of consciousness per se, but it reflects *self-awareness*, that is, the knowledge concerning self at a second level. In evolutionary terms, it is necessary as an assessment which individuals do, concerning their place in social hierarchy. While consciousness places the person in place, time, and the others (self), self-esteem places the person in relationship to others and in the *scale of social hierarchy* (self-environment interaction). One approach could be that it constitutes a kind of self-assessment in order to better select targets for resource allocation. The result is the so-called assortative mating, a term which refers to the mating with a person of the opposite sex but with similar qualities.

There is a debate concerning whether the brain is capable of fully understanding itself and, as a consequence, whether consciousness could be scientifically studied. Indeed, some philosophers, e.g., Colin McGinn (born 1950), believe that there are limits to human cognitive capacities which preclude the comprehension of what consciousness really is, while others like Searle and Nagel disagree. A third approach, suggested by the Nobel laurate Gerald Maurice Edelman (1929–2014), is that consciousness is simply the outcome of the computational workings of the association areas of the brain, and in this frame it could be the focus of scientific research (Edelman 1978, 1989; Edelman and Changeux 2001; Edelman 1992); however this statement does not answer the question pertaining to understanding.

Neurobiology of Consciousness

- Self-esteem.
  - It is not part of consciousness per se.
  - It reflects self-awareness.
  - It is the assessment which individuals do, concerning their place in social hierarchy.
- Is the brain capable of fully understanding itself?
- Could consciousness be scientifically studied?
- Astronomical scale of the complexity concerning the functional anatomy and physiology of the brain.
- Limited number of experimental trials we can conduct.
- Limited number of subjects and observations we can gather.
- The study of these phenomena is impossible with the standard reductionist approach science normally follows.

The great practical problem is the *astronomical scale of the complexity* concerning the functional anatomy and physiology of the brain. A measure of this complexity could be shown by the following: Let's take not the whole human brain but only one of the two optic nerves. This part of the nervous system contains between 770,000 and 1.7 million nerve fibers (Jonas et al. 1992), and to make calculations easy, let's assume it is 1 million. Now if each of these neurons can take only two distinct states (e.g., yes/no, fire/silent, etc.) which is a very simplistic approach if one takes into consideration the neuronal function, then the total number of states the optic nerve can take is $2^{1,000,000}$. This number is larger than $10^{250,000}$ which should be compared with 10^{80}–10^{97} which correspond to estimations of the number of all elementary particles of the visible universe.

The scale of complexity described above, considered together with the limited number of experimental trials we can conduct, and the limited number of subjects and observations we can gather, *makes impossible the study of these phenomena* with the standard reductionist approach we currently can afford which science normally follows when studying events.

Currently, we lack the tools, both theoretical and also operational, to dig deeper into these mechanisms and to understand them, far less to extrapolate subjective properties (consciousness) from the properties of objects (interconnected nerve cells).

While, eventually, we do have a general idea on how the brain works, from the atomic and the molecular level to the level of social behavior and the abstract expression of ideas, we lack the specific knowledge that bridges the different steps and levels in this complete system. We lack the precise knowledge on how the incoming sensory information is broken, coded, and stored and how a complex process leads to decision-making. We lack the knowledge on how electrical currents and chemical reactions, through an extremely organized super-complexity, produce consciousness, the sense of individuality, and, even more important in humans, the realization of the inevitable future death. This realization and the fear that follows are, according to many philosophers, the driving force of creativity (Fig. 15.1).

Fig. 15.1 The realization and the subsequent fear of death are the driving forces for creativity. Picture of a 1483 copy for Pope Sixtus IV of the book "De rerum natura" (On the nature of things), by Titus Lucretius Carus (c. 99-55 B.C.), Vatican Library

15.3 The Concept of Free Will

The concept of *free will* concerns an old debate on the very nature of human existence. While a basic definition could be that it refers to the ability of humans to choose between different possible actions unimpeded (Omoregie 2015), the details and the essence of this human characteristic remain obscure. The fundamental existential question *whether humans have real control over their actions* and to which extent exists since antiquity and occupied the thoughts of Plato, Aristotle, Epictetus, and especially of the Stoics with conflicting opinions (Nagel 1989; Searle 2013; Bobzien 1998). Several philosophers, including Arthur Schopenhauer (1788–1860), base their concept of free will on the ability of the mind to make decisions based on reason, thus implying that passions are those that limit free will (Schopenhauer 1960).

Therefore, the concept of free will has emerged more as a *social and theological/religious need* to define the responsibilities of the individual toward the society and God, rather than as the product of scientific curiosity. Its roots are social and emerged so as to serve the social needs and more specifically to determine the relationship of the individual with the other members of the society, its rights and obligations as obedience to the rules of the community, and with a subsequent punishment in case these rules are violated. It is essentially a criterion whether corrective measures should be applied or what other action from the side of the society is considered to be more appropriate. Lay person opinion is clearly supportive of the presence of free will, since this is what is in accord with the everyday human internal experience (Caruso 2012; Lamont 1969). This is probably the result of the mere consciousness of behavior, although the process of decision is often more or less "automatic" and subconscious (Spinoza 2017). Essentially it reflects a false sensation or seeming experience (Hume 1999; Clark 1999).

Such a *social construct* is unlikely to fit within the rules of scientific discussion, and subsequently it might not be possible to give a scientific definition or explanation concerning what "free will" stands for.

Free Will

- The ability of humans to choose between different possible actions unimpeded.
- Existential question whether humans have real control over their actions.
- The ability of the mind to make decisions based on reason, which implies that passions are those that limit free will.
- Social and theological/religious need to define the responsibilities of the individual toward the society and god, rather than as the product of scientific research.
- At the dangerous border between morality and disease, morality definitely prerequisites the existence of free will.
- It is impossible to create a system of morality without the concept of free will.
- The very existence of society per se is based on this.

The problem in the relationship between mind-brain and free will, their significance, their historical roots, and their importance for philosophy, religion, and the society is all extremely important for psychiatry and psychology. It is important for all professionals working in mental health to understand mental phenomena, as well as the limits of their understanding and the consequences of these limits pertaining to the way we perceive normal brain function as well as mental disorder and the way we manage the patients suffering from them and their families. However, it is also extremely important to *know our own limits* when we try to understand behaviors which are often unusual but do not come from individuals with a mental disorder per se. In other words, our understanding stands at the *dangerous border between morality and disease*, which according to Plato does not exist, since he thought that no healthy human can be immoral and vice versa.

The law often demands the assessment of capacities related to it and often in a bimodal (yes/no) way and in the frame of forensic psychiatry (Stone 2008). Free will is located at the core of morality. At the same time, both in lay culture and in philosophical thinking, free will is more or less taken for granted, and it is placed at the core of morality. *Morality prerequisites the existence of free will*, that is, persons are responsible for their actions, and therefore their actions characterize them and are also the source of important positive or negative consequences from the side of the others as individuals but also from the society as a whole (Morse 2007; Van Inwagen 1986; Pereboom 2009). It is impossible to create a system of morality without free will (Ravven 2013). To be more pragmatic, the whole societal structure of moral responsibility, law and order, reward and punishment, and even the very societal structure itself relies on the assumption that there is accountability because of the existence of free will during decision-making and choice making.

15.4 Determinism Vs. Randomness

At the roots of the mind-brain problem is the question whether nature and its laws are deterministic or not. This has been put through as the core issue by William James (Viney 1986). However, what exactly could the alternative to determinism be and what consequences this could have for the topic under discussion is elusive, since either randomness or any condition which determines the response by disengaging it from the stimulus does not imply the existence of free will per se but simply a dependence on other processes, or on the contrary a complete randomness, whose relationship with free will has not even been discussed yet.

Determinism Vs. Randomness

- Determinism and randomness in the laws of nature:
  - Both do not support the existence of free will.
  - The essence of free will does not lie in the complete independence from the stimulus or the needs.
- In spite of its complexity, the natural world has been considered to be a deterministic system.
- Modern physics leaves space for the existence of randomness.
- True randomness is not identical with inability to predict.

The basic assumption of determinism is that for each event there is some theory or system of laws such that the occurrence of that event is derivable from those laws together with initial conditions (Berofsky 1995). It suggests that only one course of events is possible. The question was of course whether nature constitutes a deterministic system. Philosophy denies considering the world through this prism. The view scientists had on the characteristics of natural laws was until recently based on the advances of the seventeenth, eighteenth, and nineteenth centuries and especially on the work of Sir Isaac Newton (1643–1727; Fig. 15.2) and the elaboration with the development of specific mathematical functions to calculate the variables related

Fig. 15.2 Left: Sir Isaac Newton (1643–1727). Right: Pierre-Simon, marquis de Laplace (1749–1827)

to physical events by Pierre-Simon, marquis de Laplace (1749–1827; Fig. 15.2). According to this view, the nature is deterministic with strict laws which mediate the cause-result effect (Laplace 1995). The same cause would always produce the same result. In spite of its complexity, the world was viewed as a deterministic system.

An allegory of this concept could be a chess match; no matter how complex could be and no matter how enormous the variants and options to play, at the end of the day, there is a finite number of combinations (though astronomical in magnitude), and with players with a sufficient process power (like modern mainframe computers), the chess game collapses into the question which player will make the first move; this player will eventually win. Although, for the average human, chess seems to be a very creative game to play, in reality it is absolutely deterministic. Since 1997 when Deep Blue, a giga-computer of that time, defeated for the first time the world champion Garry Kasparov, computers have a stable superiority vs. human players with an increasing lead as years pass (Fig. 15.3). The impressive thing is not the superiority of computers in this particular field—the opposite is, meaning that the human brain with a very limited capacity to perform arithmetic calculations (see Chap. 11.6) is able to achieve such a level of efficacy on the basis of analyses of essentially unknown nature. These analyses are probably based on *heuristics* and are utilized in order to arrive at fast decisions when faced with complex and demanding problems and in an "economic" way. Heuristics are cognitive shortcuts, and the whole method is based on the comparison of the current situation the individual faces with a template situation which is already known. This often demands a simplified and straightforward interpretation of a complex reality and the

Fig. 15.3 Photo from the historical victory of the "Deep Blue" computer versus Garry Kasparov on May 11, 1997

development of abstract mental representations which are generalized at a second phase and are called schemata. These schemata organize knowledge and channel the information processing, but often they lead to the development of a generalized network of beliefs concerning groups of people or situations. If this comparison leads to a successful solution, then the schema leads to adaptation. If not, it leads to an error or to a problematic stereotype and bias.

Throwing dice, on the other hand (Fig. 15.4), produces "random" results, impossible to predict and determine, but this is largely because you need sophisticated equipment to monitor the variables determining the action of throwing the dice and calculating speed and trajectory. At the end of the day, playing is impossible since everything could be pre-calculated, predicted, and "determined."

A radical shift in the understanding of nature came with quantum mechanics which suggested an "exotic" view of nature, not compatible with the everyday human experience, but quantum non-deterministic randomness, if it exists at all, is just as bad as determinism for free will (Suarez and Adams 2013). At any case, there is the deterministic approach championed by Albert Einstein (1879–1955; Fig. 15.5) and also in accord with the ideas of Charles Darwin and Stephen Hawking (Barlow 1958; Einstein 2006; Hawking 1994). The famous phrase by Einstein that *God does not throw dice* expresses the idea that hidden variables and laws explain any situation in a deterministic way. This means that what we observe and perceive as random is simply the result of our gap in knowledge and of our inability to calculate with precision the result because of high complexity. In sharp contrast, there is the Copenhagen interpretation proposed by Niels Bohr (1885–1962; Fig. 15.5) and Werner Heisenberg (1901–1976) which suggests that physical systems have only probabilities not specific properties.

If for a moment it is hypothesized that nature includes phenomena which are characterized by randomness, and mental phenomena are based on interaction which are partially random in character, even this idea does not add anything to the debate. The rejection of determinism leads to a partial randomness in decision-making, but the essence of free will is not the complete independence of the decision from the stimulus or the needs. Indeed, randomness would have been the

Fig. 15.4 "Randomness" vs. "determinism": dice and chess

Fig. 15.5 Niels Bohr (left) with Albert Einstein (right) in December 1925 at the house of Paul Ehrenfest in Leiden

answer if there was a need for compete disengagement between the stimulus and the response, but such a decision would have been also coercive (in a different way), and in no case it would be the expression of free will.

15.5 Physics, Mind, and the Brain

The most popular solutions to the determinism problem were built upon the traditional dualistic approach and contain *meta-physical or super-natural elements* by suggesting the existence of a "mind" which defies natural laws (Sluyter 2003).

On the other hand, some authors suggested that "exotic" quantum physics would serve for the understanding of the way the human brain functions, although it is doubtful they understand sufficiently either of them or what kind of real results could precisely come from a combined system concerning the issue of mental function and free will.

Physics, Mind, and the Brain

- Authors suggested that "exotic" quantum physics would serve for the understanding of the way the human brain functions.
- They suggest that genuine indeterminacy at the quantum level and at the right places could play a crucial role.
- In an "interplay between chance and lawfulness in the brain," "non-Turing machines" could probably serve as an operational model of the brain, but it is highly unlikely that they will serve the solving of the "free will" riddle.
- The models based on advanced and hypothetical physics do not seem to add anything strong or important to the debate.

They suggest that genuine indeterminacy at the quantum level and at the right places could play a crucial role (Gisin 2013; Acin and Masanes 2016; Suarez 2013; Brassard and Raymond-Robichaud 2013; Koch 2009), and in the words of Martin Heisenberg, they will constitute "an interplay between chance and lawfulness in the brain" (Heisenberg 2013). One of the most important such theories is that of Sir Roger Penrose (born 1931) and Stuart Hameroff (born 1947) who independently developed theories to link consciousness to quantum events. They suggest that quantum fluctuations in the brain could be amplified by deterministic chaos and lead to behavioral choices which could be not predictable (Penrose 1989, 1996; Penrose 1994; Penrose and Hameroff 1995; Hameroff and Penrose 1996). Their ORCH-OR (orchestrated objective reduction) theory locates these quantum events in the microtubules of the neurons (Penrose and Hameroff 1995). However this model is based on hypothetical physical theories that do not yet exist and assumes local quantum effects; we don't know they exist, and even if they do exist, their relationship to major mental events is unknown, and there is no clue on how to connect them (Grush and Churchland 1995).

Taken all the above together, the author's opinion is that *non-Turing machines* could probably serve as an operational model of brain function, but it is highly unlikely they will serve the solving of the "free will" riddle, simply because their functioning is also guided by rules and limitations. Additionally, the supporting models based on advanced and hypothetical physics do not seem to add anything strong or important to the debate.

15.6 Neurobiology and Free Will

In terms of experimental psychology, the most discussed experiment on free will was conducted by Benjamin Libet (1916–2007) and his colleagues in the 1980s (Libet et al. 1983; Libet 2010), but today we know that its results are limited if any value at all, since they are based on superficial and simplified assumptions which are full of conceptual mistakes which were based on the poor knowledge of the time concerning the way the brain works.

Neurobiology and Free Will

- The discussion in terms of physics and biology does not fit with the deeper essence of the concept of free will.
- Fundamentally of social origin, meaning, value, and functioning.
- Demands and entity which:
  - Is based on a code of values which is independent of bodily needs.
  - Has the absolute capacity to arrive at decisions and to execute actions.
- Three key points that cannot be neglected at all:
  - Human mental functions depend absolutely on the status of the brain.
    Maturity (developmental stage).
    Health.
    Natural degeneration (aging).
    The effect of psychoactive substances.
  - There are very serious proof concerning the role of the brain but none concerning the existence of a meta-physical or supernatural dimension of the mind.
- "Free will" or "choice based on reason":
  - End result of an astronomical number of natural interactions.
  - The scale of complexity makes impossible the study of these phenomena with the standard scientific reductionist approach.
  - Each moment in time is objectively and not poetically unique and resembles no other before or after.
  - Each combination of the values of an infinite number of variables is unique and will never exist again.
  - Every experience of decision is new, and clearly it is different from all the previous decisions.
- The uniqueness in experience creates the subjective experience of "choice" and gives birth to the idea of "free will".

The fundamental conceptual problem when the relationship between neurobiology and free will is to be discussed is definitely the fact that, irrespective of whether our world is deterministic or not, the discussion in terms of biology and physics is not relevant with the deeper essence of the concept of free will, which has a fundamentally social origin, meaning, value, and functionality.

This deeper concept of free will more or less demands an entity which is based on a code of values which is independent of bodily needs of any kind, and this entity has the absolute capacity to arrive at decisions and to execute actions on the basis of that code, even in sharp and complete contrast with the needs of the body, brain function included. And this is because to the extent it depends on brain function, to the same extent it is bound and not free.

The question how it is possible and so complex living systems like the brain came to evolve from simple organic molecules, and they are beyond our capacity to understand, remains unanswered, and it is linked to the phenomenon of life. To be precise, it is exactly the same question.

Although, on the basis of the experiments by Stanley Miller and Harold Urey in 1952, we know a possible way through which organic molecules emerged from inorganic matter, the processes that led to life are a completely different story, they are absolutely unknown, and all the hypotheses proposed so far are rather weak. The most basic and unanswered question concerns the wider mechanism via which the phenomenon of life manages to overcome the tendency of the universe toward an increased entropy and destruction of order and organization. The clarification of these mechanisms will constitute a dramatic progress, also concerning our efforts to understand how the human brain works.

The possibility that life came from processes that occurred by chance has been calculated as a probability so low that it would have taken approximately 10^{50} years, while the age of the universe is only 10^{10} (13.8 billion) years. Such a view suggests that life on earth is an exception at the sideway of natural laws; it emerged only once, in one point, and it is certain it hasn't emerged anywhere else, both in the visible and the invisible universe. The main argument against the emergence of life by chance is the fact that the age of the earth is only 4.5 billion years and life had emerged already 3.5 billion years ago, probably even much earlier, that is, almost immediately with the establishment of environmental conditions which were very close to contemporary ones.

On the other hand, there are at least three key points that cannot be neglected at all:

- The first is that the human mental functions, including that of the capacity to arrive to mature decisions, depend absolutely on the *status of the brain*, that is, the maturity (lower capacity during infancy/childhood), health (brain disorders, e.g., stroke affects it), and also natural degeneration (aging).
- The second concerns the *effect psychoactive substances have* (e.g., alcohol) on decision-making per se, as well as on behavior, and the most interesting of all is that this phenomenon does not have an "all-or-none" character.
- The third point is that while on one hand there are very strong clues and sometimes evidence that the human mental function is based and maybe it is exhausted in events inside the brain, which follow the laws of nature, *there is no scientific sign of any kind* that there is some super-natural aspect of the mind. The question whether supernatural processes can be perceived via physical means and functions, since the moment they are being perceived they fall into the domain of nature, is a philosophical question which cannot be answered.

After taking into consideration the above, and as we are definitely obliged to accept an exclusively neurobiological approach to mind, the behavior, the internal experience, and free will, one should also reply to the questions and the *contradictions that emerge*, especially concerning free will, and have led philosophical thinking to a dead end since centuries (Fig. 15.6).

As our world is most likely deterministic in deep principle, and although this probably is unimportant, the experience of "free will" or more appropriately of *choice based on reason* is probably the result of the complexity of the function of the human mind. Each decision is the end product of an astronomical number of

Fig. 15.6 As far as it concerns "free will," philosophical thinking faces a definite dead end since centuries

material interactions of unimaginable complexity. A measure of this complexity can be shown with the following example concerning the optic nerve, which has already been described above; however it is important to work with it again in more depth and to further discuss some particularly important details.

Let's take again only one of the two optic nerves. This part of the nervous system contains between 770,000 and 1.7 million nerve fibers (Jonas et al. 1992), and to make calculations easy, let's again assume it is 1 million (Fig. 15.7). Now if each of these neurons can take only two distinct states (e.g., yes/no, fire/silent, etc.) which is a very simplistic approach if one takes into consideration the neuronal function, then the total number of states the optic nerve can take is $2^{1,000,000}$. This number is larger than $10^{250,000}$ which should be compared with 10^{80}–10^{97} which correspond to estimations of the number of all elementary particles of the visible universe. Going deeper in this model, we should note that the human optic nerve could carry signals reflecting the perception of almost ten million different colors. Interestingly the wealth in color perception increases only 23 times ($2^{23} = 8,388,608$) the complexity of the system in comparison to black and white vision. This is unexpected for the naïve reader, but nonlinear mathematics often manifest similar surprises. Essentially, these mathematics are more complex variants of *chessboard mathematical problems* and suggest that the complexity of the network rather than the internal variability of its units contribute more to the resulting overall complexity of the system. This simply means that the interplay among neurons rather than the processes inside the neurons is what determines the magnitude of complexity. This observation somewhat justifies the ideas of property dualism and functionalism, but not to their full extent. It also implies that those substances acting on the connections among neurons might have a stronger effect on brain function in comparison to those with an intracellular effect.

The scale of complexity described above, considered together with the limited number of experimental trials we can conduct, and the limited number of subjects

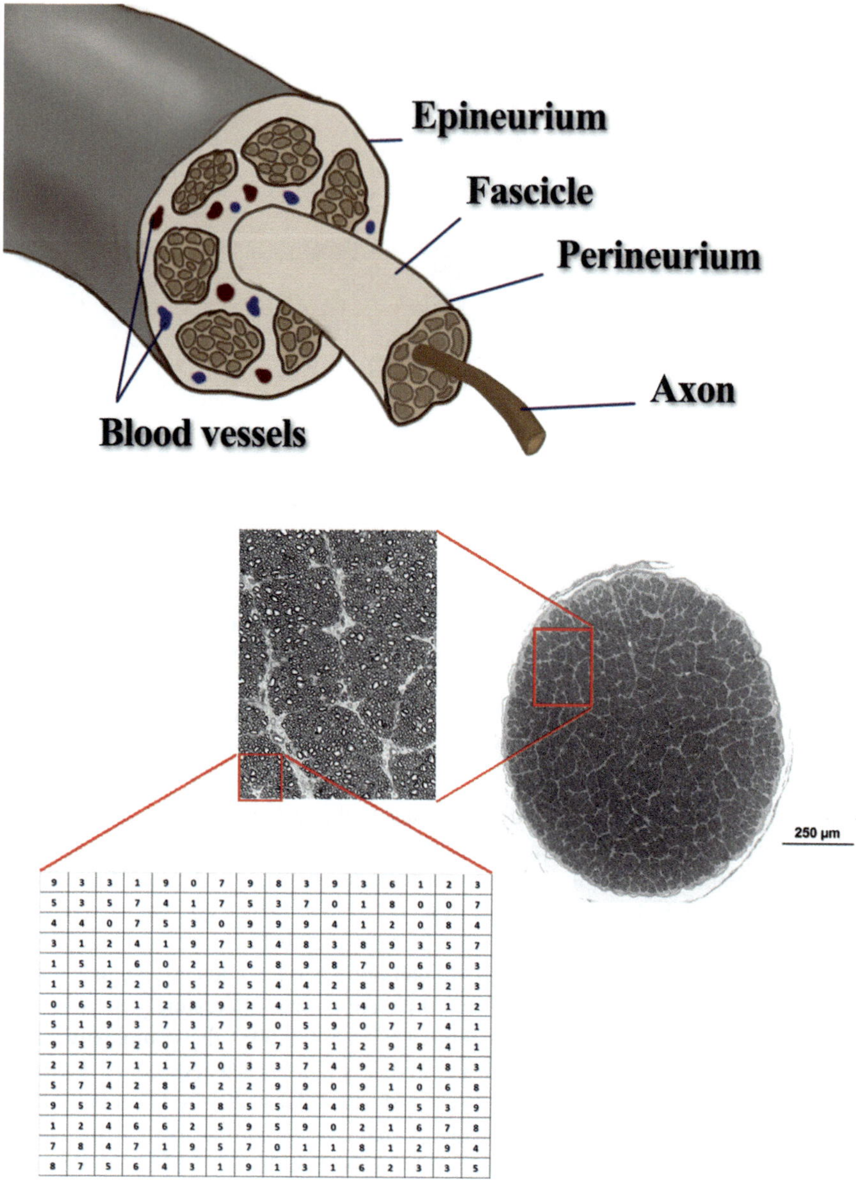

Fig. 15.7 Cross-section of nerve and reference to a "chessboard problem". (Microscope image by Whiting et al. (2014) Experimental Eye Research. 125:164-172)

and observations we can gather, makes impossible the study of these phenomena with the standard reductionist approach we can currently afford, which science normally follows when studying events. Currently we lack the tools, both theoretical and operational, to dig deeper into these mechanisms and to understand them, far less to extrapolate subjective properties (consciousness) from the properties of objects (interconnected nerve cells).

But coming back to the human subjective experience, *no interaction and no product of interactions in the brain are identical to a previous or to a future one*. This is not only a poetic expression, but it also absolutely reflects reality. Each moment in time is objectively and not poetically unique and resembles no other before or after. Each combination of the values of an infinite number of variables is unique and will never exist again.

In this sense, every experience of decision is new, and clearly it is different from all the previous decisions. This is not only in the cosmic sense but also at the practical-psychological level. Each time I make a decision concerning the "same" situation (which is never the same as explained before but more or less it is experienced as "similar"), I always consider past experiences and choices. The different constellation of situations this experience of decision-making is based on makes it unique and also makes it informed in a much different way from past ones. This uniqueness in experience creates the *subjective experience of "choice" and gives birth to the idea of "free will."*

This suggestion is in absolute accord with our current knowledge concerning the way the human brain works, and it is probably the only such existing hypothesis. It has been proposed for the first time by Ludwig Wittgenstein (1989–1951, Fig. 15.8) who in his early writings considered consciousness and free will as mere epiphenomena, a kind of "illusions" which emerge from the way the human brain works as a subjective experience (Wittgenstein 1958; Bolton 1979; Wittgenstein 1969). In his

Fig. 15.8 Ludwig Wittgenstein (1989–1951)

later writings, he was less clear on this matter. However, further discussion on the philosophical debate concerning, e.g., *eliminativism* or emergentism, is beyond the aims of the current book.

References

Acin A, Masanes L (2016) Certified randomness in quantum physics. Nature 540(7632):213–219. https://doi.org/10.1038/nature20119

Baars B (1993) Cognitive theory of consciousness. Cambridge University Press

Barlow N (1958) The autobiography of Charles Darwin 1809–1882. With the original omissions restored. Edited and with appendix and notes by his grand-daughter Nora Barlow. Collins, London

Berofsky B (1995) Determinism. In: Audi R (ed) The Cambridge dictionary of philosophy. Cambridge University Press, Cambridge, pp 199–200

Bobzien S (1998) Determinism and freedom in stoic philosophy. Oxford University Press, Clarendon

Bolton D (1979) An approach to Wittgenstein's philosophy. The Macmillan Press LtD, London

Brassard G, Raymond-Robichaud P (2013) Can free will emerge from determinism in quantum theory? In: Suarez A, Adams P (eds) Is science compatible with free will? Springer, New York, NY

Caruso G (2012) Free will and consciousness: a determinist account of the illusion of free will. Lexington Books, Landham, MD

Clark TW (1999) Fear of mechanism. A compatibilist critique of "The Volitional Brain". J Conscious Stud 6(8-9):279–293

De Sousa A (2013) Towards an integrative theory of consciousness: part 1 (neurobiological and cognitive models). Mens Sana Monogr 11(1):100–150. https://doi.org/10.4103/0973-1229.109335

Descartes R, Cottingham J, Stoothoff R, Murdochtrans D (1984) The philosophical writings of René Descartes (1641). Cambridge University Press, Cambridge

Eccles JC (1992) Evolution of consciousness. Proc Natl Acad Sci U S A 89(16):7320–7324. https://doi.org/10.1073/pnas.89.16.7320

Edelman G (1978) The mindful brain: cortical organization and the group-selective theory of higher brain function. MIT Press

Edelman G (1989) The remembered present: a biological theory of consciousness. Basic Books, New York

Edelman G (1992) Bright air, brilliant fire: on the matter of the mind. Basic books, New York

Edelman G, Changeux J (2001) The brain. Transaction Publishers, Somerset, NJ

Einstein A (2006) The world as I see it. Filiquarian Publishing

Gaulin S, McBurney DH (2003) Evolutionary psychology. Prentice Hall, Upper Saddle River, NJ

Gisin N (2013) Are there quantum effects coming from outside space–time? Nonlocality, free will and "no many-worlds". In: Suarez A, Adams P (eds) Is science compatible with free will? Springer, New York, NY

Grush R, Churchland PS (1995) Gaps in Penrose's toiling. J Conscious Stud 2(1):10–29

Hameroff SR, Penrose R (1996) Conscious events as orchestrated space-time selections. J Conscious Stud 3(1):36–53

Hawking S (1994) Black holes and baby universes and other essays. Bantam, New York

Heisenberg M (2013) The origin of freedom in animal behaviour. In: Suarez A, Adams P (eds) Is science compatible with free will? Springer, New York, NY

Herron J, Freeman S (2013) Evolutionary analysis, 5th edn. Pearson, Hoboken, NJ

Hume D (1999) Enquiry concerning human understanding. In: Tom L (ed) Beauchamp. Oxford University Press, Oxford/New York

Jonas JB, Schmidt AM, Muller-Bergh JA, Schlotzer-Schrehardt UM, Naumann GO (1992) Human optic nerve fiber count and optic disc size. Invest Ophthalmol Vis Sci 33(6):2012–2018
Koch C (2009) Free will, physics, biology, and the brain. In: Murphy N, Ellis G, O 'Connor T (eds) Downward causation and the neurobiology of free will. Springer Verlag, Berlin, pp 31–52
Lamont C (1969) Freedom of choice affirmed. Beacon Press, Boston
Laplace P (1995) Pierre-Simon Laplace philosophical essay on probabilities (Translated from the fifth French edition of 1825 with notes by the translator). Springer-Verlag, New York
Libet B (2010) Unconscious cerebral initiative and the role of conscious will in voluntary action. Behav Brain Sci 8(4):529–539. https://doi.org/10.1017/s0140525x00044903
Libet B, Gleason CA, Wright EW, Pearl DK (1983) Time of conscious intention to act in relation to onset of cerebral activity (readiness-potential). The unconscious initiation of a freely voluntary act. Brain 106(Pt 3):623–642. https://doi.org/10.1093/brain/106.3.623
Morse SJ (2007) The non-problem of free will in forensic psychiatry and psychology. Behav Sci Law 25(2):203–220. https://doi.org/10.1002/bsl.744
Nagel T (1989) Freedom. Oxford University Press, The View From Nowhere
Nichols S, Grantham T (2000) Adaptive complexity and phenomenal consciousness. Philos Sci 67(4):648–670
Omoregie J (2015) Freewill: the degree of freedom within. AuthorHouseUK, Bloomington, IN
Penrose R (1989) The emperor''s new mind: concerning computers, minds, and the laws of physics. Oxford University Press, New York, NY
Penrose R (1994) Mechanisms, microtubules, and the mind. J Conscious Stud 1:241–249
Penrose R (1996) Shadows of the mind: a search for the missing science of consciousness. Oxford University Press
Penrose R, Hameroff S (1995) What 'Gaps'? Reply to Grush and Churchland. J Conscious Stud 2:99–112
Pereboom D (2009) Living without free will. Cambridge studies in philosophy. Cambridge University Press, Cambridge. https://doi.org/10.1017/cbo9780511498824
Ravven H (2013) The self beyond itself: an alternative history of ethics, the new brain sciences, and the myth of free will. The New Press, New York
Schopenhauer A (1960) Essay on the freedom of the will. Dover Publications
Searle J (2013) "The problem of free will". Freedom and neurobiology: reflections on free will, language, and political power. Columbia University Press, New York
Sluyter A (2003) Neo-environmental determinism, intellectual damage control, and nature/society science. Antipode 35(4):813–817. https://doi.org/10.1046/j.1467-8330.2003.00354.x
Spinoza B (2017) Ethics. Loki's Publishing
Stone AA (2008) The ethical boundaries of forensic psychiatry: a view from the ivory tower. Bull Am Acad Psychiatry Law 12:209-19, 1984. J Am Acad Psychiatry Law 36(2):167–174
Suarez A (2013) Free will and nonlocality at detection as basic principles of quantum physics. In: Suarez A, Adams P (eds) Is science compatible with free will? Springer, New York, NY
Suarez A, Adams P (eds) (2013) Is science compatible with free will? Springer, New York, NY
Van Inwagen P (1986) An essay on free will, vol 34. Oxford University Press, Oxford; vol 135
Viney DW (1986) James,William, on free will and determinism. J Mind Behav 7(4):555–565
Wittgenstein L (1958) Philosophical investigations. Blackwell, Oxford
Wittgenstein L (1969) On certainty. Harper, New York

Temperament, Personality, and Character

16

Contents

16.1 Basic Concepts and Definitions

The in-depth description of *temperament*, *personality*, and *character* is beyond the scope of the present chapter and constitutes a very complex issue (Fountoulakis and Kaprinis 2006; Fountoulakis et al. 2016); however, a brief discussion concerning them is necessary, since they constitute the principal expression of brain function.

The concept of personality reflects differences among individuals concerning behavior, and individual behavior manifests high heterogeneity. This means that each person reacts and behaves differently from another person when facing the same stimulus (Sugiyama 2005; Nettle 2006; Ozer and Benet-Martinez 2006; Thornhill and Gangestad 2008). There are significant differences in terms of mating behaviors (Gangestad and Simpson 1990), the pursuing of social status, and the preference for a specific position in social hierarchy (Hawley 1999; Lund et al. 2007) as well as in the ability for the detection of deception and cheating (Ekman et al. 1999; Buss and Duntley 2008).

K. N. Fountoulakis, *The Human Connectome*,
https://doi.org/10.1007/978-3-031-10351-3_16

Temperament, Personality, and Character

- Reflect differences among individuals concerning behavior.
- Human individual behavior manifests high heterogeneity.
- Have been observed also in nonhuman species.
- Temperament.
- Personality.
- Character.
- The differences in personality constitute an expression of evolutionary adaptation at the collective level, and this is the reason it manifests primarily in social species.

These behaviors are historically classified under the umbrella of terms like "temperament," "personality," and "character," and they constitute the most fundamental systems which control and regulate the way people interact with their environment. This heterogeneity is not simply quantitative. At least concerning the plain observer, differences in behavior between two individuals seem to have a *qualitative nature*, and it is of absolutely different level from the differences that exist among individuals concerning, e.g., memory, intelligence, or other neurocognitive functions. These differences are so huge that they *tend to be characteristic of the person* (thus the term "personality") although there seem to be some wider common characteristics and facets.

These individual differences have been observed also in nonhuman species (Wolf et al. 2007; Gosling 2001), but the question whether animals have these characteristics continues to constitute a matter of debate, although it is certain there are some individual differences pertaining to behavior also in animals (Freeman and Gosling 2010; Wilsson 1997; Cloninger 1994; D'Eath et al. 2009; Martin and Reale 2008; Reale et al. 2007; King et al. 2006; Whitney 1970). These differences seem to manifest especially in social species (Penke et al. 2007; Perilloux and Buss 2008).

The extent and the "qualitative differentiation" of these individual differences constitute an oxymoron for the evolutionary approach, because heterogeneity is considered to be the substrate on which natural selection acts, rather than the final product of natural selection per se (Gaulin and McBurney 2003). A general belief is that natural selection reduces rather than facilitates individual differences; however, from a reverse point of view in this case, the differences in personality among individuals could be considered to constitute alternative strategies for the solving of different problems that tend to repeatedly appear and in this way the survival is secured at the group level, as the group is facing a changing environment, against which the isolated individual is impossible to be adequately fit, in psychological terms, to cope with all the challenges alone and by him/herself (Buss 1996; Denissen and Penke 2008a, b; Hawley 1999; Nettle 2006). In this sense, differences in personality constitute an expression of evolutionary adaptation at the collective level, and this is the reason it manifests primarily in social species.

16.2 Models of Human Personality

The so-called personality traits are defined as relatively stable dimensions that describe exactly these differences among individuals in relationship to behavior, affectivity, and neurocognitive function (Nettle 2011; Bouchard Jr. and Loehlin 2001). These characteristics manifest themselves already during childhood and they change only a little with advancing age, but they are influenced, triggered, and attenuated by both internal and external stimuli (Bouchard Jr. and Loehlin 2001; Rothbart et al. 2000; Woodward et al. 2001; Schwartz et al. 2003a, b, 2010).

There are some *basic models* for these characteristics, including the following:

- *The model of Hans Eysenck* which includes three traits (Eysenck and Eysenck 1962, 1964, 1967, 1968, 1969, 1972, 1977; Eysenck et al. 1974, 1976, 1977).
- *The model of Raymond Cattell* with 16 (Cattel et al. 1970).
- *The Big Five model* with five (Shock et al. 1984; Costa and McCrae 1985, 1990; McCrae et al. 2005; McCrae and Costa Jr. 1987).
- *The model of Robert Cloninger* which puts four temperaments and three characters in a hierarchy (Cloninger 1987; Constantino et al. 2002; Cloninger et al. 1993, 1999; Josefsson et al. 2012; Cloninger and Gilligan 1987; Cloninger and Kedia 2011).

Research has shown that in essence, these models reflect parts of a wider hierarchical model, with two trait super-factors at the top, four to seven intermediate temperaments, and more than 15 characters at the base (Fountoulakis and Gonda 2019, 2020; Akiskal et al. 2005; Cohen 1999; Maremmani et al. 2005).

It should be mentioned that to a large degree, our knowledge on human behavior and personality is based on self-reporting, and subsequently on the subjective sense, the person has on how he/she experiences him/herself, but also the behavior of others toward him/her. This is far away from an objective registration of personality and behavior; it is essentially a registration of inner experience. Having this objection in mind, the basic structure of brain function that governs the behavior which is defined by the terms temperament-personality-character seems to be as follows (Fig. 16.1). The term "emotional" is used here in a generic way to denote mood-affective-emotional mechanisms of any level (Fountoulakis and Gonda 2019, 2020):

At the top of the pyramid, there seems to be two fundamental trait super-characteristics:

- *Self (S):* which concerns the sense of identity in both affective and cognitive-mental terms.
- The *self-environment interaction (SEI)*, which concerns the existence and application of moral rules on social behavior, with emotional contribution and support. It is similar to "agreeableness" which relates to the differences pertaining to motivation toward collaboration and positive investment on others vs. egoistic behavior when conflicts concerning resources are in place, as well as with the persistence of fulfilling the target under distracting conditions. It also includes aspects of insight (Denissen and Penke 2008a; Nettle 2006, 2007, 2011).

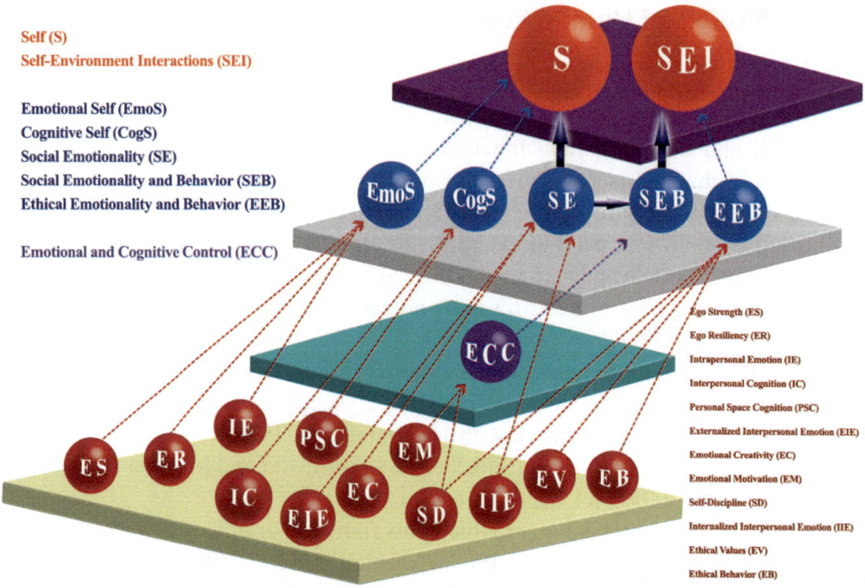

Fig. 16.1 The basic hierarchical structure of mental functioning which is defined under the terms temperament-personality-character. (From Fountoulakis et al Ann Gen Psychiatry 2019;18:21 και Fountoulakis et al. Int J Psychiatry Clin Pract:2020;1-14)

At the middle level, the following traits are present:

- *Emotional self (EmoS)*, which to a significant degree identifies with the mirror concept of *neuroticism*. This trait refers to those affective processes that concern the inner experience and the needs of the individual, and they constitute the individual capable of being psychologically balanced in a subconscious and automatic (not laborious) way. In contrast, neuroticism is related to the intensity of affects and stress management; it reflects the variability of the levels of negative affects, including fear, worry, guilt, or grief; and it is a powerful prognostic factor of psychiatric morbidity, and especially of depression (Denissen and Penke 2008a; Gurven et al. 2014; Nettle 2006; Haselton and Nettle 2006).
- *Cognitive self (CogS)* which largely identifies with the concept of *psychoticism* according to Eysenck and concerns the way of thinking through which the individual deals with his/her self and the others and understands the environment.
- *Social emotionality (SE)* is a trait that concerns the mechanisms of emotions that mediate social functioning and social life and serve their demands (attachment, aesthetics, ideas, values, etc.) It manifests significant resemblance to *openness* which refers to the sense of high-level mental activity with deep and transparent consciousness; high ability for novelty; advanced problem-solving skills; elaboration in the spiritual, artistic, and creative domains; capacity in the processing of incomplete information; as well as intrinsic curiosity which enhances mental ability (Denissen and Penke 2008a; Nettle and Clegg 2006).

- *Emotional and cognitive control (ECC)* refers to the general ability for the control of behavior and impulsivity through affective as well as cognitive mechanisms. It probably constitutes a variation of *extroversion*, which concerns the activation vs. inhibition of impulses and the predisposition toward hierarchy and possible leadership and authority as well as toward dominance and submission (Denissen and Penke 2008a).
- *Ethical emotionality and behavior (EEB)* concerns the ability to control behavior and impulsivity through emotional as well as cognitive mechanisms but with a special emphasis on moral and ethical values, which constitute the core of the functioning of this trait. It resembles the concept of *conscientiousness* which is implicated in behavior pertaining to work, self-control, satisfaction postponing, work intensity, monitoring of the non-accomplishment of targets, and executive regulation in the domain of achievements and in the ability for reliable work and the continuous commitment as well as reliability and reputation (Buss 1991; Nettle 2011).
- *Social emotionality and behavior (SEB)* is an undifferentiated trait which seems to bridge self with the self-environment interaction and includes elements of agreeableness, novelty seeking, reward, and extraversion.

At the base of the pyramid, there is a significant number of features; some of them constitute traits, while others are probably less stable. They include *ego resilience (ER)* which shares common elements with *harm avoidance* according to Cloninger, *ego strength (ES)* which shares elements with *reward dependence* and *perseveration* according to Cloninger, *intrapersonal emotionality (IE)*, *personal space cognition (PSC)*, *interpersonal cognition (IC)*, *emotional creativity (EC)* which together with *emotional motives (EM)* overlap with the concept of *novelty seeking* according to Cloninger, the *externalized (EIE)* and the *internalized interpersonal emotions (IIE)*, *self-discipline (SD)*, *ethical values (EV)*, and *ethical behavior (EB)*.

Models of Human Personality

- Personality traits.
  - Manifest already since very young age.
  - Change only a little with advancing age.
  - Influenced, triggered, and attenuated by both internal and external stimuli.
- Models of personality traits.
  - Hans Eysenck (three traits).
  - Raymond Cattell (16).
  - Big-five (five).
  - Robert Cloninger (four temperaments and three characters).
- Combined model.
  - Two super-factors at the top.
  - Four to seven intermediate temperaments.
  - Over 15 characters at the base.
- Emotional and logical analyses coexist in functional units and systems and are rarely separated.

An important conclusion which is derived from the above is that affective and logical analyses coexist in functional units and systems and are rarely separated. This observation explains by itself many features of human behavior.

16.3 Neurobiology of Human Personality

The models which were analyzed previously provide with a sense concerning which are the brain structures that possibly take part in the regulation of personality-related individual differences.

The definition of temperament itself implies a central position for biology (Buss and Plomin 1984; Rothbart 2016). On the contrary, personality is supposed to be determined to a greater degree by social influences (MacDonald 2012). However, the first biological influence which is known already since the antiquity concerns the *effect of gender*. This effect is obvious even in primates, since, for example, it has been reported that in rhesus macaques, males are more aggressive, with fewer social relationships, more impulsive, more risk taking, and with a higher mortality rate (Mehlman et al. 1994, 1997; Higley and Linnoila 1997; Higley et al. 1991, 1992).

Neurobiology of Human Personality

- The definition of temperament implies a central position for biology.
- Personality is supposed to be determined to a greater degree by social influences.
- Effect of gender.
- Participate neuronal circuits that concern:
  - Sustaining of attention.
  - Executive functions.
  - Abstract thinking.
  - Affective investment.
  - Reward and habit formation.
  - Impulse inhibition.
  - Controlling of behavioral manifestations.
- Serotonin and norepinephrine play a role in behavioral control.
- Dopamine is crucial for novelty seeking and reward-dependent behavior.
- 30–60% of the personality trait variance is due to genetic factors.
  - Novelty seeking and DRD4 gene.
  - Harm avoidance and 5-HTTLPR gene.

The analysis of personality characteristics makes clear that the neuronal circuits that participate (Fig. 16.2) concern among other functions the sustaining of attention, executive functions, the ability to abstract thinking, the affective investment, reward and habit formation, impulse inhibition, and the controlling of behavioral manifestations (Verweij et al. 2010; Cloninger et al. 1993; Gillespie et al. 2003).

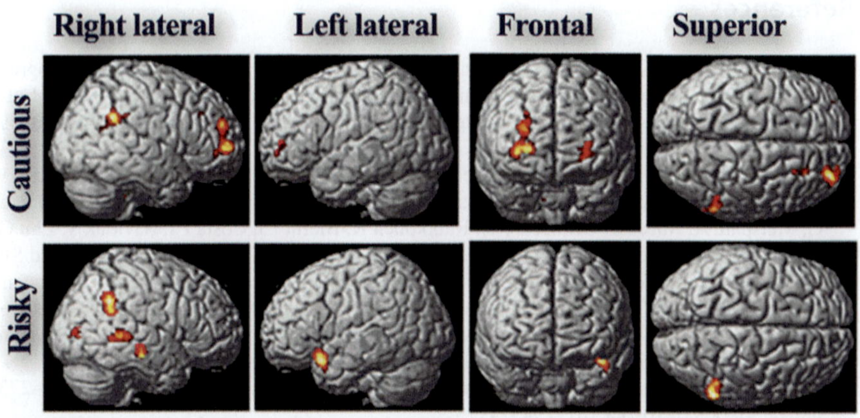

Fig. 16.2 Brain activity in typical developing adolescents. Up: During cautious behavior. Down: During risk-taking behaviors (Credit: National Institute on Drug Abuse, National Institutes of Health)

This makes granted the role of major neurotransmitters, but gives few information concerning their specific role. On the basis of the action psychopharmacological substances as well as illegal substances have, it seems that serotonin and norepinephrine play a role in behavioral control, while dopamine is crucial for novelty seeking and reward-dependent behavior.

Today, it is clear that personality traits and temperament have a biological substrate and therefore are *heritable* to a significant, but maybe each to a different, degree. Twin and adoption studies suggest that *30–60%* of the personality trait variance is due to genetic factors (Workman and Reader 2014; Verweij et al. 2012; Johnson et al. 2008; Polderman et al. 2015; Keller and Miller 2006; Gillespie et al. 2001, 2003; Keller et al. 2005). Additionally, their nature is definitely *polygenic* with a large number of possible but small additive influences or interactions among the various alleles, and in this way, they contribute to different phenotypical manifestations which however are very difficult to be traced (Penke and Jokela 2016; Plomin et al. 1994, 2008; Nettle 2006; Genetics of Personality Consortium et al. 2015; South et al. 2018; Power and Pluess 2015; Verweij et al. 2010, 2012; Gonda et al. 2009; Cloninger and Zwir 2018; Keller et al. 2005; Munafo and Flint 2011; Service et al. 2012; van den Berg et al. 2016; Amin et al. 2013; de Moor et al. 2012).

The most robust data so far concern the DRD4 gene which codes the D4 dopamine receptor and is related to novelty seeking, but its contribution is at the magnitude of 3% (Munafo et al. 2008; Golimbet et al. 2007). It manifests a significant geographical differentiation with impressive differences reported by studies which compare various populations (Chen et al. 1999; Gören 2016; Matthews and Butler 2011). Additionally, there are contradictory results pertaining to the relationship of harm avoidance and the 5-HTTLPR gene (Munafo et al. 2009; Minelli et al. 2011).

References

Akiskal HS, Akiskal KK, Haykal RF, Manning JS, Connor PD (2005) TEMPS-A: progress towards validation of a self-rated clinical version of the temperament evaluation of the Memphis, Pisa, Paris, and San Diego autoquestionnaire. J Affect Disord 85(1–2):3–16. https://doi.org/10.1016/j.jad.2004.12.001

Amin N, Hottenga JJ, Hansell NK, Janssens AC, de Moor MH, Madden PA, Zorkoltseva IV, Penninx BW, Terracciano A, Uda M, Tanaka T, Esko T, Realo A, Ferrucci L, Luciano M, Davies G, Metspalu A, Abecasis GR, Deary IJ, Raikkonen K, Bierut LJ, Costa PT, Saviouk V, Zhu G, Kirichenko AV, Isaacs A, Aulchenko YS, Willemsen G, Heath AC, Pergadia ML, Medland SE, Axenovich TI, de Geus E, Montgomery GW, Wright MJ, Oostra BA, Martin NG, Boomsma DI, van Duijn CM (2013) Refining genome-wide linkage intervals using a meta-analysis of genome-wide association studies identifies loci influencing personality dimensions. Eur J Hum Genet 21(8):876–882. https://doi.org/10.1038/ejhg.2012.263

Bouchard TJ Jr, Loehlin JC (2001) Genes, evolution, and personality. Behav Genet 31(3):243–273. https://doi.org/10.1023/a:1012294324713

Buss DM (1991) Evolutionary personality psychology. Annu Rev Psychol 42:459–491. https://doi.org/10.1146/annurev.ps.42.020191.002331

Buss DM (1996) Social adaptation and five major factors of personality. In: Wiggins JS (ed) The five-factor model of personality: theoretical perspectives. Guilford, New York, pp 180–207

Buss DM, Duntley JD (2008) Adaptations for exploitation. Group Dyn 12:53–62

Buss AH, Plomin R (1984) Temperament: early developing personality traits. Erlbaum, Hillsdale, NJ

Cattel R, Eber H, Tatsuoka M (1970) The handbook for the sixteen personality factor questionnaire. Institute for Personality and Ability Testing, Champain, IL

Chen C, Burton M, Greenberger E, Dmitrieva J (1999) Population migration and the variation of dopamine D4 receptor (DRD4) allele frequencies around the globe. Evol Hum Behav 20(5):309–324. https://doi.org/10.1016/s1090-5138(99)00015-x

Cloninger CR (1987) A systematic method for clinical description and classification of personality variants. A proposal. Arch Gen Psychiatry 44(6):573–588. https://doi.org/10.1001/archpsyc.1987.01800180093014

Cloninger CR (1994) The genetic structure of personality and learning: a phylogenetic model. Clin Genet 46(1 Spec No):124–137. https://doi.org/10.1111/j.1399-0004.1994.tb04214.x

Cloninger CR, Gilligan SB (1987) Neurogenetic mechanisms of learning: a phylogenetic perspective. J Psychiatr Res 21(4):457–472. https://doi.org/10.1016/0022-3956(87)90094-x

Cloninger CR, Kedia S (2011) The phylogenesis of human personality: identifying the precursors of cooperation, altruism, and well-being. In: Sussman RW, Cloninger CR (eds) The origins of cooperation and altruism. Developments in primatology: progress and prospects, 1st edn. Springer, New York, pp 63–110

Cloninger CR, Zwir I (2018) What is the natural measurement unit of temperament: single traits or profiles? Philos Trans R Soc Lond B Biol Sci 373(1744):20170163. https://doi.org/10.1098/rstb.2017.0163

Cloninger CR, Svrakic DM, Przybeck TR (1993) A psychobiological model of temperament and character. Arch Gen Psychiatry 50(12):975–990. https://doi.org/10.1001/archpsyc.1993.01820240059008

Cloninger C, Svrakic D, Bayon C, Przybeck T (1999) Measurement of psychopathology as variants of personality. In: Cloninger C (ed) Personality and psychopathology. American Psychiatric Press Inc., Washington DC, pp 33–65

Cohen P (1999) Personality development in childhood: old and new findings. In: Cloninger C (ed) Personality and psychopathology. American Psychiatric Press Inc, Washington DC, pp 101–127

Constantino JN, Cloninger CR, Clarke AR, Hashemi B, Przybeck T (2002) Application of the seven-factor model of personality to early childhood. Psychiatry Res 109(3):229–243. https://doi.org/10.1016/s0165-1781(02)00008-2

Costa PT, McCrae RR (1985) The NEO personality inventory manual. Psychological Assessment Resources, Odessa, FL

Costa P, McCrae R (1990) Personality disorders and the five-factor model of personality. J Personal Disord 4:362–371

de Moor MH, Costa PT, Terracciano A, Krueger RF, de Geus EJ, Toshiko T, Penninx BW, Esko T, Madden PA, Derringer J, Amin N, Willemsen G, Hottenga JJ, Distel MA, Uda M, Sanna S, Spinhoven P, Hartman CA, Sullivan P, Realo A, Allik J, Heath AC, Pergadia ML, Agrawal A, Lin P, Grucza R, Nutile T, Ciullo M, Rujescu D, Giegling I, Konte B, Widen E, Cousminer DL, Eriksson JG, Palotie A, Peltonen L, Luciano M, Tenesa A, Davies G, Lopez LM, Hansell NK, Medland SE, Ferrucci L, Schlessinger D, Montgomery GW, Wright MJ, Aulchenko YS, Janssens AC, Oostra BA, Metspalu A, Abecasis GR, Deary IJ, Raikkonen K, Bierut LJ, Martin NG, van Duijn CM, Boomsma DI (2012) Meta-analysis of genome-wide association studies for personality. Mol Psychiatry 17(3):337–349. https://doi.org/10.1038/mp.2010.128

D'Eath RB, Roehe R, Turner SP, Ison SH, Farish M, Jack MC, Lawrence AB (2009) Genetics of animal temperament: aggressive behaviour at mixing is genetically associated with the response to handling in pigs. Animal 3(11):1544–1554. https://doi.org/10.1017/S1751731109990528

Denissen JJA, Penke L (2008a) Motivational individual reaction norms underlying the five-factor model of personality: first steps towards a theory-based conceptual framework. J Res Pers 42:1285–1302

Denissen JJA, Penke L (2008b) Neuroticism predicts reactions to cues of social inclusion. Eur J Personal 22:497–517

Ekman P, O'Sullivan M, Frank MG (1999) A few can catch a liar. Psychol Sci 10:363–366

Eysenck HJ, Eysenck SB (1962) A factorial study of an interview-questionnaire. J Clin Psychol 18:286–290. https://doi.org/10.1002/1097-4679(196207)18:3<286::aid-jclp2270180311>3.0.co;2-u

Eysenck SB, Eysenck HJ (1964) An improved short questionnaire for the measurement of extraversion and neuroticism. Life Sci (1962) 3:1103–1109. https://doi.org/10.1016/0024-3205(64)90125-0

Eysenck SB, Eysenck HJ (1967) Physiological reactivity to sensory stimulation as a measure of personality. Psychol Rep 20(1):45–46. https://doi.org/10.2466/pr0.1967.20.1.45

Eysenck SB, Eysenck HJ (1968) The measurement of psychoticism: a study of factor stability and reliability. Br J Soc Clin Psychol 7(4):286–294. https://doi.org/10.1111/j.2044-8260.1968.tb00571.x

Eysenck SB, Eysenck HJ (1969) Scores on three personality variables as a function of age, sex and social class. Br J Soc Clin Psychol 8(1):69–76. https://doi.org/10.1111/j.2044-8260.1969.tb00588.x

Eysenck SB, Eysenck HJ (1972) The questionnaire measurement of psychoticism. Psychol Med 2(1):50–55. https://doi.org/10.1017/s0033291700045608

Eysenck SB, Eysenck HJ (1977) The place of impulsiveness in a dimensional system of personality description. Br J Soc Clin Psychol 16(1):57–68. https://doi.org/10.1111/j.2044-8260.1977.tb01003.x

Eysenck SB, Eysenck HJ, Shaw L (1974) The modification of personality and lie scale scores by special 'honesty' instructions. Br J Soc Clin Psychol 13(1):41–50. https://doi.org/10.1111/j.2044-8260.1974.tb00876.x

Eysenck SB, White O, Eysenck HJ (1976) Personality and mental illness. Psychol Rep 39(3):1011–1022. https://doi.org/10.2466/pr0.1976.39.3.1011

Eysenck SB, Adelaja O, Eysenck HJ (1977) A comparative study of personality in Nigerian and English subjects. J Soc Psychol 102(Second Half):171–178. https://doi.org/10.1080/00224545.1977.9713261

Fountoulakis KN, Gonda X (2019) Modeling human temperament and character on the basis of combined theoretical approaches. Ann General Psychiatry 18:21. https://doi.org/10.1186/s12991-019-0247-1

Fountoulakis KN, Gonda X (2020) Sex differences in maturation and aging of human personality on the basis of a recently developed complex hierarchical model of temperament and character. Int J Psychiatry Clin Pract 26:58. https://doi.org/10.1080/13651501.2020.1804941

Fountoulakis KN, Kaprinis GS (2006) Personality disorders: new data versus old concepts. Curr Opin Psychiatry 19(1):90–94. https://doi.org/10.1097/01.yco.0000196158.98540.b6
Fountoulakis KN, Gonda X, Koufaki I, Hyphantis T, Cloninger CR (2016) The role of temperament in the Etiopathogenesis of bipolar Spectrum illness. Harv Rev Psychiatry 24(1):36–52. https://doi.org/10.1097/HRP.0000000000000077
Freeman HD, Gosling SD (2010) Personality in nonhuman primates: a review and evaluation of past research. Am J Primatol 72(8):653–671. https://doi.org/10.1002/ajp.20833
Gangestad SW, Simpson JA (1990) Toward an evolutionary history of female sociosexual variation. J Pers 58(1):69–96. https://doi.org/10.1111/j.1467-6494.1990.tb00908.x
Gaulin S, McBurney DH (2003) Evolutionary psychology. Prentice Hall, Upper Saddle River, NJ
Genetics of Personality Consortium, de Moor MH, van den Berg SM, Verweij KJ, Krueger RF, Luciano M, Arias Vasquez A, Matteson LK, Derringer J, Esko T, Amin N, Gordon SD, Hansell NK, Hart AB, Seppala I, Huffman JE, Konte B, Lahti J, Lee M, Miller M, Nutile T, Tanaka T, Teumer A, Viktorin A, Wedenoja J, Abecasis GR, Adkins DE, Agrawal A, Allik J, Appel K, Bigdeli TB, Busonero F, Campbell H, Costa PT, Davey Smith G, Davies G, de Wit H, Ding J, Engelhardt BE, Eriksson JG, Fedko IO, Ferrucci L, Franke B, Giegling I, Grucza R, Hartmann AM, Heath AC, Heinonen K, Henders AK, Homuth G, Hottenga JJ, Iacono WG, Janzing J, Jokela M, Karlsson R, Kemp JP, Kirkpatrick MG, Latvala A, Lehtimaki T, Liewald DC, Madden PA, Magri C, Magnusson PK, Marten J, Maschio A, Medland SE, Mihailov E, Milaneschi Y, Montgomery GW, Nauck M, Ouwens KG, Palotie A, Pettersson E, Polasek O, Qian Y, Pulkki-Raback L, Raitakari OT, Realo A, Rose RJ, Ruggiero D, Schmidt CO, Slutske WS, Sorice R, Starr JM, St Pourcain B, Sutin AR, Timpson NJ, Trochet H, Vermeulen S, Vuoksimaa E, Widen E, Wouda J, Wright MJ, Zgaga L, Porteous D, Minelli A, Palmer AA, Rujescu D, Ciullo M, Hayward C, Rudan I, Metspalu A, Kaprio J, Deary IJ, Raikkonen K, Wilson JF, Keltikangas-Jarvinen L, Bierut LJ, Hettema JM, Grabe HJ, van Duijn CM, Evans DM, Schlessinger D, Pedersen NL, Terracciano A, McGue M, Penninx BW, Martin NG, Boomsma DI (2015) Meta-analysis of genome-wide association studies for neuroticism, and the polygenic association with major depressive disorder. JAMA Psychiat 72(7):642–650. https://doi.org/10.1001/jamapsychiatry.2015.0554
Gillespie NA, Johnstone SJ, Boyce P, Heath AC, Martin NG (2001) The genetic and environmental relationship between the interpersonal sensitivity measure (IPSM) and the personality dimensions of Eysenck and Cloninger. Pers Indiv Differ 31(7):1039–1051. https://doi.org/10.1016/s0191-8869(00)00200-2
Gillespie NA, Cloninger CR, Heath AC, Martin NG (2003) The genetic and environmental relationship between Cloninger's dimensions of temperament and character. Pers Indiv Differ 35(8):1931–1946. https://doi.org/10.1016/s0191-8869(03)00042-4
Golimbet VE, Alfimova MV, Gritsenko IK, Ebstein RP (2007) Relationship between dopamine system genes and extraversion and novelty seeking. Neurosci Behav Physiol 37(6):601–606. https://doi.org/10.1007/s11055-007-0058-8
Gonda X, Fountoulakis KN, Rihmer Z, Lazary J, Laszik A, Akiskal KK, Akiskal HS, Bagdy G (2009) Towards a genetically validated new affective temperament scale: a delineation of the temperament phenotype of 5-HTTLPR using the TEMPS-A. J Affect Disord 112(1–3):19–29. https://doi.org/10.1016/j.jad.2008.03.012
Gören E (2016) The biogeographic origins of novelty-seeking traits. Evol Hum Behav 37(6):456–469. https://doi.org/10.1016/j.evolhumbehav.2016.04.005
Gosling SD (2001) From mice to men: what can we learn about personality from animal research? Psychol Bull 127(1):45–86. https://doi.org/10.1037/0033-2909.127.1.45
Gurven M, C Von Rueden, Stieglitz J, Kaplan H, Rodriguez DE (2014) The evolutionary fitness of personality traits in a small-scale subsistence society. Evol Hum Behav 35(1). https://doi.org/10.1016/j.evolhumbehav.2013.09.002
Haselton MG, Nettle D (2006) The paranoid optimist: an integrative evolutionary model of cognitive biases. Personal Soc Psychol Rev 10(1):47–66. https://doi.org/10.1207/s15327957pspr1001_3
Hawley PH (1999) The ontogenesis of social dominance: a strategy-based evolutionary perspective. Dev Rev 19:97–132

Higley JD, Linnoila M (1997) A nonhuman primate model of excessive alcohol intake. Personality and neurobiological parallels of type I- and type II-like alcoholism. Recent Dev Alcohol 13:191–219

Higley JD, Suomi SJ, Linnoila M (1991) CSF monoamine metabolite concentrations vary according to age, rearing, and sex, and are influenced by the stressor of social separation in rhesus monkeys. Psychopharmacology 103(4):551–556. https://doi.org/10.1007/BF02244258

Higley JD, Mehlman PT, Taub DM, Higley SB, Suomi SJ, Vickers JH, Linnoila M (1992) Cerebrospinal fluid monoamine and adrenal correlates of aggression in free-ranging rhesus monkeys. Arch Gen Psychiatry 49(6):436–441. https://doi.org/10.1001/archpsyc.1992.01820060016002

Johnson AM, Vernon PA, Feiler AR (2008) Behavioral genetic studies of personality: an introduction and review of the results of 50+ years of research. In: Boyle GJ, Matthews G, Saklofske DH (eds) The SAGE handbook of personality theory and assessment, 1st edn. SAGE Publications, Los Angeles, CA, pp 145–173

Josefsson K, Jokela M, Cloninger CR, Hintsanen M, Salo J, Hintsa T, Pulkki-Raback L, Keltikangas-Jarvinen L (2012) Maturity and change in personality: developmental trends of temperament and character in adulthood. Dev Psychopathol 25:713

Keller MC, Miller G (2006) Resolving the paradox of common, harmful, heritable mental disorders: which evolutionary genetic models work best? Behav Brain Sci 29(4):385–404; discussion 405-352. https://doi.org/10.1017/S0140525X06009095

Keller MC, Coventry WL, Heath AC, Martin NG (2005) Widespread evidence for non-additive genetic variation in Cloninger's and Eysenck's personality dimensions using a twin plus sibling design. Behav Genet 35(6):707–721. https://doi.org/10.1007/s10519-005-6041-7

King DA, Schuehle Pfeiffer CE, Randel RD, Welsh TH Jr, Oliphint RA, Baird BE, Curley KO Jr, Vann RC, Hale DS, Savell JW (2006) Influence of animal temperament and stress responsiveness on the carcass quality and beef tenderness of feedlot cattle. Meat Sci 74(3):546–556. https://doi.org/10.1016/j.meatsci.2006.05.004

Lund OCH, Tamnes CK, Moestue C, Buss DM, Vollrath M (2007) Tactics of hierarchy negotiation. J Res Pers 41:25–44

MacDonald K (2012) Temperament and evolution. In: Zentner M, Shiner RL (eds) Handbook of temperament. Guilford Press, New York, pp 273–296

Maremmani I, Akiskal H, Signoretta S, Liguori A, Perugi G, Cloninger C (2005) The relationship of Kraepelian affective temperaments (as measured by TEMPS-I) to the tridimensional personality questionnaire (TPQ). J Affect Disord 85:17–27

Martin JG, Reale D (2008) Animal temperament and human disturbance: implications for the response of wildlife to tourism. Behav Process 77(1):66–72. https://doi.org/10.1016/j.beproc.2007.06.004

Matthews LJ, Butler PM (2011) Novelty-seeking DRD4 polymorphisms are associated with human migration distance out-of-Africa after controlling for neutral population gene structure. Am J Phys Anthropol 145(3):382–389. https://doi.org/10.1002/ajpa.21507

McCrae RR, Costa PT Jr (1987) Validation of the five-factor model of personality across instruments and observers. J Pers Soc Psychol 52(1):81–90. https://doi.org/10.1037//0022-3514.52.1.81

McCrae RR, Costa PT Jr, Martin TA (2005) The NEO-PI-3: a more readable revised NEO personality inventory. J Pers Assess 84(3):261–270. https://doi.org/10.1207/s15327752jpa8403_05

Mehlman PT, Higley JD, Faucher I, Lilly AA, Taub DM, Vickers J, Suomi SJ, Linnoila M (1994) Low CSF 5-HIAA concentrations and severe aggression and impaired impulse control in nonhuman primates. Am J Psychiatry 151(10):1485–1491. https://doi.org/10.1176/ajp.151.10.1485

Mehlman PT, Higley JD, Fernald BJ, Sallee FR, Suomi SJ, Linnoila M (1997) CSF 5-HIAA, testosterone, and sociosexual behaviors in free-ranging male rhesus macaques in the mating season. Psychiatry Res 72(2):89–102. https://doi.org/10.1016/s0165-1781(97)00084-x

Minelli A, Bonvicini C, Scassellati C, Sartori R, Gennarelli M (2011) The influence of psychiatric screening in healthy populations selection: a new study and meta-analysis of functional 5-HTTLPR and rs25531 polymorphisms and anxiety-related personality traits. BMC Psychiatry 11:50. https://doi.org/10.1186/1471-244X-11-50

Munafo MR, Flint J (2011) Dissecting the genetic architecture of human personality. Trends Cogn Sci 15(9):395–400. https://doi.org/10.1016/j.tics.2011.07.007

Munafo MR, Yalcin B, Willis-Owen SA, Flint J (2008) Association of the dopamine D4 receptor (DRD4) gene and approach-related personality traits: meta-analysis and new data. Biol Psychiatry 63(2):197–206. https://doi.org/10.1016/j.biopsych.2007.04.006

Munafo MR, Freimer NB, Ng W, Ophoff R, Veijola J, Miettunen J, Jarvelin MR, Taanila A, Flint J (2009) 5-HTTLPR genotype and anxiety-related personality traits: a meta-analysis and new data. Am J Med Genet B 150B(2):271–281. https://doi.org/10.1002/ajmg.b.30808

Nettle D (2006) The evolution of personality variation in humans and other animals. Am Psychol 61(6):622–631. https://doi.org/10.1037/0003-066X.61.6.622

Nettle D (2007) Traits and trade-offs are an important tier. Am Psychol 62(9):1074–1075. https://doi.org/10.1037/0003-066X.62.9.1074

Nettle D (2011) Evolutionary perspectives on the five factor model of personallity. In: Buss DM, Hawley PH (eds) The evolution of personality and individual differences. Oxford University Press, Oxford, p xix, 498 p

Nettle D, Clegg H (2006) Schizotypy, creativity and mating success in humans. Proc Biol Sci 273(1586):611–615. https://doi.org/10.1098/rspb.2005.3349

Ozer DJ, Benet-Martinez V (2006) Personality and the prediction of consequential outcomes. Annu Rev Psychol 57:401–421. https://doi.org/10.1146/annurev.psych.57.102904.190127

Penke L, Jokela M (2016) The evolutionary genetics of personality revisited. Curr Opin Psychol 7:104–109. https://doi.org/10.1016/j.copsyc.2015.08.021

Penke L, Denissen JJA, Miller GF (2007) The evolutionary genetics of personality. Eur J Personal 21:549–587

Perilloux C, Buss DM (2008) Breaking up romantic relationships: costs experienced and coping strategies deployed. Evol Psychol 6:164–181

Plomin R, McClearn GE, Smith DL, Vignetti S, Chorney MJ, Chorney K, Venditti CP, Kasarda S, Thompson LA, Detterman DK et al (1994) DNA markers associated with high versus low IQ: the IQ quantitative trait loci (QTL) project. Behav Genet 24(2):107–118. https://doi.org/10.1007/BF01067815

Plomin R, DeFries JC, McClearn GE, Rutter M (2008) Behavioral genetics. Freeman, New York

Polderman TJ, Benyamin B, de Leeuw CA, Sullivan PF, van Bochoven A, Visscher PM, Posthuma D (2015) Meta-analysis of the heritability of human traits based on fifty years of twin studies. Nat Genet 47(7):702–709. https://doi.org/10.1038/ng.3285

Power RA, Pluess M (2015) Heritability estimates of the big five personality traits based on common genetic variants. Transl Psychiatry 5:e604. https://doi.org/10.1038/tp.2015.96

Reale D, Reader SM, Sol D, McDougall PT, Dingemanse NJ (2007) Integrating animal temperament within ecology and evolution. Biol Rev Camb Philos Soc 82(2):291–318. https://doi.org/10.1111/j.1469-185X.2007.00010.x

Rothbart MK (2016) Temperament, development, and personality. Curr Dir Psychol Sci 16(4):207–212. https://doi.org/10.1111/j.1467-8721.2007.00505.x

Rothbart MK, Ahadi SA, Evans DE (2000) Temperament and personality: origins and outcomes. J Pers Soc Psychol 78(1):122–135. https://doi.org/10.1037//0022-3514.78.1.122

Schwartz CE, Wright CI, Shin LM, Kagan J, Rauch SL (2003a) Inhibited and uninhibited infants "grown up": adult amygdalar response to novelty. Science 300(5627):1952–1953. https://doi.org/10.1126/science.1083703

Schwartz CE, Wright CI, Shin LM, Kagan J, Whalen PJ, McMullin KG, Rauch SL (2003b) Differential amygdalar response to novel versus newly familiar neutral faces: a functional MRI probe developed for studying inhibited temperament. Biol Psychiatry 53(10):854–862. https://doi.org/10.1016/s0006-3223(02)01906-6

Schwartz CE, Kunwar PS, Greve DN, Moran LR, Viner JC, Covino JM, Kagan J, Stewart SE, Snidman NC, Vangel MG, Wallace SR (2010) Structural differences in adult orbital and ventromedial prefrontal cortex predicted by infant temperament at 4 months of age. Arch Gen Psychiatry 67(1):78–84. https://doi.org/10.1001/archgenpsychiatry.2009.171

Service SK, Verweij KJ, Lahti J, Congdon E, Ekelund J, Hintsanen M, Raikkonen K, Lehtimaki T, Kahonen M, Widen E, Taanila A, Veijola J, Heath AC, Madden PA, Montgomery GW, Sabatti

C, Jarvelin MR, Palotie A, Raitakari O, Viikari J, Martin NG, Eriksson JG, Keltikangas-Jarvinen L, Wray NR, Freimer NB (2012) A genome-wide meta-analysis of association studies of Cloninger's temperament scales. Transl Psychiatry 2(5):e116. https://doi.org/10.1038/tp.2012.37

Shock NW, Greulich RC, Andres R, Arenberg D, Costa PT, Lakatta EG (1984) Normal human aging: the Baltimore longitudinal study of aging. National Institutes of Health, Bethesda, MD

South SC, Jarnecke AM, Vize CE (2018) Sex differences in the big five model personality traits: a behavior genetics exploration. J Res Pers 74:158–165. https://doi.org/10.1016/j.jrp.2018.03.002

Sugiyama L (2005) Physical attractiveness in adaptationist perspective. In: Buss D (ed) The handbook of evolutionary psychology. Wiley, New York, pp 292–342

Thornhill R, Gangestad SW (2008) The evolutionary biology of human female sexuality. Oxford University Press, New York

van den Berg SM, de Moor MH, Verweij KJ, Krueger RF, Luciano M, Arias Vasquez A, Matteson LK, Derringer J, Esko T, Amin N, Gordon SD, Hansell NK, Hart AB, Seppala I, Huffman JE, Konte B, Lahti J, Lee M, Miller M, Nutile T, Tanaka T, Teumer A, Viktorin A, Wedenoja J, Abdellaoui A, Abecasis GR, Adkins DE, Agrawal A, Allik J, Appel K, Bigdeli TB, Busonero F, Campbell H, Costa PT, Smith GD, Davies G, de Wit H, Ding J, Engelhardt BE, Eriksson JG, Fedko IO, Ferrucci L, Franke B, Giegling I, Grucza R, Hartmann AM, Heath AC, Heinonen K, Henders AK, Homuth G, Hottenga JJ, Iacono WG, Janzing J, Jokela M, Karlsson R, Kemp JP, Kirkpatrick MG, Latvala A, Lehtimaki T, Liewald DC, Madden PA, Magri C, Magnusson PK, Marten J, Maschio A, Mbarek H, Medland SE, Mihailov E, Milaneschi Y, Montgomery GW, Nauck M, Nivard MG, Ouwens KG, Palotie A, Pettersson E, Polasek O, Qian Y, Pulkki-Raback L, Raitakari OT, Realo A, Rose RJ, Ruggiero D, Schmidt CO, Slutske WS, Sorice R, Starr JM, St Pourcain B, Sutin AR, Timpson NJ, Trochet H, Vermeulen S, Vuoksimaa E, Widen E, Wouda J, Wright MJ, Zgaga L, Generation S, Porteous D, Minelli A, Palmer AA, Rujescu D, Ciullo M, Hayward C, Rudan I, Metspalu A, Kaprio J, Deary IJ, Raikkonen K, Wilson JF, Keltikangas-Jarvinen L, Bierut LJ, Hettema JM, Grabe HJ, Penninx BW, van Duijn CM, Evans DM, Schlessinger D, Pedersen NL, Terracciano A, McGue M, Martin NG, Boomsma DI (2016) Meta-analysis of genome-wide association studies for extraversion: findings from the genetics of personality consortium. Behav Genet 46(2):170–182. https://doi.org/10.1007/s10519-015-9735-5

Verweij KJ, Zietsch BP, Medland SE, Gordon SD, Benyamin B, Nyholt DR, McEvoy BP, Sullivan PF, Heath AC, Madden PA, Henders AK, Montgomery GW, Martin NG, Wray NR (2010) A genome-wide association study of Cloninger's temperament scales: implications for the evolutionary genetics of personality. Biol Psychol 85(2):306–317. https://doi.org/10.1016/j.biopsycho.2010.07.018

Verweij KJ, Yang J, Lahti J, Veijola J, Hintsanen M, Pulkki-Raback L, Heinonen K, Pouta A, Pesonen AK, Widen E, Taanila A, Isohanni M, Miettunen J, Palotie A, Penke L, Service SK, Heath AC, Montgomery GW, Raitakari O, Kahonen M, Viikari J, Raikkonen K, Eriksson JG, Keltikangas-Jarvinen L, Lehtimaki T, Martin NG, Jarvelin MR, Visscher PM, Keller MC, Zietsch BP (2012) Maintenance of genetic variation in human personality: testing evolutionary models by estimating heritability due to common causal variants and investigating the effect of distant inbreeding. Evolution 66(10):3238–3251. https://doi.org/10.1111/j.1558-5646.2012.01679.x

Whitney G (1970) Timidity and fearfulness of laboratory mice: an illustration of problems in animal temperament. Behav Genet 1(1):77–85. https://doi.org/10.1007/BF01067373

Wilsson E (1997) Maternal effects on behavior of juvenile & adult dogs, Doctoral Thesis. Stockholm University, Department of Zoology

Wolf M, van Doorn GS, Leimar O, Weissing FJ (2007) Life-history trade-offs favour the evolution of animal personalities. Nature 447(7144):581–584. https://doi.org/10.1038/nature05835

Woodward SA, McManis MH, Kagan J, Deldin P, Snidman N, Lewis M, Kahn V (2001) Infant temperament and the brainstem auditory evoked response in later childhood. Dev Psychol 37(4):533–538

Workman L, Reader W (2014) Evolutionary psychology: an introduction, Third edition. edn. Cambridge University Press, Cambridge

Differences Between Genders in the Organization and the Functioning of the Brain

17

Contents

17.1 Differences Between Genders in Observed Behavior

Differences between the two genders are widespread in nature (Fig. 17.1) and concern a variety of characteristics, including not only the external appearance but also inclinations and skills, as well as the way of thinking and the overall behavior.

While differences between genders in terms of behavior are clear, the differences concerning the psychological characteristics, the temperament and the personality, from which eventually behavior comes from, are less concrete. It is important here to discuss an important methodological-statistical detail. Two populations (in the current case the two genders) might manifest little or even an undetectable difference and certainly without essential significance in the mean concerning a specific characteristic, but this small difference in the mean could produce dramatic differences at the extremes of the distribution and as we move away from the mean. This is especially true for rare manifestations. For example, the difference in mean somatic aggressiveness between males and females is small, and the probability the average male to commit murder is not different from that of the average female.

K. N. Fountoulakis, *The Human Connectome*,
https://doi.org/10.1007/978-3-031-10351-3_17

Fig. 17.1 Couple of Mandarin ducks (or *Aix galericulata*), at the Martin Mere, Lancashire, UK. The differences between the male (left) and the female (right) are impressive (photo by Francis C. Franklin/ CC-BY-SA-3.0)

However, the total of incarcerated for violent crimes and especially for murder are males, and this is because murder is an extremely rare event, which is possibly committed by individuals with extreme characteristics or an extreme combination of them. In contrast, as we move to everyday behaviors, the difference between genders is greater and easily observed in mean scores (e.g., in driving behavior).

The question whether observed differences are the product of biology or on the contrary are the result mainly of social pressure has not been answered with absolute certainty; however, in nature and in primates, *clear differences in the behavior between genders* are observed. The exact role of this dimorphism and its significance in a number of issues, sexual orientation and preferences included, is a matter of debate with important moral-ideological and political dimensions (Rieber and Sigusch 1979).

Differences Between Genders in Observed Behavior

- Culturally induced
- Robust effect of an underlying biological substrate
- Anisogamy
- Caring of the offspring
  - Leads to the allocation of roles in the two genders
  - Sexual behavior
  - Aggression
  - Social and interpersonal behavior
- Emotional reactions
- Response to stress
- Visuospatial tasks
- Verbal tasks
- Memory

- Sexual arousal threshold
- Different type of reaction to cheating
- Different prevalence of neuropsychiatric disorders
- General attitude/mood
- Psychomotor activity
- Body posture and gestures
- Way of thinking

There are reports arguing that any such difference observed is *culturally induced*. Margaret Mead promoted the concept of the "Blank Slate," which enjoyed wide acceptance and in a way is the continuation of the thoughts of Aristotle (384–322 π.X.), as well as of the "tabula rasa" which had been conceptualized by John Locke (1632–1704). This approach reflects a kind of radical equality at all levels and aspects, and this was extending beyond political and social equality, and eventually considered all individuals not only equal but *essentially identical* in all terms. It was also implying that all individual differences were the product of social forces and not of inherent structural characteristics (Pinker 2002; Rose 2001; Mead 1928; Shankman 2009; Freeman 1983). This topic will be further discussed in detail in Chap. 19; however, it should be noted here that the overwhelming majority of the behaviors characteristic of gender are stable in all human groups around the world and are similar to those observed by analogy in the animal kingdom. These observations support the *robust effect of a biological substrate* (Allen and Gorski 1990; Giedd et al. 1996; Pakkenberg and Gundersen 1997; Schlaepfer et al. 1995; Swaab and Fliers 1985; Witelson et al. 1995).

Caring for the offspring has a central role in the lives of humans, and the same holds true for nature. This leads to the allocation of roles in the two genders, so that specific needs pertaining to parenthood could be satisfied. Many of them are absolutely biological. The differences between genders manifest themselves mainly through sexual behavior and the expression of aggression, but also in various aspects of social and interpersonal behavior. However, in these latter areas, the differences are not that great, and variability is significant. The largest differences concern *sexuality* and *aggressiveness*.

Less obvious but relatively proven differences between genders concerning behavior and mental functions are the following:

- Differences have been reported concerning the response to *emotionally* loaded images of faces; however, the data manifest inconsistencies (Sacher et al. 2013; Aleman and Swart 2008; Fischer et al. 2004; Lee et al. 2002, 2005; McClure et al. 2004; Schirmer et al. 2004; Schroeder et al. 2004; Fusar-Poli et al. 2009). It seems that males tend to manifest more intense responses to threatening stimuli, and they more often respond with physical violence (Kret and De Gelder 2012). Studies in animals have shown that male rats improved their performance under *stressful* conditions, while females manifested an impairment (Shors 2001).

- Already since infancy, boys outperform girls in *visuospatial tests*, and more specifically in tests concerning the handling of objects (Miller and Halpern 2014) as well as in tests of targeted motor skills, reaction times, and concentration (Sacher et al. 2013; Roivainen 2011). While both males and females are able to the same degree to learn to solve a spatial problem, males can learn more easily, but females seem to be able to recall more details of the process. The two genders seem to recruit different strategies which could potentially affect performance under specific circumstances (Beiko et al. 2004; Perrot-Sinal et al. 1996). The difference in the pattern of brain areas which are activated suggests that concerning visuospatial function, males probably mainly utilize more specialized circuits for this aim (parietal lobe), while females recruit less specialized and analytical systems (frontal lobe) (Gur et al. 2000; Hugdahl et al. 2006).
- Females manifest higher capacity of *recalling information from memory*, especially with social context (Cahill 2006), while they also manifest higher ability of recalling lists of words (Miller and Halpern 2014). This higher recall ability seems to be generalized (Sacher et al. 2013; Roivainen 2011) and concerns mainly *verbal* (de Lima Xavier et al. 2019) and to a lesser degree also nonverbal tasks, for example, memory of place of objects (Li 2014; Wallentin 2009).
- The presence of a *lower threshold for sexual excitement in males* is clear, and males are therefore more prone to pornography, prostitution, and paraphilias.
- Following the previous, there is a *very large anisogamy*, which comes from the fact that males are capable of producing millions of sperm cells, while on the contrary, females can produce only a limited number of eggs. This leads the two genders to adopt different sexual behaviors. Additionally, *the female carries the fetus* for a duration of 9 months, during which she needs global support. Also the newborn stays at first with the mother, under her protection and care. In 1972, Robert Trivers (1943–) developed the *theory of parental investments* (Trivers and Willard 1973; Trivers 1972) and proposed that there are different levels of parental investment between genders.
- Additionally, in general, females respond more negatively to *emotional cheating*, while males respond more to *sexual cheating*, particularly because paternity could not be proven until recently (Galperin et al. 2013; Buss 1995; Buss et al. 1992).
- In order for the human fetus to fully develop, as the fetuses of other mammals and the great apes do, a gestation period of 18–21 months would be necessary (Cooke et al. 1999; Weiner et al. 2008; Charnov and Ernest 2006; Wells et al. 2012; Washburn 1973, 1978, 1982; Washburn and McCown 1972; Dunsworth et al. 2012). The reason is irrelevant (for an analysis, see also Chap. 19); the fact is that *the human newborn is extremely unprepared to survive by itself alone* in nature, and demands significant care and parental investment, thus pushing toward the establishing of stable long-term relationships of mating. This makes humans much more oriented toward monogamy in comparison to primates and the other mammals. *Social rules* do not permit the open expression of deviances from monogamy, especially for women, which means that at least part of the differences between genders in sexual behavior is socially imposed.

- There are important differences concerning the manifestation of *neurologic and psychiatric disorders,* with depression being twice as usual in females, anorexia being found almost exclusively in females, and Parkinson's, Alzheimer's, and Huntington's diseases and amyotrophic lateral sclerosis (ALS), dyslexia, and attention deficit hyperactivity disorder (ADHD) being more frequent in males (Abel et al. 2010; Baron-Cohen et al. 2005; Cahill 2006; De Vries and Simerly 2002; Forger 2009; Jazin and Cahill 2010; McCarthy and Arnold 2011; McCarthy et al. 2009; Morris et al. 2004; Simerly 2002). Overall, the manifestation of *developmental psychiatric disorders* is much more frequent in males, while the manifestation of mental disorders during adulthood is more frequent in females (Beery and Zucker 2011).
- Apart from the above large differences between genders, there are reports on less well-studied and documented differences in general attitude/mood, psychomotor activity, somatic posture and gestures, emotionality, and the way of thinking.

17.2 Structural Brain Differences Between Genders

17.2.1 Basic Biological Differences

Basic Biological Differences Between Genders

- Hormonological profile
- Anisogamy
- Males are physically larger
  - Tendency of males to be competitive and have often violent behavior
- Females mature sexually earlier than males
  - Prefer males older than themselves
- Sexual dimorphism
- Sexual differential maturation
- Males:
  - Commit 90% of murders
  - The total of sexual offences
  - They constitute the overwhelming majority of incarcerated for violent crimes

Before someone considers the structural differences between genders (Fig. 17.2), we should have in mind that between them *four biological facts* are important:

- The *hormonological profile* is quite different between genders, to the extent it is possible to identify gender on the basis of laboratory testing alone.
- There is *great anisogamy*, because of the fact that males are in the position to produce millions of sperm cells, while females only a limited number of eggs, a fact that leads to different sexual behaviors between genders as analyzed above.

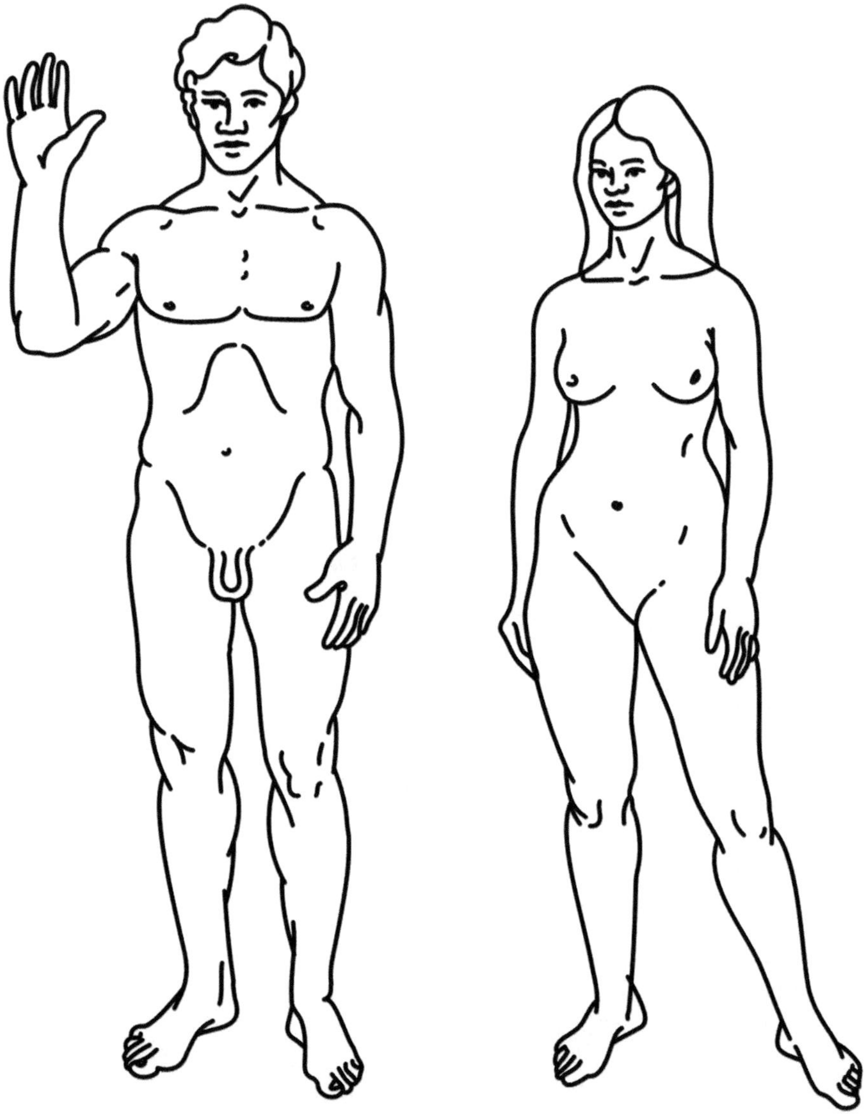

Fig. 17.2 Man and woman as depicted on the plate of the spaceship Pioneer 10

- The third concerns, as mentioned and analyzed above, the estimation that for the human fetus to be fully developed the same way the fetuses of other mammals and the great apes do, a gestation period of 18–21 months would be necessary (Cooke et al. 1999; Weiner et al. 2008; Charnov and Ernest 2006; Wells et al. 2012; Washburn 1973, 1978, 1982; Washburn and McCown 1972; Dunsworth et al. 2012). Many of the needs that are consequences of this fact have their corresponding analogs in primates also, which suggests that there is a *robust biological determinant* apart from social pressure.

- On average, males are *physically larger and with a tendency to be competitive and have often violent behavior*. Females mature sexually earlier than males, and they often prefer males older than themselves. These correspond to a pattern of *sexual dimorphism* and *sexual differential maturation* (Foley and Lewin 2013; Lewin 2009). Of course, this is enhanced culturally but it seems based on inherent traits. There is not a single culture in which the images of genders show something opposite, with females being culturally expected to be more violent than males. Additionally, the competition among men is possibly more intense during late adolescence and early adulthood, and this could explain the *particularly high rates of criminality* in these age groups (Rose 2001). Approximately 90% of murders and almost all sexual offences are committed by men (Spratt 2000), which constitute the overwhelming majority of incarcerated for violent crimes.

17.2.2 Whole Brain Volume

As males are physically larger than females, this seems to be the case also concerning the *volume and the weight of the brain*. Males have on average 8–13% larger brain than females (Ruigrok et al. 2014; Cahill 2006; Chen et al. 2007; Allen et al. 2003; Carne et al. 2006; Gur et al. 1999; Paus et al. 1996; Peters et al. 1998), although there are opposing opinions in the literature (Blatter et al. 1996; Cosgrove et al. 2007; Courchesne et al. 2000; De Vries 2004; Ruigrok et al. 2014). It is possible that the inconsistencies in the literature could probably be due in part because of the fact that the results are also influenced by the developmental phase (McCarthy and Konkle 2005). It is possible that the need for keeping an equivalent level of functioning with a smaller volume leads the female brain to manifest a larger gray-to-white matter ratio (Sacher et al. 2013). However, it seems that this difference in brain volume exists even after correction in terms of whole body volume, which means there is a *genuine* and *qualitative* and not simply a proportional quantitative difference between genders (Allen et al. 2003; Cosgrove et al. 2007; Luders et al. 2002; Nopoulos et al. 2000; Shin et al. 2005).

17.2.3 Regional Differences

Studies concerning differences between genders in specific brain regions are even more *contradictory* and do not lead to clear conclusions (Ruigrok et al. 2014; Baron-Cohen et al. 2005; Cosgrove et al. 2007; Goldstein et al. 2001; Hsu et al. 2008; Schmithorst et al. 2008; Szeszko et al. 2003; Guillamon et al. 2016).

17.2.3.1 Hypothalamus

The *hypothalamus* is the brain area in which most differences between genders have been identified, and these differences seem to relate to differences in reproductive behavior, gender identity, and sexual orientation (James 2005; Swaab et al. 1992;

Swaab and Hofman 1995; Kindon et al. 1996; Paredes and Baum 1995). Most findings concern the *medial preoptic area* of the hypothalamus and its nucleus (SDN-POA) (Swaab and Hofman 1984; Hofman and Swaab 1989; Swaab and Fliers 1985; Pilgrim and Reisert 1992; Yahr et al. 1994; McKenna 1998, 1999, 2000).

One of the early studies in the field reported that the group of the four *interstitial nuclei of the anterior hypothalamus (INAH)* is double in volume in males in comparison to females, and this is why many researchers label this area as the sexually dimorphic nucleus (SDN-POA) (Allen et al. 1989). More recent studies have shown that dimorphism concerns rather specifically the No3 nucleus (INAH3) of the area (Allen et al. 1989; LeVay 1991; Byne et al. 2000; Swaab et al. 2001).

The variability of reports includes the finding that the *suprachiasmatic nucleus* (SCN) which is related to circadian rhythms is more elongated in females and more spherical in males, while both the volume and the number of neurons of the *sexually dimorphic nucleus of the preoptic area* are double in size in males in comparison to females (Hofman and Swaab 1991). Especially those neurons of the nucleus that secret the vasoactive peptide (VIP) are found in larger numbers in males (Zhou et al. 1995b; Swaab and Fliers 1985; Swaab et al. 1994). The differences observed in this nucleus might constitute the reason for the differences found in circadian rhythms between males and females, with higher rates of slow-wave sleep, lower rates of stage 1 sleep, as well as shorter duration of unbounded circadian rhythms in females (van Hilten et al. 1994; Wever 1984).

Brain Regional Differences Between Genders

(The gender with largest volume is in the parenthesis)

- Hypothalamus
  - Group of the four interstitial nuclei of the anterior hypothalamus (INAH) (males)
  - Suprachiasmatic nucleus (SCN) (males)
- Limbic system and basal ganglia
  - Nucleus caudatus (females)
  - Hippocampus (females)
  - Globus pallidus (males)
  - Amygdala (males)
- Mesencephalon
  - Left inferior and right medial temporal gyrus (males)
  - Right occipital lingual gyrus (males)
  - Cerebellum (males)
  - Anterior commissure (females)
  - Greater lateralization of volume left in males and right in females
- Posterior brain and medulla oblongata
  - Darkly staining posteromedial part of the bed nucleus of the stria terminalis (BST-DSPM) (males)
  - Central nucleus of the bed nucleus of the stria terminalis (BST-DSPM) (males)

17.2.3.2 Limbic System and Basal Ganglia

The *nucleus caudatus* and the *hippocampus* are larger in females and the *globus pallidus* and the *amygdala* in males (Giedd et al. 1997). But concerning the hippocampus specifically, the relatively larger volume in females is under debate (Tan et al. 2016), while in males, if the total brain volume is taken into consideration, the difference concerning the amygdala which is reported to be 10% larger disappears (Marwha et al. 2017).

Males have larger volumes at the mesencephalon, the left inferior and right medial temporal gyrus, the right occipital lingual gyrus, and the cerebellum (Chen et al. 2007). The size of the anterior commissure probably manifests sexual dimorphism, and it is thicker with more fibers in females, in spite of the inconsistencies in the literature (Allen and Gorski 1991; Demeter et al. 1988; Martin 1985; Botez-Marquard and Botez 1992; Risse et al. 1978).

In general, males manifest a tendency for a *left lateralization* of systems (amygdala, hippocampus, and anterior parahippocampal gyrus), while females for a *right lateralization* (frontal lobe, medial and inferior frontal lobe, anterior cingulate, and occipital cortex) (Ruigrok et al. 2014). However, a meta-analysis found no significant differences in lateralization (Sommer et al. 2008).

17.2.3.3 Posterior Brain and Medulla Oblongata

In males, the darkly staining posteromedial part (Allen and Gorski 1990) and the central nucleus of the bed nucleus of the stria terminalis (BST-DSPM) (Zhou et al. 1995a) manifest larger volumes.

17.2.4 Differences in Gray Matter

The simultaneous finding of increased total volume in males and the *higher ratio of gray to total volume in females* are important for the understanding of the functional needs of the brain, but it also reveals the methodological problems that arise when trying to compare brain structures (Witte et al. 2010; Sacher et al. 2013; Allen et al. 2003; Goldstein et al. 2001; Gur et al. 1999; Luders et al. 2005). This difference is reflected also in the different thickness of gray matter in females, especially in the anterior dorsal, the posterior, and the ventral cingulate gyrus and the right inferior parietal lobule (Chen et al. 2007), but maybe also in the left auditory and related to language cortex as well as in the primary visual and visuospatial association areas (Brun et al. 2009). However, the total gray matter volume is smaller in females (Gur et al. 1999)

17.2.5 Differences in White Matter and Connectivity

Older studies have shown differences between males and females concerning white but not gray matter, and this is in accord with what has already been discussed above concerning the difference in total brain volume and the gray-to-whole brain volume ratio (Filipek et al. 1994; Passe et al. 1997; Gur et al. 1999). In general, a *smaller white matter volume in females* is reported (Hsu et al. 2008; Lemaitre et al. 2005) however not all studies report this (Smith et al. 2007; Bourisly et al. 2017). Here again the developmental phase plays a crucial role, with the white matter volume to increase much more in males during adolescence (Perrin et al. 2009), while the developmental trajectories seem to significantly differ concerning the connections between the amygdala and the lateral prefrontal cortex (Zuo et al. 2010).

More important seem to be the reports concerning differences in white matter in specific brain areas. These findings might imply that maybe there are qualitative differences between genders in connectivity among the various brain regions (Gong et al. 2011). These differences refer to regions that concern *emotional and motor control as well as higher mental functions*, including the thalamus, the cingulate gyrus, the corpus callosum, the occipitofrontal fasciculus, the under the parahippocampal gyrus white matter, the internal capsule, the white matter that corresponds to the middle frontal, the cingulate and the posterior central gyri, the hippocampus, the isle, the periaqueductal gray matter, as well as the fusiform (middle occipitotemporal) gyrus (Sacher et al. 2013; Berglund et al. 2006; Savic et al. 2005; Savic and Lindstrom 2008; Zhou et al. 2014; Kong et al. 2010; Chou et al. 2011; Menzler et al. 2011; Westerhausen et al. 2003, 2004; Liu et al. 2010; Bourisly et al. 2017; Biswal et al. 2010; Huster et al. 2009; Inano et al. 2011; Kanaan et al. 2014; Rametti et al. 2011a; Schoonheim et al. 2014; Takao et al. 2014; van Hemmen et al. 2017). Concerning the corpus callosum, significant differences are reported in the thickness and microstructure, especially in the trunk and the rostrum, with females manifesting larger volume and density (Sacher et al. 2013; Westerhausen et al. 2011). In contrast, some studies showed a higher density and volume in the dorsal and ventral optical tract in females (Keller and Menon 2009) as well as only in the trunk and the genu of the corpus callosum (Westerhausen et al. 2011).

It has been reported that *males manifest a denser connectivity in the right and females in the left hemisphere* (Sacher et al. 2013; Tian et al. 2011).

It seems that these differences cannot be explained solely on the basis of the difference in the whole brain volume (Bourisly et al. 2017) and led many authors to postulate that there are important gender-determined differences in the functional organization of the human brain (Yan et al. 2011); however, one should be cautious in the way he/she understands this statement.

17.2.6 Relationship to Sexual Orientation

Before someone proceeds to discuss a sensitive issue like the differences in brain structure and function between genders, and the role this organization plays in sexual orientation, and the agreement between the biological with the experienced and self-determined sex, he/she should make a clear analysis of the sociopolitical and medical frame that determines these issues and should stress that the *acceptance of diversity rather than the demand for similarity* is the foundation for the respect of human rights.

Homosexuality was removed from the list of mental disorders already since the 1970s in a pioneering move of the American Psychiatric Association. Its previous consideration as a mental disorder was based on the dominant at that time psychoanalytic literature (Friedman and Downey 1998) and was imposed by social and legal pressure. It is important to note how revolutionary this act was, since the decriminalization of homosexuality has not been completed in the USA and in all their states before the early years of the twenty-first century, and it is still considered a penal offence in many Middle Eastern and Asian countries. In DSM-I, homosexuality had been classified under the umbrella of "sociopathic personality disorder," while in DSM-II, it had been recognized as an independent nosological entity. Its removal from nosology was a personal and clear choice by Robert Spitzer (1932–2015), while the American Psychiatric Association has established an award in the memory of John Fryer (1937–2003), who in 1972 made a sensational speech during the works of the annual APA congress while concealing his identity by wearing a mask and a wig as Dr. H. Anonymous and starting his speech with the phrase "I am homosexual. I am a Psychiatrist" (Lieberman 2016; Lieberman and First 2016). Removing of homosexuality from DSM was also due to the decision to follow a more empirical-scientific approach and to abandon the theoretical-hypothetical etiological approaches (Decker 2013; Bayer and Spitzer 1985). On the other hand, the classification of mental disorders by the World Health Organization followed this path only with its tenth edition (ICD-10) in 1990, and this delay was because of the need for a wider international consensus. In the new ICD-11 classification, all disorders related to sexuality are classified in a separate chapter, outside the chapter of mental disorders, and the presence of dysphoria or dysfunction for any diagnosis is a prerequisite (Reed et al. 2016).

In the literature, there is a limited number of reports concerning the presence of structural differences in the brains between heterosexual and homosexual men (LeVay 1991; Swaab and Hofman 1990) as well as between transgender and other men (Zhou et al. 1995a; Guillamon et al. 2016; Kreukels and Guillamon 2016; Swaab 2004, 2007).

It has been reported that homosexual men manifest smaller volume (thus closer to the size found in women) in the anterior hypothalamus and more specifically in the interstitial nuclei (INAH) (LeVay 1991), thinner cortex in the cuneus, and smaller volume of the thalamus, in comparison to heterosexual men (Abe et al. 2014). It is interesting that there are no differences in neurocognition between homosexual and heterosexual women (Rahman and Wilson 2003; Rahman et al. 2004). One study reported double volume of the suprachiasmatic nucleus (SCN) with double number of neurons in the brains of homosexual men who died of AIDS, but without any other findings in the rest of the hypothalamus (Hofman and Swaab 1991; Swaab and Hofman 1990).

Findings from the brain MRIs of transgender females showed that they had typical for females putamen of the lentiform nucleus, which was larger in volume in comparison to non-transgender males (Saleem and Rizvi 2017) and similarly typical central part of the bed nucleus of the stria terminalis (BNST) and of No3 interstitial nucleus of the anterior hypothalamus (INAH-3)(Guillamon et al. 2016).

Recent studies in transgender and homosexual men and women in comparison to heterosexuals showed that *transsexualism is probably related to differences in brain areas related with the perception of self and body, while sexual orientation is related with lower sexual differentiation* (Burke et al. 2017). There is a discrepancy concerning the role of homosexuality in transgender persons pertaining to the differences which are observed in brain structures and the instability of the findings (Rametti et al. 2011a, b; Kranz et al. 2014; Burke et al. 2018; Feusner et al. 2016, 2017; Manzouri et al. 2017). According to some studies, the contrast between self-defined and biological gender is related to the thickness of gray matter (Luders et al. 2012; Manzouri et al. 2017; Zubiaurre-Elorza et al. 2013) and its total volume (Hoekzema et al. 2015; Luders et al. 2009; Simon et al. 2013) as well as with the structural connectivity among brain areas (Hahn et al. 2015, 2016; Kranz et al. 2014; Rametti et al. 2011a, b; Burke et al. 2018; Feusner et al. 2016, 2017; Manzouri et al. 2017), with a reverse sexual dimorphism. In this frame, the existence of a fronto-occipito-parietal dissociation between the processing of the perception of self and the perception of familiarity and identification with own body has been proposed (Burke et al. 2018; Feusner et al. 2016, 2017; Manzouri et al. 2017; Sarubbo et al. 2013; Rametti et al. 2011a, b; Blanke et al. 2015; Ionta et al. 2014; Hahn et al. 2015; Zubiaurre-Elorza et al. 2013; Giummarra et al. 2011; Longo et al. 2010; Tsakiris 2010).

The prevailing opinion is that this is a consequence of the effects of sexual hormones, mainly during gestation but also during the perinatal period (Balthazart 2011; Bradley and Zucker 1990; Hines 2011; Savic et al. 2010; Pesaresi et al. 2015; Burke et al. 2018).

17.3 Differences in Neurobiological Function of the Brain Between Genders

The human brain includes a number of structures and functions that manifest *sexual dimorphism*, whose origin should be traced back in the developmental phases and should be considered to be a result of organization. In general, it is believed that this dimorphism is organized or imprinted through hormonal signals, and mainly during the early two periods during which a sexually dimorphic hormonal peak from the gonads is observed, that is, during *gestation and the perinatal period*. From adolescence on, sex hormones modify the functioning of already formulated neuronal networks (Swaab et al. 1992; de Zegher et al. 1992; Halpern et al. 1998; van Amelsvoort et al. 2001).

The most important dimorphic sexual differences in the brain are the following:

- Although it is reported that in principle, under steady state, the differences in brain function are minimal or they do not exist at all (Weissman-Fogel et al. 2010), there are reports concerning a higher activation of the total brain during steady state in females (Devous Sr. et al. 1986; Gur et al. 1982; Baxter Jr. et al. 1987), especially in the basal forebrain (Andreason et al. 1994). The findings are contradictory in general, and conclusions are difficult (Azari et al. 1992; Kuhl et al. 1982), partially because of the changes caused by the menstrual cycle in females (Reiman et al. 1996).

- Healthy males are reported to have up to 50% higher *serotonin synthesis* in comparison to healthy females, which must be noted that they manifest double rates of depression in comparison to males (Nishizawa et al. 1997). The levels of the metabolite *5-hydroxyindoleacetic acid (5-HIAA)* in the hypothalamus are lower in males (Gottfries et al. 1974), while the functioning of the serotonin transporter in the synaptic cleft is reported to be increased in females, but it is possible that this is reversed under conditions of stress (Jovanovic et al. 2008; Staley et al. 2001; Cosgrove et al. 2007). Increased activation of 5-HT_{1A} has been reported in females (Parsey et al. 2002) and lower activation of 5-HT_{2A} (Biver et al. 1996), but these observations have not been confirmed (Adams et al. 2004; Frokjaer et al. 2008; Stein et al. 2008; Rabiner et al. 2002; Meyer et al. 1999; Rosier et al. 1996).
- Similarly, the reports on differences concerning the dopaminergic system are also inconsistent (Best et al. 2005; Hietala et al. 1994; Kaasinen et al. 2001; van Dyck et al. 2002). There are some observations concerning an increased dopaminergic tone in the striatum (Laakso et al. 2002; Lavalaye et al. 2000; Mozley et al. 2001; Staley et al. 2001) and higher dopamine receptor density in the rest of the brain in females (Wong et al. 1988).
- A *more robust right lateralization of executive function for males and left for females* is reported (Koscik et al. 2010), and a similar report exists for the same type of lateralized activation in relationship with memory tasks (Cahill 2003; Cahill et al. 2004; Canli et al. 2002). One report restricts the sexually more robust lateralization in the amygdala functioning and to the affective components of memory alone (Mackiewicz et al. 2006), which is a finding that other authors consider to be false and an artifact because of the variations in the amygdala anatomy (Sergerie et al. 2008). Another report found a larger activation in the prefrontal and limbic areas for females and a more diffused brain activation in males (Hill et al. 2014).
- Visuospatial tasks cause higher activation in the right hemisphere in males and in the left in females (Gur et al. 2000; Clements et al. 2006). Additionally, this is related with the activation of different areas in the superior parietal lobule in males and the inferior frontal cortex in females. This later finding suggests a differential way of processing during visuospatial analysis between genders, with a *more fuzzy way in males and a more laborious digital way in females* (Hugdahl et al. 2006).
- Emotional processing in males includes the increased activation of the amygdala bilaterally and of the orbital frontal cortex, while in females it includes the increased activation of the globus pallidus, the putamen, the thalamus bilaterally, the cingulate gyrus, and the mesencephalon (Sacher et al. 2013). Especially, the viewing of pictures with sexual content in comparison to neutral images caused a significant increase in the amygdala and the hypothalamus bilaterally and the right cerebellum in males, which was evident even after controlling for the

degree of sexual arousal (Hamann et al. 2004). As mentioned above, especially for the amygdala, many authors suggest that its anatomical variations are the real reason behind the findings (Sergerie et al. 2008).

- It is interesting that the provocation of jealousness in males causes increased activation in the amygdala and the hypothalamus, while in females in the posterior superior temporal sulcus (Takahashi et al. 2006)

Differences in Neurobiological Function of the Brain Between Genders

- Under steady state, the differences in brain function are minimal or they do not exist at all
- In males up to 50% higher serotonin synthesis
- The levels of 5-hydroxyindoleacetic acid (5-HIAA) in the hypothalamus are lower in males
- The functioning of the serotonin transporter increased in females
- Inconsistent reports concerning the dopaminergic system
- More robust lateralization right for males and left for females of:
  - Executive functions
  - Memory
  - Visuospatial tasks
- Emotional processing
  - Increased activation of the amygdala bilaterally (males)
  - Orbital frontal cortex (males)
  - Increased activation of the globus pallidus (females)
  - Putamen (females)
  - Thalamus bilaterally (females)
  - Cingulate gyrus (females)
  - Mesencephalon (females)
- Provocation of jealousness causes increased activation in the:
  - Amygdala and the hypothalamus (males)
  - Posterior superior temporal sulcus (females)

17.4 The Role of Sexual Hormones and Genes in the Development and Functioning of the Brain

Even in primary cell cultures of neurons, there are sex-related differences (Carruth et al. 2002; Nunez and McCarthy 2009). The *effect of sex hormones on brain development* is considered by many authors as certain and robust; however, it is not studied to a satisfactory extent, and many of its special characteristics continue to remain unknown. It is also certain that genes exert an effect as well, but their effect is also understudied.

It is clear that these hormones' levels in the blood of adults vary significantly (Gladue et al. 1984; Zhou et al. 1995a); however, there are data from animal studies suggesting that they continue having an effect even during adulthood. For example, the *castration* of adult animals leads to an impressive loss of volume in those brain areas that manifest sexual dimorphism, and on the contrary, the administration of male hormones partially but not completely reverses dimorphism (Bloch and Gorski 1988; Commins and Yahr 1984; Panzica et al. 1987; DeVoogd and Nottebohm 1981; Devoogd et al. 1985).

In humans, changes in behavior, but also in mental functions, have been observed after the *administration of hormones for purposes of gender change or other therapeutical reasons* (Korol 2004; Van Goozen et al. 1994, 1995), and generally there are some data supporting that also in humans, sex hormones have an effect on regional brain volume and also on behavior, although to a much smaller extent in comparison to animals, as dimorphism in humans is much less pronounced (Allen and Gorski 1990; Giedd et al. 1996; Pakkenberg and Gundersen 1997; Schlaepfer et al. 1995; Swaab and Fliers 1985; Witelson et al. 1995; Pfaff et al. 2009). The hormonal effect seems to be stronger during adolescence (Sisk and Zehr 2005). The effect of sex hormones and their fluctuations on various receptors have been well documented during *menstruation* and *menopause* (Smith et al. 1998, 2001; Yoshida et al. 2000; Zubieta et al. 1999), and this documentation has been particularly strong concerning the serotonin receptors (Frokjaer et al. 2010; Kugaya et al. 2003; Moses et al. 2000; Moses-Kolko et al. 2011)

The sex-related genes seem to play an important role on the secondary characteristics of genders, including social behavior, reproduction, and nociception (perception of pain) (Arnold and Chen 2009; Budefeld et al. 2008; Dewing et al. 2006), and it seems that the same allele manifests different functioning depending on whether it is located on the specific chromosome (either X or Y) that comes from the father or the mother, a phenomenon called parental imprinting (Gregg et al. 2010).

References

Abe C, Johansson E, Allzen E, Savic I (2014) Sexual orientation related differences in cortical thickness in male individuals. PLoS One 9(12):e114721. https://doi.org/10.1371/journal.pone.0114721

Abel KM, Drake R, Goldstein JM (2010) Sex differences in schizophrenia. Int Rev Psychiatry 22(5):417–428. https://doi.org/10.3109/09540261.2010.515205

Adams KH, Pinborg LH, Svarer C, Hasselbalch SG, Holm S, Haugbol S, Madsen K, Frokjaer V, Martiny L, Paulson OB, Knudsen GM (2004) A database of [(18)F]-altanserin binding to 5-HT(2A) receptors in normal volunteers: normative data and relationship to physiological and demographic variables. NeuroImage 21(3):1105–1113. https://doi.org/10.1016/j.neuroimage.2003.10.046

Aleman A, Swart M (2008) Sex differences in neural activation to facial expressions denoting contempt and disgust. PLoS one 3(11):e3622. https://doi.org/10.1371/journal.pone.0003622

Allen LS, Gorski RA (1990) Sex difference in the bed nucleus of the stria terminalis of the human brain. J Comp Neurol 302(4):697–706. https://doi.org/10.1002/cne.903020402

Allen LS, Gorski RA (1991) Sexual dimorphism of the anterior commissure and massa intermedia of the human brain. J Comp Neurol 312(1):97–104. https://doi.org/10.1002/cne.903120108

Allen LS, Hines M, Shryne JE, Gorski RA (1989) Two sexually dimorphic cell groups in the human brain. J Neurosci 9(2):497–506. https://doi.org/10.1523/JNEUROSCI.09-02-00497.1989

Allen JS, Damasio H, Grabowski TJ, Bruss J, Zhang W (2003) Sexual dimorphism and asymmetries in the gray-white composition of the human cerebrum. NeuroImage 18(4):880–894. https://doi.org/10.1016/s1053-8119(03)00034-x

Andreason PJ, Zametkin AJ, Guo AC, Baldwin P, Cohen RM (1994) Gender-related differences in regional cerebral glucose metabolism in normal volunteers. Psychiatry Res 51(2):175–183. https://doi.org/10.1016/0165-1781(94)90037-x

Arnold AP, Chen X (2009) What does the "four core genotypes" mouse model tell us about sex differences in the brain and other tissues? Front Neuroendocrinol 30(1):1–9. https://doi.org/10.1016/j.yfrne.2008.11.001

Azari NP, Rapoport SI, Grady CL, DeCarli C, Haxby JV, Schapiro MB, Horwitz B (1992) Gender differences in correlations of cerebral glucose metabolic rates in young normal adults. Brain Res 574(1-2):198–208. https://doi.org/10.1016/0006-8993(92)90817-s

Balthazart J (2011) Minireview: Hormones and human sexual orientation. Endocrinology 152(8):2937–2947. https://doi.org/10.1210/en.2011-0277

Baron-Cohen S, Knickmeyer RC, Belmonte MK (2005) Sex differences in the brain: implications for explaining autism. Science 310(5749):819–823. https://doi.org/10.1126/science.1115455

Baxter LR Jr, Mazziotta JC, Phelps ME, Selin CE, Guze BH, Fairbanks L (1987) Cerebral glucose metabolic rates in normal human females versus normal males. Psychiatry Res 21(3):237–245. https://doi.org/10.1016/0165-1781(87)90028-x

Bayer R, Spitzer RL (1985) Neurosis, psychodynamics, and DSM-III. A history of the controversy. Arch Gen Psychiatry 42(2):187–196. https://doi.org/10.1001/archpsyc.1985.01790250081011

Beery AK, Zucker I (2011) Sex bias in neuroscience and biomedical research. Neurosci Biobehav Rev 35(3):565–572. https://doi.org/10.1016/j.neubiorev.2010.07.002

Beiko J, Lander R, Hampson E, Boon F, Cain DP (2004) Contribution of sex differences in the acute stress response to sex differences in water maze performance in the rat. Behav Brain Res 151(1-2):239–253. https://doi.org/10.1016/j.bbr.2003.08.019

Berglund H, Lindstrom P, Savic I (2006) Brain response to putative pheromones in lesbian women. Proc Natl Acad Sci U S A 103(21):8269–8274. https://doi.org/10.1073/pnas.0600331103

Best SE, Sarrel PM, Malison RT, Laruelle M, Zoghbi SS, Baldwin RM, Seibyl JP, Innis RB, van Dyck CH (2005) Striatal dopamine transporter availability with [123I]beta-CIT SPECT is unrelated to gender or menstrual cycle. Psychopharmacology (Berl) 183(2):181–189. https://doi.org/10.1007/s00213-005-0158-5

Biswal BB, Mennes M, Zuo XN, Gohel S, Kelly C, Smith SM, Beckmann CF, Adelstein JS, Buckner RL, Colcombe S, Dogonowski AM, Ernst M, Fair D, Hampson M, Hoptman MJ, Hyde JS, Kiviniemi VJ, Kotter R, Li SJ, Lin CP, Lowe MJ, Mackay C, Madden DJ, Madsen KH, Margulies DS, Mayberg HS, McMahon K, Monk CS, Mostofsky SH, Nagel BJ, Pekar JJ, Peltier SJ, Petersen SE, Riedl V, Rombouts SA, Rypma B, Schlaggar BL, Schmidt S, Seidler RD, Siegle GJ, Sorg C, Teng GJ, Veijola J, Villringer A, Walter M, Wang L, Weng XC, Whitfield-Gabrieli S, Williamson P, Windischberger C, Zang YF, Zhang HY, Castellanos FX, Milham MP (2010) Toward discovery science of human brain function. Proc Natl Acad Sci U S A 107(10):4734–4739. https://doi.org/10.1073/pnas.0911855107

Biver F, Lotstra F, Monclus M, Wikler D, Damhaut P, Mendlewicz J, Goldman S (1996) Sex difference in 5HT2 receptor in the living human brain. Neurosci Lett 204(1-2):25–28. https://doi.org/10.1016/0304-3940(96)12307-7

Blanke O, Slater M, Serino A (2015) Behavioral, neural, and computational principles of bodily self-consciousness. Neuron 88(1):145–166. https://doi.org/10.1016/j.neuron.2015.09.029

Blatter DD, Bigler ED, Johnson CS, Anderson C, Gale SD (1996) A normative database from magnetic resonance imaging. In: Neuroimaging I. Human brain function. Springer, Boston, MA. https://doi.org/10.1007/978-1-4899-1701-0_4

Bloch GJ, Gorski RA (1988) Cytoarchitectonic analysis of the SDN-POA of the intact and gonadectomized rat. J Comp Neurol 275(4):604–612. https://doi.org/10.1002/cne.902750408

Botez-Marquard T, Botez MI (1992) Visual memory deficits after damage to the anterior commissure and right fornix. Arch Neurol 49(3):321–324. https://doi.org/10.1001/archneur.1992.00530270141032

Bourisly AK, Gejo G, Hayat AA, Alsarraf L, Dashti FM, Di Paola M (2017) White matter sexual dimorphism of the adult human brain. Transl Neurosci 8(1):49–53. https://doi.org/10.1515/tnsci-2017-0009

Bradley SJ, Zucker KJ (1990) Gender identity disorder and psychosexual problems in children and adolescents. Can J Psychiatry 35(6):477–486. https://doi.org/10.1177/070674379003500603

Brun CC, Lepore N, Luders E, Chou YY, Madsen SK, Toga AW, Thompson PM (2009) Sex differences in brain structure in auditory and cingulate regions. Neuroreport 20(10):930–935. https://doi.org/10.1097/wnr.0b013e32832c5e65

Budefeld T, Grgurevic N, Tobet SA, Majdic G (2008) Sex differences in brain developing in the presence or absence of gonads. Dev Neurobiol 68(7):981–995. https://doi.org/10.1002/dneu.20638

Burke SM, Manzouri AH, Savic I (2017) Structural connections in the brain in relation to gender identity and sexual orientation. Sci Rep 7(1):17954. https://doi.org/10.1038/s41598-017-17352-8

Burke SM, Manzouri AH, Dhejne C, Bergstrom K, Arver S, Feusner JD, Savic-Berglund I (2018) Testosterone effects on the brain in transgender men. Cereb Cortex 28(5):1582–1596. https://doi.org/10.1093/cercor/bhx054

Buss DM (1995) Psychological sex differences. Origins through sexual selection. Am Psychol 50(3):164–168; discussion 169-171. https://doi.org/10.1037/0003-066x.50.3.164

Buss DM, Larsen RJ, Westen D, Semmelroth J (1992) Sex differences in jealousy: evolution, physiology, and psychology. Psychol Sci 3(4):251–255

Byne W, Lasco MS, Kemether E, Shinwari A, Edgar MA, Morgello S, Jones LB, Tobet S (2000) The interstitial nuclei of the human anterior hypothalamus: an investigation of sexual variation in volume and cell size, number and density. Brain Res 856(1-2):254–258. https://doi.org/10.1016/s0006-8993(99)02458-0

Cahill L (2003) Sex-related influences on the neurobiology of emotionally influenced memory. Ann N Y Acad Sci 985(1):163–173. https://doi.org/10.1111/j.1749-6632.2003.tb07080.x

Cahill L (2006) Why sex matters for neuroscience. Nat Rev Neurosci 7(6):477–484. https://doi.org/10.1038/nrn1909

Cahill L, Uncapher M, Kilpatrick L, Alkire MT, Turner J (2004) Sex-related hemispheric lateralization of amygdala function in emotionally influenced memory: an FMRI investigation. Learn Mem 11(3):261–266. https://doi.org/10.1101/lm.70504

Canli T, Desmond JE, Zhao Z, Gabrieli JD (2002) Sex differences in the neural basis of emotional memories. Proc Natl Acad Sci U S A 99(16):10789–10794. https://doi.org/10.1073/pnas.162356599

Carne RP, Vogrin S, Litewka L, Cook MJ (2006) Cerebral cortex: an MRI-based study of volume and variance with age and sex. J Clin Neurosci 13(1):60–72. https://doi.org/10.1016/j.jocn.2005.02.013

Carruth LL, Reisert I, Arnold AP (2002) Sex chromosome genes directly affect brain sexual differentiation. Nat Neurosci 5(10):933–934. https://doi.org/10.1038/nn922

Charnov EL, Ernest SK (2006) The offspring-size/clutch-size trade-off in mammals. Am Nat 167(4):578–582. https://doi.org/10.1086/501141

Chen X, Sachdev PS, Wen W, Anstey KJ (2007) Sex differences in regional gray matter in healthy individuals aged 44-48 years: a voxel-based morphometric study. NeuroImage 36(3):691–699. https://doi.org/10.1016/j.neuroimage.2007.03.063

Chou KH, Cheng Y, Chen IY, Lin CP, Chu WC (2011) Sex-linked white matter microstructure of the social and analytic brain. NeuroImage 54(1):725–733. https://doi.org/10.1016/j.neuroimage.2010.07.010

Clements AM, Rimrodt SL, Abel JR, Blankner JG, Mostofsky SH, Pekar JJ, Denckla MB, Cutting LE (2006) Sex differences in cerebral laterality of language and visuospatial processing. Brain Lang 98(2):150–158. https://doi.org/10.1016/j.bandl.2006.04.007

Commins D, Yahr P (1984) Adult testosterone levels influence the morphology of a sexually dimorphic area in the Mongolian gerbil brain. J Comp Neurol 224(1):132–140. https://doi.org/10.1002/cne.902240112

Cooke BM, Tabibnia G, Breedlove SM (1999) A brain sexual dimorphism controlled by adult circulating androgens. Proc Natl Acad Sci U S A 96(13):7538–7540. https://doi.org/10.1073/pnas.96.13.7538

Cosgrove KP, Mazure CM, Staley JK (2007) Evolving knowledge of sex differences in brain structure, function, and chemistry. Biol Psychiatry 62(8):847–855. https://doi.org/10.1016/j.biopsych.2007.03.001

Courchesne E, Chisum HJ, Townsend J, Cowles A, Covington J, Egaas B, Harwood M, Hinds S, Press GA (2000) Normal brain development and aging: quantitative analysis at in vivo MR imaging in healthy volunteers. Radiology 216(3):672–682. https://doi.org/10.1148/radiology.216.3.r00au37672

de Lima Xavier L, Hanekamp S, Simonyan K (2019) Sexual dimorphism within brain regions controlling speech production. Front Neurosci 13:795. https://doi.org/10.3389/fnins.2019.00795

De Vries GJ (2004) Minireview: sex differences in adult and developing brains: compensation, compensation, compensation. Endocrinology 145(3):1063–1068. https://doi.org/10.1210/en.2003-1504

De Vries GJ, Simerly RB (2002) Anatomy, development, and function of sexually dimorphic neural circuits in the mammalian brain. In: Hormones, brain and behavior. Elsevier, Boston. https://doi.org/10.1016/b978-012532104-4/50066-4

de Zegher F, Devlieger H, Veldhuis JD (1992) Pulsatile and sexually dimorphic secretion of luteinizing hormone in the human infant on the day of birth. Pediatr Res 32(5):605–607. https://doi.org/10.1203/00006450-199211000-00025

Decker HS (2013) The making of DSM-III: a diagnostic manual's conquest of American Psychiatry. Oxford University Press, New York

Demeter S, Ringo JL, Doty RW (1988) Morphometric analysis of the human corpus callosum and anterior commissure. Hum Neurobiol 6(4):219–226

DeVoogd T, Nottebohm F (1981) Gonadal hormones induce dendritic growth in the adult avian brain. Science 214(4517):202–204. https://doi.org/10.1126/science.7280692

Devoogd TJ, Nixdorf B, Nottebohm F (1985) Synaptogenesis and changes in synaptic morphology related to acquisition of a new behavior. Brain research 329(1-2):304–308. https://doi.org/10.1016/0006-8993(85)90539-6

Devous MD Sr, Stokely EM, Chehabi HH, Bonte FJ (1986) Normal distribution of regional cerebral blood flow measured by dynamic single-photon emission tomography. J Cereb Blood Flow Metab 6(1):95–104. https://doi.org/10.1038/jcbfm.1986.12

Dewing P, Chiang CW, Sinchak K, Sim H, Fernagut PO, Kelly S, Chesselet MF, Micevych PE, Albrecht KH, Harley VR, Vilain E (2006) Direct regulation of adult brain function by the male-specific factor SRY. Curr Biol 16(4):415–420. https://doi.org/10.1016/j.cub.2006.01.017

Dunsworth HM, Warrener AG, Deacon T, Ellison PT, Pontzer H (2012) Metabolic hypothesis for human altriciality. Proc Natl Acad Sci U S A 109(38):15212–15216. https://doi.org/10.1073/pnas.1205282109

Feusner JD, Dervisic J, Kosidou K, Dhejne C, Bookheimer S, Savic I (2016) Female-to-male transsexual individuals demonstrate different own body identification. Arch Sex Behav 45(3):525–536. https://doi.org/10.1007/s10508-015-0596-z

Feusner JD, Lidstrom A, Moody TD, Dhejne C, Bookheimer SY, Savic I (2017) Intrinsic network connectivity and own body perception in gender dysphoria. Brain Imaging Behav 11(4):964–976. https://doi.org/10.1007/s11682-016-9578-6

Filipek PA, Richelme C, Kennedy DN, Caviness VS Jr (1994) The young adult human brain: an MRI-based morphometric analysis. Cereb Cortex 4(4):344–360. https://doi.org/10.1093/cercor/4.4.344

Fischer H, Sandblom J, Herlitz A, Fransson P, Wright CI, Backman L (2004) Sex-differential brain activation during exposure to female and male faces. Neuroreport 15(2):235–238. https://doi.org/10.1097/00001756-200402090-00004

Foley R, Lewin R (2013) Principles of human evolution. Wiley, London

Forger NG (2009) The organizational hypothesis and final common pathways: sexual differentiation of the spinal cord and peripheral nervous system. Horm Behav 55(5):605–610. https://doi.org/10.1016/j.yhbeh.2009.03.008

Freeman D (1983) Margaret Mead and Samoa. Harvard University Press, Cambridge

Friedman RC, Downey JI (1998) Psychoanalysis and the model of homosexuality as psychopathology: a historical overview. Am J Psychoanal 58(3):249–270. https://doi.org/10.1023/a:1022583223820

Frokjaer VG, Mortensen EL, Nielsen FA, Haugbol S, Pinborg LH, Adams KH, Svarer C, Hasselbalch SG, Holm S, Paulson OB, Knudsen GM (2008) Frontolimbic serotonin 2A receptor binding in healthy subjects is associated with personality risk factors for affective disorder. Biol Psychiatry 63(6):569–576. https://doi.org/10.1016/j.biopsych.2007.07.009

Frokjaer VG, Erritzoe D, Juul A, Nielsen FA, Holst K, Svarer C, Madsen J, Paulson OB, Knudsen GM (2010) Endogenous plasma estradiol in healthy men is positively correlated with cerebral cortical serotonin 2A receptor binding. Psychoneuroendocrinology 35(9):1311–1320. https://doi.org/10.1016/j.psyneuen.2010.03.002

Fusar-Poli P, Placentino A, Carletti F, Landi P, Allen P, Surguladze S, Benedetti F, Abbamonte M, Gasparotti R, Barale F, Perez J, McGuire P, Politi P (2009) Functional atlas of emotional faces processing: a voxel-based meta-analysis of 105 functional magnetic resonance imaging studies. J Psychiatry Neurosci 34(6):418–432

Galperin A, Haselton MG, Frederick DA, Poore J, von Hippel W, Buss DM, Gonzaga GC (2013) Sexual regret: evidence for evolved sex differences. Arch Sex Behav 42(7):1145–1161. https://doi.org/10.1007/s10508-012-0019-3

Giedd JN, Vaituzis AC, Hamburger SD, Lange N, Rajapakse JC, Kaysen D, Vauss YC, Rapoport JL (1996) Quantitative MRI of the temporal lobe, amygdala, and hippocampus in normal human development: ages 4-18 years. J Comp Neurol 366(2):223–230. https://doi.org/10.1002/(SICI)1096-9861(19960304)366:2<223::AID-CNE3>3.0.CO;2-7

Giedd JN, Castellanos FX, Rajapakse JC, Vaituzis AC, Rapoport JL (1997) Sexual dimorphism of the developing human brain. Prog Neuropsychopharmacol Biol Psychiatry 21(8):1185–1201. https://doi.org/10.1016/s0278-5846(97)00158-9

Giummarra MJ, Bradshaw JL, Nicholls ME, Hilti LM, Brugger P (2011) Body integrity identity disorder: deranged body processing, right fronto-parietal dysfunction, and phenomenological experience of body incongruity. Neuropsychol Rev 21(4):320–333. https://doi.org/10.1007/s11065-011-9184-8

Gladue BA, Green R, Hellman RE (1984) Neuroendocrine response to estrogen and sexual orientation. Science 225(4669):1496–1499. https://doi.org/10.1126/science.6089349

Goldstein JM, Seidman LJ, Horton NJ, Makris N, Kennedy DN, Caviness VS Jr, Faraone SV, Tsuang MT (2001) Normal sexual dimorphism of the adult human brain assessed by in vivo magnetic resonance imaging. Cereb Cortex 11(6):490–497. https://doi.org/10.1093/cercor/11.6.490

Gong G, He Y, Evans AC (2011) Brain connectivity: gender makes a difference. Neuroscientist 17(5):575–591. https://doi.org/10.1177/1073858410386492

Gottfries CG, Roos BE, Winblad B (1974) Determination of 5-hydroxytryptamine, 5-hydroxyindoleacetic acid and homovanillic acid in brain tissue from an autopsy material. Acta Psychiatr Scand 50(5):496–507. https://doi.org/10.1111/j.1600-0447.1974.tb09711.x

Gregg C, Zhang J, Butler JE, Haig D, Dulac C (2010) Sex-specific parent-of-origin allelic expression in the mouse brain. Science 329(5992):682–685. https://doi.org/10.1126/science.1190831

Guillamon A, Junque C, Gomez-Gil E (2016) A Review of the Status of Brain Structure Research in Transsexualism. Arch Sex Behav 45(7):1615–1648. https://doi.org/10.1007/s10508-016-0768-5

Gur RC, Gur RE, Obrist WD, Hungerbuhler JP, Younkin D, Rosen AD, Skolnick BE, Reivich M (1982) Sex and handedness differences in cerebral blood flow during rest and cognitive activity. Science 217(4560):659–661. https://doi.org/10.1126/science.7089587

Gur RC, Turetsky BI, Matsui M, Yan M, Bilker W, Hughett P, Gur RE (1999) Sex differences in brain gray and white matter in healthy young adults: correlations with cognitive performance. J Neurosci 19(10):4065–4072. https://doi.org/10.1523/JNEUROSCI.19-10-04065.1999

Gur RC, Alsop D, Glahn D, Petty R, Swanson CL, Maldjian JA, Turetsky BI, Detre JA, Gee J, Gur RE (2000) An fMRI study of sex differences in regional activation to a verbal and a spatial task. Brain Lang 74(2):157–170. https://doi.org/10.1006/brln.2000.2325

Hahn A, Kranz GS, Kublbock M, Kaufmann U, Ganger S, Hummer A, Seiger R, Spies M, Winkler D, Kasper S, Windischberger C, Swaab DF, Lanzenberger R (2015) Structural Connectivity Networks of Transgender People. Cereb Cortex 25(10):3527–3534. https://doi.org/10.1093/cercor/bhu194

Hahn A, Kranz GS, Sladky R, Kaufmann U, Ganger S, Hummer A, Seiger R, Spies M, Vanicek T, Winkler D, Kasper S, Windischberger C, Swaab DF, Lanzenberger R (2016) Testosterone affects language areas of the adult human brain. Hum Brain Mapp 37(5):1738–1748. https://doi.org/10.1002/hbm.23133

Halpern CT, Udry JR, Suchindran C (1998) Monthly measures of salivary testosterone predict sexual activity in adolescent males. Arch Sex Behav 27(5):445–465. https://doi.org/10.1023/a:1018700529128

Hamann S, Herman RA, Nolan CL, Wallen K (2004) Men and women differ in amygdala response to visual sexual stimuli. Nat Neurosci 7(4):411–416. https://doi.org/10.1038/nn1208

Hietala J, West C, Syvalahti E, Nagren K, Lehikoinen P, Sonninen P, Ruotsalainen U (1994) Striatal D2 dopamine receptor binding characteristics in vivo in patients with alcohol dependence. Psychopharmacology (Berl) 116(3):285–290. https://doi.org/10.1007/BF02245330

Hill AC, Laird AR, Robinson JL (2014) Gender differences in working memory networks: a BrainMap meta-analysis. Biol Psychol 102:18–29. https://doi.org/10.1016/j.biopsycho.2014.06.008

Hines M (2011) Prenatal endocrine influences on sexual orientation and on sexually differentiated childhood behavior. Front Neuroendocrinol 32(2):170–182. https://doi.org/10.1016/j.yfrne.2011.02.006

Hoekzema E, Schagen SE, Kreukels BP, Veltman DJ, Cohen-Kettenis PT, Delemarre-van de Waal H, Bakker J (2015) Regional volumes and spatial volumetric distribution of gray matter in the gender dysphoric brain. Psychoneuroendocrinology 55:59–71. https://doi.org/10.1016/j.psyneuen.2015.01.016

Hofman MA, Swaab DF (1989) The sexually dimorphic nucleus of the preoptic area in the human brain: a comparative morphometric study. J Anat 164:55–72

Hofman MA, Swaab DF (1991) Sexual dimorphism of the human brain: myth and reality. Exp Clin Endocrinol 98(2):161–170. https://doi.org/10.1055/s-0029-1211113

Hsu JL, Leemans A, Bai CH, Lee CH, Tsai YF, Chiu HC, Chen WH (2008) Gender differences and age-related white matter changes of the human brain: a diffusion tensor imaging study. NeuroImage 39(2):566–577. https://doi.org/10.1016/j.neuroimage.2007.09.017

Hugdahl K, Thomsen T, Ersland L (2006) Sex differences in visuo-spatial processing: an fMRI study of mental rotation. Neuropsychologia 44(9):1575–1583. https://doi.org/10.1016/j.neuropsychologia.2006.01.026

Huster RJ, Westerhausen R, Kreuder F, Schweiger E, Wittling W (2009) Hemispheric and gender related differences in the midcingulum bundle: a DTI study. Hum Brain Mapp 30(2):383–391. https://doi.org/10.1002/hbm.20509

Inano S, Takao H, Hayashi N, Abe O, Ohtomo K (2011) Effects of age and gender on white matter integrity. AJNR Am J Neuroradiol 32(11):2103–2109. https://doi.org/10.3174/ajnr.A2785

Ionta S, Martuzzi R, Salomon R, Blanke O (2014) The brain network reflecting bodily self-consciousness: a functional connectivity study. Soc Cogn Affect Neurosci 9(12):1904–1913. https://doi.org/10.1093/scan/nst185

James WH (2005) Biological and psychosocial determinants of male and female human sexual orientation. J Biosoc Sci 37(5):555–567. https://doi.org/10.1017/S0021932004007059

Jazin E, Cahill L (2010) Sex differences in molecular neuroscience: from fruit flies to humans. Nature reviews Neuroscience 11(1):9–17. https://doi.org/10.1038/nrn2754

Jovanovic H, Lundberg J, Karlsson P, Cerin A, Saijo T, Varrone A, Halldin C, Nordstrom AL (2008) Sex differences in the serotonin 1A receptor and serotonin transporter binding in the human brain measured by PET. NeuroImage 39(3):1408–1419. https://doi.org/10.1016/j.neuroimage.2007.10.016

Kaasinen V, Nagren K, Hietala J, Farde L, Rinne JO (2001) Sex differences in extrastriatal dopamine d(2)-like receptors in the human brain. Am J Psychiatry 158(2):308–311. https://doi.org/10.1176/appi.ajp.158.2.308

Kanaan RA, Chaddock C, Allin M, Picchioni MM, Daly E, Shergill SS, McGuire PK (2014) Gender influence on white matter microstructure: a tract-based spatial statistics analysis. PLoS One 9(3):e91109. https://doi.org/10.1371/journal.pone.0091109

Keller K, Menon V (2009) Gender differences in the functional and structural neuroanatomy of mathematical cognition. NeuroImage 47(1):342–352. https://doi.org/10.1016/j.neuroimage.2009.04.042

Kindon HA, Baum MJ, Paredes RJ (1996) Medial preoptic/anterior hypothalamic lesions induce a female-typical profile of sexual partner preference in male ferrets. Horm Behav 30(4):514–527. https://doi.org/10.1006/hbeh.1996.0055

Kong J, Tu PC, Zyloney C, Su TP (2010) Intrinsic functional connectivity of the periaqueductal gray, a resting fMRI study. Behav Brain Res 211(2):215–219. https://doi.org/10.1016/j.bbr.2010.03.042

Korol DL (2004) Role of estrogen in balancing contributions from multiple memory systems. Neurobiol Learn Mem 82(3):309–323. https://doi.org/10.1016/j.nlm.2004.07.006

Koscik T, Bechara A, Tranel D (2010) Sex-related functional asymmetry in the limbic brain. Neuropsychopharmacology 35(1):340–341. https://doi.org/10.1038/npp.2009.122

Kranz GS, Hahn A, Kaufmann U, Kublbock M, Hummer A, Ganger S, Seiger R, Winkler D, Swaab DF, Windischberger C, Kasper S, Lanzenberger R (2014) White matter microstructure in transsexuals and controls investigated by diffusion tensor imaging. J Neurosci 34(46):15466–15475. https://doi.org/10.1523/JNEUROSCI.2488-14.2014

Kret ME, De Gelder B (2012) A review on sex differences in processing emotional signals. Neuropsychologia 50(7):1211–1221. https://doi.org/10.1016/j.neuropsychologia.2011.12.022

Kreukels BP, Guillamon A (2016) Neuroimaging studies in people with gender incongruence. Int Rev Psychiatry 28(1):120–128. https://doi.org/10.3109/09540261.2015.1113163

Kugaya A, Epperson CN, Zoghbi S, van Dyck CH, Hou Y, Fujita M, Staley JK, Garg PK, Seibyl JP, Innis RB (2003) Increase in prefrontal cortex serotonin 2A receptors following estrogen treatment in postmenopausal women. Am J Psychiatry 160(8):1522–1524. https://doi.org/10.1176/appi.ajp.160.8.1522

Kuhl DE, Metter EJ, Riege WH, Phelps ME (1982) Effects of human aging on patterns of local cerebral glucose utilization determined by the [18F]fluorodeoxyglucose method. J Cereb Blood Flow Metab 2(2):163–171. https://doi.org/10.1038/jcbfm.1982.15

Laakso A, Vilkman H, Bergman J, Haaparanta M, Solin O, Syvalahti E, Salokangas RK, Hietala J (2002) Sex differences in striatal presynaptic dopamine synthesis capacity in healthy subjects. Biol Psychiatry 52(7):759–763. https://doi.org/10.1016/s0006-3223(02)01369-0

Lavalaye J, Booij J, Reneman L, Habraken JB, van Royen EA (2000) Effect of age and gender on dopamine transporter imaging with [123I]FP-CIT SPET in healthy volunteers. Eur J Nucl Med 27(7):867–869. https://doi.org/10.1007/s002590000279

Lee TM, Liu HL, Hoosain R, Liao WT, Wu CT, Yuen KS, Chan CC, Fox PT, Gao JH (2002) Gender differences in neural correlates of recognition of happy and sad faces in humans assessed by functional magnetic resonance imaging. Neurosci Lett 333(1):13–16. https://doi.org/10.1016/s0304-3940(02)00965-5

Lee TM, Liu HL, Chan CC, Fang SY, Gao JH (2005) Neural activities associated with emotion recognition observed in men and women. Mol Psychiatry 10(5):450–455. https://doi.org/10.1038/sj.mp.4001595

Lemaitre H, Crivello F, Grassiot B, Alperovitch A, Tzourio C, Mazoyer B (2005) Age- and sex-related effects on the neuroanatomy of healthy elderly. NeuroImage 26(3):900–911. https://doi.org/10.1016/j.neuroimage.2005.02.042

LeVay S (1991) A difference in hypothalamic structure between heterosexual and homosexual men. Science 253(5023):1034–1037. https://doi.org/10.1126/science.1887219
Lewin R (2009) Human evolution: an illustrated introduction, 5th edn. Wiley
Li R (2014) Why women see differently from the way men see? A review of sex differences in cognition and sports. J Sport Health Sci 3(3):155–162. https://doi.org/10.1016/j.jshs.2014.03.012
Lieberman JA (2016) Shrinks: the untold story of psychiatry. Back Bay Books
Lieberman J, First M (2016) Robert Spitzer: a psychiatrist for the ages. Neuropsychopharmacology 41(13):3124. https://doi.org/10.1038/npp.2016.30
Liu F, Vidarsson L, Winter JD, Tran H, Kassner A (2010) Sex differences in the human corpus callosum microstructure: a combined T2 myelin-water and diffusion tensor magnetic resonance imaging study. Brain Res 1343:37–45. https://doi.org/10.1016/j.brainres.2010.04.064
Longo MR, Azanon E, Haggard P (2010) More than skin deep: body representation beyond primary somatosensory cortex. Neuropsychologia 48(3):655–668. https://doi.org/10.1016/j.neuropsychologia.2009.08.022
Luders E, Steinmetz H, Jancke L (2002) Brain size and grey matter volume in the healthy human brain. Neuroreport 13(17):2371–2374. https://doi.org/10.1097/01.wnr.0000049603.85580.da
Luders E, Narr KL, Thompson PM, Woods RP, Rex DE, Jancke L, Steinmetz H, Toga AW (2005) Mapping cortical gray matter in the young adult brain: effects of gender. NeuroImage 26(2):493–501. https://doi.org/10.1016/j.neuroimage.2005.02.010
Luders E, Sanchez FJ, Gaser C, Toga AW, Narr KL, Hamilton LS, Vilain E (2009) Regional gray matter variation in male-to-female transsexualism. NeuroImage 46(4):904–907. https://doi.org/10.1016/j.neuroimage.2009.03.048
Luders E, Sanchez FJ, Tosun D, Shattuck DW, Gaser C, Vilain E, Toga AW (2012) Increased Cortical Thickness in Male-to-Female Transsexualism. J Behav Brain Sci 2(3):357–362. https://doi.org/10.4236/jbbs.2012.23040
Mackiewicz KL, Sarinopoulos I, Cleven KL, Nitschke JB (2006) The effect of anticipation and the specificity of sex differences for amygdala and hippocampus function in emotional memory. Proc Natl Acad Sci U S A 103(38):14200–14205. https://doi.org/10.1073/pnas.0601648103
Manzouri A, Kosidou K, Savic I (2017) Anatomical and functional findings in female-to-male transsexuals: testing a new hypothesis. Cereb Cortex 27(2):998–1010. https://doi.org/10.1093/cercor/bhv278
Martin A (1985) A qualitative limitation on visual transfer via the anterior commissure. Evidence from a case of callosal agenesis. Brain 108(Pt 1):43–63. https://doi.org/10.1093/brain/108.1.43
Marwha D, Halari M, Eliot L (2017) Meta-analysis reveals a lack of sexual dimorphism in human amygdala volume. NeuroImage 147:282–294. https://doi.org/10.1016/j.neuroimage.2016.12.021
McCarthy MM, Arnold AP (2011) Reframing sexual differentiation of the brain. Nat Neurosci 14(6):677–683. https://doi.org/10.1038/nn.2834
McCarthy MM, Konkle AT (2005) When is a sex difference not a sex difference? Front Neuroendocrinol 26(2):85–102. https://doi.org/10.1016/j.yfrne.2005.06.001
McCarthy MM, de Vries GJ, Forger NG (2009) Sexual differentiation of the brain: mode, mechanisms, and meaning. Hormones, brain and behavior. Elsevier, Boston. https://doi.org/10.1016/b978-008088783-8.00054-1
McClure EB, Monk CS, Nelson EE, Zarahn E, Leibenluft E, Bilder RM, Charney DS, Ernst M, Pine DS (2004) A developmental examination of gender differences in brain engagement during evaluation of threat. Biol Psychiatry 55(11):1047–1055. https://doi.org/10.1016/j.biopsych.2004.02.013
McKenna KE (1998) Central control of penile erection. Int J Impot Res 10(Suppl 1):S25–S34
McKenna KE (1999) Central nervous system pathways involved in the control of penile erection. Annu Rev Sex Res 10:157–183
McKenna KE (2000) Some proposals regarding the organization of the central nervous system control of penile erection. Neurosci Biobehav Rev 24(5):535–540. https://doi.org/10.1016/s0149-7634(00)00021-x
Mead M (1928) Coming of age in Samoa. William Morrow and Co

Menzler K, Belke M, Wehrmann E, Krakow K, Lengler U, Jansen A, Hamer HM, Oertel WH, Rosenow F, Knake S (2011) Men and women are different: diffusion tensor imaging reveals sexual dimorphism in the microstructure of the thalamus, corpus callosum and cingulum. NeuroImage 54(4):2557–2562. https://doi.org/10.1016/j.neuroimage.2010.11.029

Meyer JH, Kapur S, Houle S, DaSilva J, Owczarek B, Brown GM, Wilson AA, Kennedy SH (1999) Prefrontal cortex 5-HT2 receptors in depression: an [18F]setoperone PET imaging study. Am J Psychiatry 156(7):1029–1034. https://doi.org/10.1176/ajp.156.7.1029

Miller DI, Halpern DF (2014) The new science of cognitive sex differences. Trends in cognitive sciences 18(1):37–45. https://doi.org/10.1016/j.tics.2013.10.011

Morris JA, Jordan CL, Breedlove SM (2004) Sexual differentiation of the vertebrate nervous system. Nat Neurosci 7(10):1034–1039. https://doi.org/10.1038/nn1325

Moses EL, Drevets WC, Smith G, Mathis CA, Kalro BN, Butters MA, Leondires MP, Greer PJ, Lopresti B, Loucks TL, Berga SL (2000) Effects of estradiol and progesterone administration on human serotonin 2A receptor binding: a PET study. Biol Psychiatry 48(8):854–860. https://doi.org/10.1016/s0006-3223(00)00967-7

Moses-Kolko EL, Price JC, Shah N, Berga S, Sereika SM, Fisher PM, Coleman R, Becker C, Mason NS, Loucks T, Meltzer CC (2011) Age, sex, and reproductive hormone effects on brain serotonin-1A and serotonin-2A receptor binding in a healthy population. Neuropsychopharmacology 36(13):2729–2740. https://doi.org/10.1038/npp.2011.163

Mozley LH, Gur RC, Mozley PD, Gur RE (2001) Striatal dopamine transporters and cognitive functioning in healthy men and women. Am J Psychiatry 158(9):1492–1499. https://doi.org/10.1176/appi.ajp.158.9.1492

Nishizawa S, Benkelfat C, Young SN, Leyton M, Mzengeza S, de Montigny C, Blier P, Diksic M (1997) Differences between males and females in rates of serotonin synthesis in human brain. Proc Natl Acad Sci U S A 94(10):5308–5313. https://doi.org/10.1073/pnas.94.10.5308

Nopoulos P, Flaum M, O'Leary D, Andreasen NC (2000) Sexual dimorphism in the human brain: evaluation of tissue volume, tissue composition and surface anatomy using magnetic resonance imaging. Psychiatry Res 98(1):1–13. https://doi.org/10.1016/s0925-4927(99)00044-x

Nunez JL, McCarthy MM (2009) Resting intracellular calcium concentration, depolarizing Gamma-Aminobutyric Acid and possible role of local estradiol synthesis in the developing male and female hippocampus. Neuroscience 158(2):623–634. https://doi.org/10.1016/j.neuroscience.2008.09.061

Pakkenberg B, Gundersen HJ (1997) Neocortical neuron number in humans: effect of sex and age. J Comp Neurol 384(2):312–320. https://doi.org/10.1002/(sici)1096-9861(19970728)384:2<312::aid-cne10>3.0.co;2-k

Panzica GC, Viglietti-Panzica C, Calacagni M, Anselmetti GC, Schumacher M, Balthazart J (1987) Sexual differentiation and hormonal control of the sexually dimorphic medial preoptic nucleus in the quail. Brain Res 416(1):59–68. https://doi.org/10.1016/0006-8993(87)91496-x

Paredes RG, Baum MJ (1995) Altered sexual partner preference in male ferrets given excitotoxic lesions of the preoptic area/anterior hypothalamus. J Neurosci 15(10):6619–6630

Parsey RV, Oquendo MA, Simpson NR, Ogden RT, Van Heertum R, Arango V, Mann JJ (2002) Effects of sex, age, and aggressive traits in man on brain serotonin 5-HT1A receptor binding potential measured by PET using [C-11]WAY-100635. Brain Res 954(2):173–182. https://doi.org/10.1016/s0006-8993(02)03243-2

Passe TJ, Rajagopalan P, Tupler LA, Byrum CE, MacFall JR, Krishnan KR (1997) Age and sex effects on brain morphology. Prog Neuropsychopharmacol Biol Psychiatry 21(8):1231–1237. https://doi.org/10.1016/s0278-5846(97)00160-7

Paus T, Otaky N, Caramanos Z, MacDonald D, Zijdenbos A, D'Avirro D, Gutmans D, Holmes C, Tomaiuolo F, Evans AC (1996) In vivo morphometry of the intrasulcal gray matter in the human cingulate, paracingulate, and superior-rostral sulci: hemispheric asymmetries, gender differences and probability maps. J Comp Neurol 376(4):664–673. https://doi.org/10.1002/(SICI)1096-9861(19961223)376:4<664::AID-CNE12>3.0.CO;2-M

Perrin JS, Leonard G, Perron M, Pike GB, Pitiot A, Richer L, Veillette S, Pausova Z, Paus T (2009) Sex differences in the growth of white matter during adolescence. NeuroImage 45(4):1055–1066. https://doi.org/10.1016/j.neuroimage.2009.01.023

Perrot-Sinal TS, Kostenuik MA, Ossenkopp KP, Kavaliers M (1996) Sex differences in performance in the Morris water maze and the effects of initial nonstationary hidden platform training. Behav Neurosci 110(6):1309–1320. https://doi.org/10.1037//0735-7044.110.6.1309

Pesaresi M, Soon-Shiong R, French L, Kaplan DR, Miller FD, Paus T (2015) Axon diameter and axonal transport: In vivo and in vitro effects of androgens. NeuroImage 115:191–201. https://doi.org/10.1016/j.neuroimage.2015.04.048

Peters M, Jancke L, Staiger JF, Schlaug G, Huang Y, Steinmetz H (1998) Unsolved problems in comparing brain sizes in Homo sapiens. Brain Cogn 37(2):254–285. https://doi.org/10.1006/brcg.1998.0983

Pfaff DW, Arnold AP, Etgen AM, Fahrbach SE, Rubin RT (eds) (2009) Hormones, brain and behavior. Elsevier, Boston. https://doi.org/10.1016/b978-008088783-8.09002-1

Pilgrim C, Reisert I (1992) Differences between male and female brains--developmental mechanisms and implications. Horm Metab Res 24(8):353–359. https://doi.org/10.1055/s-2007-1003334

Pinker S (2002) The blank slate: the modern denial of human nature. Viking, New York, NY

Rabiner EA, Messa C, Sargent PA, Husted-Kjaer K, Montgomery A, Lawrence AD, Bench CJ, Gunn RN, Cowen P, Grasby PM (2002) A database of [(11)C]WAY-100635 binding to 5-HT(1A) receptors in normal male volunteers: normative data and relationship to methodological, demographic, physiological, and behavioral variables. NeuroImage 15(3):620–632. https://doi.org/10.1006/nimg.2001.0984

Rahman Q, Wilson GD (2003) Large sexual-orientation-related differences in performance on mental rotation and judgement of line orientation tasks. Neuropsychology 17(1):25–31. https://doi.org/10.1037/0894-4105.17.1.25

Rahman Q, Wilson GD, Abrahams S (2004) Biosocial factors, sexual orientation and neurocognitive functioning. Psychoneuroendocrinology 29(7):867–881. https://doi.org/10.1016/S0306-4530(03)00154-9

Rametti G, Carrillo B, Gomez-Gil E, Junque C, Segovia S, Gomez A, Guillamon A (2011a) White matter microstructure in female to male transsexuals before cross-sex hormonal treatment. A diffusion tensor imaging study. J Psychiatr Res 45(2):199–204. https://doi.org/10.1016/j.jpsychires.2010.05.006

Rametti G, Carrillo B, Gomez-Gil E, Junque C, Zubiarre-Elorza L, Segovia S, Gomez A, Guillamon A (2011b) The microstructure of white matter in male to female transsexuals before cross-sex hormonal treatment. A DTI study. J Psychiatr Res 45(7):949–954. https://doi.org/10.1016/j.jpsychires.2010.11.007

Reed GM, Drescher J, Krueger RB, Atalla E, Cochran SD, First MB, Cohen-Kettenis PT, Arango-de Montis I, Parish SJ, Cottler S, Briken P, Saxena S (2016) Disorders related to sexuality and gender identity in the ICD-11: revising the ICD-10 classification based on current scientific evidence, best clinical practices, and human rights considerations. World Psychiatry 15(3):205–221. https://doi.org/10.1002/wps.20354

Reiman EM, Armstrong SM, Matt KS, Mattox JH (1996) The application of positron emission tomography to the study of the normal menstrual cycle. Hum Reprod 11(12):2799–2805. https://doi.org/10.1093/oxfordjournals.humrep.a019214

Rieber I, Sigusch V (1979) Psychosurgery on sex offenders and sexual "deviants" in West Germany. Arch Sex Behav 8(6):523–527. https://doi.org/10.1007/BF01541419

Risse GL, LeDoux J, Springer SP, Wilson DH, Gazzaniga MS (1978) The anterior commissure in man: functional variation in a multisensory system. Neuropsychologia 16(1):23–31. https://doi.org/10.1016/0028-3932(78)90039-8

Roivainen E (2011) Gender differences in processing speed: a review of recent research. Learn Individ Differ 21(2):145–149. https://doi.org/10.1016/j.lindif.2010.11.021

Rose S (2001) Revisiting evolutionary psychology and psychiatry. Br J Psychiatry 179:558. https://doi.org/10.1192/bjp.179.6.558-a

Rosier A, Dupont P, Peuskens J, Bormans G, Vandenberghe R, Maes M, de Groot T, Schiepers C, Verbruggen A, Mortelmans L (1996) Visualisation of loss of 5-HT2A receptors with age in healthy volunteers using [18F]altanserin and positron emission tomographic imaging. Psychiatry Res 68(1):11–22. https://doi.org/10.1016/s0925-4927(96)02806-5

Ruigrok AN, Salimi-Khorshidi G, Lai MC, Baron-Cohen S, Lombardo MV, Tait RJ, Suckling J (2014) A meta-analysis of sex differences in human brain structure. Neurosci Biobehav Rev 39(100):34–50. https://doi.org/10.1016/j.neubiorev.2013.12.004

Sacher J, Neumann J, Okon-Singer H, Gotowiec S, Villringer A (2013) Sexual dimorphism in the human brain: evidence from neuroimaging. Magn Reson Imaging 31(3):366–375. https://doi.org/10.1016/j.mri.2012.06.007

Saleem F, Rizvi SW (2017) Transgender associations and possible etiology: a literature review. Cureus 9(12):e1984. https://doi.org/10.7759/cureus.1984

Sarubbo S, De Benedictis A, Maldonado IL, Basso G, Duffau H (2013) Frontal terminations for the inferior fronto-occipital fascicle: anatomical dissection, DTI study and functional considerations on a multi-component bundle. Brain Struct Funct 218(1):21–37. https://doi.org/10.1007/s00429-011-0372-3

Savic I, Lindstrom P (2008) PET and MRI show differences in cerebral asymmetry and functional connectivity between homo- and heterosexual subjects. Proc Natl Acad Sci U S A 105(27):9403–9408. https://doi.org/10.1073/pnas.0801566105

Savic I, Berglund H, Lindstrom P (2005) Brain response to putative pheromones in homosexual men. Proc Natl Acad Sci U S A 102(20):7356–7361. https://doi.org/10.1073/pnas.0407998102

Savic I, Garcia-Falgueras A, Swaab DF (2010) Sexual differentiation of the human brain in relation to gender identity and sexual orientation. In: Savic I (ed) Sex differences in the human brain, their underpinnings and implications. Elsevier, Amsterdam. https://doi.org/10.1016/b978-0-444-53630-3.00004-x

Schirmer A, Zysset S, Kotz SA, Yves von Cramon D (2004) Gender differences in the activation of inferior frontal cortex during emotional speech perception. NeuroImage 21(3):1114–1123. https://doi.org/10.1016/j.neuroimage.2003.10.048

Schlaepfer TE, Harris GJ, Tien AY, Peng L, Lee S, Pearlson GD (1995) Structural differences in the cerebral cortex of healthy female and male subjects: a magnetic resonance imaging study. Psychiatry Res 61(3):129–135. https://doi.org/10.1016/0925-4927(95)02634-a

Schmithorst VJ, Holland SK, Dardzinski BJ (2008) Developmental differences in white matter architecture between boys and girls. Hum Brain Mapp 29(6):696–710. https://doi.org/10.1002/hbm.20431

Schoonheim MM, Vigeveno RM, Rueda Lopes FC, Pouwels PJ, Polman CH, Barkhof F, Geurts JJ (2014) Sex-specific extent and severity of white matter damage in multiple sclerosis: implications for cognitive decline. Hum Brain Mapp 35(5):2348–2358. https://doi.org/10.1002/hbm.22332

Schroeder U, Hennenlotter A, Erhard P, Haslinger B, Stahl R, Lange KW, Ceballos-Baumann AO (2004) Functional neuroanatomy of perceiving surprised faces. Hum Brain Mapp 23(4):181–187. https://doi.org/10.1002/hbm.20057

Sergerie K, Chochol C, Armony JL (2008) The role of the amygdala in emotional processing: a quantitative meta-analysis of functional neuroimaging studies. Neurosci Biobehav Rev 32(4):811–830. https://doi.org/10.1016/j.neubiorev.2007.12.002

Shankman P (2009) The Trashing of Margaret Mead: the anatomy of an anthropological myth. University of Wisconsin Press, Madison

Shin YW, Kim DJ, Ha TH, Park HJ, Moon WJ, Chung EC, Lee JM, Kim IY, Kim SI, Kwon JS (2005) Sex differences in the human corpus callosum: diffusion tensor imaging study. Neuroreport 16(8):795–798. https://doi.org/10.1097/00001756-200505310-00003

Shors TJ (2001) Acute stress rapidly and persistently enhances memory formation in the male rat. Neurobiology of learning and memory 75(1):10–29. https://doi.org/10.1006/nlme.1999.3956

Simerly RB (2002) Wired for reproduction: organization and development of sexually dimorphic circuits in the mammalian forebrain. Annu Rev Neurosci 25(1):507–536. https://doi.org/10.1146/annurev.neuro.25.112701.142745

Simon L, Kozak LR, Simon V, Czobor P, Unoka Z, Szabo A, Csukly G (2013) Regional grey matter structure differences between transsexuals and healthy controls--a voxel based morphometry study. PLoS One 8(12):e83947. https://doi.org/10.1371/journal.pone.0083947

Sisk CL, Zehr JL (2005) Pubertal hormones organize the adolescent brain and behavior. Front Neuroendocrinol 26(3-4):163–174. https://doi.org/10.1016/j.yfrne.2005.10.003

Smith YR, Zubieta JK, del Carmen MG, Dannals RF, Ravert HT, Zacur HA, Frost JJ (1998) Brain opioid receptor measurements by positron emission tomography in normal cycling women: relationship to luteinizing hormone pulsatility and gonadal steroid hormones. J Clin Endocrinol Metab 83(12):4498–4505. https://doi.org/10.1210/jcem.83.12.5351

Smith YR, Minoshima S, Kuhl DE, Zubieta JK (2001) Effects of long-term hormone therapy on cholinergic synaptic concentrations in healthy postmenopausal women. J Clin Endocrinol Metab 86(2):679–684. https://doi.org/10.1210/jcem.86.2.7222

Smith CD, Chebrolu H, Wekstein DR, Schmitt FA, Markesbery WR (2007) Age and gender effects on human brain anatomy: a voxel-based morphometric study in healthy elderly. Neurobiol Aging 28(7):1075–1087. https://doi.org/10.1016/j.neurobiolaging.2006.05.018

Sommer IE, Aleman A, Somers M, Boks MP, Kahn RS (2008) Sex differences in handedness, asymmetry of the planum temporale and functional language lateralization. Brain Res 1206:76–88. https://doi.org/10.1016/j.brainres.2008.01.003

Spratt D (2000) Sex differences in the brain. J Neuroendocrinol 12(7):597–598. https://doi.org/10.1046/j.1365-2826.2000.012007597.x

Staley JK, Krishnan-Sarin S, Zoghbi S, Tamagnan G, Fujita M, Seibyl JP, Maciejewski PK, O'Malley S, Innis RB (2001) Sex differences in [123I]beta-CIT SPECT measures of dopamine and serotonin transporter availability in healthy smokers and nonsmokers. Synapse 41(4):275–284. https://doi.org/10.1002/syn.1084

Stein P, Savli M, Wadsak W, Mitterhauser M, Fink M, Spindelegger C, Mien LK, Moser U, Dudczak R, Kletter K, Kasper S, Lanzenberger R (2008) The serotonin-1A receptor distribution in healthy men and women measured by PET and [carbonyl-11C]WAY-100635. Eur J Nucl Med Mol Imaging 35(12):2159–2168. https://doi.org/10.1007/s00259-008-0850-x

Swaab DF (2004) Sexual differentiation of the human brain: relevance for gender identity, transsexualism and sexual orientation. Gynecol Endocrinol 19(6):301–312. https://doi.org/10.1080/09513590400018231

Swaab DF (2007) Sexual differentiation of the brain and behavior. Best Pract Res Clin Endocrinol Metab 21(3):431–444. https://doi.org/10.1016/j.beem.2007.04.003

Swaab DF, Fliers E (1985) A sexually dimorphic nucleus in the human brain. Science 228(4703):1112–1115. https://doi.org/10.1126/science.3992248

Swaab DF, Hofman MA (1984) Sexual differentiation of the human brain. A historical perspective. In: De Vries GJ et al (eds) Sex differences in the brain: relation between structure and function, Progress in brain research, vol 61. Elsevier, Amsterdam, pp 361–374

Swaab DF, Hofman MA (1990) An enlarged suprachiasmatic nucleus in homosexual men. Brain Res 537(1-2):141–148. https://doi.org/10.1016/0006-8993(90)90350-k

Swaab DF, Hofman MA (1995) Sexual differentiation of the human hypothalamus in relation to gender and sexual orientation. Trends Neurosci 18(6):264–270

Swaab DF, Gooren LJG, Hofman MA (1992) The human hypothalamus in relation to gender and sexual orientation. In: Swaab DF, Hofman MA, Mirmiran M, Ravid R, Van Leeuwen FW (eds) The human hypothalamus in health and disease, Progress in brain research, vol 93. Elsevier, Amsterdam, pp 205–215

Swaab DF, Zhou JN, Ehlhart T, Hofman MA (1994) Development of vasoactive intestinal polypeptide neurons in the human suprachiasmatic nucleus in relation to birth and sex. Brain Res Dev Brain Res 79(2):249–259. https://doi.org/10.1016/0165-3806(94)90129-5

Swaab DF, Chung WC, Kruijver FP, Hofman MA, Ishunina TA (2001) Structural and functional sex differences in the human hypothalamus. Horm Behav 40(2):93–98. https://doi.org/10.1006/hbeh.2001.1682

Szeszko PR, Vogel J, Ashtari M, Malhotra AK, Bates J, Kane JM, Bilder RM, Frevert T, Lim K (2003) Sex differences in frontal lobe white matter microstructure: a DTI study. Neuroreport 14(18):2469–2473. https://doi.org/10.1097/00001756-200312190-00035

Takahashi H, Matsuura M, Yahata N, Koeda M, Suhara T, Okubo Y (2006) Men and women show distinct brain activations during imagery of sexual and emotional infidelity. NeuroImage 32(3):1299–1307. https://doi.org/10.1016/j.neuroimage.2006.05.049

Takao H, Hayashi N, Ohtomo K (2014) Sex dimorphism in the white matter: fractional anisotropy and brain size. J Magn Reson Imaging 39(4):917–923. https://doi.org/10.1002/jmri.24225

Tan A, Ma W, Vira A, Marwha D, Eliot L (2016) The human hippocampus is not sexually-dimorphic: meta-analysis of structural MRI volumes. NeuroImage 124(Pt A):350–366. https://doi.org/10.1016/j.neuroimage.2015.08.050

Tian L, Wang J, Yan C, He Y (2011) Hemisphere- and gender-related differences in small-world brain networks: a resting-state functional MRI study. NeuroImage 54(1):191–202. https://doi.org/10.1016/j.neuroimage.2010.07.066

Trivers RL (1972) Parental investment and sexual selection. In: Campbell B (ed) Sexual Selection and the Descent of Man, 1871-1971. Aldine-Atherton, Chicago, pp 136–179

Trivers RL, Willard DE (1973) Natural selection of parental ability to vary the sex ratio of offspring. Science 179(4068):90–92. https://doi.org/10.1126/science.179.4068.90

Tsakiris M (2010) My body in the brain: a neurocognitive model of body-ownership. Neuropsychologia 48(3):703–712. https://doi.org/10.1016/j.neuropsychologia.2009.09.034

van Amelsvoort T, Compton J, Murphy D (2001) In vivo assessment of the effects of estrogen on human brain. Trends Endocrinol Metab 12(6):273–276. https://doi.org/10.1016/s1043-2760(01)00422-2

van Dyck CH, Seibyl JP, Malison RT, Laruelle M, Zoghbi SS, Baldwin RM, Innis RB (2002) Age-related decline in dopamine transporters: analysis of striatal subregions, nonlinear effects, and hemispheric asymmetries. Am J Geriatr Psychiatry 10(1):36–43. https://doi.org/10.1097/00019442-200201000-00005

Van Goozen SH, Cohen-Kettenis PT, Gooren LJ, Frijda NH, Van de Poll NE (1994) Activating effects of androgens on cognitive performance: causal evidence in a group of female-to-male transsexuals. Neuropsychologia 32(10):1153–1157. https://doi.org/10.1016/0028-3932(94)90099-x

Van Goozen SH, Cohen-Kettenis PT, Gooren LJ, Frijda NH, Van de Poll NE (1995) Gender differences in behaviour: activating effects of cross-sex hormones. Psychoneuroendocrinology 20(4):343–363. https://doi.org/10.1016/0306-4530(94)00076-x

van Hemmen J, Saris IMJ, Cohen-Kettenis PT, Veltman DJ, Pouwels PJW, Bakker J (2017) Sex differences in white matter microstructure in the human brain predominantly reflect differences in sex hormone exposure. Cereb Cortex 27(5):2994–3001. https://doi.org/10.1093/cercor/bhw156

van Hilten B, Hoff JI, Middelkoop HA, van der Velde EA, Kerkhof GA, Wauquier A, Kamphuisen HA, Roos RA (1994) Sleep disruption in Parkinson's disease. Assessment by continuous activity monitoring. Arch Neurol 51(9):922–928. https://doi.org/10.1001/archneur.1994.00540210094018

Wallentin M (2009) Putative sex differences in verbal abilities and language cortex: a critical review. Brain and language 108(3):175–183. https://doi.org/10.1016/j.bandl.2008.07.001

Washburn SL (1973) The promise of primatology. Am J Phys Anthropol 38(2):177–182. https://doi.org/10.1002/ajpa.1330380206

Washburn SL (1978) The evolution of man. Sci Am 239(3):194–198, 201-192, 204 passim. https://doi.org/10.1038/scientificamerican0978-194

Washburn SL (1982) Human evolution. Perspectives in biology and medicine 25(4):583–602. https://doi.org/10.1353/pbm.1982.0072

Washburn SL, McCown ER (1972) Evolution of human behavior. Soc Biol 19(2):163–170. https://doi.org/10.1080/19485565.1972.9987980

Weiner S, Monge J, Mann A (2008) Bipedalism and parturition: an evolutionary imperative for cesarean delivery? Clin Perinatol 35(3):469–478, ix. https://doi.org/10.1016/j.clp.2008.06.003

Weissman-Fogel I, Moayedi M, Taylor KS, Pope G, Davis KD (2010) Cognitive and default-mode resting state networks: do male and female brains "rest" differently? Hum Brain Mapp 31(11):1713–1726. https://doi.org/10.1002/hbm.20968

Wells JC, DeSilva JM, Stock JT (2012) The obstetric dilemma: an ancient game of Russian roulette, or a variable dilemma sensitive to ecology? Am J Phys Anthropol 149(Suppl 55):40–71. https://doi.org/10.1002/ajpa.22160

Westerhausen R, Walter C, Kreuder F, Wittling RA, Schweiger E, Wittling W (2003) The influence of handedness and gender on the microstructure of the human corpus callosum: a diffusion-tensor magnetic resonance imaging study. Neuroscience letters 351(2):99–102. https://doi.org/10.1016/j.neulet.2003.07.011

Westerhausen R, Kreuder F, Dos Santos Sequeira S, Walter C, Woerner W, Wittling RA, Schweiger E, Wittling W (2004) Effects of handedness and gender on macro- and microstructure of the corpus callosum and its subregions: a combined high-resolution and diffusion-tensor MRI study. Brain Res Cogn Brain Res 21(3):418–426. https://doi.org/10.1016/j.cogbrainres.2004.07.002

Westerhausen R, Kompus K, Dramsdahl M, Falkenberg LE, Gruner R, Hjelmervik H, Specht K, Plessen K, Hugdahl K (2011) A critical re-examination of sexual dimorphism in the corpus callosum microstructure. NeuroImage 56(3):874–880. https://doi.org/10.1016/j.neuroimage.2011.03.013

Wever RA (1984) Sex differences in human circadian rhythms: intrinsic periods and sleep fractions. Experientia 40(11):1226–1234. https://doi.org/10.1007/BF01946652

Witelson SF, Glezer II, Kigar DL (1995) Women have greater density of neurons in posterior temporal cortex. J Neurosci 15(5 Pt 1):3418–3428. https://doi.org/10.1523/JNEUROSCI.15-05-03418.1995

Witte AV, Savli M, Holik A, Kasper S, Lanzenberger R (2010) Regional sex differences in grey matter volume are associated with sex hormones in the young adult human brain. NeuroImage 49(2):1205–1212. https://doi.org/10.1016/j.neuroimage.2009.09.046

Wong DF, Broussolle EP, Wand G, Villemagne V, Dannals RF, Links JM, Zacur HA, Harris J, Naidu S, Braestrup C et al (1988) In vivo measurement of dopamine receptors in human brain by positron emission tomography. Age and sex differences. Ann N Y Acad Sci 515:203–214. https://doi.org/10.1111/j.1749-6632.1988.tb32986.x

Yahr P, Finn PD, Hoffman NW, Sayag N (1994) Sexually dimorphic cell groups in the medial preoptic area that are essential for male sex behavior and the neural pathways needed for their effects. Psychoneuroendocrinology 19(5-7):463–470. https://doi.org/10.1016/0306-4530(94)90033-7

Yan C, Gong G, Wang J, Wang D, Liu D, Zhu C, Chen ZJ, Evans A, Zang Y, He Y (2011) Sex- and brain size-related small-world structural cortical networks in young adults: a DTI tractography study. Cereb Cortex 21(2):449–458. https://doi.org/10.1093/cercor/bhq111

Yoshida T, Kuwabara Y, Sasaki M, Fukumura T, Ichimiya A, Takita M, Ogomori K, Ichiya Y, Masuda K (2000) Sex-related differences in the muscarinic acetylcholinergic receptor in the healthy human brain--a positron emission tomography study. Ann Nucl Med 14(2):97–101. https://doi.org/10.1007/BF02988587

Zhou JN, Hofman MA, Gooren LJ, Swaab DF (1995a) A sex difference in the human brain and its relation to transsexuality. Nature 378(6552):68–70. https://doi.org/10.1038/378068a0

Zhou JN, Hofman MA, Swaab DF (1995b) VIP neurons in the human SCN in relation to sex, age, and Alzheimer's disease. Neurobiol Aging 16(4):571–576. https://doi.org/10.1016/0197-4580(95)00043-e

Zhou W, Yang X, Chen K, Cai P, He S, Jiang Y (2014) Chemosensory communication of gender through two human steroids in a sexually dimorphic manner. Curr Biol 24(10):1091–1095. https://doi.org/10.1016/j.cub.2014.03.035

Zubiaurre-Elorza L, Junque C, Gomez-Gil E, Segovia S, Carrillo B, Rametti G, Guillamon A (2013) Cortical thickness in untreated transsexuals. Cereb Cortex 23(12):2855–2862. https://doi.org/10.1093/cercor/bhs267

Zubieta JK, Dannals RF, Frost JJ (1999) Gender and age influences on human brain mu-opioid receptor binding measured by PET. Am J Psychiatry 156(6):842–848. https://doi.org/10.1176/ajp.156.6.842

Zuo XN, Kelly C, Di Martino A, Mennes M, Margulies DS, Bangaru S, Grzadzinski R, Evans AC, Zang YF, Castellanos FX, Milham MP (2010) Growing together and growing apart: regional and sex differences in the lifespan developmental trajectories of functional homotopy. J Neurosci 30(45):15034–15043. https://doi.org/10.1523/JNEUROSCI.2612-10.2010

Genetics of Brain Structure and Function 18

Contents

18.1 Human Genetics

Before we proceed with the describing of brain genetics and their relationship to brain function, it is important to list and analyze in brief the basic principles and the basic terms of genetics, since the average mental health professional is not familiar with the precise meaning of concepts and the framework.

18.1.1 Basic Human Genetics

Humans are *diploid organisms*. They have 23 pairs of chromosomes (Fig. 18.1), of which 22 pairs are somatic chromosomes and 1 pair is sex chromosome (female, XX; male, XY). The *chromosomes* are made of DNA chains (Fig. 18.2), and they contain all the genetic information of the organism. The DNA is made of two strands of nucleotides, composed by four main nitrogenous nucleotide bases, two purines

K. N. Fountoulakis, *The Human Connectome*,
https://doi.org/10.1007/978-3-031-10351-3_18

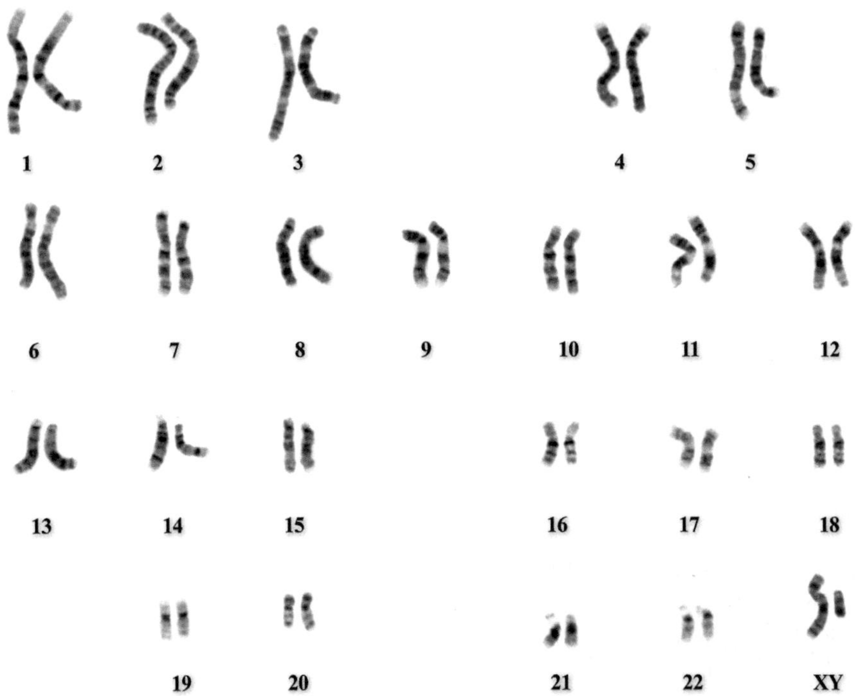

Fig. 18.1 Male karyotype (from National Human Genome Research Institute - http://www.genome.gov/glossary/resources/karyotype.pdf)

(adenine-A and guanine-G), and two pyrimidines (thymine-T and cytosine-C). The two strands form a *double helix*, held by coupling of nucleotides between the strands; adenine pairs with thymine and guanine with cytosine.

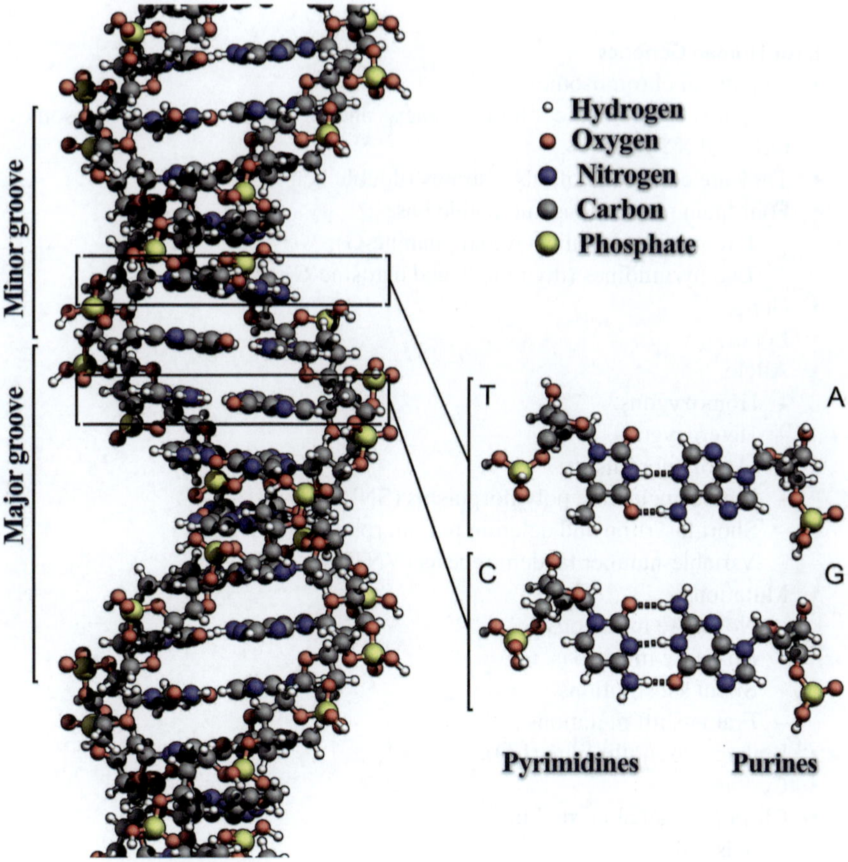

Fig. 18.2 The DNA double helix (image by Richard Wheeler after permission)

Basic Human Genetics

- 23 pairs of chromosomes
- 22 pairs are somatic chromosomes, and 1 pair is sex chromosome (XX and XY)
- They are composed of DNA strains (double helix).
- Four main nitrogenous nucleotide bases:
  - Two purines (adenine-A and guanine-G).
  - Two pyrimidines (thymine-T and cytosine-C).
- Gene.
- Locus.
- Allele:
  - Homozygous.
  - Heterozygous.
- DNA polymorphisms:
  - Single nucleotide polymorphisms (SNPs).
  - Short insertion and deletion polymorphisms (INDELs).
  - Variable-number tandem repeats (VNTRs).
- Mutations:
  - Nonsense mutations.
  - Missense mutations.
  - Silent substitutions.
  - Frame-shift mutations.
- Linkage disequilibrium (LD).
- Codon.
- Chromosomal abnormalities:
  - Trisomy.
  - Monosomy.
  - Mosaicism.
- Breakage of chromosomes:
  - Deletion.
  - Inversion.
  - Duplication of chromosome segment.
  - Translocation of segments from one chromosome to another (nonhomologous) chromosome.
    - Balanced.
    - Unbalanced.
- Unstable DNA.
- Mitochondrial genome.

The basic nomenclature is as follows:

- *Gene* is a DNA segment which codes a specific protein. The sum of genes is called *genome*. It is estimated that in humans there are approximately 20,000 genes coding proteins and the total number of genes is over 100,000.
- *Locus* is a term which describes a sequence at a defined position on a chromosome.
- *Allele* is called the variation in the nucleotide sequence of a particular gene among different persons. This variation results in a significant number of different phenotypic expressions among humans. Because chromosomes are paired, each person carries two alleles of a gene, of which one has paternal and the other maternal origin. When the person carries two same alleles, then it is homozygous, while when the alleles are different, the person is heterozygous for this particular gene.
- *DNA polymorphisms* are those alleles with at least 1% prevalence in the general population. They are the result of DNA damage, mutations, or incorrect DNA replication. The most common are:
  - *Single nucleotide polymorphisms (SNPs)*: They are caused by a variation in a single nucleotide at a specific position.
  - *Short insertion and deletion polymorphisms (INDELs)*: They correspond to insertions or deletions of up to 50 nucleotides at a single locus.
  - *Variable-number tandem repeats (VNTRs)*: They are formed by genetic elements repeated in tandem arrays.
- *Mutation* is a permanent change in the nucleotide sequence, and it is the result from errors during DNA replication or other types of damage to DNA. Repair of damages can also lead to mutations as also can the *insertion* or *deletion* of segments of DNA due to mobile genetic elements.
  - *Nonsense mutations* lead to the synthesis of a truncated protein with a dramatic effect on the functionality of the protein.
  - *Missense mutations* change the specific amino acid coded by a single codon (non-synonymous substitution).
  - *Silent substitutions* do not alter the amino acid coded by a codon (synonymous substitutions).
  - *Frame-shift mutations* are small insertions or deletions in coding sequences; however, no matter the length, they can cause dramatic consequences because they shift the entire downstream sequence and cause it to be read out of frame. This results in the synthesis of a completely different polypeptide.
- *Linkage disequilibrium (LD)* refers to nonrandom association of alleles at two or more loci since neighboring alleles tend to co-segregate and be inherited together.
- *Codon* is the unit of information in DNA or RNA which encodes an individual amino acid consisting of three adjacent bases.
- *Chromosomal abnormalities* include either the numerical increase or decrease in chromosomal number.
  - *Trisomy*: Increase as the result of the existence of an extra copy of a single chromosome (e.g., trisomy 21 or Down's syndrome).

- *Monosomy*: Decrease can happen when an entire chromosome is missing. This is not compatible with life, with the exception of females with Turner's syndrome.
- *Mosaicism*: In case of a failure of a pair of homologous chromosomes to separate during cell division in embryogenesis, some cells in the body will have a normal number of chromosomes, while others will not. The consequences vary dramatically, depending on the proportion of abnormal cells.
- *Breakage of chromosomes*:

  Deletion refers to the situation when a large stretch of DNA is lost.

  Inversion refers to the case when a piece of chromosome breaks and becomes reattached in the opposite orientation. In this case there is no loss of DNA; however, individual genes may be interrupted.

  Duplication of a segment of chromosome.

  Translocation of segments from one chromosome to another (nonhomologous) chromosome. They are subdivided into:
  - *Balanced.*
  - *Unbalanced.*

- *Unstable DNA*: DNA sequences containing repetitive sequences are often unstable and are prone to extremely high rates of mutation.
- *Mitochondrial genome*: Mitochondria contain their own small genome. It consists of around 16,500 bases encoding a small number of proteins. Mitochondrial DNA is almost always and exclusively transmitted maternally. It has a very high mutation rate.

18.1.2 Patterns of Inheritance

Patterns of Inheritance

- Single-gene inheritance.
- Atypical patterns of Mendelian inheritance:
- Anticipation.
- Imprinting.
- Chromosomal anomalies and changes.
- Polygenic inheritance.
- Multifactorial and polygenic liability threshold models.
- Epigenetic inheritance.

- *Single-gene inheritance*: This is the typical way according to Mendel's laws.
- *Atypical patterns of Mendelian inheritance*: Often, single-gene characteristics and disorders can show an irregular pattern of inheritance due to incomplete penetrance, which refers to the probability the particular phenotype will be manifested.
- *Anticipation* refers to the manifestation of greater clinical severity and earlier age at onset in subsequent generations.

- *Imprinting* refers to the variability of the phenotypic manifestation on the basis of whether it originates from the mother or father, meaning that it depends on whether the genetic information is located on a chromosome that came from a specific parent.
- *Chromosomal anomalies and changes* including additional or absence of a complete chromosome or parts of it.
- *Polygenic inheritance*: Most human characteristics are distributed continuously and normally in the general population. The more genes are implicated in the manifestation of a characteristic, the more its manifestation resembles a normal distribution.
- *Multifactorial and polygenic liability threshold models*: While even disorders with complex genetic origin are either present or absent, their severity could vary widely probably on the basis of some kind of an underlying liability to the disorder which is continuously distributed in the general population. This liability is determined by a number of genes, and beyond a certain threshold of severity, the disorder is present. Subthreshold liability is present in the family members of patients in higher frequency in comparison with the general population. Often more complex genetic backgrounds exist, including the presence of multiple liabilities being necessary for the manifestation of a single disorder.
- *Epigenetic inheritance*: It refers to the presence of mechanisms by which variation in the rate at which a gene is expressed can be inherited in the absence of a change in the primary DNA sequence (Fig. 18.3). One mechanism involves methylation of bases at the end of genes making the gene silent. Imprinting is

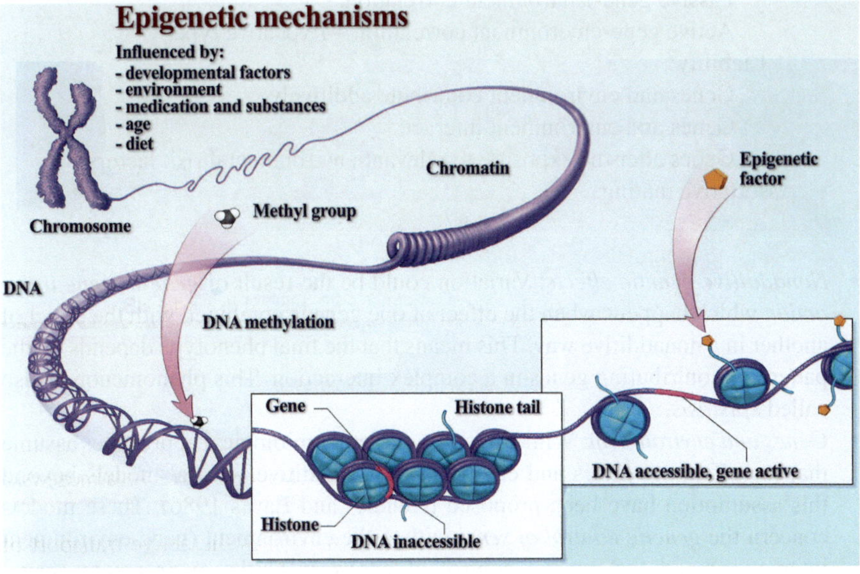

Fig. 18.3 Graphical representation of epigenetic mechanisms (image from US National Institutes of Health)

another mechanism suggesting that the transmitting parent determines the functioning of the gene because of different state in the surrounding DNA between parents. Other cause of epigenetic changes is translocation since the precise chromosomal localization of the gene could affect its functioning (position effect).

18.1.3 Human Variation and Its Components

Phenotypic variation is the observed variation of a specific phenotype in a specified population (VP). This variation is explained partially by *genetic factors* (Vg) and partially by the *environment* (Ve). The environmentally explained variance can be further subdivided into *common or shared environmental variance* (Vc) and *non-shared environmental variance* (Ve).

Human Variation and Its Components

- Genetic factors (Vg).
- Environment (Ve):
  - Common or shared environmental variance (Vc).
  - Non-shared environmental variance (Ve).
- Nonadditive genetic effects—Epistasis.
- Genes and environment (nature vs. nurture).
- Gen-environment interaction:
  - Gene-environment correlation:
    Passive gene-environment correlation.
    Active gene-environment correlation—Evocative type.
  - Liability:
    Genes and environment contribute additively.
    Genes and environment interact.
    Genes alter the exposure to relevant environmental risk factors.
- Assortative mating.

- *Nonadditive genetic effects*: Variation could be the result of *gene-to-gene interaction* which happens when the effect of one gene is combined with the effect of another in a nonadditive way. This means that the final phenotype depends on the pattern of contributing genes in a complex interaction. This phenomenon is also called *epistasis*.
- *Genes and environment*: While most genetic-epidemiological approaches assume that the effects of genes and environment are additive, several models beyond this assumption have been proposed (Kendler and Eaves 1986). These models concern the *genetic control of sensitivity* to the environment (gene-environment interaction) and the genetic control of exposure to the environment (gene-environment correlation).

- **Gene-environment interaction**: There is a widely accepted idea that genetic factors (nature) and environmental factors (nurture) interact with each other (Caspi et al. 2002; Reif et al. 2007; Cicchetti et al. 2012). The exact nature of this interaction is a matter of debate which does not always obey the laws and the methods of science, because it is influenced by ideological, sociopolitical, and religious currents. However, a number of different ways of possible interaction have been proposed:
  - *Gene-environment correlation* corresponds to the exposure to the environment. It includes:

    Passive gene-environment correlation occurs when parents pass on the genes that increase the likelihood of a particular trait, and additionally they also provide an environment during childhood that is correlated with this particular parental genetic make-up.

    Active gene-environment correlation develops when the specific inherited trait selectively shapes parts of or the whole environment to fit the manifestation of this particular trait. This type is also called "evocative" because it actively evokes responses from the environment (Jaffee and Price 2008).
  - *Liability* corresponds to the likelihood a person will manifest a specific phenotype under the influence of the environment. The variation in liability itself may be caused by both genetic and environmental influences (Eaves 1984). There are three major models to explain the observed variation in liability, but they should not be considered to be mutually exclusive since mixed models also exist in reality (Kendler and Eaves 1986):

    Genes and environment contribute *additively* to liability of a disorder.

    Genes and environment *interact* with either genes controlling sensitivity to the environment or on the contrary with the environment controlling gene expression.

    Genes and environment are correlated with genes *altering the exposure* to relevant environmental risk factors.
- *Assortative mating*: While the quantitative genetics theory assumes random mating in principle, this might not be the case in the real world, where often like may be attracted to like. This is called *positive assortative mating*, while the opposite is called *negative assortative mating*. There is evidence that assortative mating is often the case concerning a number of human traits including body type, intelligence, antisocial traits, and emotionality, and there is some evidence that social conditions further enhance assortative mating behaviors. The overall result is that a similar genetic background of both parents leads to *increased genetic load of these traits in the offspring* and also to the *accumulation of modified environment elements from generation to generation*. There are voices today that sound the alarm concerning this phenomenon, which might lead to an increase and further widening of the social gap with possible dangerous consequences for social coherence but even for democracy per se.

18.1.4 Methods in Behavior Genetics Research

Methods in Behavior Genetics Research

- Linkage studies.
- Association studies.
- Family studies.
- Twin studies.
- Adoption studies:
  - Adoptee's method.
  - Familial method.
- Animal studies.

The basic techniques for the study of genetic components include a variety of methods whose description is outside the scope of the current text. The design of studies includes the following:

- *Linkage studies* examine the co-inheritance of a trait/disorder and set of genetic markers within families; the assumption is that if the disease/trait susceptibility locus and the marker locus are located close together on a chromosome, they will tend to be cotransmitted more frequently than expected by chance.
- *Association studies* compare a group of affected individuals with a control group consisting of people without the trait/disorder.
- The *family study design* focuses on comparing the rate of a trait or a disorder in relatives of probands compared with the rate among relatives of subjects without the trait/disorder or with some other specific trait/disorder.
- *Twin studies*: They concern the comparison of affected vs. nonaffected monozygotic or dizygotic twins. Monozygotic twins share 100% of their genes while dizygotic only 50% on average.
- *Adoption studies* are typically used together with twin studies and aim to disentangle genetic from environmental influence.
- The *adoptee's method* investigates similarities between the adoptee and their biological and adoptive parents.
- The *familial method* compares non-biological siblings who are raised in the same household.
- *Animal studies* are based on animal models simulating human behavior, and the selective breeding or genetic manipulation allows studying gene effects.

18.2 Brain and Behavior Genetics

Although the literature cannot provide a comprehensive synthesis of results into a single mechanism for the various brain functions and the way they translate into observable behaviors, there is a bulk of data linking genetic markers with phenotypic expressions. Of course, the brain, as it happens with all the body, is determined by a large number of genes which regulate from basic structure and differentiation of cells to the synthesis of thousands of different substances. Such an in-depth and detailed description of the genetics of the brain is outside the scope of the current book, which will be limited to those fundamental genetic knowledge that help in the understanding of brain function and its expression in behavior and mental health.

18.2.1 The Most Important Genetic Markers Identified So Far

Οι κυριότεροι γενετικοί δείκτες που έχουν αναγνωρισθεί ως σήμερα-Γονίδια

- Serotonin transporter (SERT ή 5HTT), chromosome 17 (17q11.1-q12):
  - Polymorphism 5-HTTLPR—14 different alleles.
  - "Long" (L):
    - LA.
    - LG.
  - "Short" (S)
  - STin2.12.
- Serotonin receptor 1A (5-HTR1A), chromosome 5 (5q11.2–13):
  - Allele G.
- Serotonin receptor 1B (5-HTR1B), chromosome 6 (6q14.1):
  - Several polymorphisms.
  - G861C SNP (rs6296).
  - Allele G.
- Serotonin receptor 2A (5-HTR2A), chromosome 13 (13q14-q21):
  - Haplotype 102T/-1438A.
  - Haplotype 102C/-1438G.
  - rs6314, rs7322347, rs643627 (A/G), and rs594242 (C/G),
- Serotonin receptor 2B (5-HTR2B), chromosome 2 (2q36.3-q37.1):
  - Stop codon (C20T, Q20*).
- Serotonin receptor 2C (5-HTR2C), chromosome X (Xq23):
  - SNP (rs6318) in gene's coding region (C68G).
- The tryptophan hydroxylase isoform 1 (TPH1), chromosome 11 (11p15.1):
  - Allele 218A (U).
  - Allele 218C (L).
- Dopamine D2 receptor, chromosome 11 (11q22-q23):
  - TaqI restriction fragment length polymorphism (RFLP).
  - Alleles A1 και A2.
- Dopamine D4 receptor, chromosome 11 (11p15.5):
  - Alleles 2R, 4R, and 7R more frequent.
  - Alleles 3R, 5R, 6R, and 8R rarely.
  - rs1800955,
  - Long (L) and short (S) allele.

- MAO-A, short arm of X chromosome (Xp11.3):
  - MAOA-H.
  - MAOA-L.
- Catechol-o-methyltransferase gene (COMT), chromosome 22 (22q11.21):
  - SNP in exon 4 (rs4680, G472A transition):
    Valine is substituted by methionine at codon:
    - 158 (Met158Val)
    - 108 (Met108Val).
  - COMT-LL (Met/Met).
  - COMT-LH (Met/Val).
  - COMT-HH (Val/Val).
- Brain-derived neurotrophic factor (BDNF), chromosome 11 (11p14.1):
  - SNP (rs6265), G196A, (Val66Met).
- Cholinergic receptor nicotinic alpha 4 subunit gene (CHRNA4), chromosome 20 (20q13.2-13.3):
  - SNP (rs1044396) in exon 5.
- Four genes that code the AMPA receptor subunits.
- A single NR1 gene which gives eight different isoforms of NR1 subunits in NMDA receptors.
- Four genes coding the NR2 family (NR2A-NR2D) subunits.

- *The serotonin transporter (SERT or 5HTT)* is coded for by the gene "solute carrier family 6, member 4" (SLC6A4), located on chromosome 17 (17q11.1-q12) (Nakamura et al. 2000). The polymorphism "5-HTTLPR" includes up to 14 different alleles (Nakamura et al. 2000), of which the most common and best studied are the "long" (L) and the "short" (S) (Heils et al. 1996; Hariri and Weinberger 2003). Homozygosity for the long variant (L/L) was associated with resulting in higher amounts of serotonin transporter and thus higher serotonin reuptake (Heils et al. 1996; Cadoret et al. 2003; Greenberg et al. 1999; Lesch et al. 1996; Collier et al. 1996). The L allele is divided into LA and LG because of the presence of an SNP. The LG is similar to the S allele leading to the low functioning of the SERT (Hu et al. 2006). Another SLC6A4 gene polymorphism is STin2.12 and is a more potent positive transcriptional regulator (MacKenzie and Quinn 1999).
- *The serotonin receptor 1A (5-HTR1A)* is encoded by a gene located at chromosome 5 (5q11.2-13) (Kobilka et al. 1987). The allele G is associated with a higher receptor expression, leading to increased negative feedback inhibition in raphe nuclei serotonergic neurons and subsequently to decreased serotonergic activity (Wu and Comings 1999; Lemonde et al. 2003).
- *The serotonin receptor 1B (5-HTR1B)* is coded for by a short, intronless gene located at chromosome 6 (6q14.1), and several gene polymorphisms exist. The G861C SNP (rs6296) affects receptor binding, and homozygosity for the G allele was associated with a higher receptor binding compared with G/C heterozygous genotype (Sanders et al. 2002; Huang et al. 1999).
- *The serotonin receptor 2A (5-HTR2A)* is encoded by a gene located at chromosome 13 (13q14-q21). The haplotype 102T/-1438A is associated with increased

serotonin binding compared with haplotype 102C/-1438G (Turecki et al. 1999). Other SNPs include rs6314, rs7322347, rs643627 (A/G), and rs594242 (C/G).

- *Serotonin receptor 2B (5-HTR2B)* is encoded by a gene located at chromosome 2 (2q36.3-q37.1). There is a functional stop codon (C20T, Q20*), most probably limited to the Finnish population, causing RNA decay and 5-HTR2B expression blockage (Bonaventure et al. 2002).
- *Serotonin receptor 2C (5-HTR2C)* is encoded by a gene located at X chromosome (Xq23). There is a SNP (rs6318) in gene's coding region, C68G.
- *The tryptophan hydroxylase isoform 1 (TPH1)*, is coded by a gene, located at chromosome 11 (11p15.1) (Zill et al. 2007; Gutknecht et al. 2009). The allele 218A is also referred to as "upper" or U allele and constitutes the minor allele, while 218C is also referred to as "lower" or L allele. The U allele was associated with decreased 5-HTR2A density (Ono et al. 2002).
- *The dopamine 2 receptor (D2R)* is encoded by a gene located at chromosome 11 (11q22-q23). There is a TaqI restriction fragment length polymorphism (RFLP), giving rise to alleles A1 and A2 which are of unknown significance (Grandy et al. 1989). There is a possible association between A1 allele and reduced D2 receptor's density and activity (Noble et al. 1997; Pohjalainen et al. 1998) and also an association between A1 allele and increased dopamine synthesis, perhaps due to decreased D2R expression (Laakso et al. 2005).
- *Dopamine 4 receptor (DRD4)* is encoded by a gene at chromosome 11 (11p15.5). Different alleles lead to different receptor products. Allele frequencies vary greatly between different populations. It was shown, that allele 4R is the most commonly found, together with 7R and 2R, while alleles 3R, 5R, 6R, and 8R are rare. Alleles up to 5R are considered "short," while the rest, including allele 7R, are considered "long" (Lichter et al. 1993; Chang et al. 1996; Ding et al. 2002). Allele 7R was associated with decreased in vitro gene expression and receptor binding compared with short alleles (Asghari et al. 1994, 1995). Another DRD4 gene polymorphism (**rs1800955)** was shown to lead to a 40% lower transcription levels compared with C allele (Ronai et al. 2001). Lastly, there is a tandem 120 bp duplication located 1.2 kb upstream from the initiation codon, giving rise to long (L) and short (S) alleles, with the S allele associated with increased transcriptional activity compared with L allele (D'Souza et al. 2004).
- *The MAO-A gene* is linked to the short arm of X chromosome (Xp11.3) (Hunter 2010). There are high- (MAOA-H) and low-activity MAO-A gene variants (MAOA-L). MAOA-H is associated with higher while MAOA-L with reduced MAO-A expression and therefore increased synaptic serotonin levels (Sabol et al. 1998; Buckholtz and Meyer-Lindenberg 2008; Guo et al. 2008).
- *The catechol-o-methyltransferase (COMT) gene* is located at chromosome 22 (22q11.21). There is a functional SNP in exon 4 (rs4680, G472A transition), resulting in an amino acid change. Valine is substituted by methionine at codon 158 of membrane-bound COMT (Met158Val) and at codon 108 of soluble COMT (Met108Val). Polymorphism G472A was shown to decrease enzyme's thermostability and therefore activity three- to fourfold. Specifically, genetically polymorphic COMT enzyme activity may be low (COMT-LL, corresponding to Met/Met genotype), intermediate (COMT-LH, corresponding to Met/Val geno-

type), or high (COMT-HH, corresponding to Val/Val genotype). Altogether, carriers of the more active form of the enzyme display less dopamine for neurotransmission (Stahl 2003; Lachman et al. 1996).

- *Brain-derived neurotrophic factor (BDNF)* is coded by a gene located at chromosome 11 (11p14.1). There is a common functional SNP (rs6265), G196A, resulting in an amino acid substitution (valine is substituted by methionine at codon 66, Val66Met). The more common G allele codes for valine, while A allele codes for methionine. The G196A A/A genotype was associated with decreased BDNF neuronal secretion (Egan et al. 2003).
- *The cholinergic receptor nicotinic alpha 4 subunit gene (CHRNA4).* Nicotinic acetylcholine receptors constitute a receptor family, whose members are formed by diverse combinations of five different subunits (α1-α10 and β2-β4). Neuronal receptor $\alpha 4\beta 2$ is the main receptor found in mammalian brain (Gotti et al. 2009). Subunit α4 is encoded by CHRNA4 gene, located at chromosome 20 (20q13.2–13.3). There is a synonymous SNP (rs1044396) in exon five, a C-to-T transition, giving rise to two corresponding alleles. There is evidence that it affects the receptor's sensitivity (Eggert et al. 2015).
- There are four genes that code the *AMPA receptor subunits* (Hollmann et al. 1989).
- There is a single NR1 gene which gives eight different isoforms of *NR1 subunits in NMDA receptors* (Hollmann et al. 1993; Sugihara et al. 1992). In contrast, there are *four genes coding the NR2 family* (NR2A–NR2D) subunits, and therefore it is evident that the properties of the NMDA receptor depend on the specific NR2 subunits which participate in the heteromer (Buller et al. 1994; Monyer et al. 1992, 1994; Williams 1993).

18.2.2 Phenotypes and Endophenotypes

A number of behaviors have a strong genetic component. In fact, most adult behavior has a genetic background even though it is strongly shaped by experience and the environment. However, there are some behaviors, e.g., the smile of babies, which could be considered as almost completely genetically determined.

Phenotypes and Endophenotypes

- The "big five" personality dimensions, heritability 40–60%:
  - SERT S allele (both S/S and S/L genotype).
  - C(-1019)G polymorphism of HTR1A receptor gene.
  - G196A polymorphism of BDNF gene.
  - CHRNA4 rs1044396.
  - COMT-LL.
- Novelty seeking, heritability 36%:
  - 7R allele of the DRD4 48 bp VNTR
  - 48 bp VNTR long allele

  - Short 2R and 5R alleles.
  - C-521 T SNP C/C.
  - COMT-HH.
- Sensation seeking, heritability 30–60%:
  - 7R allele
- Harm avoidance, heritability 36%:
  - In females, COMT-LL, COMT-LH, and COMT-HH.
  - BDNF G196A C allele.
- Impulsivity, heritability 45–50%:
  - MAOA-L.
  - 5-HTTLPR S/S, L/S, L/L
  - HTR1A G/G, G/C, C/C.
  - 5-HTR2B Q20
- Aggressive behavior, genetic component 40–60% of variability:
  - MAOA-L.
  - HTTLPR (S/S, S/LG, and LG/LG).
  - 5-HTR2A SNP T/A (rs7322347)
  - TPH1 218A, 779A.
  - DRD2 A1/A2.
  - COMT G472A (especially the COMT-HH genotype).
- Antisocial behavior, genetic component 40–60% of variability:
  - MAOA-L and history of abuse.
- Psychopathy, genetic component 50–80% of variability.
- Suicidality, heritability 55%:
  - A number of gene polymorphisms concerning the serotonergic system.
  - SERT 218C.
  - TPH1 779C.
  - COMT-LL.

A number of behaviors have been studied, and although results are rather inconclusive and only indicative, they provide valuable insight. The following concern nonclinical populations.

- *The big five personality dimensions* (neuroticism, introversion-extraversion, agreeableness, conscientiousness, and openness to experience) (Costa and McCrae 1985; McCrae et al. 2005) manifest a heritability of around 40–60% (Borkenau et al. 2001; Jang et al. 1996). The SERT S allele (both S/S and S/L genotype) was associated with neuroticism (Lesch et al. 1996) and especially the S allele with trait anxiety, affective temperament (depressive, cyclothymic, irritable, and anxious), guilt, hostility, and somatization (Gonda et al. 2009). Some results were positive concerning the relationship of HTR1A C(-1019)G polymorphism with anxiety- and depression-related personality traits, such as neuroticism and harm avoidance (Strobel et al. 2003; Serretti et al. 2009). BDNF G196A polymorphism explained 1.9% of the variance of trait anxiety, which was significantly higher in G/G carriers (Lang et al. 2005). Another report suggested that the G allele explained 4% of genetic variance for anxiety, depression, self-consciousness, and vulnerability (Sen et al. 2003). The A/A genotype was related with extraversion, but only in females (Itoh et al. 2004), while a recent genome-wide association study confirmed the association between the G196A polymorphism and extraversion (Terracciano et al. 2010). Lastly, a study of maltreated children showed that CHRNA4 rs1044396 moderated the effects of maltreatment on childhood personality outcome. Children with the T/T genotype displayed higher levels of neuroticism when they had a background of childhood maltreatment, but on the contrary, non-maltreated children with T/T genotype displayed lower levels of neuroticism and higher levels of openness to experience (Grazioplene et al. 2013). However, there is also a number of negative or equivocal reports concerning all the above (Ball et al. 1997; Schinka et al. 2004; Koller et al. 2006; Lang et al. 2005). Lastly, there was some evidence for an association between COMT-LL genotype and neuroticism in females, which was drawn though from a study employing a selected population (high or low scorers) (Eley et al. 2003) and the BDNF G/G genotype and high neuroticism levels (Frustaci et al. 2008).

- *Novelty seeking* is a behavioral trait according to the theory of Robert Cloninger (Cloninger 1987; Constantino et al. 2002; Cloninger et al. 1993, 1999), and heritability accounts for 36% of its variance (Heiman et al. 2004). Polymorphism HTR1A G-1438A was not associated with novelty seeking (Nomura et al. 2006), but the 7R allele of the DRD4 48 bp VNTR allele might be (Ebstein et al. 1996). Data support an association between 48 bp VNTR long allele carriers and novelty seeking (Benjamin et al. 1996; Ono et al. 1997; Tomitaka et al. 1999; Lee et al. 2003a; Strobel et al. 1999). Findings were contradictory for the 5R (Tsuchimine et al. 2009) but positive for the short 2R and 5R alleles (Ekelund et al. 1999). There are also negative reports (Kim et al. 2006a; Lusher et al. 2001; Baron 1998; Mitsuyasu et al. 2001; Tochigi et al. 2006), including meta-analyses (Kluger et al. 2002). Two other meta-analyses reported conflicting results concerning the long repeat alleles and high novelty seeking (Schinka et al. 2002; Munafo et al. 2008). A meta-analysis confirmed an association between *DRD4* C-521 T SNP C/C genotype and high novelty seeking (Schinka et al. 2002). Another one reported that T allele carriers (C/T or T/T genotype) displayed lower levels of novelty seeking, while C-521T SNP accounted for 2% of the phenotypic variance (Munafo et al. 2008). COMT-HH might relate to high novelty seeking (Reuter and Hennig 2005; Tsai et al. 2004b; Lang et al. 2007); however, two recent studies were negative concerning both COMT and BDNF polymorphisms (Kang et al. 2010; Tsai et al. 2004a).
- A similar construct is *sensation seeking* (McCourt et al. 1993; Zuckerman et al. 1972), and it has been shown to be highly heritable in both sexes with the genetic component explaining 30–60% of variance (Stoel et al. 2006). Sensation seeking was found to correlate with the 7R (Campbell et al. 2010; Dmitrieva et al. 2011).
- *Harm avoidance* is essentially the opposite of novelty seeking, and its heritability accounts for 36% of the variance (Heiman et al. 2004). The data are rather

negative concerning the association of HTR1A gene alleles and harm avoidance (Kazantseva et al. 2008; Strobel et al. 2003; Koller et al. 2006; Serretti et al. 2009; Nomura et al. 2006). Females genotyped as COMT-LL showed the lowest harm avoidance levels, whereas COMT-LH genotype was associated with intermediate and COMT-HH with higher harm avoidance levels (Kim et al. 2006b). A meta-analysis on BDNF G196A polymorphisms was negative (Tsai et al. 2004a); however; two later studies were positive concerning the C allele and high harm avoidance (Markett et al. 2011; Bey et al. 2016).

- *Impulsivity* has heritability around 45–50% (Pedersen et al. 1988; Hur and Bouchard Jr. 1997) with its role increasing with age. This is also the case with risk-taking behavior whose heritability seems to increase with age, but only in males (Anokhin et al. 2009). MAOA-L allele was reported to relate to high impulsivity (Manuck et al. 2000) as was the 5-HTTLPR S/S genotype (Sakado et al. 2003; Lee et al. 2003b). It seems that S/S carriers relate to higher levels of impulsivity compared with L/S carriers displaying intermediate and L/L carriers displaying lower levels of impulsiveness, while, additionally, the effect of S allele on the expression of impulsivity are sex-dependent, with males being more likely to exhibit impulsiveness (Walderhaug et al. 2010). Again negative reports exist (Lage et al. 2011). HTR1A gene G/G carriers displayed significantly higher impulsivity levels, compared with G/C and C/C carriers (Benko et al. 2010; Nomura et al. 2006). In a study of a Finnish population comprised of violent criminal offenders, 5-HTR2B Q20* carriers showed no cognitive deficits and committed crimes mediated by high impulsivity levels (Bevilacqua et al. 2010).
- *Aggressive behavior* constitutes a broad phenotype, and in general, heritability might be more pronounced during adulthood with its genetic component probably accounting for 40–60% of the variance (Craig and Halton 2009; Miles and Carey 1997; Rhee and Waldman 2002). The literature suggest a positive relationship between aggressiveness and the MAOA-L allele (Manuck et al. 2000; Eisenberger et al. 2007; Reif et al. 2007; McDermott et al. 2009). One study reported a significant association between HTTLPR genotypes with low-expressing alleles (S/S, S/LG, and LG/LG) and aggressive behavior in children aged 5–15 years (Beitchman et al. 2006). L/L genotype was considered a protective factor against manifestation of violence in adults with a high childhood adverse environment index (Reif et al. 2007). The allele S and S/S genotype were associated with violent behavior, explaining 5% of the variance of violent behavior (Retz et al. 2004). An association was reported between 5-HTR2A SNP T/A (rs7322347) and aggressive traits, with T/T genotype carriers displaying more aggressive traits compared with allele A carriers (Banlaki et al. 2015). Homozygosity for TPH1 218A allele was associated with higher aggression and more intense tendency towards unprovoked anger in males (Manuck et al. 1999), as well as with higher levels of proactive aggression (Hennig et al. 2005). Both TPH1 218A and 779A were associated with trait anger, state anger, and anger temperament (Rujescu et al. 2002). Heterozygous DRD2 A1/A2 but not homozygous A1/A1 genotype in adolescents and young adults was related to serious and violent delinquency but only in males (Guo et al. 2007). This is called "het-

erosis," a condition in which it is heterozygous rather than homozygous individuals manifesting a trait to a greater or lesser extent (Comings and MacMurray 2000). Other studies suggested the presence of A1 genotype, either as A1/A1 homozygosity or A1/A2 heterozygosity, to be associated with aggressive-violent behavior (Chen et al. 2005). In general, COMT G472A polymorphism (in particular COMT-HH genotype) may be associated with aggressive traits (Calati et al. 2011; Kulikova et al. 2008).

- *Antisocial behavior* constitutes a broad phenotype and includes the manifestation of aggressive and criminal behavior as well as delinquency and psychopathy since early adolescence (Baker et al. 2007), and it is more common in males (Craig and Halton 2009). Interestingly, genetic factors seem to be stronger in females during childhood, but this disappears later, and in adults the effects of genetic and environmental factors have a similar effect on both sexes (Jacobson et al. 2002). Overall, the genetic component was shown to account for 40–60% of the variance (Fergusson et al. 2011; Gunter et al. 2010). Male MAOA-L carriers with a background of childhood maltreatment or other early-life traumatic events were more prompt to the development of antisocial behavior compared with male MAOA-H carriers (Caspi et al. 2002; Frazzetto et al. 2007; Aslund et al. 2011). However, there are also negative reports (Haberstick et al. 2005; Huizinga et al. 2006; Weder et al. 2009; Prichard et al. 2007; Jacob et al. 2005; Widom and Brzustowicz 2006). Two meta-analyses supported the relationship of MAOA-L and antisocial behavior with a special importance of childhood maltreatment especially for males (Kim-Cohen et al. 2006; Ficks and Waldman 2014; Byrd and Manuck 2014). The data concerning the role of 5-HTTLPR genotype are conflicting (Liao et al. 2004; Davidge et al. 2004), but a meta-analysis was positive (with some reservations) concerning the role of the S allele (Ficks and Waldman 2014).
- *Psychopathy* is another but not identical term with antisocial behavior and refers to a personality trait characterized by a lack of empathy and remorse. Its genetic component explains 50–80% of variance (Gunter et al. 2010; Retz et al. 2004).
- *Suicidality* seems to be influenced by a genetic component independent of the genetic contribution to mental disorders. Many genes are reported to contribute, with the heritability of serious suicidality to be around 55% (Clayden et al. 2012; Mann et al. 2001), with more important the effect of a number of serotonergic gene polymorphisms (Arango et al. 2003). A meta-analysis was negative concerning the role of MAOA alleles (Clayden et al. 2012). There data are contradictory on the role of the 5-HTTLPR alleles (Courtet et al. 2001; Du et al. 1999; Anguelova et al. 2003), while meta-analyses supported the role of the S allele for attempters only (Anguelova et al. 2003; Clayden et al. 2012; Jernej et al. 2004). Other meta-analyses were difficult to interpret (Lin and Tsai 2004; Li and He 2007). The role of 5-HTR1A allele G is unclear (Lemonde et al. 2003; Sawiniec et al. 2007; Wasserman et al. 2006; Gonzalez-Castro et al. 2013b), and the results of three meta-analyses were negative (Angles et al. 2012; Clayden et al. 2012; Gonzalez-Castro et al. 2013b). A number of other negative studies on the role of 5-HTR1A polymorphisms also exist (Anguelova et al. 2003; Gonzalez-Castro

et al. 2013b). The data on 5-HTR1B polymorphisms are also contradictory (New et al. 2001; Nishiguchi et al. 2001; Rujescu et al. 2003c; Stefulj et al. 2004b), with meta-analysis returning negative results (Kia-Keating et al. 2007; Clayden et al. 2012). Studies on 5-HTR2A polymorphisms were also inconclusive (Zhang et al. 1997; Gonzalez-Castro et al. 2013a; Bondy et al. 2000; Anguelova et al. 2003; Giegling et al. 2006), with negative meta-analyses (Anguelova et al. 2003; Li et al. 2006; Gonzalez-Castro et al. 2013a; Wang et al. 2015) except for one concerning the role of G-1438A (Li et al. 2006). One study on 5-HTR2C polymorphisms was negative (Stefulj et al. 2004a). Studies on TPH1 polymorphisms were negative (Clayden et al. 2012; Serretti et al. 2009; Rujescu et al. 2003b; Turecki et al. 2001; Zalsman et al. 2001). There are some data on the relation of the SERT 218C with suicidal behavior (Jernej et al. 2004), confirmed also by meta-analyses (Rujescu et al. 2003b; Bellivier et al. 2004), although the findings could be valid to attempts alone (Clayden et al. 2012). There are some evidence for an association between TPH1 779C allele and non-impulsive suicide attempts (Nielsen et al. 1998), but results for the A779C and A218C polymorphisms are controversial or negative (Lalovic and Turecki 2002; Saetre et al. 2010; Li and He 2006; Clayden et al. 2012). The COMT-LL allele was reported to relate to violent but not to nonviolent suicide attempts in a German psychiatric population, independent of diagnosis, compared with healthy controls (Rujescu et al. 2003a). Although negative reports exist (Russ et al. 2000; Tovilla-Zarate et al. 2011), a meta-analysis was positive with a probable moderating effect identified for gender (Kia-Keating et al. 2007). Three other meta-analyses were negative (Calati et al. 2011; Tovilla-Zarate et al. 2011; Clayden et al. 2012). One meta-analysis suggested that the low-functioning A allele of BDNF constituted a risk allele for the manifestation of suicidal behavior (Zai et al. 2012), but a second meta-analysis was negative (Clayden et al. 2012).

18.3 Conclusions from Genetic Studies

Association studies within the research field of behavioral genetics have provided in most cases *controversial or inconsistent results* when nonclinical populations were studied. The reasons for this include ethnicity with differences in genetic background (Li and He 2007; Lea and Chambers 2007; Hunter 2010; Brunner et al. 1993a, b; Kelsoe 2010; Zai et al. 2012), gender bias in the study sample (Tsai et al. 2004a; Kulikova et al. 2008; Manuck et al. 2000; Alia-Klein et al. 2008; Kim-Cohen et al. 2006; Derringer et al. 2010; Ficks and Waldman 2014; Lin and Tsai 2004), age (Lusher et al. 2001; Lang et al. 2005), and socioeconomic or cultural differences (Campbell et al. 2010; Kang et al. 2010), as well as a number of other methodological issues (Munafo et al. 2008; Olivier and van Oorschot 2005; Mann 1998; Clayden et al. 2012; Schinka et al. 2004) including issues of shared pathophysiology between behavioral traits and mental disorders (Singh et al. 2012; Williams et al. 2007; Gonzalez-Castro et al. 2013b; Lin and Tsai 2004; Saetre et al. 2010) together with their high comorbidity (Schild et al. 2013). Environmental factors also act as confounding variables (Caspi et al. 2002; Derringer et al. 2010; Hart and Marmorstein 2009).

Associating a single gene polymorphism with behaviors and personality traits was proven a non-accomplishable task until now. Both human behavior and personality are of complex origin, the result of genes interacting with each other, as well as with environmental factors. Altogether, studies of the association between specific alleles and behaviors/personality traits have mostly contributed to exclude the theory of a single gene's major contribution. *Missing heritability* has been attributed to variants, possibly rare variants with greater effects, which have not been discovered yet; however, they exert a robust effect. Gene-gene interactions at an astronomical scale of complexity might also lay behind many of human behavior characteristics (Zuk et al. 2012; Manolio et al. 2009).

References

Alia-Klein N, Goldstein RZ, Kriplani A, Logan J, Tomasi D, Williams B, Telang F, Shumay E, Biegon A, Craig IW, Henn F, Wang GJ, Volkow ND, Fowler JS (2008) Brain monoamine oxidase A activity predicts trait aggression. J Neurosci 28(19):5099–5104. https://doi.org/10.1523/JNEUROSCI.0925-08.2008

Angles MR, Ocana DB, Medellin BC, Tovilla-Zarate C (2012) No association between the HTR1A gene and suicidal behavior: a meta-analysis. Braz J Psychiatry 34(1):38–42

Anguelova M, Benkelfat C, Turecki G (2003) A systematic review of association studies investigating genes coding for serotonin receptors and the serotonin transporter: II. Suicidal behavior. Mol Psychiatry 8(7):646–653. https://doi.org/10.1038/sj.mp.4001336

Anokhin AP, Golosheykin S, Grant J, Heath AC (2009) Heritability of risk-taking in adolescence: a longitudinal twin study. Twin Res Hum Genet 12(4):366–371. https://doi.org/10.1375/twin.12.4.366

Arango V, Huang YY, Underwood MD, Mann JJ (2003) Genetics of the serotonergic system in suicidal behavior. J Psychiatr Res 37(5):375–386. https://doi.org/10.1016/s0022-3956(03)00048-7

Asghari V, Schoots O, van Kats S, Ohara K, Jovanovic V, Guan HC, Bunzow JR, Petronis A, Van Tol HH (1994) Dopamine D4 receptor repeat: analysis of different native and mutant forms of the human and rat genes. Mol Pharmacol 46(2):364–373

Asghari V, Sanyal S, Buchwaldt S, Paterson A, Jovanovic V, Van Tol HH (1995) Modulation of intracellular cyclic AMP levels by different human dopamine D4 receptor variants. J Neurochem 65(3):1157–1165. https://doi.org/10.1046/j.1471-4159.1995.65031157.x

Aslund C, Nordquist N, Comasco E, Leppert J, Oreland L, Nilsson KW (2011) Maltreatment, MAOA, and delinquency: sex differences in gene-environment interaction in a large population-based cohort of adolescents. Behav Genet 41(2):262–272. https://doi.org/10.1007/s10519-010-9356-y

Baker LA, Jacobson KC, Raine A, Lozano DI, Bezdjian S (2007) Genetic and environmental bases of childhood antisocial behavior: a multi-informant twin study. J Abnorm Psychol 116(2):219–235. https://doi.org/10.1037/0021-843X.116.2.219

Ball D, Hill L, Freeman B, Eley TC, Strelau J, Riemann R, Spinath FM, Angleitner A, Plomin R (1997) The serotonin transporter gene and peer-rated neuroticism. Neuroreport 8(5):1301–1304. https://doi.org/10.1097/00001756-199703240-00048

Banlaki Z, Elek Z, Nanasi T, Szekely A, Nemoda Z, Sasvari-Szekely M, Ronai Z (2015) Polymorphism in the serotonin receptor 2a (HTR2A) gene as possible predisposal factor for aggressive traits. PLoS One 10(2):e0117792. https://doi.org/10.1371/journal.pone.0117792

Baron M (1998) Mapping genes for personality: is the saga sagging? Mol Psychiatry 3(2):106–108. https://doi.org/10.1038/sj.mp.4000377

Beitchman JH, Baldassarra L, Mik H, De Luca V, King N, Bender D, Ehtesham S, Kennedy JL (2006) Serotonin transporter polymorphisms and persistent, pervasive childhood aggression. Am J Psychiatry 163(6):1103–1105. https://doi.org/10.1176/ajp.2006.163.6.1103

Bellivier F, Chaste P, Malafosse A (2004) Association between the TPH gene A218C polymorphism and suicidal behavior: a meta-analysis. Am J Med Genet B Neuropsychiatr Genet 124B(1):87–91. https://doi.org/10.1002/ajmg.b.20015

Benjamin J, Li L, Patterson C, Greenberg BD, Murphy DL, Hamer DH (1996) Population and familial association between the D4 dopamine receptor gene and measures of novelty seeking. Nat Genet 12(1):81–84. https://doi.org/10.1038/ng0196-81

Benko A, Lazary J, Molnar E, Gonda X, Tothfalusi L, Pap D, Mirnics Z, Kurimay T, Chase D, Juhasz G, Anderson IM, Deakin JFW, Bagdy G (2010) Significant association between the C(−1019)G functional polymorphism of the HTR1A gene and impulsivity. Am J Med Genet B Neuropsychiatr Genet 153B(2):592–599. https://doi.org/10.1002/ajmg.b.31025

Bevilacqua L, Doly S, Kaprio J, Yuan Q, Tikkanen R, Paunio T, Zhou Z, Wedenoja J, Maroteaux L, Diaz S, Belmer A, Hodgkinson CA, Dell'osso L, Suvisaari J, Coccaro E, Rose RJ, Peltonen L, Virkkunen M, Goldman D (2010) A population-specific HTR2B stop codon predisposes to severe impulsivity. Nature 468(7327):1061–1066. https://doi.org/10.1038/nature09629

Bey K, Lennertz L, Markett S, Petrovsky N, Gallinat J, Grunder G, Spreckelmeyer KN, Wienker TF, Mobascher A, Dahmen N, Thuerauf N, Kornhuber J, Kiefer F, Toliat MR, Nurnberg P, Winterer G, Wagner M (2016) Replication of the association between CHRNA4 rs1044396 and harm avoidance in a large population-based sample. Eur Neuropsychopharmacol 26(1):150–155. https://doi.org/10.1016/j.euroneuro.2015.11.007

Bonaventure P, Guo H, Tian B, Liu X, Bittner A, Roland B, Salunga R, Ma XJ, Kamme F, Meurers B, Bakker M, Jurzak M, Leysen JE, Erlander MG (2002) Nuclei and subnuclei gene expression profiling in mammalian brain. Brain Res 943(1):38–47. https://doi.org/10.1016/s0006-8993(02)02504-0

Bondy B, Kuznik J, Baghai T, Schule C, Zwanzger P, Minov C, de Jonge S, Rupprecht R, Meyer H, Engel RR, Eisenmenger W, Ackenheil M (2000) Lack of association of serotonin-2A receptor gene polymorphism (T102C) with suicidal ideation and suicide. Am J Med Genet 96(6):831–835. https://doi.org/10.1002/1096-8628(20001204)96:6<831::aid-ajmg27>3.0.co;2-k

Borkenau P, Riemann R, Angleitner A, Spinath FM (2001) Genetic and environmental influences on observed personality: evidence from the German observational study of adult twins. J Pers Soc Psychol 80(4):655–668. https://doi.org/10.1037//0022-3514.80.4.655

Brunner HG, Nelen M, Breakefield XO, Ropers HH, van Oost BA (1993a) Abnormal behavior associated with a point mutation in the structural gene for monoamine oxidase A. Science 262(5133):578–580. https://doi.org/10.1126/science.8211186

Brunner HG, Nelen MR, van Zandvoort P, Abeling NG, van Gennip AH, Wolters EC, Kuiper MA, Ropers HH, van Oost BA (1993b) X-linked borderline mental retardation with prominent behavioral disturbance: phenotype, genetic localization, and evidence for disturbed monoamine metabolism. Am J Hum Genet 52(6):1032–1039

Buckholtz JW, Meyer-Lindenberg A (2008) MAOA and the neurogenetic architecture of human aggression. Trends Neurosci 31(3):120–129. https://doi.org/10.1016/j.tins.2007.12.006

Buller AL, Larson HC, Schneider BE, Beaton JA, Morrisett RA, Monaghan DT (1994) The molecular basis of NMDA receptor subtypes: native receptor diversity is predicted by subunit composition. J Neurosci 14(9):5471–5484

Byrd AL, Manuck SB (2014) MAOA, childhood maltreatment, and antisocial behavior: meta-analysis of a gene-environment interaction. Biol Psychiatry 75(1):9–17. https://doi.org/10.1016/j.biopsych.2013.05.004

Cadoret RJ, Langbehn D, Caspers K, Troughton EP, Yucuis R, Sandhu HK, Philibert R (2003) Associations of the serotonin transporter promoter polymorphism with aggressivity, attention deficit, and conduct disorder in an adoptee population. Compr Psychiatry 44(2):88–101. https://doi.org/10.1053/comp.2003.50018

Calati R, Porcelli S, Giegling I, Hartmann AM, Moller HJ, De Ronchi D, Serretti A, Rujescu D (2011) Catechol-o-methyltransferase gene modulation on suicidal behavior and personality traits: review, meta-analysis and association study. J Psychiatr Res 45(3):309–321. https://doi.org/10.1016/j.jpsychires.2010.07.004

Campbell BC, Dreber A, Apicella CL, Eisenberg DT, Gray PB, Little AC, Garcia JR, Zamore RS, Lum JK (2010) Testosterone exposure, dopaminergic reward, and sensation-seeking in young men. Physiol Behav 99(4):451–456. https://doi.org/10.1016/j.physbeh.2009.12.011

Caspi A, McClay J, Moffitt TE, Mill J, Martin J, Craig IW, Taylor A, Poulton R (2002) Role of genotype in the cycle of violence in maltreated children. Science 297(5582):851–854. https://doi.org/10.1126/science.1072290

Chang FM, Kidd JR, Livak KJ, Pakstis AJ, Kidd KK (1996) The world-wide distribution of allele frequencies at the human dopamine D4 receptor locus. Hum Genet 98(1):91–101. https://doi.org/10.1007/s004390050166

Chen TJ, Blum K, Mathews D, Fisher L, Schnautz N, Braverman ER, Schoolfield J, Downs BW, Comings DE (2005) Are dopaminergic genes involved in a predisposition to pathological aggression? Hypothesizing the importance of "super normal controls" in psychiatricgenetic research of complex behavioral disorders. Med Hypotheses 65(4):703–707. https://doi.org/10.1016/j.mehy.2005.04.037

Cicchetti D, Rogosch FA, Thibodeau EL (2012) The effects of child maltreatment on early signs of antisocial behavior: genetic moderation by tryptophan hydroxylase, serotonin transporter, and monoamine oxidase A genes. Dev Psychopathol 24(3):907–928. https://doi.org/10.1017/S0954579412000442

Clayden RC, Zaruk A, Meyre D, Thabane L, Samaan Z (2012) The association of attempted suicide with genetic variants in the SLC6A4 and TPH genes depends on the definition of suicidal behavior: a systematic review and meta-analysis. Transl Psychiatry 2:e166. https://doi.org/10.1038/tp.2012.96

Cloninger CR (1987) A systematic method for clinical description and classification of personality variants. A proposal. Arch Gen Psychiatry 44(6):573–588. https://doi.org/10.1001/archpsyc.1987.01800180093014

Cloninger CR, Svrakic DM, Przybeck TR (1993) A psychobiological model of temperament and character. Arch Gen Psychiatry 50(12):975–990. https://doi.org/10.1001/archpsyc.1993.01820240059008

Cloninger C, Svrakic D, Bayon C, Przybeck T (1999) Measurement of psychopathology as variants of personality. In: Cloninger C (ed) Personality and psychopathology. American Psychiatric Press Inc., Washington DC, pp 33–65

Collier DA, Stober G, Li T, Heils A, Catalano M, Di Bella D, Arranz MJ, Murray RM, Vallada HP, Bengel D, Muller CR, Roberts GW, Smeraldi E, Kirov G, Sham P, Lesch KP (1996) A novel functional polymorphism within the promoter of the serotonin transporter gene: possible role in susceptibility to affective disorders. Mol Psychiatry 1(6):453–460

Comings DE, MacMurray JP (2000) Molecular heterosis: a review. Mol Genet Metab 71(1–2):19–31. https://doi.org/10.1006/mgme.2000.3015

Constantino JN, Cloninger CR, Clarke AR, Hashemi B, Przybeck T (2002) Application of the seven-factor model of personality to early childhood. Psychiatry Res 109(3):229–243. https://doi.org/10.1016/s0165-1781(02)00008-2

Costa PT, McCrae RR (1985) The NEO personality inventory manual. Psychological Assessment Resources, Odessa, FL

Courtet P, Baud P, Abbar M, Boulenger JP, Castelnau D, Mouthon D, Malafosse A, Buresi C (2001) Association between violent suicidal behavior and the low activity allele of the serotonin transporter gene. Mol Psychiatry 6(3):338–341. https://doi.org/10.1038/sj.mp.4000856

Craig IW, Halton KE (2009) Genetics of human aggressive behaviour. Hum Genet 126(1):101–113. https://doi.org/10.1007/s00439-009-0695-9

Davidge KM, Atkinson L, Douglas L, Lee V, Shapiro S, Kennedy JL, Beitchman JH (2004) Association of the serotonin transporter and 5HT1Dbeta receptor genes with extreme, persistent and pervasive aggressive behaviour in children. Psychiatr Genet 14(3):143–146. https://doi.org/10.1097/00041444-200409000-00004

Derringer J, Krueger RF, Irons DE, Iacono WG (2010) Harsh discipline, childhood sexual assault, and MAOA genotype: an investigation of main and interactive effects on diverse clinical externalizing outcomes. Behav Genet 40(5):639–648. https://doi.org/10.1007/s10519-010-9358-9

Ding YC, Chi HC, Grady DL, Morishima A, Kidd JR, Kidd KK, Flodman P, Spence MA, Schuck S, Swanson JM, Zhang YP, Moyzis RK (2002) Evidence of positive selection acting at the human dopamine receptor D4 gene locus. Proc Natl Acad Sci U S A 99(1):309–314. https://doi.org/10.1073/pnas.012464099

Dmitrieva J, Chen C, Greenberger E, Ogunseitan O, Ding YC (2011) Gender-specific expression of the DRD4 gene on adolescent delinquency, anger and thrill seeking. Soc Cogn Affect Neurosci 6(1):82–89. https://doi.org/10.1093/scan/nsq020

D'Souza UM, Russ C, Tahir E, Mill J, McGuffin P, Asherson PJ, Craig IW (2004) Functional effects of a tandem duplication polymorphism in the 5'flanking region of the DRD4 gene. Biol Psychiatry 56(9):691–697. https://doi.org/10.1016/j.biopsych.2004.08.008

Du L, Faludi G, Palkovits M, Demeter E, Bakish D, Lapierre YD, Sotonyi P, Hrdina PD (1999) Frequency of long allele in serotonin transporter gene is increased in depressed suicide victims. Biol Psychiatry 46(2):196–201. https://doi.org/10.1016/s0006-3223(98)00376-x

Eaves LJ (1984) The resolution of genotype x environment interaction in segregation analysis of nuclear families. Genet Epidemiol 1(3):215–228. https://doi.org/10.1002/gepi.1370010302

Ebstein RP, Novick O, Umansky R, Priel B, Osher Y, Blaine D, Bennett ER, Nemanov L, Katz M, Belmaker RH (1996) Dopamine D4 receptor (D4DR) exon III polymorphism associated with the human personality trait of novelty seeking. Nat Genet 12(1):78–80. https://doi.org/10.1038/ng0196-78

Egan MF, Kojima M, Callicott JH, Goldberg TE, Kolachana BS, Bertolino A, Zaitsev E, Gold B, Goldman D, Dean M, Lu B, Weinberger DR (2003) The BDNF val66met polymorphism affects activity-dependent secretion of BDNF and human memory and hippocampal function. Cell 112(2):257–269. https://doi.org/10.1016/s0092-8674(03)00035-7

Eggert M, Winterer G, Wanischeck M, Hoda JC, Bertrand D, Steinlein O (2015) The nicotinic acetylcholine receptor alpha 4 subunit contains a functionally relevant SNP haplotype. BMC Genet 16:46. https://doi.org/10.1186/s12863-015-0204-1

Eisenberger NI, Way BM, Taylor SE, Welch WT, Lieberman MD (2007) Understanding genetic risk for aggression: clues from the brain's response to social exclusion. Biol Psychiatry 61(9):1100–1108. https://doi.org/10.1016/j.biopsych.2006.08.007

Ekelund J, Lichtermann D, Jarvelin MR, Peltonen L (1999) Association between novelty seeking and the type 4 dopamine receptor gene in a large Finnish cohort sample. Am J Psychiatry 156(9):1453–1455. https://doi.org/10.1176/ajp.156.9.1453

Eley TC, Tahir E, Angleitner A, Harriss K, McClay J, Plomin R, Riemann R, Spinath F, Craig I (2003) Association analysis of MAOA and COMT with neuroticism assessed by peers. Am J Med Genet B Neuropsychiatr Genet 120B(1):90–96. https://doi.org/10.1002/ajmg.b.20046

Fergusson DM, Boden JM, Horwood LJ, Miller AL, Kennedy MA (2011) MAOA, abuse exposure and antisocial behaviour: 30-year longitudinal study. Br J Psychiatry 198(6):457–463. https://doi.org/10.1192/bjp.bp.110.086991

Ficks CA, Waldman ID (2014) Candidate genes for aggression and antisocial behavior: a meta-analysis of association studies of the 5HTTLPR and MAOA-uVNTR. Behav Genet 44(5):427–444. https://doi.org/10.1007/s10519-014-9661-y

Frazzetto G, Di Lorenzo G, Carola V, Proietti L, Sokolowska E, Siracusano A, Gross C, Troisi A (2007) Early trauma and increased risk for physical aggression during adulthood: the moderating role of MAOA genotype. PLoS One 2(5):e486. https://doi.org/10.1371/journal.pone.0000486

Frustaci A, Pozzi G, Gianfagna F, Manzoli L, Boccia S (2008) Meta-analysis of the brain-derived neurotrophic factor gene (BDNF) Val66Met polymorphism in anxiety disorders and anxiety-related personality traits. Neuropsychobiology 58(3–4):163–170. https://doi.org/10.1159/000182892

Giegling I, Hartmann AM, Moller HJ, Rujescu D (2006) Anger- and aggression-related traits are associated with polymorphisms in the 5-HT-2A gene. J Affect Disord 96(1–2):75–81. https://doi.org/10.1016/j.jad.2006.05.016

Gonda X, Fountoulakis KN, Juhasz G, Rihmer Z, Lazary J, Laszik A, Akiskal HS, Bagdy G (2009) Association of the s allele of the 5-HTTLPR with neuroticism-related traits and temperaments in a psychiatrically healthy population. Eur Arch Psychiatry Clin Neurosci 259(2):106–113. https://doi.org/10.1007/s00406-008-0842-7

Gonzalez-Castro TB, Tovilla-Zarate C, Juarez-Rojop I, Pool Garcia S, Velazquez-Sanchez MP, Genis A, Nicolini H, Lopez Narvaez L (2013a) Association of the 5HTR2A gene with suicidal behavior: case-control study and updated meta-analysis. BMC Psychiatry 13:25. https://doi.org/10.1186/1471-244X-13-25

Gonzalez-Castro TB, Tovilla-Zarate CA, Juarez-Rojop I, Pool Garcia S, Genis A, Nicolini H, Lopez Narvaez L (2013b) Association of 5HTR1A gene variants with suicidal behavior: case-control study and updated meta-analysis. J Psychiatr Res 47(11):1665–1672. https://doi.org/10.1016/j.jpsychires.2013.04.011

Gotti C, Clementi F, Fornari A, Gaimarri A, Guiducci S, Manfredi I, Moretti M, Pedrazzi P, Pucci L, Zoli M (2009) Structural and functional diversity of native brain neuronal nicotinic receptors. Biochem Pharmacol 78(7):703–711. https://doi.org/10.1016/j.bcp.2009.05.024

Grandy DK, Litt M, Allen L, Bunzow JR, Marchionni M, Makam H, Reed L, Magenis RE, Civelli O (1989) The human dopamine D2 receptor gene is located on chromosome 11 at q22-q23 and identifies a TaqI RFLP. Am J Hum Genet 45(5):778–785

Grazioplene RG, Deyoung CG, Rogosch FA, Cicchetti D (2013) A novel differential susceptibility gene: CHRNA4 and moderation of the effect of maltreatment on child personality. J Child Psychol Psychiatry 54(8):872–880. https://doi.org/10.1111/jcpp.12031

Greenberg BD, Tolliver TJ, Huang SJ, Li Q, Bengel D, Murphy DL (1999) Genetic variation in the serotonin transporter promoter region affects serotonin uptake in human blood platelets. Am J Med Genet 88(1):83–87

Gunter TD, Vaughn MG, Philibert RA (2010) Behavioral genetics in antisocial spectrum disorders and psychopathy: a review of the recent literature. Behav Sci Law 28(2):148–173. https://doi.org/10.1002/bsl.923

Guo G, Roettger ME, Shih JC (2007) Contributions of the DAT1 and DRD2 genes to serious and violent delinquency among adolescents and young adults. Hum Genet 121(1):125–136. https://doi.org/10.1007/s00439-006-0244-8

Guo G, Ou XM, Roettger M, Shih JC (2008) The VNTR 2 repeat in MAOA and delinquent behavior in adolescence and young adulthood: associations and MAOA promoter activity. Eur J Hum Genet 16(5):626–634. https://doi.org/10.1038/sj.ejhg.5201999

Gutknecht L, Kriegebaum C, Waider J, Schmitt A, Lesch KP (2009) Spatio-temporal expression of tryptophan hydroxylase isoforms in murine and human brain: convergent data from Tph2 knockout mice. Eur Neuropsychopharmacol 19(4):266–282. https://doi.org/10.1016/j.euroneuro.2008.12.005

Haberstick BC, Lessem JM, Hopfer CJ, Smolen A, Ehringer MA, Timberlake D, Hewitt JK (2005) Monoamine oxidase a (MAOA) and antisocial behaviors in the presence of childhood and adolescent maltreatment. Am J Med Genet B Neuropsychiatr Genet 135B(1):59–64. https://doi.org/10.1002/ajmg.b.30176

Hariri AR, Weinberger DR (2003) Functional neuroimaging of genetic variation in serotonergic neurotransmission. Genes Brain Behav 2(6):341–349. https://doi.org/10.1046/j.1601-1848.2003.00048.x

Hart D, Marmorstein NR (2009) Neighborhoods and genes and everything in between: understanding adolescent aggression in social and biological contexts. Dev Psychopathol 21(3):961–973. https://doi.org/10.1017/S0954579409000510

Heils A, Teufel A, Petri S, Stober G, Riederer P, Bengel D, Lesch KP (1996) Allelic variation of human serotonin transporter gene expression. J Neurochem 66(6):2621–2624. https://doi.org/10.1046/j.1471-4159.1996.66062621.x

Heiman N, Stallings MC, Young SE, Hewitt JK (2004) Investigating the genetic and environmental structure of Cloninger's personality dimensions in adolescence. Twin Res 7(5):462–470. https://doi.org/10.1375/1369052042335313

Hennig J, Reuter M, Netter P, Burk C, Landt O (2005) Two types of aggression are differentially related to serotonergic activity and the A779C TPH polymorphism. Behav Neurosci 119(1):16–25. https://doi.org/10.1037/0735-7044.119.1.16

Hollmann M, O'Shea-Greenfield A, Rogers SW, Heinemann S (1989) Cloning by functional expression of a member of the glutamate receptor family. Nature 342(6250):643–648. https://doi.org/10.1038/342643a0

Hollmann M, Boulter J, Maron C, Beasley L, Sullivan J, Pecht G, Heinemann S (1993) Zinc potentiates agonist-induced currents at certain splice variants of the NMDA receptor. Neuron 10(5):943–954. https://doi.org/10.1016/0896-6273(93)90209-a

Hu XZ, Lipsky RH, Zhu G, Akhtar LA, Taubman J, Greenberg BD, Xu K, Arnold PD, Richter MA, Kennedy JL, Murphy DL, Goldman D (2006) Serotonin transporter promoter gain-of-function genotypes are linked to obsessive-compulsive disorder. Am J Hum Genet 78(5):815–826. https://doi.org/10.1086/503850

Huang YY, Grailhe R, Arango V, Hen R, Mann JJ (1999) Relationship of psychopathology to the human serotonin1B genotype and receptor binding kinetics in postmortem brain tissue. Neuropsychopharmacology 21(2):238–246. https://doi.org/10.1016/S0893-133X(99)00030-5

Huizinga D, Haberstick BC, Smolen A, Menard S, Young SE, Corley RP, Stallings MC, Grotpeter J, Hewitt JK (2006) Childhood maltreatment, subsequent antisocial behavior, and the role of monoamine oxidase A genotype. Biol Psychiatry 60(7):677–683. https://doi.org/10.1016/j.biopsych.2005.12.022

Hunter P (2010) The psycho gene. EMBO Rep 11(9):667–669. https://doi.org/10.1038/embor.2010.122

Hur YM, Bouchard TJ Jr (1997) The genetic correlation between impulsivity and sensation seeking traits. Behav Genet 27(5):455–463. https://doi.org/10.1023/a:1025674417078

Itoh K, Hashimoto K, Kumakiri C, Shimizu E, Iyo M (2004) Association between brain-derived neurotrophic factor 196 G/a polymorphism and personality traits in healthy subjects. Am J Med Genet B Neuropsychiatr Genet 124B(1):61–63. https://doi.org/10.1002/ajmg.b.20078

Jacob CP, Muller J, Schmidt M, Hohenberger K, Gutknecht L, Reif A, Schmidtke A, Mossner R, Lesch KP (2005) Cluster B personality disorders are associated with allelic variation of monoamine oxidase a activity. Neuropsychopharmacology 30(9):1711–1718. https://doi.org/10.1038/sj.npp.1300737

Jacobson KC, Prescott CA, Kendler KS (2002) Sex differences in the genetic and environmental influences on the development of antisocial behavior. Dev Psychopathol 14(2):395–416. https://doi.org/10.1017/s0954579402002110

Jaffee SR, Price TS (2008) Genotype-environment correlations: implications for determining the relationship between environmental exposures and psychiatric illness. Psychiatry 7(12):496–499. https://doi.org/10.1016/j.mppsy.2008.10.002

Jang KL, Livesley WJ, Vernon PA (1996) Heritability of the big five personality dimensions and their facets: a twin study. J Pers 64(3):577–591. https://doi.org/10.1111/j.1467-6494.1996.tb00522.x

Jernej B, Stefulj J, Hranilovic D, Balija M, Skavic J, Kubat M (2004) Intronic polymorphism of tryptophan hydroxylase and serotonin transporter: indication for combined effect in predisposition to suicide. J Neural Transm (Vienna) 111(6):733–738. https://doi.org/10.1007/s00702-003-0114-7

Kang JI, Song DH, Namkoong K, Kim SJ (2010) Interaction effects between COMT and BDNF polymorphisms on boredom susceptibility of sensation seeking traits. Psychiatry Res 178(1):132–136. https://doi.org/10.1016/j.psychres.2010.04.001

Kazantseva AV, Gaysina DA, Faskhutdinova GG, Noskova T, Malykh SB, Khusnutdinova EK (2008) Polymorphisms of the serotonin transporter gene (5-HTTLPR, A/G SNP in 5-HTTLPR, and STin2 VNTR) and their relation to personality traits in healthy individuals from Russia. Psychiatr Genet 18(4):167–176. https://doi.org/10.1097/YPG.0b013e328304deb8

Kelsoe JR (2010) Behavioural neuroscience: a gene for impulsivity. Nature 468(7327):1049–1050. https://doi.org/10.1038/4681049a

Kendler KS, Eaves LJ (1986) Models for the joint effect of genotype and environment on liability to psychiatric illness. Am J Psychiatry 143(3):279–289. https://doi.org/10.1176/ajp.143.3.279

Kia-Keating BM, Glatt SJ, Tsuang MT (2007) Meta-analyses suggest association between COMT, but not HTR1B, alleles, and suicidal behavior. Am J Med Genet B Neuropsychiatr Genet 144B(8):1048–1053. https://doi.org/10.1002/ajmg.b.30551

Kim SJ, Kim YS, Kim CH, Lee HS (2006a) Lack of association between polymorphisms of the dopamine receptor D4 and dopamine transporter genes and personality traits in a Korean population. Yonsei Med J 47(6):787–792. https://doi.org/10.3349/ymj.2006.47.6.787

Kim SJ, Kim YS, Kim SY, Lee HS, Kim CH (2006b) An association study of catechol-O-methyltransferase and monoamine oxidase A polymorphisms and personality traits in Koreans. Neurosci Lett 401(1–2):154–158. https://doi.org/10.1016/j.neulet.2006.03.006

Kim-Cohen J, Caspi A, Taylor A, Williams B, Newcombe R, Craig IW, Moffitt TE (2006) MAOA, maltreatment, and gene-environment interaction predicting children's mental health: new evidence and a meta-analysis. Mol Psychiatry 11(10):903–913. https://doi.org/10.1038/sj.mp.4001851

Kluger AN, Siegfried Z, Ebstein RP (2002) A meta-analysis of the association between DRD4 polymorphism and novelty seeking. Mol Psychiatry 7(7):712–717. https://doi.org/10.1038/sj.mp.4001082

Kobilka BK, Frielle T, Collins S, Yang-Feng T, Kobilka TS, Francke U, Lefkowitz RJ, Caron MG (1987) An intronless gene encoding a potential member of the family of receptors coupled to guanine nucleotide regulatory proteins. Nature 329(6134):75–79. https://doi.org/10.1038/329075a0

Koller G, Bondy B, Preuss UW, Zill P, Soyka M (2006) The C(−1019)G 5-HT1A promoter polymorphism and personality traits: no evidence for significant association in alcoholic patients. Behav Brain Funct 2:7. https://doi.org/10.1186/1744-9081-2-7

Kulikova MA, Maluchenko NV, Timofeeva MA, Shlepzova VA, Schegolkova JV, Sysoeva OV, Ivanitsky AM, Tonevitsky AG (2008) Effect of functional catechol-O-methyltransferase Val158Met polymorphism on physical aggression. Bull Exp Biol Med 145(1):62–64. https://doi.org/10.1007/s10517-008-0006-9

Laakso A, Pohjalainen T, Bergman J, Kajander J, Haaparanta M, Solin O, Syvalahti E, Hietala J (2005) The A1 allele of the human D2 dopamine receptor gene is associated with increased activity of striatal L-amino acid decarboxylase in healthy subjects. Pharmacogenet Genomics 15(6):387–391. https://doi.org/10.1097/01213011-200506000-00003

Lachman HM, Papolos DF, Saito T, Yu YM, Szumlanski CL, Weinshilboum RM (1996) Human catechol-O-methyltransferase pharmacogenetics: description of a functional polymorphism and its potential application to neuropsychiatric disorders. Pharmacogenetics 6(3):243–250. https://doi.org/10.1097/00008571-199606000-00007

Lage GM, Malloy-Diniz LF, Matos LO, Bastos MA, Abrantes SS, Correa H (2011) Impulsivity and the 5-HTTLPR polymorphism in a non-clinical sample. PLoS One 6(2):e16927. https://doi.org/10.1371/journal.pone.0016927

Lalovic A, Turecki G (2002) Meta-analysis of the association between tryptophan hydroxylase and suicidal behavior. Am J Med Genet 114(5):533–540. https://doi.org/10.1002/ajmg.10518

Lang UE, Hellweg R, Kalus P, Bajbouj M, Lenzen KP, Sander T, Kunz D, Gallinat J (2005) Association of a functional BDNF polymorphism and anxiety-related personality traits. Psychopharmacology 180(1):95–99. https://doi.org/10.1007/s00213-004-2137-7

Lang UE, Bajbouj M, Sander T, Gallinat J (2007) Gender-dependent association of the functional catechol-O-methyltransferase Val158Met genotype with sensation seeking personality trait. Neuropsychopharmacology 32(9):1950–1955. https://doi.org/10.1038/sj.npp.1301335

Lea R, Chambers G (2007) Monoamine oxidase, addiction, and the "warrior" gene hypothesis. N Z Med J 120(1250):U2441

Lee HJ, Lee HS, Kim YK, Kim L, Lee MS, Jung IK, Suh KY, Kim S (2003a) D2 and D4 dopamine receptor gene polymorphisms and personality traits in a young Korean population. Am J Med Genet B Neuropsychiatr Genet 121B(1):44–49. https://doi.org/10.1002/ajmg.b.20054

Lee JH, Kim HT, Hyun DS (2003b) Possible association between serotonin transporter promoter region polymorphism and impulsivity in Koreans. Psychiatry Res 118(1):19–24. https://doi.org/10.1016/s0165-1781(03)00065-9

Lemonde S, Turecki G, Bakish D, Du L, Hrdina PD, Bown CD, Sequeira A, Kushwaha N, Morris SJ, Basak A, Ou XM, Albert PR (2003) Impaired repression at a 5-hydroxytryptamine 1A receptor gene polymorphism associated with major depression and suicide. J Neurosci 23(25):8788–8799

Lesch KP, Bengel D, Heils A, Sabol SZ, Greenberg BD, Petri S, Benjamin J, Muller CR, Hamer DH, Murphy DL (1996) Association of anxiety-related traits with a polymorphism in the serotonin transporter gene regulatory region. Science 274(5292):1527–1531. https://doi.org/10.1126/science.274.5292.1527

Li D, He L (2006) Further clarification of the contribution of the tryptophan hydroxylase (TPH) gene to suicidal behavior using systematic allelic and genotypic meta-analyses. Hum Genet 119(3):233–240. https://doi.org/10.1007/s00439-005-0113-x

Li D, He L (2007) Meta-analysis supports association between serotonin transporter (5-HTT) and suicidal behavior. Mol Psychiatry 12(1):47–54. https://doi.org/10.1038/sj.mp.4001890

Li D, Duan Y, He L (2006) Association study of serotonin 2A receptor (5-HT2A) gene with schizophrenia and suicidal behavior using systematic meta-analysis. Biochem Biophys Res Commun 340(3):1006–1015. https://doi.org/10.1016/j.bbrc.2005.12.101

Liao DL, Hong CJ, Shih HL, Tsai SJ (2004) Possible association between serotonin transporter promoter region polymorphism and extremely violent crime in Chinese males. Neuropsychobiology 50(4):284–287. https://doi.org/10.1159/000080953

Lichter JB, Barr CL, Kennedy JL, Van Tol HH, Kidd KK, Livak KJ (1993) A hypervariable segment in the human dopamine receptor D4 (DRD4) gene. Hum Mol Genet 2(6):767–773. https://doi.org/10.1093/hmg/2.6.767

Lin PY, Tsai G (2004) Association between serotonin transporter gene promoter polymorphism and suicide: results of a meta-analysis. Biol Psychiatry 55(10):1023–1030. https://doi.org/10.1016/j.biopsych.2004.02.006

Lusher JM, Chandler C, Ball D (2001) Dopamine D4 receptor gene (DRD4) is associated with novelty seeking (NS) and substance abuse: the saga continues. Mol Psychiatry 6(5):497–499. https://doi.org/10.1038/sj.mp.4000918

MacKenzie A, Quinn J (1999) A serotonin transporter gene intron 2 polymorphic region, correlated with affective disorders, has allele-dependent differential enhancer-like properties in the mouse embryo. Proc Natl Acad Sci U S A 96(26):15251–15255. https://doi.org/10.1073/pnas.96.26.15251

Mann JJ (1998) The neurobiology of suicide. Nat Med 4(1):25–30. https://doi.org/10.1038/nm0198-025

Mann JJ, Brent DA, Arango V (2001) The neurobiology and genetics of suicide and attempted suicide: a focus on the serotonergic system. Neuropsychopharmacology 24(5):467–477. https://doi.org/10.1016/S0893-133X(00)00228-1

Manolio TA, Collins FS, Cox NJ, Goldstein DB, Hindorff LA, Hunter DJ, McCarthy MI, Ramos EM, Cardon LR, Chakravarti A, Cho JH, Guttmacher AE, Kong A, Kruglyak L, Mardis E, Rotimi CN, Slatkin M, Valle D, Whittemore AS, Boehnke M, Clark AG, Eichler EE, Gibson G, Haines JL, Mackay TF, McCarroll SA, Visscher PM (2009) Finding the missing heritability of complex diseases. Nature 461(7265):747–753. https://doi.org/10.1038/nature08494

Manuck SB, Flory JD, Ferrell RE, Dent KM, Mann JJ, Muldoon MF (1999) Aggression and anger-related traits associated with a polymorphism of the tryptophan hydroxylase gene. Biol Psychiatry 45(5):603–614. https://doi.org/10.1016/s0006-3223(98)00375-8

Manuck SB, Flory JD, Ferrell RE, Mann JJ, Muldoon MF (2000) A regulatory polymorphism of the monoamine oxidase-A gene may be associated with variability in aggression, impulsivity, and central nervous system serotonergic responsivity. Psychiatry Res 95(1):9–23. https://doi.org/10.1016/s0165-1781(00)00162-1

Markett S, Montag C, Reuter M (2011) The nicotinic acetylcholine receptor gene CHRNA4 is associated with negative emotionality. Emotion 11(2):450–455. https://doi.org/10.1037/a0021784

McCourt WF, Gurrera RJ, Cutter HS (1993) Sensation seeking and novelty seeking. Are they the same? J Nerv Ment Dis 181(5):309–312. https://doi.org/10.1097/00005053-199305000-00006
McCrae RR, Costa PT Jr, Martin TA (2005) The NEO-PI-3: a more readable revised NEO personality inventory. J Pers Assess 84(3):261–270. https://doi.org/10.1207/s15327752jpa8403_05
McDermott R, Tingley D, Cowden J, Frazzetto G, Johnson DD (2009) Monoamine oxidase A gene (MAOA) predicts behavioral aggression following provocation. Proc Natl Acad Sci U S A 106(7):2118–2123. https://doi.org/10.1073/pnas.0808376106
Miles DR, Carey G (1997) Genetic and environmental architecture of human aggression. J Pers Soc Psychol 72(1):207–217. https://doi.org/10.1037//0022-3514.72.1.207
Mitsuyasu H, Hirata N, Sakai Y, Shibata H, Takeda Y, Ninomiya H, Kawasaki H, Tashiro N, Fukumaki Y (2001) Association analysis of polymorphisms in the upstream region of the human dopamine D4 receptor gene (DRD4) with schizophrenia and personality traits. J Hum Genet 46(1):26–31. https://doi.org/10.1007/s100380170120
Monyer H, Sprengel R, Schoepfer R, Herb A, Higuchi M, Lomeli H, Burnashev N, Sakmann B, Seeburg PH (1992) Heteromeric NMDA receptors: molecular and functional distinction of subtypes. Science 256(5060):1217–1221. https://doi.org/10.1126/science.256.5060.1217
Monyer H, Burnashev N, Laurie DJ, Sakmann B, Seeburg PH (1994) Developmental and regional expression in the rat brain and functional properties of four NMDA receptors. Neuron 12(3):529–540. https://doi.org/10.1016/0896-6273(94)90210-0
Munafo MR, Yalcin B, Willis-Owen SA, Flint J (2008) Association of the dopamine D4 receptor (DRD4) gene and approach-related personality traits: meta-analysis and new data. Biol Psychiatry 63(2):197–206. https://doi.org/10.1016/j.biopsych.2007.04.006
Nakamura M, Ueno S, Sano A, Tanabe H (2000) The human serotonin transporter gene linked polymorphism (5-HTTLPR) shows ten novel allelic variants. Mol Psychiatry 5(1):32–38. https://doi.org/10.1038/sj.mp.4000698
New AS, Gelernter J, Goodman M, Mitropoulou V, Koenigsberg H, Silverman J, Siever LJ (2001) Suicide, impulsive aggression, and HTR1B genotype. Biol Psychiatry 50(1):62–65. https://doi.org/10.1016/s0006-3223(01)01108-8
Nielsen DA, Virkkunen M, Lappalainen J, Eggert M, Brown GL, Long JC, Goldman D, Linnoila M (1998) A tryptophan hydroxylase gene marker for suicidality and alcoholism. Arch Gen Psychiatry 55(7):593–602. https://doi.org/10.1001/archpsyc.55.7.593
Nishiguchi N, Shirakawa O, Ono H, Nishimura A, Nushida H, Ueno Y, Maeda K (2001) No evidence of an association between 5HT1B receptor gene polymorphism and suicide victims in a Japanese population. Am J Med Genet 105(4):343–345. https://doi.org/10.1002/ajmg.1347
Noble EP, Gottschalk LA, Fallon JH, Ritchie TL, Wu JC (1997) D2 dopamine receptor polymorphism and brain regional glucose metabolism. Am J Med Genet 74(2):162–166
Nomura M, Kusumi I, Kaneko M, Masui T, Daiguji M, Ueno T, Koyama T, Nomura Y (2006) Involvement of a polymorphism in the 5-HT2A receptor gene in impulsive behavior. Psychopharmacology 187(1):30–35. https://doi.org/10.1007/s00213-006-0398-z
Olivier B, van Oorschot R (2005) 5-HT1B receptors and aggression: a review. Eur J Pharmacol 526(1–3):207–217. https://doi.org/10.1016/j.ejphar.2005.09.066
Ono Y, Manki H, Yoshimura K, Muramatsu T, Mizushima H, Higuchi S, Yagi G, Kanba S, Asai M (1997) Association between dopamine D4 receptor (D4DR) exon III polymorphism and novelty seeking in Japanese subjects. Am J Med Genet 74(5):501–503
Ono H, Shirakawa O, Kitamura N, Hashimoto T, Nishiguchi N, Nishimura A, Nushida H, Ueno Y, Maeda K (2002) Tryptophan hydroxylase immunoreactivity is altered by the genetic variation in postmortem brain samples of both suicide victims and controls. Mol Psychiatry 7(10):1127–1132. https://doi.org/10.1038/sj.mp.4001150
Pedersen NL, Plomin R, McClearn GE, Friberg L (1988) Neuroticism, extraversion, and related traits in adult twins reared apart and reared together. J Pers Soc Psychol 55(6):950–957. https://doi.org/10.1037//0022-3514.55.6.950
Pohjalainen T, Rinne JO, Nagren K, Lehikoinen P, Anttila K, Syvalahti EK, Hietala J (1998) The A1 allele of the human D2 dopamine receptor gene predicts low D2 receptor availability in healthy volunteers. Mol Psychiatry 3(3):256–260. https://doi.org/10.1038/sj.mp.4000350

Prichard ZM, Jorm AF, Mackinnon A, Easteal S (2007) Association analysis of 15 polymorphisms within 10 candidate genes for antisocial behavioural traits. Psychiatr Genet 17(5):299–303. https://doi.org/10.1097/YPG.0b013e32816ebc9e

Reif A, Rosler M, Freitag CM, Schneider M, Eujen A, Kissling C, Wenzler D, Jacob CP, Retz-Junginger P, Thome J, Lesch KP, Retz W (2007) Nature and nurture predispose to violent behavior: serotonergic genes and adverse childhood environment. Neuropsychopharmacology 32(11):2375–2383. https://doi.org/10.1038/sj.npp.1301359

Retz W, Retz-Junginger P, Supprian T, Thome J, Rosler M (2004) Association of serotonin transporter promoter gene polymorphism with violence: relation with personality disorders, impulsivity, and childhood ADHD psychopathology. Behav Sci Law 22(3):415–425. https://doi.org/10.1002/bsl.589

Reuter M, Hennig J (2005) Association of the functional catechol-O-methyltransferase VAL158MET polymorphism with the personality trait of extraversion. Neuroreport 16(10):1135–1138. https://doi.org/10.1097/00001756-200507130-00020

Rhee SH, Waldman ID (2002) Genetic and environmental influences on antisocial behavior: a meta-analysis of twin and adoption studies. Psychol Bull 128(3):490–529

Ronai Z, Barta C, Guttman A, Lakatos K, Gervai J, Staub M, Sasvari-Szekely M (2001) Genotyping the -521C/T functional polymorphism in the promoter region of dopamine D4 receptor (DRD4) gene. Electrophoresis 22(6):1102–1105. https://doi.org/10.1002/1522-2683()22:6<1102::AID-ELPS1102>3.0.CO;2-3

Rujescu D, Giegling I, Bondy B, Gietl A, Zill P, Moller HJ (2002) Association of anger-related traits with SNPs in the TPH gene. Mol Psychiatry 7(9):1023–1029. https://doi.org/10.1038/sj.mp.4001128

Rujescu D, Giegling I, Gietl A, Hartmann AM, Moller HJ (2003a) A functional single nucleotide polymorphism (V158M) in the COMT gene is associated with aggressive personality traits. Biol Psychiatry 54(1):34–39. https://doi.org/10.1016/s0006-3223(02)01831-0

Rujescu D, Giegling I, Sato T, Hartmann AM, Moller HJ (2003b) Genetic variations in tryptophan hydroxylase in suicidal behavior: analysis and meta-analysis. Biol Psychiatry 54(4):465–473. https://doi.org/10.1016/s0006-3223(02)01748-1

Rujescu D, Giegling I, Sato T, Moller HJ (2003c) Lack of association between serotonin 5-HT1B receptor gene polymorphism and suicidal behavior. Am J Med Genet B Neuropsychiatr Genet 116B(1):69–71. https://doi.org/10.1002/ajmg.b.10732

Russ MJ, Lachman HM, Kashdan T, Saito T, Bajmakovic-Kacila S (2000) Analysis of catechol-O-methyltransferase and 5-hydroxytryptamine transporter polymorphisms in patients at risk for suicide. Psychiatry Res 93(1):73–78. https://doi.org/10.1016/s0165-1781(00)00128-1

Sabol SZ, Hu S, Hamer D (1998) A functional polymorphism in the monoamine oxidase A gene promoter. Hum Genet 103(3):273–279. https://doi.org/10.1007/s004390050816

Saetre P, Lundmark P, Wang A, Hansen T, Rasmussen HB, Djurovic S, Melle I, Andreassen OA, Werge T, Agartz I, Hall H, Terenius L, Jonsson EG (2010) The tryptophan hydroxylase 1 (TPH1) gene, schizophrenia susceptibility, and suicidal behavior: a multi-centre case-control study and meta-analysis. Am J Med Genet B Neuropsychiatr Genet 153B(2):387–396. https://doi.org/10.1002/ajmg.b.30991

Sakado K, Sakado M, Muratake T, Mundt C, Someya T (2003) A psychometrically derived impulsive trait related to a polymorphism in the serotonin transporter gene-linked polymorphic region (5-HTTLPR) in a Japanese nonclinical population: assessment by the Barratt impulsiveness scale (BIS). Am J Med Genet B Neuropsychiatr Genet 121B(1):71–75. https://doi.org/10.1002/ajmg.b.20063

Sanders AR, Duan J, Gejman PV (2002) DNA variation and psychopharmacology of the human serotonin receptor 1B (HTR1B) gene. Pharmacogenomics 3(6):745–762. https://doi.org/10.1517/14622416.3.6.745

Sawiniec J, Borkowski K, Ginalska G, Lewandowska-Stanek H (2007) Association between 5-hydroxytryptamine 1A receptor gene polymorphism and suicidal behavior. Przegl Lek 64(4–5):208–211

Schild AH, Pietschnig J, Tran US, Voracek M (2013) Genetic association studies between SNPs and suicidal behavior: a meta-analytical field synopsis. Prog Neuro-Psychopharmacol Biol Psychiatry 46:36–42. https://doi.org/10.1016/j.pnpbp.2013.06.014

Schinka JA, Letsch EA, Crawford FC (2002) DRD4 and novelty seeking: results of meta-analyses. Am J Med Genet 114(6):643–648. https://doi.org/10.1002/ajmg.10649

Schinka JA, Busch RM, Robichaux-Keene N (2004) A meta-analysis of the association between the serotonin transporter gene polymorphism (5-HTTLPR) and trait anxiety. Mol Psychiatry 9(2):197–202. https://doi.org/10.1038/sj.mp.4001405

Sen S, Nesse RM, Stoltenberg SF, Li S, Gleiberman L, Chakravarti A, Weder AB, Burmeister M (2003) A BDNF coding variant is associated with the NEO personality inventory domain neuroticism, a risk factor for depression. Neuropsychopharmacology 28(2):397–401. https://doi.org/10.1038/sj.npp.1300053

Serretti A, Calati R, Giegling I, Hartmann AM, Moller HJ, Rujescu D (2009) Serotonin receptor HTR1A and HTR2C variants and personality traits in suicide attempters and controls. J Psychiatr Res 43(5):519–525. https://doi.org/10.1016/j.jpsychires.2008.06.005

Singh JP, Volavka J, Czobor P, Van Dorn RA (2012) A meta-analysis of the Val158Met COMT polymorphism and violent behavior in schizophrenia. PLoS One 7(8):e43423. https://doi.org/10.1371/journal.pone.0043423

Stahl SM (2003) Deconstructing psychiatric disorders, part 1. Genotypes, symptom phenotypes, and endophenotypes. J Clin Psychiatry 64(9):982–983. https://doi.org/10.4088/jcp.v64n0901

Stefulj J, Buttner A, Kubat M, Zill P, Balija M, Eisenmenger W, Bondy B, Jernej B (2004a) 5HT-2C receptor polymorphism in suicide victims. Association studies in German and Slavic populations. Eur Arch Psychiatry Clin Neurosci 254(4):224–227. https://doi.org/10.1007/s00406-004-0482-5

Stefulj J, Buttner A, Skavic J, Zill P, Balija M, Eisenmenger W, Bondy B, Jernej B (2004b) Serotonin 1B (5HT-1B) receptor polymorphism (G861C) in suicide victims: association studies in German and Slavic population. Am J Med Genet B Neuropsychiatr Genet 127B(1):48–50. https://doi.org/10.1002/ajmg.b.20174

Stoel RD, De Geus EJ, Boomsma DI (2006) Genetic analysis of sensation seeking with an extended twin design. Behav Genet 36(2):229–237. https://doi.org/10.1007/s10519-005-9028-5

Strobel A, Wehr A, Michel A, Brocke B (1999) Association between the dopamine D4 receptor (DRD4) exon III polymorphism and measures of novelty seeking in a German population. Mol Psychiatry 4(4):378–384. https://doi.org/10.1038/sj.mp.4000535

Strobel A, Gutknecht L, Rothe C, Reif A, Mossner R, Zeng Y, Brocke B, Lesch KP (2003) Allelic variation in 5-HT1A receptor expression is associated with anxiety- and depression-related personality traits. J Neural Transm (Vienna) 110(12):1445–1453. https://doi.org/10.1007/s00702-003-0072-0

Sugihara H, Moriyoshi K, Ishii T, Masu M, Nakanishi S (1992) Structures and properties of seven isoforms of the NMDA receptor generated by alternative splicing. Biochem Biophys Res Commun 185(3):826–832. https://doi.org/10.1016/0006-291x(92)91701-q

Terracciano A, Sanna S, Uda M, Deiana B, Usala G, Busonero F, Maschio A, Scally M, Patriciu N, Chen WM, Distel MA, Slagboom EP, Boomsma DI, Villafuerte S, Sliwerska E, Burmeister M, Amin N, Janssens AC, van Duijn CM, Schlessinger D, Abecasis GR, Costa PT Jr (2010) Genome-wide association scan for five major dimensions of personality. Mol Psychiatry 15(6):647–656. https://doi.org/10.1038/mp.2008.113

Tochigi M, Hibino H, Otowa T, Kato C, Marui T, Ohtani T, Umekage T, Kato N, Sasaki T (2006) Association between dopamine D4 receptor (DRD4) exon III polymorphism and neuroticism in the Japanese population. Neurosci Lett 398(3):333–336. https://doi.org/10.1016/j.neulet.2006.01.020

Tomitaka M, Tomitaka S, Otuka Y, Kim K, Matuki H, Sakamoto K, Tanaka A (1999) Association between novelty seeking and dopamine receptor D4 (DRD4) exon III polymorphism in Japanese subjects. Am J Med Genet 88(5):469–471. https://doi.org/10.1002/(sici)1096-8628(19991015)88:5<469::aid-ajmg6>3.0.co;2-f

Tovilla-Zarate C, Juarez-Rojop I, Ramon-Frias T, Villar-Soto M, Pool-Garcia S, Medellin BC, Genis Mendoza AD, Narvaez LL, Humberto N (2011) No association between COMT val158met polymorphism and suicidal behavior: meta-analysis and new data. BMC Psychiatry 11:151. https://doi.org/10.1186/1471-244X-11-151

Tsai SJ, Hong CJ, Yu YW, Chen TJ (2004a) Association study of a brain-derived neurotrophic factor (BDNF) Val66Met polymorphism and personality trait and intelligence in healthy young females. Neuropsychobiology 49(1):13–16. https://doi.org/10.1159/000075333

Tsai SJ, Hong CJ, Yu YW, Chen TJ (2004b) Association study of catechol-O-methyltransferase gene and dopamine D4 receptor gene polymorphisms and personality traits in healthy young chinese females. Neuropsychobiology 50(2):153–156. https://doi.org/10.1159/000079107

Tsuchimine S, Yasui-Furukori N, Kaneda A, Saito M, Sugawara N, Kaneko S (2009) Minor genetic variants of the dopamine D4 receptor (DRD4) polymorphism are associated with novelty seeking in healthy Japanese subjects. Prog Neuro-Psychopharmacol Biol Psychiatry 33(7):1232–1235. https://doi.org/10.1016/j.pnpbp.2009.07.005

Turecki G, Briere R, Dewar K, Antonetti T, Lesage AD, Seguin M, Chawky N, Vanier C, Alda M, Joober R, Benkelfat C, Rouleau GA (1999) Prediction of level of serotonin 2A receptor binding by serotonin receptor 2A genetic variation in postmortem brain samples from subjects who did or did not commit suicide. Am J Psychiatry 156(9):1456–1458. https://doi.org/10.1176/ajp.156.9.1456

Turecki G, Zhu Z, Tzenova J, Lesage A, Seguin M, Tousignant M, Chawky N, Vanier C, Lipp O, Alda M, Joober R, Benkelfat C, Rouleau GA (2001) TPH and suicidal behavior: a study in suicide completers. Mol Psychiatry 6(1):98–102. https://doi.org/10.1038/sj.mp.4000816

Walderhaug E, Herman AI, Magnusson A, Morgan MJ, Landro NI (2010) The short (S) allele of the serotonin transporter polymorphism and acute tryptophan depletion both increase impulsivity in men. Neurosci Lett 473(3):208–211. https://doi.org/10.1016/j.neulet.2010.02.048

Wang JY, Jia CX, Lian Y, Sun SH, Lyu M, Wu A (2015) Association of the HTR2A 102T/C polymorphism with attempted suicide: a meta-analysis. Psychiatr Genet 25(4):168–177. https://doi.org/10.1097/YPG.0000000000000091

Wasserman D, Geijer T, Sokolowski M, Rozanov V, Wasserman J (2006) The serotonin 1A receptor C(−1019)G polymorphism in relation to suicide attempt. Behav Brain Funct 2:14. https://doi.org/10.1186/1744-9081-2-14

Weder N, Yang BZ, Douglas-Palumberi H, Massey J, Krystal JH, Gelernter J, Kaufman J (2009) MAOA genotype, maltreatment, and aggressive behavior: the changing impact of genotype at varying levels of trauma. Biol Psychiatry 65(5):417–424. https://doi.org/10.1016/j.biopsych.2008.09.013

Widom CS, Brzustowicz LM (2006) MAOA and the "cycle of violence:" childhood abuse and neglect, MAOA genotype, and risk for violent and antisocial behavior. Biol Psychiatry 60(7):684–689. https://doi.org/10.1016/j.biopsych.2006.03.039

Williams K (1993) Ifenprodil discriminates subtypes of the N-methyl-D-aspartate receptor: selectivity and mechanisms at recombinant heteromeric receptors. Mol Pharmacol 44(4):851–859

Williams HJ, Owen MJ, O'Donovan MC (2007) Is COMT a susceptibility gene for schizophrenia? Schizophr Bull 33(3):635–641. https://doi.org/10.1093/schbul/sbm019

Wu S, Comings DE (1999) A common C-1018G polymorphism in the human 5-HT1A receptor gene. Psychiatr Genet 9(2):105–106. https://doi.org/10.1097/00041444-199906000-00010

Zai CC, Manchia M, De Luca V, Tiwari AK, Chowdhury NI, Zai GC, Tong RP, Yilmaz Z, Shaikh SA, Strauss J, Kennedy JL (2012) The brain-derived neurotrophic factor gene in suicidal behaviour: a meta-analysis. Int J Neuropsychopharmacol 15(8):1037–1042. https://doi.org/10.1017/S1461145711001313

Zalsman G, Frisch A, King RA, Pauls DL, Grice DE, Gelernter J, Alsobrook J, Michaelovsky E, Apter A, Tyano S, Weizman A, Leckman JF (2001) Case control and family-based studies of tryptophan hydroxylase gene A218C polymorphism and suicidality in adolescents. Am J Med Genet 105(5):451–457. https://doi.org/10.1002/ajmg.1406

Zhang HY, Ishigaki T, Tani K, Chen K, Shih JC, Miyasato K, Ohara K, Ohara K (1997) Serotonin2A receptor gene polymorphism in mood disorders. Biol Psychiatry 41(7):768–773. https://doi.org/10.1016/S0006-3223(96)00160-6

Zill P, Buttner A, Eisenmenger W, Moller HJ, Ackenheil M, Bondy B (2007) Analysis of tryptophan hydroxylase I and II mRNA expression in the human brain: a post-mortem study. J Psychiatr Res 41(1–2):168–173. https://doi.org/10.1016/j.jpsychires.2005.05.004

Zuckerman M, Bone RN, Neary R, Mangelsdorff D, Brustman B (1972) What is the sensation seeker? Personality trait and experience correlates of the sensation-seeking scales. J Consult Clin Psychol 39(2):308–321. https://doi.org/10.1037/h0033398

Zuk O, Hechter E, Sunyaev SR, Lander ES (2012) The mystery of missing heritability: genetic interactions create phantom heritability. Proc Natl Acad Sci U S A 109(4):1193–1198. https://doi.org/10.1073/pnas.1119675109

Psychoneurobiology, Brain, and Society (Sociobiology) 19

Contents

19.1 Evolutionary Social Biopsychology

Evolutionary social psychology is a rather recently developed scientific field and tries *to understand the complex area of social behavior in the frame of evolutionary psychology and biology* (Santrock 2005; Schaller et al. 2006). Essentially, it constitutes the expansion of evolutionary psychology in the social domain.

The humankind, throughout its historical course, was living in organized small groups of hunter-gatherers (Fig. 19.1) and only during the recent two to three millennia in larger groups like tribes. Only recently, since only a century, large homogenous tribal groups of the level of tens of millions of people appeared, while currently we are experiencing the era of mega-states with populations of hundreds of millions. In this environment, human psychological function tries to help humans to adapt but also to further evolve.

In this frame, social cognition and Theory of Mind is a field of research concerning the *neurocognitive basis of social behavior*. This refers to the ability of individuals to process and interpret stimuli relevant to the behavior of others, that is, their intentions, desires, and abilities, but also concerning the prediction of their future behavior. Collectively, this is called *"attribution"* and can be ascribed an internal (personality, character, etc.) or external (environmental) locus (Reisenzein 2015; Dunfield and Johnson 2015; Schreiber 2012; Seidel et al. 2010; Santiago and Tarantino 2002; Andrews 2001; Klin 2000; Block and Funder 1986; Kruglanski 1986).

K. N. Fountoulakis, *The Human Connectome*,
https://doi.org/10.1007/978-3-031-10351-3_19

Fig. 19.1 The humankind during all its history until recently was living in organized small groups of hunter-gatherers (Photo by Hans Splinter)

A number of *biases* and errors in the attribution process have been described (see also Chap. 8), and probably all of them are the product of evolution. These biases and errors include:

- The tendency to overestimate the role of personality and underestimate the role of situations.
- To attribute dispositional causes for successes and failures.
- Blaming victims for their suffering.
- False memory of having predicted events.
- Overestimation of true predictions after the outcome is known.

Confirmation bias leads to search for information or interpretations that confirm preconceptions and disregard the others. All these protect the person from feeling vulnerable and mortal. It seems there is a kind of "white lies" everybody tells himself as a defense mechanism in order to keep psychological wellbeing, and maybe these "adjustment lies" are not functioning in depression (Andrews 2001).

However, *biases* are not identical with *errors*. Biases can help quick adaptation, but they could constitute errors under specific conditions. This is because the human brain utilizes heuristics in order to arrive at fast decisions to complex and demanding problems. If this comparison leads to a successful solution, then the bias or the schema leads to adaptation; if not, it leads to an error or to a problematic stereotype and prejudice.

Another important concept is that of "*social influence*" which refers to the persuasive effects people have on each other and includes:

- *Conformity* (act or think like other members of a group).
- *Compliance* (change in behavior due to a request or suggestion from another person).
- *Obedience* (change in behavior as a result of a direct command from another person).

An interesting form of social influence is the so called self-fulfilling prophecy which refers to the situation when a prediction is made, the person's behavior actually causes it to happen, e.g., when expecting hostility from others, the behavior of the person itself actually causes it, while it did not preexist.

Evolutionary Social Biopsychology

- Neurocognitive basis of social behavior.
- Biases and errors:
  - Tendency to overestimate the role of personality and underestimate the role of situations.
  - To attribute dispositional causes for successes and failures.
  - Blaming victims for their suffering.
  - False memory of having predicted events.
  - Overestimation of true predictions after the outcome is known.
- Heuristics—Fast decisions in complex and demanding problems.
- Social influence:
  - Conformity.
  - Compliance.
  - Obedience.
- Self-fulfilling prophecy.
- Crowd behavior.
- Many common characteristics link human civilizations from around the world no matter how isolated they are from each other:
  - Systems of recognition and treatment of individuals according to degree of kinship.
  - Some kind of marital bond.
  - More frequently polygyny.
  - Presence of status hierarchy.
  - Separation into castes and groups.

Apart from the interaction between individuals and between individuals and groups, groups also interact with each other since they possess a distinct identity, rules to follow, and solidarity among members. A related phenomenon is the *behavior of crowds* which often leads to deindividuation, a term reflecting a state of altered self-awareness caused by feelings of anonymity (including large crowds, disguise, and online anonymity). This is associated with uninhibited and maybe dangerous behavior.

To put social psychology in an evolutionary perspective is both challenging and rewarding. An essential first observation is that there seems to be a lot of common features that link human cultures from around the world irrespective of how isolated they are from each other and similarities are more than differences (Brown 1991; Rosch 1973). For example, in all human cultures and societies, there are systems to recon kinship and treating individuals according to kinship status (Daly et al. 1997). While this is not the case for the vast majority of mammals, all human societies have some kind of marital bond for the sharing of parenting (Daly and Wilson 1983; Broude 1994; Geary 1998). These *cultural similarities* exist alongside many *cultural variations*, which often seem very peculiar like the mating customs of the aboriginal tribe of Tiwi in Australia who manifest an interesting interplay between general human mating preferences and a particular social ecology (Hart and Pillig 1960).

Another example is that historically, in most cultures, there is *polygyny* (one man and more than one wife), while a few permits *polyandry* (one woman and more than one husband). It is interesting that in spite of this, *modern societies are monogamous*. It is standard procedure that when biologists find variations across species in behavior, they search for correlations with ecological factors (Alcock 2001). In general, polyandry, though rare, can be explained by the presence of an environment with limited resources. In such environments (e.g., in the high Himalayas), it needs more than one male to support a female and her offspring successfully. In such places, often brothers marry the same woman (Platek and Shackelford 2007; Salmon and Shackelford 2008). On the contrary, in places rich in resources, the opposite phenomenon, that is, polygyny, is observed. Extreme polygyny is manifested in harems, and they are associated with societies with strict hierarchy in a rich environment (Crook and Crook 1988). However, other factors apart from the physical environment play a role in the quality and characteristics of the marital bond. One such factor is the ratio of males to females which can change because of war, migration, and other similar causes. In excess of females, later marriage, more divorce, and permissive sexual norms are observed. In excess of males, males are committed to more stable monogamy (Gangestad and Buss 1994; Kenrick et al. 2003).

It is interesting to note that in the great apes, all kinds of social organization are seen, from monogamy in gibbons to unimale polygyny in orangutans and gorillas and to multimale polygyny (or polygynandry) in chimpanzees where a group of males defends a group of females and their offspring (Foley and Lewin 2013; Lewin 2009).

In all human cultures, also the presence of *status hierarchies* is a basic characteristic of society with separations of casts and groups and variable restrictions and rules (Brown 1991). Anthropological and archaeological data suggest that originally, humans were living in small groups of hunter-gatherers with the size of 50–80 individuals on average and the members of each group were biologically related. These groups were characterized by less strict social hierarchy, members knew very well each other, and they were connected with (actual or fictive) bonds of kinship and were occupying large territories in an exclusive way (Barnard 1999; Maryanski and Turner 1992).

Within these groups, the genetic relatedness together with the long-term reciprocal exchanges usually developed trust and cooperation rather than market-like reciprocal exchange (Fiske 1992). Stigmas concerning disability or disease probably reflect fear of threat not only to the psychology of the individual but also to the collective group welfare (Kurzban and Leary 2001; Neuberg et al. 2000).

Also, an *unknown individual* not belonging to the group was considered to be a potential enemy, and although cordial exchange relationships with other groups were in place, outsiders were always considered to be a threat in terms of stealing, kidnapping females, rape, or homicide (Chagnon 1988; Radcliffe-Brown 1913). Therefore, it seems that evolution could have made humans to be cognitively inclined to divide other people into *"with us"* and *"not with us"* and to perceive the latter as a source of danger and threat (Krebs and Denton 1997; Wilson 1978). However, since outsiders often did not constitute a threat but, on the contrary, they constituted opportunities for trade and development, a flexible response system would be more adaptive in comparison to a rigid one that would reject all outsiders. Such a flexible system would recognize the true value of the outsider and its place in the complex environment of the group (Kenrick 1994; Kenrick et al. 1994).

19.2 Sociobiology

Sociobiology attempts to predict social behavior by utilizing the tools and theories of evolutionary psychology (Haig 2002). Social life is the preferred mode of living in many species in nature. They prefer to live in groups and manifest characteristics of social life, some of them very complex, but the complexity of human societies is exceptional. This complexity is such that many individuals find it difficult to adjust to its demands and subsequently are concerned more with keeping up with the rules and laws of the society than with survival itself, which these rules are supposed to serve. According to Jean-Jacques Rousseau (1712–1778), "social institutions are those that best know how to denature man, to take his absolute existence from him in order to give him a relative one and transport the I into the common unity" (Rousseau 1979).

Through the history of human societies, and since the strong pro-social teachings of Socrates (470–399 BC) and Confucius (551–479 BC), a number of theories evolved, which essentially were attacking the very concept of social organization itself, from the philosophical ideas of Jean-Jacques Rousseau to the anarchist political ideology of William Godwin (1756–1836) and Mikhail Alexandrovich Bakunin (1814–1876).

But one of the most fascinating ideas which appeared in the twentieth century was that of the *"Noble Savage"* which enjoyed wide acceptance, especially in the frame of the civil rights movement after WWII. It referred to an ideal human being, living in peace with nature and his neighbors, with no negative thoughts or feelings. It was preached by Margaret Mead (1901–1978) who claimed that she had found a tribe in Samoa with an ideal psychological and social behavior and lacked things like jealousy or rigid sex roles (Mead 1928). Her reports were proved to be completely false and biased (Shankman 2009; Freeman 1983). However, this idea was appealing, since it suggested that humans are inherently good and moral and modern civilization was the cause of all negative things (a convenient external locus). This rejection of civilization and also of the strict technocratic science is considered by many authors to be a consequence of the two world wars and the horror they embedded in collective consciousness with dreadful images strongly connected with technological advances. This is of course very close to the ideas of Jean-Jacques Rousseau and his 1762 book *Emil* (Rousseau 1979).

Sociobiology

- Social life is the preferred mode of living in many species in nature.
- The complexity of human societies is exceptional.
- The idea that the modern civilization is the cause of all negative things is widely distributed.
- Rejection of civilization and also of the strict technocratic science is considered by many authors to be a consequence of the two world wars and the horror they embedded in collective consciousness with dreadful images strongly connected with technological advances.
- Blank slate.
  - Fundamentally antiscientific and anti-biological.
  - Suggests that all people as individuals are able to learn anything with an equal amount of effort.
  - There are no biological predispositions or inclinations.
  - There are only psychological preferences and choices.
  - Individuals are not only equal but essentially identical in all aspects including biology.
  - Individual differences were the product of social forces.
- Scientific research has proven that:
  - Humans have a strong predisposition to learn some behaviors over others.
  - Individual differences clearly do exist.
  - This in not simply the result of social forces or personal preference.
  - There are no data supporting consistent differences among groups of people that in this way could justify any kind of racism.
- The evolution of human behavior took place mainly during the Pleistocene.

- Contemporary behaviors constitute more abstract and refined versions of basic adaptation strategies.
- In social life:
  - There is a constant conflict between the needs of the one and the needs of the many.
  - Through inclusive fitness, the survival and adaptation of the group and the society as a whole increase.
  - Reciprocity refers to the behaviors with which individuals exchange favors.
  - Altruism and other non-selfish behaviors.
  - Ability for empathy.
  - Ability to detect cheating.
- Parenthood:
  - The human fetus would need gestation period of 18–21 months to be fully developed.
  - Such a long gestation period is impossible in humans.
  - The newborn and the baby demand many months of care and protection.
  - As a result, humans are much more monogamous than primates.
- Sexual dimorphism, differences between genders:
  - Physical.
  - Behavioral.
  - In mating behavior.
  - Impulsivity in males.
  - Increased physical aggressiveness in males.
  - Increased social cognition and sociality in females.

Around the same time Margaret Mead was developing her ideas, another concept, that of the *"Blank Slate,"* was widespread. This suggested that all people were able to learn anything with an equal amount of effort and that there were *no biological predispositions or inclinations, only psychological preferences and choices*. This reflected a concept of *radical equality* at all levels and aspects, which took the step from civil rights equality to consider individuals being not only equal but essentially identical in all aspects. This was also suggesting that all individual differences were the product of social forces rather than inherent traits. This has its roots in Aristotle (384–322 BC), but it was shaped with the concept of "tabula rasa" by John Locke (1632–1704), although in a different frame and purpose. However, the modern concept has important modern political implications. According to the modern concept of the "Blank Slate," in cases of state failure or in non-state societies, people organize in group for protection, and concepts like revenge and honor become extremely important (Pinker 2002; Rose 2001). This is in sharp contrast to the very influential collectivist anarchism theories developed by Mikhail Alexandrovich Bakunin.

Beyond theoretical, ideological, and philosophical preferences and opinions, scientific research has proved that humans have a strong predisposition to learn some behaviors over others and that *individual differences* clearly do exist and this is not simply the result of social forces or personal preference (Buss 2001). What has also been clearly shown is that there are no data supporting consistent differences among groups of people that in this way could justify any kind of racism. Of course, there is the opinion that the approach which considers similarity to be the basis of equality is racist in itself, and as a consequence, the activistic promotion of similarity essentially hides a deeply racist mentality. On the contrary, the *acceptance (not tolerance) of difference* should be promoted.

Therefore, it is pretty evident that as every living organism on Earth has a repertoire of "hard-wired" behaviors and instincts which serve survival, so do humans. For all species, these traits are evolutionary developed and determined. For humans, this evolution took place mainly during the Pleistocene, and contemporary behaviors constitute more abstract and refined versions of basic adaptation strategies (e.g., friendship evolved from sharing of resources) (Kenrick et al. 2003). All aspects of human mental life were developed during that period, and their properties, advantages, and problems can be traced back to then, from differences in cognitive function between sexes to the selective and representative but imprecise way memories are retrieved (Klein et al. 2002).

Social life is characterized by a way of life and behaviors which are often sharply different from a solitude way of life. Thus, there is a constant conflict between the needs of the one and the needs of the many, and this is a problem evolutionary sociobiology needs to address, but it is often very difficult. The behaviors and restrictions when being part of a group might reduce that reproductive success or induce harm or even death to the individual. Eventually however, through inclusive fitness, the survival and adaptation of the group and the society as a whole increase.

At this point, it is interesting to mention the experiments in rats by John B. Calhoun during the 1950s, 1960s, and 1970s, as well as those by Bruce Alexander during the 1970s. Calhoun built the "rat utopia," a place for rats to live which provided the animals with the best possible conditions for food, recreation, and reproduction, and as a result, he unexpectedly observed a decay in their behavior with a collapse of adaptive behaviors and the emergence of incomprehensible and unsuitable for the situation behaviors, e.g., cannibalism, sexual behavior disorders, and a decrease in the total social dynamism of the population. The absolutely unexpected finding appeared during the larger Universe 25 experiment when the rat population, instead of prospering and increasing, within 2 years reached extinction (Calhoun 1962, 1970, 1973). These observations are very difficult to interpret, they probably reflect the lack of social Darwinism mechanisms, and it is unclear whether conclusions could be carried to humans.

Alexander built what was called "park" or "rat paradise" and studied their social behavior as well as their tendency to use and abuse substances in comparison to conditions of loneliness in a cage (Hadaway et al. 1979; Gage and Sumnall 2019; Alexander et al. 1981). The results of that study made a huge impression both in intellectuals and in the public as they were showing the decisive role of the environment on the development of certain behaviors. However, research that followed provided with equivocal and inconsistent results and failed to confirm Alexander's theory (Solinas et al. 2008; Khoo 2020; Petrie 1996). Both theories were extremely appealing and have secured their place in popular culture as it is usual that the most appealing and socio-ideologically accepted theories are discussed more.

The term *"reciprocity"* refers to the behaviors with which individuals exchange favors. The issue of *altruism* and other non-selfish or self-harming behaviors in the frame of inclusive fitness theory has been described above. Issues pertaining mating and polygamy have also been discussed above. The ability for empathy is a rapid way to induce cooperation and to facilitate reciprocal altruism (Burt and Trivers 2006; Dawkins 1976). The same holds for communication and *"cheating."* Reciprocity is essential for humans; failure to reciprocate reduces, while on the contrary the reliable reciprocation increases social reputation. Especially for the study of reciprocal systems and cheating, the game theory paradigm known as the "prisoner's dilemma" tries to model the development of cooperation among independently acting egoists (Barash 2003b).

It has been estimated that in order for the human fetus to be fully develop like the fetuses of other mammals and great apes, a gestation period of 18–21 months would be necessary. The reason why it has been restricted to only 9 months has been considered to be a side effect of upright bipedal walking which leads to a small pelvis size which in turn limits the size of the fetus the human female can carry (Weiner et al. 2008; Charnov and Ernest 2006). This was called the *"obstetrical dilemma,"* a term coined in 1960, by Sherwood Larned Washburn (1911–2000) (Wells et al. 2012; Washburn 1973, 1978, 1982; Washburn and McCown 1972). Another theory suggests that it is the result of limits in the metabolic burden the mother can take in favor of the fetus, and subsequently, this puts a limit to how large and energetically expensive a fetus can be (metabolic crossover hypothesis) (Dunsworth et al. 2012). No matter what the cause is, the fact remains *that the human newborn is extremely unprepared to survive by itself in nature* and demands significant care and parental investment, thus pushing toward stable long-term mating relationships. This makes humans much more monogamous oriented than primates and other mammals (Fig. 19.2).

Observations are consistent with the belief that humans are mildly polygamous by nature and this is a stable cross-cultural characteristic. Serial monogamy is a situation in-between monogamy and polygamy (Barash and Lipton 2002; Platek and Shackelford 2007; Salmon and Shackelford 2008). Social norms do not permit departures from monogamy to manifest openly, especially for women. In spite of this, departures from monogamy are present for both sexes universally (Barash and Lipton 2002). This is also true for previously considered monogamous species, as DNA testing proved that 10–80% of offspring come from a father which is not the male social partner of the couple. This might also was the case for prehistoric humans as well, but in modern human societies, the rate is probably below 4% (Bellis et al. 2005). The globalization which started with the colonial era brought a gradual social homogenization, and while before, almost 90% of societies were polygynous, currently, the vast majority are monogamous, and monogamy is also a feature, among others, that determines whether a society is modern or antiquated.

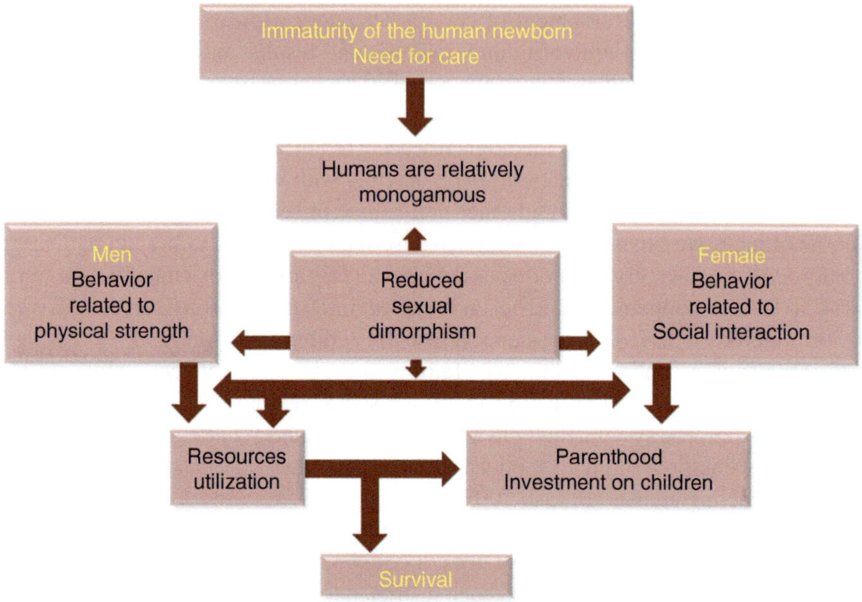

Fig. 19.2 The fact that the human newborn is extremely unprepared to survive by itself in nature makes humans much more monogamous oriented than primates and other mammals

Polygyny was socially more dominant in the past, and it is related to a number of differences between sexes. On average, men are physically larger and with a tendency toward competitive and often violent behavior. Females mature sexually earlier than men and often prefer men older than themselves. All these are also found in mammals where polygyny is the rule since through natural selection, males obtain characteristics which help them to compete with other males, with higher strength, status, and chances for success with advancing age, while females reproduce early and more frequently. These correspond to a pattern of *"sexual dimorphism"* and *"sexual bimaturism."* In monogamous species like gibbons, males and females are of the same size. In gorillas which live in unimale polygamy, males are 30% bigger (Foley and Lewin 2013; Lewin 2009).

Studies of old societies who accepted polygyny suggest that nearly all women were mated and reproductive. On the contrary, there was significant variability in men, with some men being non-reproductive bachelors, most being monogamous, and a few having a harem. This points to another fact that although the physical tendency is toward polygamy, humans are clearly capable of monogamy, at least at the social level. Even their physical inclination toward polygamy seems to be less strong in comparison to most mammals. In evolutionary terms, as described above, this could be explained by the fact that the human offspring is helpless at birth and takes several years to grow up, and this is more pronounced in comparison to the offspring of other mammals. Subsequently, it needs more postnatal care and parental investment from both parents for several years until it grows up.

On the other hand, the tendency of females to seek multiple sexual partners is difficult to explain. It was also difficult to document until recently, because of the secretive way it was conducted in all species. This tendency occurs in spite of risking even violent behaviors from the side of the male and the high risk of abandonment. The probable explanation is that females try to improve the genetic quality of offspring, while at the same time they keep the advantages of stable parenting through social monogamy.

The above are in accord with the presence of a *lower threshold for sexual excitement in men* which therefore are more susceptible to pornography, prostitution, and paraphilias. They are also more eager to engage in sexual activities with strangers, and they pay more emphasis on the physical attributes of the partner since these reflect fertility status. They pay less attention to intellectual attributes and are jealous and possessive. On the contrary, females seem to be more concern with male access to recourses. The mechanism for these choices is largely unconscious, at least concerning the deeper evolutionary goal, and the differences between genders are relative and not absolute, while there are some indications that in the modern society, they are much attenuated.

Men are more inclined to physical aggression and violence than women, and much of their behavior is based on the show-off or physical strength. This is of course culturally enhanced, but it is based on inherent traits. There is no culture in which the sex images are the opposite and women are culturally expected to be more violent than men, which simply means that it is widely accepted that this is impossible. An argument in favor of the biological origin of these differences in behavior is that around the world, the overwhelming majority of murders are committed by men, while also the incarcerated for violent delinquency are also males in overwhelming majority.

In nature, physical aggression and violence are often linked with mating, even in the great apes. While in human society and culture, rape is linked to moral and political issues, in the animal kingdom, rape is probably the strategy followed by the otherwise socially and sexually unsuccessful individuals. In humans, the "domination hypothesis," introduced by Susan Brownmiller (1935–), suggests that rape is not sexually motivated, but instead, it is a conscious process of intimidation by which all men keep all women in a state of fear. This was a highly controversial theory in a clear feminist politico-ideological frame (Lalumiere et al. 2005), and one of the key arguments in favor was that no zoologist ever observed animals raping in their natural habitat. However, this is far from true. At that time, such evidence was available and in the following decades accumulated (Alcock 2001). Randy Thornhill (1944–) and Greg Palmer argued against this hypothesis and pointed out that the improved understanding of what motivates rape rather than ideological approaches are likely help prevent rape. They argued that sexuality is the motive behind rape since a disproportionate number of victims are very young women and suggested that rape is either an adaptation or a byproduct of adaptive traits such as sexual desire and aggressiveness (Thornhill and Palmer 2000; Lalumiere et al. 2005; Figueredo et al. 2011). Furthermore, competition among males is probably more intensive in late adolescence and young adulthood, and this might explain the particularly high crime rates in these age groups (Rose 2001).

Parenting is a complex situation and task and demands a number of somatic but also intellectual abilities from the side of parents. Conflicts between parents but also among parents and offspring are frequent, and from an evolutionary perspective, this is also because their evolutionary goals and agendas are not identical. Parents share only half of their genetic material with each offspring which acts as a limiting factor in parental investment. The area of parent-offspring conflict is predictable and concerns mainly the tendency of the offspring to seek and obtain more investment than the parent has chosen to provide. In a reversal of situations, similar conflicts may occur when the offspring shows significantly less than expected inclination toward the parent and preference for other members of the family. The psychological dynamics within a family were the focus of psychological, sociological, and anthropological studies, and the conclusion is that they could be very intense as well as unconscious. Competition with same-sex parent and adjustment to opposite-sex parent can be considered under this view, as they can, even sophisticated psychoanalytic concepts like the "Oedipus concept."

A more complex social and psychological issue is *step-parenting*. Among animals, step-parenting is rare, and when it occurs, it might hide different aims from the side of the step-parent. When created experimentally, these families are very dysfunctional, and even murders happen. Infanticide and neglect of step-children are the rule rather than the exception (Power 1975; Hasegawa and Hiraiwa 1980; Barash 2003a).

While *most human step-parents are clearly able to function and invest in their step-children*, which is an important element especially in societies with high divorce rates, in evolutionary psychology, the term "Cinderella effect" (Daly and Wilson 2005) refers to the alleged higher incidence of child abuse and mistreatment by step-parents in comparison to biological parents. Of course, it refers to the famous Cinderella fairy tale, and it is believed to be a direct effect of competition concerning mating and parental investment as described above. There is a wide belief that step-families are emotionally conflicted more frequently in comparison to biological families. Non-biological parents have an inherent disinclination to invest in unrelated children. It is both interesting and disturbing that there are some data suggesting that even the accidental injury of children is higher when a step-parent exists but not in single-parent families when one of biological parents has left (Tooley et al. 2006).

One of the sensitive and delicate issues to deal with in a sociobiological perspective is *religion and religiosity*. Currently, there is a controversy whether they are a consequence of evolved psychological adaptations or is a byproduct of other cognitive adaptations (Beit-Hallahmi 2012; D'Onofrio et al. 1999; Fountoulakis et al. 2008; Rossano 2006). It is to be noted however that primitive religious-like ceremonies have been observed in chimpanzees (Harrod 2014).

An issue which recently started to be the focus of interest and research is the fact that the more culturally and technologically advanced a society is, the more it demands biologically determined secondary abilities, especially neurocognitive, from its members. This seems to be even more intensified and pronounced with the informatics and electronics revolution of the last few decades (Geary 1995) and creates new challenges since human societies try to be simultaneously competitive but also inclusive and caring. With the fast rate the modern human society develops, maybe the greatest challenge in the near future would be the incorporation to society of those individuals which, although are normal in terms of psychological and cognitive aspects, are less capable to keep up and follow the developments of a fast-changing environment.

The history of the humankind teaches that humans are capable for the best as well as for the worst; it is extremely easy for them to succumb to their darkest passions, but all the more often they emerge victorious, thus overturning the pessimistic prognostics of their nature.

After all

> We are just an advanced breed of monkeys on a minor planet of a very average star. But we can understand the universe. That makes us something very special.—Stephen Hawking (1942–2018)

(Fig. 19.3).

Fig. 19.3 Stephen Hawking (1942–2018) (Photo by Tom Pilston)

References

Alcock J (2001) The triumph of sociobiology. Oxford University Press, New York
Alexander BK, Beyerstein BL, Hadaway PF, Coambs RB (1981) Effect of early and later colony housing on oral ingestion of morphine in rats. Pharmacol Biochem Behav 15(4):571–576. https://doi.org/10.1016/0091-3057(81)90211-2
Andrews PW (2001) The psychology of social chess and the evolution of attribution mechanisms: explaining the fundamental attribution error. Evol Hum Behav 22(1):11–29. https://doi.org/10.1016/s1090-5138(00)00059-3
Barash D (2003a) Revolutionary biology: the new, gene-centered view of life. Transaction Publishers
Barash D (2003b) The survival game: how game theory explains the biology of cooperation and competition. Henry Holt and Company, New York
Barash D, Lipton J (2002) The myth of monogamy: fidelity and infidelity in animals and people. Henry Holt, New York
Barnard A (1999) Modem hunter-gatherers and early symbolic culture. In: Dunbar R, Knight C, Power C (eds) The evolution of culture: an interdisciplinary view. Rutgers University Press, New Brunswick, NJ
Beit-Hallahmi B (2012) Connecting biological concepts and religious behavior. Behav Brain Sci 35(2):80–81. https://doi.org/10.1017/S0140525X11000938
Bellis MA, Hughes K, Hughes S, Ashton JR (2005) Measuring paternal discrepancy and its public health consequences. J Epidemiol Community Health 59(9):749–754. https://doi.org/10.1136/jech.2005.036517
Block J, Funder DC (1986) Social roles and social perception: individual differences in attribution and error. J Pers Soc Psychol 51(6):1200–1207. https://doi.org/10.1037//0022-3514.51.6.1200
Broude GJ (1994) Marriage, family, and relationships: a cross cultural encyclopedia. ABC-CLIO, Santa Barbara, CA
Brown DE (1991) Human universals. McGraw Hill, New York
Burt A, Trivers R (2006) Genes in conflict: the biology of selfish genetic elements. Belknap Press, Cambridge, MA
Buss DM (2001) Human nature and culture: an evolutionary psychological perspective. J Pers 69(6):955–978. https://doi.org/10.1111/1467-6494.696171
Calhoun JB (1962) Population density and social pathology. Sci Am 206:139–148. https://doi.org/10.1038/scientificamerican0262-139
Calhoun JB (1970) Population density and social pathology. Calif Med 113(5):54
Calhoun JB (1973) Death squared: the explosive growth and demise of a mouse population. Proc R Soc Med 66(1 Pt 2):80–88
Chagnon NA (1988) Life histories, blood revenge, and warfare in a tribal population. Science 239(4843):985–992. https://doi.org/10.1126/science.239.4843.985
Charnov EL, Ernest SK (2006) The offspring-size/clutch-size trade-off in mammals. Am Nat 167(4):578–582. https://doi.org/10.1086/501141
Crook JH, Crook SJ (1988) Tibetan polyandry: problems of adaptation and fitness. In: Betzig L, Borgerhoff Mulder M, Turke P (eds) Human reproductive behavior: a Darwinian perspective. Cambridge University Press, Cambridge
Daly M, Wilson M (1983) Sex, evolution, and behavior, 2nd edn. Wadsworth, Belmont, CA
Daly M, Wilson M (2005) The "Cinderella effect" is no fairy tale. Trends Cogn Sci 9(11):507–508; author reply 508-510. https://doi.org/10.1016/j.tics.2005.09.007
Daly M, Salmon C, Wilson MIE (1997) Kinship: the conceptual hole in psychological studies of social cognition and close relationships. In: Simpson J, Kenrick D (eds) Evolutionary social psychology. Lawrence Erlbaum Associates, Mahnaw, NJ, pp 265–296
Dawkins R (1976) The selfish gene. Oxford University Press
D'Onofrio BM, Eaves LJ, Murrelle L, Maes HH, Spilka B (1999) Understanding biological and social influences on religious affiliation, attitudes, and behaviors: a behavior genetic perspective. J Pers 67(6):953–984. https://doi.org/10.1111/1467-6494.00079

Dunfield KA, Johnson SC (2015) Variability in social reasoning: the influence of attachment security on the attribution of goals. Front Psychol 6:1487. https://doi.org/10.3389/fpsyg.2015.01487

Dunsworth HM, Warrener AG, Deacon T, Ellison PT, Pontzer H (2012) Metabolic hypothesis for human altriciality. Proc Natl Acad Sci U S A 109(38):15212–15216. https://doi.org/10.1073/pnas.1205282109

Figueredo A, Gladden P, Hohman Z (2011) The evolutionary psychology of criminal behaviour. In: Roberts SC (ed) Applied evolutionary psychology. Oxford University Press

Fiske AP (1992) The four elementary forms of sociality: framework for a unified theory of social relations. Psychol Rev 99(4):689–723. https://doi.org/10.1037/0033-295x.99.4.689

Foley R, Lewin R (2013) Principles of human evolution. Wiley, London

Fountoulakis KN, Siamouli M, Magiria S, Kaprinis G (2008) Late-life depression, religiosity, cerebrovascular disease, cognitive impairment and attitudes towards death in the elderly: interpreting the data. Med Hypotheses 70(3):493–496. https://doi.org/10.1016/j.mehy.2007.01.093

Freeman D (1983) Margaret Mead and Samoa. Harvard University Press, Cambridge

Gage SH, Sumnall HR (2019) Rat Park: how a rat paradise changed the narrative of addiction. Addiction 114(5):917–922. https://doi.org/10.1111/add.14481

Gangestad SW, Buss DM (1994) Pathogen prevalence and human mate preferences. Ethol Sociobiol 14(2):89–96

Geary DC (1995) Reflections of evolution and culture in children's cognition. Implications for mathematical development and instruction. Am Psychol 50(1):24–37. https://doi.org/10.1037//0003-066x.50.1.24

Geary DC (1998) Male, female: the evolution of human sex differences. American Psychological Association, Washington, DC

Hadaway PF, Alexander BK, Coambs RB, Beyerstein B (1979) The effect of housing and gender on preference for morphine-sucrose solutions in rats. Psychopharmacology 66(1):87–91. https://doi.org/10.1007/BF00431995

Haig D (2002) Genomic imprinting and kinship. Rutgers University Press, New Brunswick, NJ

Harrod J (2014) The case for chimpanzee religion. J Study Religion Nat Culture 8(1):8–45

Hart CW, Pillig AR (1960) The Tiwi of North Australia. Holt, Rinehart, & Winston, New York

Hasegawa T, Hiraiwa M (1980) Social interactions of orphans observed in a free-ranging troop of Japanese monkeys. Folia Primatol (Basel) 33(1-2):129–158. https://doi.org/10.1159/000155930

Kenrick DT (1994) Evolutionary social psychology: from sexual selection to social cognition. In: Zanna M (ed) Advances in experimental social psychology, vol 26. Academic Press, San Diego, CA

Kenrick DT, Neuberg SL, Zierk KL, Krones JM (1994) Evolution and social cognition: contrast effects as a function of sex, dominance, and physical attractiveness. Personal Soc Psychol Bull 20:210–217

Kenrick DT, Li NP, Butner J (2003) Dynamical evolutionary psychology: individual decision rules and emergent social norms. Psychol Rev 110(1):3–28. https://doi.org/10.1037/0033-295x.110.1.3

Khoo SY-S (2020) Have we reproduced Rat Park? Conceptual but not direct replication of the protective effects of social and environmental enrichment in addiction. J Reprod Neuroscience 1:1318. https://doi.org/10.31885/jrn.1.2020.1318

Klein SB, Cosmides L, Tooby J, Chance S (2002) Decisions and the evolution of memory: multiple systems, multiple functions. Psychol Rev 109(2):306–329. https://doi.org/10.1037/0033-295x.109.2.306

Klin A (2000) Attributing social meaning to ambiguous visual stimuli in higher-functioning autism and Asperger syndrome: the social attribution task. J Child Psychol Psychiatry 41(7):831–846

Krebs DL, Denton K (1997) Social illusions and self-deception: the evolution of biases in person perception. In: Simpson J, Kenrick D (eds) Evolutionary social psychology. Erlbaum, Mahwah, NJ, pp 21–48

Kruglanski AW (1986) Social psychology: attribution. Science 232(4750):665–666. https://doi.org/10.1126/science.232.4750.665

Kurzban R, Leary MR (2001) Evolutionary origins of stigmatization: the functions of social exclusion. Psychol Bull 127(2):187–208. https://doi.org/10.1037/0033-2909.127.2.187
Lalumiere M, Harris G, Quinsey V, Rice M (2005) The causes of rape. American Psychological Association, Washington, DC
Lewin R (2009) Human evolution: an illustrated introduction, 5th edn. John Wiley & Sons, Hoboken
Maryanski A, Turner JH (1992) The social cage: human nature and the evolution of society. Stanford University Press, Stanford, CA
Mead M (1928) Coming of age in Samoa. William Morrow and Co.
Neuberg SL, Smith DM, Asher T (2000) Why people stigmatize: toward a biocultural framework. In: Heatherton T, Kleck R (eds) The social psychology of stigma. The Guilford Press, New York
Petrie BF (1996) Environment is not the most important variable in determining oral morphine consumption in Wistar rats. Psychol Rep 78(2):391–400. https://doi.org/10.2466/pr0.1996.78.2.391
Pinker S (2002) The blank slate: the modern denial of human nature. Viking, New York, NY
Platek S, Shackelford T (2007) Female infidelity and paternal uncertainty. Cambridge University Press, Cambridge
Power HW (1975) Mountain bluebirds: experimental evidence against altruism. Science 189(4197):142–143. https://doi.org/10.1126/science.189.4197.142
Radcliffe-Brown A (1913) Three tribes of Western Australia. J R Anthropol Inst 43:143–194
Reisenzein R (2015) On the universality of the attribution-affect model of helping. Int J Psychol 50(4):308–311. https://doi.org/10.1002/ijop.12153
Rosch EH (1973) Natural categories. Cogn Psychol 4:328–350
Rose S (2001) Revisiting evolutionary psychology and psychiatry. The British journal of psychiatry : the journal of mental science 179:558. https://doi.org/10.1192/bjp.179.6.558-a
Rossano M (2006) The religious mind and the evolution of religion. Rev Gen Psychol 10(4):346–364
Rousseau JJ (1979) Emile, or on education (Trans. Allan Bloom). Basic Books, New York
Salmon C, Shackelford T (2008) Family relationships: an evolutionary perspective. Oxford University Press, Oxford
Santiago JH, Tarantino SJ (2002) Individualism and collectivism: cultural orientation in locus of control and moral attribution under conditions of social change. Psychol Rep 91(3 Pt 2):1155–1168. https://doi.org/10.2466/pr0.2002.91.3f.1155
Santrock WJ (2005) A topical approach to life-span development, 3rd edn. McGraw-Hill, New York, NY
Schaller M, Simpson JA, Kenrick DT (2006) Evolution and social psychology. Psychology Press, New York
Schreiber D (2012) On social attribution: implications of recent cognitive neuroscience research for race, law, and politics. Sci Eng Ethics 18(3):557–566. https://doi.org/10.1007/s11948-012-9381-8
Seidel EM, Eickhoff SB, Kellermann T, Schneider F, Gur RC, Habel U, Derntl B (2010) Who is to blame? Neural correlates of causal attribution in social situations. Soc Neurosci 5(4):335–350. https://doi.org/10.1080/17470911003615997
Shankman P (2009) The trashing of Margaret Mead: the anatomy of an anthropological myth. University of Wisconsin Press, Madison
Solinas M, Chauvet C, Thiriet N, El Rawas R, Jaber M (2008) Reversal of cocaine addiction by environmental enrichment. Proc Natl Acad Sci U S A 105(44):17145–17150. https://doi.org/10.1073/pnas.0806889105
Thornhill R, Palmer C (2000) A natural history of rape: biological bases of sexual coercion. The MIT Press, Cambridge
Tooley G, Karakis M, Stokes M, Ozannesmith J (2006) Generalising the Cinderella effect to unintentional childhood fatalities. Evol Hum Behav Brain Sci 27:224–230
Washburn SL (1973) The promise of primatology. Am J Phys Anthropol 38(2):177–182. https://doi.org/10.1002/ajpa.1330380206
Washburn SL (1978) The evolution of man. Sci Am 239(3):194–198, 201–192, 204 passim. https://doi.org/10.1038/scientificamerican0978-194

Washburn SL (1982) Human evolution. Perspect Biol Med 25(4):583–602. https://doi.org/10.1353/pbm.1982.0072

Washburn SL, McCown ER (1972) Evolution of human behavior. Soc Biol 19(2):163–170. https://doi.org/10.1080/19485565.1972.9987980

Weiner S, Monge J, Mann A (2008) Bipedalism and parturition: an evolutionary imperative for cesarean delivery? Clin Perinatol 35(3):469–478, ix. https://doi.org/10.1016/j.clp.2008.06.003

Wells JC, DeSilva JM, Stock JT (2012) The obstetric dilemma: an ancient game of Russian roulette, or a variable dilemma sensitive to ecology? Am J Phys Anthropol 149(Suppl 55):40–71. https://doi.org/10.1002/ajpa.22160

Wilson E (1978) On human nature. Harvard University Press, Cambridge, MA

Acronyms 20

Contents

20.1 Acronyms to Full Names

3-MT	3-Methoxytyramine
5-HIAA	5-Hydroxyindoleacetic acid
5-HTOL	5-Hydroxytryptophol
5HTT	Serotonin transporter
ACE	Angiotensin-converting enzyme
Acetyl-CoA	Acetyl coenzyme A
ACTH	Adrenocorticotropic hormone
ADH	Antidiuretic hormone
ADHD	Attention-deficit hyperactivity disorder
ALS	Amyotrophic lateral sclerosis
AMPA	a-Amino-3-hydroxy-5-methyl-4-isoxazole propionic acid
AVP	Arginine vasopressin
BDNF	Brain-derived neurotrophic factor
BNST	Bed nucleus of the stria terminalis
BST-DSPM	Darkly staining posteromedial part of the bed nucleus of the stria terminalis
C	Cytosine
Ca2+	Calcium ions
CaM	Calcium-modulated protein calmodulin
CART	Cocaine- and amphetamine-regulated transcript
CeA	Central nucleus of the amygdala
CKK	Cholecystokinin

K. N. Fountoulakis, *The Human Connectome*,
https://doi.org/10.1007/978-3-031-10351-3_20

CLN	Caudal linear (raphe) nucleus
CNGCs	Cyclic nucleotide-gated channels
CNTF	Ciliary neurotrophic factor
COMT	Catechol-O-methyltransferase
CRH	Corticotropin-releasing hormone
CSF	Cerebrospinal fluid
DAG	Diacylglycerol
DAN	Default state network, dorsal attention network
DBH	Dopamine beta (β)-hydroxylase
D-FPN	Dorsal frontoparietal network
DHEA	Dehydroepiandrosterone
DHMA	3,4-Dihydroxymandelic acid
DHPG	3,4-Dihydroxyphenylglycol
DHPGA	3,4-Dihydroxyphenylglycolaldehyde
DMN	Default mode network
DOPAC	Dihydroxyphenylacetic acid
DRN	Dorsal raphe nucleus
EPAC	Exchange protein activated by cAMP
FEF	Frontal eye fields
FSH	Follicle-stimulating hormone
G	Guanine
GAB1	GRB2-associated-binding protein 1 or growth-factor receptor-bound protein- 2-associated binding protein 1
GABA	Gamma-aminobutyric acid
GDNF	Glial cell line-derived neurotrophic factor
GDP	Guanosine diphosphate
Gi	Inhibitory regulative G-protein
Glu	Glutamate
Gly	Glycine
GnRH	Gonadotropin-releasing hormone
GPLR	G-protein-linked receptors
GRB2	Growth factor receptor-bound protein 2
GRH	Growth hormone-releasing hormone
GRP	Gastrin-releasing peptide
Gs	Stimulative regulative G-protein
GTP	Guanosine triphosphate
GWAS	Genome-wide association study
HVA	Homovanillic acid
INAH	Interstitial nucleus of the anterior hypothalamus
INAH2 and INAH3	Preoptic anterior hypothalamus areas
INDELs	Short insertion and deletion polymorphisms
IP3	Inositol 1,4,5-trisphosphate or inositol trisphosphate
IQ	Intelligence quotient
K+	Potassium ions
KA	Kainic acid
L-DOPA	L-3,4-dihydroxyphenylalanine or L-dihydroxyphenylalanine

LD	Linkage disequilibrium
LH	Luteinizing hormone
LHRH	Luteinizing hormone-releasing hormone
LNAA	Large neutral amino acid transporter
LTD	Long-term depression
LTP	Long-term potentiation
MAO-A	Monoaminoxidase-A
MAO-B	Monoaminoxidase-B
MAP kinase	Mitogen-activated protein kinase
MCH	Melanin-concentrating hormone
M-FPN	Medial frontoparietal network
MHPG	3-Metoxy-4-hydroxyglycol
MHPGA	3-Metoxy-4-hydroxyphenylglycoaldehyde
MRN	Median raphe nucleus
MSH	Melanocyte-stimulating hormone
Na+	Sodium ions
NE	Norepinephrine
NGF	Nerve growth factor
NMDA	N-methyl-D-aspartate
NO	Nitric oxide
NPY	Neuropeptide Y
NTS	Nucleus of the solitary tract or nucleus tractus solitarius
OFQ	Orphanin FQ
OGF	Opioid growth factor
OGFr	Opioid growth factor receptor
ORCH-OR	Theory of orchestrated objective reduction
PACAP	Pituitary adenylate cyclase-activating peptide
PCD	Programmed cell death
PCP	Phencyclidin
Pgi	Nucleus paragigantocellular
PHI-27	Peptide histidine-isoleucine with 27 amino acids
PI3	Phosphoinositide 3-kinase
PIP2	Phospholipid phosphatidylinositol 4,5-bisphosphate
PKA	Protein kinase A
PLCG1	Phospholipase C, gamma 1
PNMT	Phenylethanolamine N-methyltransferase
PPTg	Pedunculopontine tegmental nuclei
PRH	Nucleus prepositus hypoglossi
PVN	Paraventricular nucleus of hypothalamus
REM sleep	Rapid eye movement sleep
RFLP	Restriction fragment length polymorphism
SDN-POA	Sexually dimorphic nucleus of the preoptic area
SERT	Serotonin transporter
SLC	Solute carrier family
SNPs	Single-nucleotide polymorphisms
T	Thymine
TGFβ	Transforming growth factor-β

TNF	Tumor necrosis factor
TRH	Thyrotropin-releasing hormone
Trk	Tropomyosin-related kinase
Trk	Tyrosine kinase receptors
VIP	Vasoactive intestinal polypeptide
VMA	Vanilylmandelic acid
VNTRs	Variable number tandem repeats
VTA	Ventral tegmental area
ANP	Atrial natriuretic peptide

20.2 Full Names to Acronyms

3-Methoxytyramine	3-MT
3-Metoxy-4-hydroxyglycol	MHPG
3-Metoxy-4-hydroxyphenylglycoaldehyde	MHPGA
3,4-Dihydroxymandelic acid	DHMA
3,4-Dihydroxyphenylglycol	DHPG
3,4-Dihydroxyphenylglycolaldehyde	DHPGA
5-Hydroxyindoleacetic acid	5-HIAA
5-Hydroxytryptophol	5-HTOL
a-Amino-3-hydroxy-5-methyl-4-isoxazole propionic acid	AMPA
Acetyl coenzyme A	Acetyl-CoA
Adrenocorticotropic hormone	ACTH
Amyotrophic lateral sclerosis	ALS
Angiotensin-converting enzyme	ACE
Antidiuretic hormone	ADH
Arginine vasopressin	AVP
Atrial natriuretic peptide	ANP
Attention-deficit hyperactivity disorder	ADHD
Bed nucleus of the stria terminalis	BNST
Brain-derived neurotrophic factor	BDNF
Calcium ions	Ca2+
Calcium-modulated protein calmodulin	CaM
Catechol-O-methyltransferase	COMT
Caudal linear (raphe) nucleus	CLN
Central nucleus of the amygdala	CeA
Cerebrospinal fluid	CSF
Cholecystokinin	CKK
Ciliary neurotrophic factor	CNTF
Cocaine- and amphetamine-regulated transcript	CART
Corticotropin-releasing hormone	CRH
Cyclic nucleotide-gated channels	CNGCs
Cytosine	C
Darkly staining posteromedial part of the bed nucleus of the stria terminalis	BST- DSPM

Default mode network	DMN
Default state network, dorsal attention network	DAN
Dehydroepiandrosterone	DHEA
Diacylglycerol	DAG
Dihydroxyphenylacetic acid	DOPAC
Dopamine beta (β)-hydroxylase	DBH
Dorsal frontoparietal network	D-FPN
Dorsal raphe nucleus	DRN
Exchange protein activated by cAMP	EPAC
Follicle-stimulating hormone	FSH
Frontal eye fields	FEF
G-protein-linked receptors	GPLR
Gamma-aminobutyric acid	GABA
Gastrin-releasing peptide	GRP
Genome-wide association study	GWAS
Glial cell line-derived neurotrophic factor	GDNF
Glutamate	Glu
Glycine	Gly
Gonadotropin-releasing hormone	GnRH
GRB2-associated-binding protein 1	GAB1
Growth factor receptor-bound protein 2	GRB2
Growth factor receptor-bound protein-2-associated binding protein 1	GAB1
Growth hormone-releasing hormone	GRH
Guanine	G
Guanosine diphosphate	GDP
Guanosine triphosphate	GTP
Homovanillic acid	HVA
Inhibitory regulative G-protein	Gi
Inositol 1,4,5-trisphosphate	IP3
Inositol trisphosphate	IP3
Intelligence quotient	IQ
Interstitial nucleus of the anterior hypothalamus	INAH
Kainic acid	KA
L-3,4-dihydroxyphenylalanine	L-DOPA
L-Dihydroxyphenylalanine	L-DOPA
Large neutral amino acid transporter	LNAA
Linkage disequilibrium	LD
Long-term depression	LTD
Long-term potentiation	LTP
Luteinizing hormone	LH
Luteinizing hormone-releasing hormone	LHRH
Medial frontoparietal network	M-FPN
Median raphe nucleus	MRN
Melanin-concentrating hormone	MCH
Melanocyte-stimulating hormone	MSH
Mitogen-activated protein kinase	MAP kinase
Monoamine oxidase-A	MAO-A

Monoamine oxidase-B	MAO-B
N-methyl-D-aspartate	NMDA
Nerve growth factor	NGF
Neuropeptide Y	NPY
Nitric oxide	NO
Norepinephrine	NE
Nucleus of the solitary tract	NTS
Nucleus paragigantocellular	PGi
Nucleus prepositus hypoglossi	PRH
Nucleus tractus solitarius	NTS
Opioid growth factor	OGF
Opioid growth factor receptor	OGFr
Orphanin FQ	OFQ
Paraventricular nucleus of hypothalamus	PVN
Pedunculopontine tegmental nuclei	PPTg
Peptide histidine-isoleucine with 27 amino acids	PHI-27
Phencyclidin	PCP
Phenylethanolamine N-methyltransferase	PNMT
Phosphoinositide 3-kinase	PI3
Phospholipase C, gamma 1	PLCG1
Phospholipid phosphatidylinositol 4,5-bisphosphate	PIP2
Pituitary adenylate cyclase-activating peptide	PACAP
Potassium ions	K^+
Preoptic anterior hypothalamus areas	INAH2 and INAH3
Programmed cell death	PCD
Protein kinase A	PKA
Rapid eye movement sleep	REM sleep
Restriction fragment length polymorphism	RFLP
Serotonin transporter	5HTT
Serotonin transporter	SERT
Sexually dimorphic nucleus of the preoptic area	SDN-POA
Short insertion and deletion polymorphisms	INDELs
Single-nucleotide polymorphisms	SNPs
Sodium ions	Na^+
Solute carrier family	SLC
Stimulative regulative G-protein	Gs
Theory of orchestrated objective reduction	ORCH-OR
Thymine	T
Thyrotropin-releasing hormone	TRH
Transforming growth factor-β	TGFβ
Tropomyosin-related kinase	Trk
Tumor necrosis factor	TNF
Tyrosine kinase receptors	Trk
Vanilylmandelic acid	VMA
Variable number tandem repeats	VNTRs
Vasoactive intestinal polypeptide	VIP
Ventral tegmental area	VTA

GPSR Compliance

The European Union's (EU) General Product Safety Regulation (GPSR) is a set of rules that requires consumer products to be safe and our obligations to ensure this.

If you have any concerns about our products, you can contact us on ProductSafety@springernature.com

In case Publisher is established outside the EU, the EU authorized representative is:

Springer Nature Customer Service Center GmbH
Europaplatz 3
69115 Heidelberg, Germany

Batch number: 10732031

Printed by Printforce, the Netherlands